About the Editors

István Barabás

István Barabás is a Professor of System Theory and Automation and Automotive Testing at the Faculty of Automotive, Mechatronic and Mechanical Engineering, Technical University of Cluj-Napoca, Romania. He coordinates the Bachelor's program in Command and Control Equipment and Systems for Road Vehicles. His main research interests include intelligent sensors, sensor fusion for automotive applications, automotive testing, powertrain concepts, energy efficiency, autonomous vehicles, and computer modeling and simulation in the automotive field. He has authored more than 150 scientific papers in conferences and journals and received the "Traian Vuia" Award from the Romanian Academy in 2024 for his scientific contributions. István Barabás serves on the editorial boards of *Periodica Polytechnica Transportation Engineering* and *Technical Review*, and is a member of in the advisory and organizing committees of several international conferences.

Calin Iclodean

Calin Iclodean is currently an Associate Professor Habil in the Department of Automotive Engineering and Transport at the Technical University of Cluj-Napoca in Romania. He holds a Ph.D. and Habilitation in Mechanical Engineering. He graduated from the Faculty of Electronics and Telecommunications, Department of Applied Electronics, at the Technical University of Cluj-Napoca in 1994, and he has over 25 years of IT&C experience. His primary research areas include electric vehicles, autonomous vehicles, fuel cell vehicles, powertrain concepts, electronic control units, in-vehicle communication networks, energy efficiency, computer modeling, and automotive simulation.

Máté Zöldy

Máté Zöldy is currently a Professor in the Department of Automotive Technologies, BME Faculty of Transportation Engineering and Vehicle Engineering, Budapest, Hungary. His primary research areas include legislation of autonomous vehicles, trajectory planning in roundabouts, refueling prediction of autonomous vehicles, and drivetrains.

AF713857

Preface

Autonomous vehicles (AVs) must function with high levels of intelligence. Beyond the functional algorithms included in their autonomous driving systems, these vehicles must be capable of determining appropriate driving actions in real time, including in safety-critical conditions. Achieving such decision-making autonomy, particularly as vehicles progress toward SAE J3016 Levels 4 and 5, requires the integration of an intelligent perception system.

The critical activities of localization, perception, prediction, and motion planning rely heavily on data collected by intelligent sensors that continuously evaluate their surroundings. As a result, the academic and industrial community working on AV development requires an up-to-date and complete review of the capabilities, limitations, benefits, and downsides of intelligent sensors for autonomous vehicles.

This knowledge directly contributes to the increased dependability, safety, and robustness of future autonomous systems.

This Reprint is a compilation of contributions from authors who are committed to advancing the field of autonomous vehicles through their research on intelligent sensory systems and associated technologies. The published works cover a wide range of topics, including RADAR, LiDAR, camera, ultrasonic, GPS/GNSS sensors, and V2V and V2X communication systems. Furthermore, the investigations look into sensor data fusion methods, object classification techniques, and different approaches to environmental perception and interpretation.

The research covered in this volume includes investigations into intelligent sensory systems for autonomous driving, paradigms and architectures of intelligent sensory networks, and comparisons of real and virtual sensory system models.

Additional research focuses on sensor integration and data fusion for autonomous decision-making, intelligent proprioceptive and exteroceptive sensors, and object recognition and classification via intelligent sensing. The usage of artificial intelligence algorithms, fog and edge computing frameworks, and cybersecurity procedures for intelligent sensory systems are also discussed, as is the crucial importance of intelligent sensors in vehicle-to-everything (V2X) communication.

Overall, the research presented advances our understanding of how intelligent sensing and perception technologies support the transition to fully autonomous, connected, and safe mobility systems.

István Barabás, Calin Iclodean, and Máté Zöldy
Guest Editors

sensors

MDPI

Editorial

Intelligent Sensors for Smart and Autonomous Vehicles

István Barabás [1], Calin Iclodean [1,*] and Máté Zöldy [2]

[1] Department of Automotive Engineering and Transports, Technical University of Cluj-Napoca, 400001 Cluj-Napoca, Romania; istvan.barabas@auto.utcluj.ro
[2] Department of Automotive Technologies, BME Faculty of Transportation Engineering and Vehicle Engineering, 1111 Budapest, Hungary; zoldy.mate@kjk.bme.hu
* Correspondence: calin.iclodean@auto.utcluj.ro

1. Introduction

Autonomous vehicles (AVs) must exhibit a high degree of intelligence; beyond the functional algorithms implemented within their autonomous driving systems, AVs are also required to independently determine appropriate actions in real time, including in safety-critical scenarios. Achieving higher levels of driving automation—specifically Levels 4 and 5 as defined by the SAE J3016 standard [1]—necessitates the integration of advanced intelligent perception systems [2,3].

Core functions such as localization, environmental perception, behavior prediction, and motion planning rely heavily on data acquired through intelligent sensors embedded within a vehicle. These sensors interpret dynamic environmental inputs and enable informed decision-making by the AV [4].

A comprehensive and up-to-date overview of the capabilities, limitations, advantages, and disadvantages of intelligent sensor technologies is essential for researchers, engineers, and stakeholders engaged in the development of autonomous vehicles. Such knowledge facilitates the creation of AV systems that are not only more reliable and robust but also significantly safer [5].

This Special Issue, entitled "Intelligent Sensors for Smart and Autonomous Vehicles", addresses the crucial role of advanced sensorial systems in the development of autonomous vehicles (AVs). As AV technologies progress toward higher levels of automation, the need for robust, accurate, and intelligent perception becomes increasingly important. Sensor systems such as RADAR, LiDAR, cameras, ultrasonic devices, GPS/GNSS modules, and Vehicle-to-Vehicle (V2V) communication units form the foundation of environmental awareness in autonomous systems [6].

The integration and fusion of data obtained from these sensors are essential for achieving reliable localization, object detection, classification, and decision-making processes. Moreover, the combination of artificial intelligence algorithms with sensorial inputs enhances a vehicle's ability to perceive, understand, and respond to complex driving scenarios [7]. Emerging paradigms such as edge and fog computing, alongside the evolution of intelligent proprioceptive and exteroceptive sensors, are shaping the architecture of next-generation AV systems [8].

In this Special Issue, recent research in the field of smart sensors is published, with a focus on increasing the performance and capabilities of these sensors, reducing expenses and constructive dimensions, advancing sensor signal fusion, and decreasing signal transmission time and processing latencies. Advances in edge artificial intelligence and sensor-specific processing enable real-time results while adhering to safety standards [9,10]. Advances in smart sensor technology have converted vehicle sensor systems from static to adaptive sensor networks based on machine learning and which are capable of complex environmental interpretation [11].

Despite significant progress in the field of smart sensors, many knowledge and technology gaps remain, including regarding the robustness of perception in adverse weather conditions and extreme events; reliable calibration and synchronization between heterogeneous sensors; the immaturity of some sensor signal fusion algorithms; insufficient volumes of multimodal datasets; latency in signal transmission, processing, and interpretation; energy efficiency factors; and implementation costs [12]. Furthermore, as connection between smart vehicles and smart environments expands, cybersecurity and data privacy issues become more prevalent [13].

Future research should concentrate on creating verifiable and validatable multimodal fusion systems that allow for real-world uncertainty while maintaining consistent performance under these unpredictable real-world settings [14]. Efforts toward achieving physics-based simulation, long-tail dataset creation, energy-efficient edge inference, and functional safety certification frameworks will be critical in reducing the gap between laboratory innovation and large-scale implementation in intelligent and autonomous vehicles on the road [15,16].

This Special Issue invites the submission of scientific contributions that explore various aspects of intelligent sensing, including conceptual and architectural models, comparisons between real and virtual systems, advanced sensor fusion techniques, cybersecurity in sensor-based systems, and the integration of intelligent sensors into V2X communication networks. By consolidating cutting-edge research in these domains, this Issue aims to support the development of autonomous vehicles that are not only more capable and efficient but also significantly safer and more trustworthy in real-world conditions.

2. Short Presentation of the Papers

In the review titled "An Approach to Modeling and Developing Virtual Sensors Used in the Simulation of Autonomous Vehicles" (contribution 1), Barabás et al. present a virtual modeling approach that facilitates the study of real-world phenomena within a simulated environment. This approach is based on a theoretical model that digitally replicates a real system, enabling the analysis and validation of vehicle behavior under various operating conditions. The complexity of the virtual model must reflect the complexity of the actual system under evaluation—being as intricate as necessary, yet as simplified as possible—to ensure the reliability of computer simulations and their validation against experimental measurements.

The virtual model of the autonomous vehicle was developed using CarMaker, version 12.0, a simulation platform created by IPG Automotive. This software is widely adopted both in the international academic community and in the automotive industry, and supports real-time simulation of elementary vehicle systems at the system level and offers an open framework for designing virtual test scenarios in domains such as autonomous driving, advanced driver assistance systems (ADASs), powertrain development, and vehicle dynamics.

This model incorporates a variety of virtual sensors, including slip angle, inertial, object detection, free-space, traffic sign recognition, lane detection, road surface, object-by-line, camera-based, and Global Navigation Satellite System (GNSS) sensors, as well as radar, LiDAR, and ultrasonic sensors. These virtual sensors can be categorized according to their response generation mechanisms: sensors based on parameters derived from measurement characteristics, sensors developed through modeling techniques, and application-specific sensors.

Medina et al., in the review "An Overview of Autonomous Parking Systems: Strategies, Challenges, and Future Directions" (contribution 2), discuss the rapid evolution of Autonomous Parking Systems (APSs), which promise significant improvements in convenience, safety, and operational efficiency. This paper critically examines current strategies in perception, path planning, and vehicle control, as well as system-level considerations such as integration, validation, and security.

Despite notable advancements driven by deep learning techniques and sophisticated sensor fusion, substantial challenges remain. The authors explore the inherent trade-offs, including balancing computational complexity with real-time performance requirements, unresolved foundational issues such as the verification of non-deterministic AI components, and the considerable difficulty of ensuring robust deployment in real-world environments characterized by diverse and unpredictable conditions—ranging from cluttered urban canyons to poorly lit and ambiguously marked parking facilities.

The review further addresses the limitations of existing technologies; the complexities involved in safety assurance within dynamic settings; the pervasive influence of cost constraints on system capabilities; and the critical, often underestimated, importance of establishing genuine user trust. The authors conclude that future research must not only develop innovative technological solutions to these challenges but also engage with the complex socio-technical dimensions necessary to fully realize the potential of Autonomous Parking Systems.

In their review titled "Recent Developments on Drivable Area Estimation: A Survey and a Functional Analysis" (contribution 3), Hortelano et al. highlight that the most advanced Autonomous Driving Systems (ADSs) currently depend heavily on the prior generation of High-Definition (HD) maps. While effective, this process is costly and requires frequent updates to reflect the dynamic nature of road environments. As an alternative, the online generation of accurate navigation maps presents a promising solution to reduce these costs and to extend the Operational Design Domains (ODDs) of modern ADSs.

The authors provide a comprehensive overview of the state of the art in drivable area estimation—an essential component for enabling ADS functionality in environments where HD maps are limited or unavailable. The study proposes a novel architectural framework that encompasses both learning- and non-learning-based methods, facilitating a systematic analysis of recent and impactful algorithms in this field.

Furthermore, the review includes practical resources for researchers and practitioners, such as an evaluation of the impact of modern sensing technologies on drivable area estimation, and a curated list of benchmark datasets relevant for performance evaluation and algorithm comparison.

Tollner et al., in their paper "How Do We Calibrate a Battery Electric Vehicle Model Based on Controller Area Network Bus Data?" (contribution 4), argue that transforming a modern vehicle into a measurement system is a highly beneficial approach, given the wide array of sensors integrated within onboard control and diagnostic systems. As these procedures are not executed by a single control unit, it is essential to transmit signal values over a communication network, enabling an external device to be connected to achieve real-world data acquisition during driving scenarios.

The primary objective of the paper is to utilize the recorded data to validate a one-Degree-Of-Freedom (1-DOF) longitudinal vehicle and powertrain model. To ensure repeatability, three urban driving routes are selected: a flat road, a road with a slight incline, and a road with a steeper gradient, each in both directions. This setup allows the drivetrain system to be tested across a wide load spectrum, including periods of extended energy recuperation. Altitude variations are captured using a Differential GPS (DGPS) system.

Based on the collected data, key vehicle and drivetrain model parameters can be calibrated, including aerodynamic drag coefficients, rolling resistance values, and drivetrain efficiency metrics. Model validation is performed by evaluating speed tracking performance and ensuring that the relative deviation in cumulative energy remains below a 10% threshold.

Ultimately, the developed model is suitable for further energy analysis or the design of control strategies. The study also presents the energy balance of the test cycles as supporting evidence for the model's accuracy.

Cap et al., in the paper "Hybrid MambaVision and Transformer-Based Architecture for 3D Lane Detection" (contribution 5), demonstrate that lane detection is a critical task in the fields of computer vision and autonomous driving. This includes identifying road markings on the road surface, not only assisting drivers in keeping their vehicles in the proper lane, but also offering essential data for advanced driver assistance systems and autonomous vehicles.

The authors provide a new framework built on the MambaVision-S-1K backbone that combines Mamba-based processing with Transformer capabilities to extract both local detail and global context from monocular pictures. This hybrid approach enables precise three-dimensional simulation of lane geometry, even when elevation varies. By replacing the typical convolutional neural network backbone with MambaVision, the proposed model dramatically enhances 3D lane detection systems. This technique achieves cutting-edge performance on the ONCE-3DLanes dataset, demonstrating its supremacy in precisely capturing lane curvature and elevation changes. These findings demonstrate the possibility of integrating enhanced backbones based on Vision Transformers into autonomous driving systems to improve lane detection robustness and reliability.

Radu et al., in their paper "Design and Validation of an Active Headrest System with Integrated Sensing in Rear-End Crash Scenarios" (contribution 6), address the significant safety challenges posed by rear-end collisions, particularly due to the high risk of whiplash injuries among vehicle occupants. The accurate simulation of occupant kinematics during such impacts is essential for the development of advanced occupant protection systems.

This paper introduces an enhanced multibody simulation model tailored specifically for rear-end crash scenarios, integrating active headrest mechanisms and sensor-based activation logic. The model features detailed representations of vehicle structural components, suspension systems, restraint systems, and occupant biomechanics, enabling precise prediction of crash dynamics and occupant responses.

The system was developed using Simscape Multibody, with components derived from CAD models and interconnected through physically realistic joints. Validation was conducted through controlled experimental crash tests, with particular emphasis placed on accurately modeling contact forces, suspension dynamics, and actuator response times within the active headrest system.

The model demonstrated high predictive accuracy, achieving a root mean square error (RMSE) of 4.19 m/s^2 and a Mean Absolute Percentage Error (MAPE) of 0.71% in head acceleration measurements during frontal impact tests, confirming the model's reliability in replicating occupant kinematics and head acceleration profiles.

This research underscores the importance of integrated sensor–actuator systems in enhancing occupant safety and establishes a robust, adaptable platform for future investigations into intelligent vehicle safety technologies.

Beles et al., in their paper "Driver Drowsiness Multi-Method Detection for Vehicles with Autonomous Driving Functions" (contribution 7), present a comprehensive approach to the development of a fuzzy-logic-based decision algorithm aimed at detecting and issuing warnings regarding driver drowsiness. The proposed algorithm integrates an analysis of ElectroOculoGraphy (EOG) signals and eye state images to assess the driver's level of alertness, with the primary goal of preventing accidents caused by reduced attentiveness.

The drowsiness detection system consists of several core components responsible for acquiring data, analyzing physiological and visual indicators, and making decisions regarding the driver's state of vigilance. When signs of drowsiness are detected, the system can issue appropriate warnings to alert the driver.

Driver drowsiness is typically manifested through a progressive decline in attention to road and traffic conditions, impaired driving performance, and increased reaction times—all of which significantly elevate the risk of accidents. In scenarios where the driver fails to respond to repeated alerts, the advanced driver assistance system (ADAS) is expected to intervene by taking control of the vehicle to ensure continued safety.

This research highlights the critical role of multi-method detection systems and fuzzy logic decision-making in enhancing safety in semi-autonomous and autonomous driving environments.

Montiel-Marín et al., in their paper "Point Cloud Painting for 3D Object Detection with Camera and Automotive 3+1D RADAR Fusion" (contribution 8), discuss the longstanding integration of RADAR and camera sensors in automotive systems since the inception of advanced driver assistance systems (ADASs). While these sensors exhibit complementary strengths and weaknesses, their potential has been relatively underexplored within learning-based object detection methods.

This work proposes a novel approach for object detection in autonomous driving by employing a geometrical and sequential sensor fusion technique that combines 3+1D RADAR data with semantic information extracted from camera imagery via point cloud painting from the perspective view. To this end, the authors adapt the PointPainting method—originally developed for LiDAR and camera data—to the fusion of 3+1D RADAR and camera sensors.

The process begins with the application of YOLOv8-seg to generate instance segmentation masks, which are subsequently projected onto the RADAR point cloud. A set of heuristic rules is introduced as a refinement step to reduce error propagation from the segmentation phase to the detection phase. The pipeline concludes with the application of PointPillars as the object detection network operating on the painted RADAR point cloud.

The proposed approach is validated using the novel View of Delft dataset, which contains 3+1D RADAR sequences captured in urban environments. Experimental results demonstrate that the fusion of RADAR and camera data significantly outperforms the RADAR-only baseline, achieving an increase in mean Average Precision (mAP) from 41.18 to 52.67, corresponding to a relative improvement of 27.9%.

Zhao et al., in their study "YOLOv7-TS: A Traffic Sign Detection Model Based on Sub-Pixel Convolution and Feature Fusion" (contribution 9), report significant advancements in deep learning-based object detection in recent years. Within this domain, traffic sign detection represents a critical subtask with substantial potential for development; however, existing object detection methods for traffic signs in real-world scenarios often suffer from challenges such as missed detections of small objects and low overall detection accuracy.

To address these limitations, the authors propose a novel traffic sign detection model named YOLOv7-Traffic Sign (YOLOv7-TS), which leverages sub-pixel convolution and features fusion techniques. Firstly, the model exploits the up-sampling capability of sub-pixel convolution by integrating the channel dimension and introduces a Feature Map Extraction Module (FMEM) to reduce channel information loss. Furthermore, a Multi-feature Interactive Fusion Network (MIFNet) is designed to enhance information interaction across all feature layers, thereby improving feature fusion effectiveness and augmenting the capability of detecting small objects. Additionally, a Deep Feature Enhancement Module

(DFEM) is incorporated to accelerate the pooling process while enriching features at the highest layer.

YOLOv7-TS is evaluated on two traffic sign datasets: CCTSDB2021 and TT100K. Compared to the baseline YOLOv7 model, YOLOv7-TS achieves notable improvements in mean Average Precision (mAP) of 3.63% and 2.68% on the respective datasets, while employing a smaller number of parameters. These results demonstrate the effectiveness of the proposed model.

Gong et al., in their paper "Identification of Driver Status Hazard Level and the System" (contribution 10), highlight that according to survey statistics, the majority of traffic accidents result from irregularities in driver behavior and status. Due to the absence of a multi-level hazardous state grading system both domestically and internationally, this paper proposes a comprehensive state grading system for real-time detection and dynamic tracking of driver status.

The system employs an OpenMV acquisition camera combined with a cradle head tracking mechanism to dynamically capture real-time images of the driver's current driving behavior. It integrates the YOLOX with the OpenPose algorithm to identify dangerous driving behaviors by detecting unsafe objects within the cabin and analyzing the driver's posture. Additionally, an improved RetinaFace face detection algorithm is combined with the Dlib feature-point algorithm to assess driver fatigue.

Experimental results demonstrate that the proposed system achieves accuracies of 95.8%, 94.5%, and 96.3% in detecting three defined driver hazard levels (R1, R2, and R3), respectively. These findings indicate that the system holds practical significance for real-time warnings against driver distraction and related unsafe behaviors.

Ghang et al., in their paper "Cyclist Orientation Estimation Using LiDAR Data" (contribution 11), emphasize the importance of predicting cyclist behavior in autonomous vehicle decision-making. When navigating real traffic environments, a cyclist's body orientation indicates their current direction of movement, while head orientation reflects their intention to observe the road situation before making subsequent maneuvers. Consequently, accurate estimation of both body and head orientations is a critical component of cyclist behavioral prediction in autonomous driving.

This research proposes two deep-neural-network-based methods for estimating cyclist orientation—encompassing both body and head—using data from Light Detection and Ranging (LiDAR) sensors. The first method represents LiDAR data through 2D images encoding reflectivity, ambience, and range information. The second method directly utilizes 3D point cloud data to represent the sensor's output. Both approaches employ ResNet50, a 50-layer convolutional neural network, for orientation classification.

The study compares the performance of these two methods to determine the most effective use of LiDAR sensor data for cyclist orientation estimation. A dedicated cyclist dataset was developed, comprising multiple cyclists exhibiting diverse body and head orientations.

Experimental results indicate that the model based on 3D point cloud data outperforms the 2D-image-based model in cyclist orientation estimation. Furthermore, within the 3D point cloud approach, the use of reflectivity information yields more accurate estimations than ambient information.

Kibii et al., in their paper "Design and Calibration of Plane Mirror Setups for Mobile Robots with a 2D-Lidar" (contribution 12), discuss the widespread use of LiDAR sensors for environmental perception in Autonomous Robot Vehicles (ARVs). They note that the Field Of View (FOV) of LiDAR sensors can be effectively reshaped by strategically positioning plane mirrors in their vicinity. Such mirror setups are particularly advantageous for enhancing ground detection capabilities in ARVs equipped with 2D LiDAR sensors.

This paper presents an overview of several geometric mirror designs and evaluates their strengths for various vehicle types. Furthermore, the authors introduce a novel, easy-to-implement calibration procedure for 2D LiDAR mirror setups, aimed at precisely determining mirror orientations and positions. This method utilizes a single flat calibration object featuring a pre-aligned, simple fiducial marker.

The study presents measurement data collected from a prototype vehicle equipped with a 2D LiDAR sensor (with a 2 m range) employing this calibration procedure. The results demonstrate that the calibrated mirror orientations achieve an accuracy of less than 0.6° within this short range, representing a significant improvement over orientation angles derived directly from CAD models. The accuracy of the resulting point cloud data improves without any notable increase in distance noise.

Based on these findings, the authors propose general guidelines for successful calibration using their method. In conclusion, a 2D LiDAR sensor combined with two plane mirrors calibrated via this approach offers a cost-effective and precise solution for robot engineers seeking to enhance the environmental perception capabilities of ARVs.

Peiris et al., in their study "Quantifying the Foregone Benefits of Intelligent Speed Assist Due to the Limited Availability of Speed Signs across Three Australian States" (contribution 13), demonstrate that Intelligent Speed Assist (ISA), which communicates speed limits to drivers via speed sign recognition cameras, is expected to yield significant road safety improvements by enhancing speed compliance. However, due to the absence of comprehensive digital speed maps and limited cellular connectivity across Australia, this study estimates the potential safety benefits foregone when ISA relies solely on physical speed signs for optimal advisory function.

The analysis begins by identifying speed-related Fatalities and Serious Injuries (FSIs) in the Australian states of Victoria, South Australia, and Queensland over the period 2013–2018. Published ISA effectiveness estimates are then applied to quantify the potential safety benefits of ISA deployment. Taking into account the presence and absence of speed signage across these states, the study estimates the foregone savings as the proportion of FSIs that would remain unprevented due to missing speed sign infrastructure.

The results indicate that annually, 27–35% of speed-related FSIs in each state are unlikely to be prevented by ISA because of inadequate speed sign coverage, corresponding to economic losses estimated at between AUD 62 million and AUD 153 million. Despite several assumptions regarding ISA fitment rates and driver acceptance, conservative estimates suggest that the consistent placement of speed signs across various road classes and remoteness levels would substantially outweigh the costs associated with current speed sign deficiencies.

This study introduces a novel methodology for quantifying the foregone benefits of ISA attributable to suboptimal road infrastructure. It provides a valuable framework for prioritizing infrastructure investments to maximize the safety gains achievable through advanced driver assistance technologies.

Author Contributions: Writing—original draft preparation, C.I.; writing—review and editing, I.B., C.I., and M.Z. All authors have read and agreed to the published version of the manuscript.

Acknowledgments: We would like to thank all of the authors for the papers they submitted to this Special Issue. We would also like to acknowledge all of the reviewers for their careful and timely reviews to help improve the quality of this Special Issue.

Conflicts of Interest: The authors declare no conflicts of interest.

List of Contributions:

1. Barabás, I.; Iclodean, C.; Beles, H.; Antonya, C.; Molea, A.; Scurt, F.B. An Approach to Modeling and Developing Virtual Sensors Used in the Simulation of Autonomous Vehicles. *Sensors* **2025**, *25*, 3338. https://doi.org/10.3390/s25113338.
2. Olmos Medina, J.S.; Maradey Lázaro, J.G.; Rassõlkin, A.; González Acuña, H. An Overview of Autonomous Parking Systems: Strategies, Challenges, and Future Directions. *Sensors* **2025**, *25*, 4328. https://doi.org/10.3390/s25144328.
3. Hortelano, J.L.; Villagrá, J.; Godoy, J.; Jiménez, V. Recent Developments on Drivable Area Estimation: A Survey and a Functional Analysis. *Sensors* **2023**, *23*, 7633. https://doi.org/10.3390/s23177633.
4. Tollner, D.; Nyerges, Á.; Jneid, M.S.; Geleta, A.; Zöldy, M. How Do We Calibrate a Battery Electric Vehicle Model Based on Controller Area Network Bus Data? *Sensors* **2024**, *24*, 4637. https://doi.org/10.3390/s24144637.
5. Cap, R.-M.; Popa, C.-A. Hybrid MambaVision and Transformer-Based Architecture for 3D Lane Detection. *Sensors* **2025**, *25*, 5729. https://doi.org/10.3390/s25185729.
6. Radu, A.I.; Tolea, B.A.; Beles, H.; Scurt, F.B.; Tusinean, A.N. Design and Validation of an Active Headrest System with Integrated Sensing in Rear-End Crash Scenarios. *Sensors* **2025**, *25*, 4291. https://doi.org/10.3390/s25144291.
7. Beles, H.; Vesselenyi, T.; Rus, A.; Mitran, T.; Scurt, F.B.; Tolea, B.A. Driver Drowsiness Multi-Method Detection for Vehicles with Autonomous Driving Functions. *Sensors* **2024**, *24*, 1541. https://doi.org/10.3390/s24051541.
8. Montiel-Marín, S.; Llamazares, Á.; Antunes, M.; Revenga, P.A.; Bergasa, L.M. Point Cloud Painting for 3D Object Detection with Camera and Automotive 3+1D RADAR Fusion. *Sensors* **2024**, *24*, 1244. https://doi.org/10.3390/s24041244.
9. Zhao, S.; Yuan, Y.; Wu, X.; Wang, Y.; Zhang, F. YOLOv7-TS: A Traffic Sign Detection Model Based on Sub-Pixel Convolution and Feature Fusion. *Sensors* **2024**, *24*, 989. https://doi.org/10.3390/s24030989.
10. Gong, J.; Zhou, S.; Ren, W. Identification of Driver Status Hazard Level and the System. *Sensors* **2023**, *23*, 7536. https://doi.org/10.3390/s23177536.
11. Chang, H.; Gu, Y.; Goncharenko, I.; Hsu, L.-T.; Premachandra, C. Cyclist Orientation Estimation Using LiDAR Data. *Sensors* **2023**, *23*, 3096. https://doi.org/10.3390/s23063096.
12. Kibii, J.E.; Dreher, A.; Wormser, P.L.; Gimpel, H. Design and Calibration of Plane Mirror Setups for Mobile Robots with a 2D-Lidar. *Sensors* **2022**, *22*, 7830. https://doi.org/10.3390/s22207830.
13. Peiris, S.; Newstead, S.; Berecki-Gisolf, J.; Fildes, B. Quantifying the Foregone Benefits of Intelligent Speed Assist Due to the Limited Availability of Speed Signs across Three Australian States. *Sensors* **2022**, *22*, 7765. https://doi.org/10.3390/s22207765.

References

1. SAE Levels of Driving Automation™ Refined for Clarity and International Audience. Available online: https://www.sae.org/standards/j3016_202104-taxonomy-definitions-terms-related-driving-automation-systems-road-motor-vehicles (accessed on 20 September 2025).
2. Wang, H.; Shao, W.; Sun, C.; Yang, K.; Cao, D.; Li, J. A Survey on an Emerging Safety Challenge for Autonomous Vehicles: Safety of the Intended Functionality. *Engineering* **2024**, *33*, 17–34. [CrossRef]
3. Banda, L.G.; Chandra, S. Natural intelligence in autonomous vehicles: Redefining the path to human-centric adoption. *J. Retail. Consum. Serv.* **2025**, *84*, 104195. [CrossRef]
4. Liu, X.; Li, J.; Ma, J.; Sun, H.; Xu, Z.; Zhang, T.; Yu, H. Deep transfer learning for intelligent vehicle perception: A survey. *Green Energy Intell. Transp.* **2023**, *2*, 100125. [CrossRef]
5. Fernández Llorca, D.; Hamon, R.; Junklewitz, H. Testing autonomous vehicles and AI: Perspectives and challenges from cybersecurity, transparency, robustness and fairness. *Eur. Transp. Res. Rev.* **2025**, *17*, 38. [CrossRef]
6. Zhang, Y.; Carballo, A.; Yang, H.; Takeda, K. Perception and sensing for autonomous vehicles under adverse weather conditions: A survey. *ISPRS J. Photogramm. Remote Sens.* **2023**, *196*, 146–177. [CrossRef]
7. Mobini Seraji, M.H.; Shaffiee Haghshenas, S.; Shaffiee Haghshenas, S. A state-of-the-art review on machine learning techniques for driving behavior analysis: Clustering and classification approaches. *Complex Intell. Syst.* **2025**, *11*, 386. [CrossRef]

8. Zhang, F.; Yu, J.; Lin, D.; Zhang, J. UnIC: Towards Unmanned Intelligent Cluster and Its Integration into Society. *Engineering* **2022**, *12*, 24–38. [CrossRef]
9. Singh, R.; Singh Gill, S. Edge AI: A survey. *Internet Things Cyber-Phys. Syst.* **2023**, *3*, 71–92. [CrossRef]
10. Baek, Y.; Bae, B.; Shin, H.; Sonnadara, C.; Cho, H.; Lin, C.Y.; Mu, Y.; Shen, C.; Shah, S.; Wang, G.; et al. Edge intelligence through in-sensor and near-sensor computing for the artificial intelligence of things. *npj Unconv. Comput.* **2025**, *2*, 25. [CrossRef]
11. Ness, S. Vehicle detection and recognition approach in smart surveillance system: A comparative analysis. *Array* **2025**, *27*, 100473. [CrossRef]
12. Zhang, Y.; Tu, C.; Gao, K. Multisensor information fusion: Future of environmental perception in intelligent vehicles. *J. Intell. Connect. Veh.* **2024**, *7*, 163–176. [CrossRef]
13. Agus Prasetio, E.; Nurliyana, C. Evaluating perceived safety of autonomous vehicle: The influence of privacy and cybersecurity to cognitive and emotional safety. *IATSS Res.* **2023**, *47*, 160–170. [CrossRef]
14. Germiniani, S.; Nicoletti, D.; Pravadelli, G. A Systematic Literature Review on Mining LTL Specifications. *IEEE Access* **2025**, *13*, 48950. [CrossRef]
15. Kumar Tyagi, A.; Aswathy, S.U. Autonomous Intelligent Vehicles (AIV): Research statements, open issues, challenges and road for future. *Int. J. Intell. Netw.* **2021**, *2*, 83–102. [CrossRef]
16. Yang, S.; Sun, B.; Guang, H.; Wang, R.; Zheng, B.; Gong, W. CHAINS: CHAIN-Based Fusion Safety System Framework for Intelligent Connected Vehicle. *Chain* **2024**, *1*, 2–45. [CrossRef]

Disclaimer/Publisher's Note: The statements, opinions and data contained in all publications are solely those of the individual author(s) and contributor(s) and not of MDPI and/or the editor(s). MDPI and/or the editor(s) disclaim responsibility for any injury to people or property resulting from any ideas, methods, instructions or products referred to in the content.

sensors

MDPI

Review

An Approach to Modeling and Developing Virtual Sensors Used in the Simulation of Autonomous Vehicles

István Barabás [1], Calin Iclodean [1,*], Horia Beles [2,*], Csaba Antonya [3], Andreia Molea [1] and Florin Bogdan Scurt [2]

1 Department of Automotive Engineering and Transports, Technical University of Cluj-Napoca Romania, Muncii Bd. 103-105, 400114 Cluj-Napoca, Romania; istvan.barabas@auto.utcluj.ro (I.B.); andreia.molea@auto.utcluj.ro (A.M.)
2 Department of Mechanical Engineering and Automotive, University of Oradea, Universitatii Str. 1, Oradea, 7, 410087 Oradea, Romania; scurt.florinbogdan@didactic.uoradea.ro
3 Department of Automotive and Transport Engineering, Transilvania University of Brasov, Eroilor Bd. 29, 500036 Brasov, Romania; antonya@unitbv.ro
* Correspondence: calin.iclodean@auto.utcluj.ro (C.I.); horia.beles@uoradea.ro (H.B.)

Abstract: A virtual model enables the study of reality in a virtual environment using a theoretical model, which is a digital image of a real model. The complexity of the virtual model must correspond to the reality of the evaluated system, becoming as complex as necessary and nevertheless as simple as possible, allowing for computer simulation results to be validated by experimental measurements. The virtual model of the autonomous vehicle was created using the CarMaker software package version 12.0, which was developed by the IPG Automotive company and is extensively used in both the international academic community and the automotive industry. The virtual model simulates the real-time operation of a vehicle's elementary systems at the system level and provides an open platform for the development of virtual test scenarios in the application areas of autonomous vehicles, ADAS, Powertrain, and vehicle dynamics. This model included the following virtual sensors: slip angle sensor, inertial sensor, object sensor, free space sensor, traffic sign sensor, line sensor, road sensor, object-by-line sensor, camera sensor, global navigation sensor, radar sensor, lidar sensor, and ultrasonic sensor. Virtual sensors can be classified based on how they generate responses: sensors that operate on parameters derived from measurement characteristics, sensors that operate on developed modeling methods, and sensors that operate on applications.

Keywords: virtual sensors; ideal sensor; Hi-Fi sensor; RSI sensor; virtual vehicle model; autonomous driving; deep learning; machine learning

1. Introduction

Sensors are electronic devices that generate electrical signals in response to various environmental stimuli [1]. Sensors' operating principles are determined by how information is recorded, and sensors can be classified as resistive, piezoelectric, capacitive, optical, magnetic, QTC (Quantum Tunneling Composite), triboelectric effect, FET (Field-Effect Transistor), and so on [2]. Solid-state sensors are integrated devices that preprocess an analog or digital signal before delivering a sensory response that embedded systems can process.

A virtual sensor is a software emulation of a physical sensor that uses real data, mathematical models, fuzzy logic, neural networks, genetic algorithms, and ML (Machine Learning) and AI (artificial intelligence) models to estimate parameter values and anticipate scenarios [3]. ML algorithms comprise two types of mechanisms: interactive learning, in which the virtual sensor identifies relevant elements in data streams, and automatic teaching, in which the human factor proactively contributes relevant elements to training the learning process [4]. Virtual sensors are data-driven prediction models that identify the physical characteristics of the system into which they are integrated, providing a viable substitute for real sensors while also being a cost-effective solution [5–9]. A key advantage of virtual sensors lies in their ability to significantly reduce overall vehicle mass—not only by replacing physical sensors, but also by eliminating the associated wiring, insulators, and connectors needed to interface with embedded systems. This weight reduction contributes to enhanced fuel efficiency or extended range in electric vehicles, ultimately resulting in lower greenhouse gas emissions. These benefits underscore both the environmental and economic advantages of virtual sensing technologies. Such considerations are evident in models developed by companies producing autonomous shuttle buses (e.g., Apollo Baidu, EasyMile, Navya), which utilize electric vehicle platforms equipped with low-power engines (20–40 kW) and batteries with limited energy capacity (20–40 kWh) [10].

The methodology presented in this review enables the development of virtual sensors by abstracting physical sensor models and integrating empirical behavioral representations, thereby allowing output variables to be directly derived from input variables relevant to the monitored process.

Virtual sensor models are algorithms or mathematical models that estimate physical quantities using input data obtained from real sensors, statistical analysis, or input data generated by AI [11]. Virtual sensors are developed using software applications and do not require any additional or specific hardware components to function. According to [12], a virtual sensor is a software-only sensor (no hardware components) that can generate signals autonomously by integrating data/signals obtained synchronously or asynchronously from physical sensors or even other virtual sensors. Virtual sensor models developed by [13] and implemented in the software architecture of real vehicles include, for example, the tire wear virtual sensor and the brake wear virtual sensor, which monitor wear, lifespan, and potential anomalies during the operation of some of the vehicle's systems and components (headlight levelling, tire pressure, tire temperature, e-motor temperature, brake temperature, suspension displacement, and so on) [14].

Virtual sensors offer a wide range of applications in vehicle active safety systems, including ABS (Antilock Braking System), AWD (All-Wheel Drive), ACC (Adaptive Cruise Control), and SAS (Smart Airbag System). The integration of virtual sensor parameters into the previously specified systems contributes to the management of the vehicle's optimal operating state in order to improve functional performance and reduce fuel/energy consumption, which is correlated with a reduction in pollutant emissions [15].

Implementing virtual sensors on real vehicles improves the accuracy of monitored data while also expanding coverage to locations where physical sensors are unavailable. When integrated into an embedded system, virtual sensors perform preprocessing, error correction, merging, and optimization of input data sets. Essentially, virtual sensors utilize an algorithm or mathematical apparatus to process input data and produce high-complexity output data sets that match specified requirements, as demonstrated by Hu et al. [16].

Unlike physical sensors, which must be added to precise positions inside a vehicle's structural architecture in order to function properly, virtual sensors depend on data sets collected from the vehicle's embedded systems. Based on this information, virtual sensors calculate specified parameters without the need for extra hardware. The development of a virtual sensor necessitates the implementation of a functional algorithm for the system under consideration, which is based on a statistical model that reliably anticipates the essential parameters being studied [17]. Because virtual sensors are made up of software components, firmware upgrades can be accomplished remotely via the OTA (Over-The-Air) approach, eliminating the need for physical interventions to remove and install these sensors.

Virtual sensors may improve data accuracy and resolution by merging information from numerous sources (other sensors, electronic control units, actuator feedback) using advanced data fusion and processing algorithms [18,19]. These sensors might be very simple or extremely sophisticated, depending on the activities and consequences they simulate: stimulus, electrical requirements, ambient environment, operational restrictions, and functional safety [20]. However, the performance of virtual sensors could decrease with time because of changes in nonlinear dynamics and the complexity of physical processes in the environment, as well as nonlinear interactions between input and output variables [6,21]. Virtual sensors increase the accessibility of data from physical sensors, facilitating collaboration at the sensor, equipment, and organizational levels (allowing service providers to offer solutions based on the same hardware), allowing for more efficient use of the same hardware resources in interconnected systems, such as IoT (Internet of Things). Virtual sensors take data acquired by physical sensors and incorporate them into complicated software applications, where they are merged with other sources of information (databases) and processed by specialized algorithms to produce meaningful results [1,22].

Figure 1 illustrates three combinations of connected virtual and real sensors.

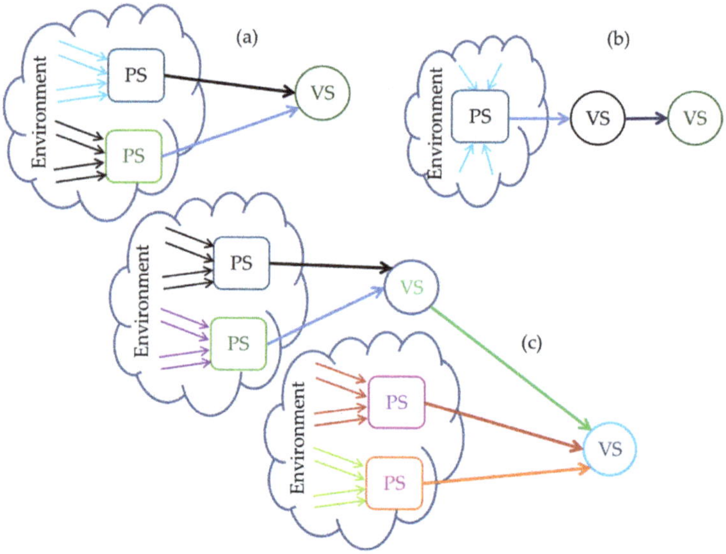

Figure 1. Different configurations of VS (virtual sensors) and PS (physical sensors). (**a**) Virtual sensors depend only on data from physical sensors (**b**) Virtual sensors depend entirely on information from other virtual sensors (**c**) Virtual sensors depend on data from both physical and virtual sensors.

(a) Virtual sensors depend only on data from physical sensors. ESC (Electronic Stability Control) uses physical sensors like gyroscopes, accelerometers, wheel speed sensors, and virtual sensors to estimate the yaw/slip angle, allowing the vehicle to maintain control in low-grip conditions or dangerous turns.
(b) Virtual sensors depend entirely on information from other virtual sensors. In the case of FCW (Forward Collision Warning) and AEB (Automatic Emergency Braking), a virtual sensor is used to predict the trajectory of the vehicle and evaluate the distance to other vehicles.
(c) Virtual sensors depend on data from both physical and virtual sensors. This configuration can be found in the DMS (Driver Monitoring System), which uses physical

sensors like a video camera and/or pressure sensors in the steering wheel and/or seat, and virtual sensors like those for estimating the driver's level of attention and detecting the intention to leave the lane.

Finally, there is a requirement for using a suitable combination of physical and virtual sensors, in addition to maintaining functional algorithms for virtual sensors up to date [23].

Tactile Mobility [24] is a platform for monitoring, processing, and storing data from specific types of physical sensors that are installed in smart and interconnected vehicles in over 70 locations globally. This platform utilizes the data to create virtual sensors that, based on recorded scenarios, generate output parameters designed to improve the safety and performance of these vehicles [24]. The Tactile Mobility platform's solution incorporates a software program into the vehicle's built-in command and control systems, improving the operating regime by delivering information on road traffic, road conditions, tire grip and condition, vehicle mass, and so on.

Another platform that enables the use of virtual sensors in the automobile sector is the Compredict Virtual Sensor Platform [25], which calibrates, verifies, and implements these sensors on a wide range of real vehicle models. Thus, the Compredict platform can generate virtual models based on cloud-stored input data for the following virtual sensor categories: suspension travel, brake wear, brake temperature, wheel force transducer, vehicle mass, strain gauge, tire wear, tire pressure, tire temperature, LV (low-voltage) battery health, HV (high-voltage) battery health, and battery anomaly.

The development of virtual sensors is accelerating; according to a market study conducted by Mordor Intelligence for the period 2025–2030 in [26], the virtual sensor market will be worth 1.37 billion USD in 2025 and 5.35 billion USD by 2030. The advancement of smart manufacturing technologies, specifically the digitization of industrial processes in addition to the digitalization and validation of real vehicle models, contributes to the development of the virtual sensor sector.

Autonomous driving is one of the most innovative and rapidly evolving technologies in the automotive industry, with the potential to transform both personal and public transportation. It offers substantial benefits in road safety, traffic efficiency, and urban mobility. Central to this technological advancement are smart sensors, which are critical to enabling autonomous vehicles to operate safely and effectively. These sensors provide environmental perception, support real-time decision-making, and allow adaptive responses to dynamic traffic and weather conditions.

However, the deployment of smart sensors in autonomous systems presents several challenges, including high costs, hardware–software integration, cybersecurity risks, and a lack of component standardization.

This review investigates the transition from physical (hardware-based) sensors to virtual (software-based) sensors, drawing on a project conducted by the authors. This project involved digitizing real sensor models and implementing their functionalities within a virtual autonomous vehicle, thereby demonstrating the potential and practical application of virtual sensing in autonomous vehicle development. Virtual sensors are critical for modern vehicles in terms of improving autonomous driving capabilities, safety, and efficiency. Table 1 shows the sensors' progression and their implementation in a vehicle's constructive architecture from level 1 to level 5 automation (according to SAE J3016^{TM}) [27,28]. The sensors used in the equipment of autonomous vehicles are constrained by their physical dimensions, mass, the necessity to be positioned in less accessible sections of the vehicle's structural architecture, and the cost of these sensors [29]. It is evident that the number of sensors increases as automation levels rise, with ultrasonic sensors and Lidar 2D/3D indicating the most significant numerical increases.

Table 1. The evolution of sensors depending on driving automation level.

Level 1		Level 2		Level 3		Level 4		Level 5 (Estimate)	
Model	Units	Model	Units	Model	Units	Model	Units	Model	Units
Ultrasonic	4	Ultrasonic	8	Ultrasonic	8	Ultrasonic	8	Ultrasonic	10
Radar long range	1	Radar long range	1	Radar long range	2	Radar long range	2	Radar long range	2
Radar short range	2	Radar short range	4	Radar short range	4	Radar short range	4	Radar short range	4
Camera mono	1	Camera mono	4	Camera mono	2	Camera mono	3	Camera mono	3
-	-	-	-	Camera stereo	1	Camera stereo	1	Camera stereo	2
-	-	-	-	Infra-red	1	Infra-red	1	Infra-red	2
-	-	-	-	Lidar 2D/3D	1	Lidar 2D/3D	4	Lidar 2D/3D	4
-	-	-	-	Global navigation	1	Global navigation	1	Global navigation	1
Total	8	Total	17	Total	20	Total	24	Total	28
2012		2016		2018		2020		Estimated by 2030	

2. Classification of the Virtual Sensor

2.1. Virtual Sensor Model

According to research teams, the key real sensors frequently utilized in the constructive architecture of autonomous vehicles, sensors that form the basis for defining the virtual sensors used in modeling the virtual vehicle model, are as follows [30]:

1. Camera sensors generate synthetic data on the recognition and classification of objects in the area [31–33], in addition to the vehicle's positioning and orientation relative to close to objects and V2V (Vehicle-to-Vehicle) communication [34] based on the VLC (Visible Light Communication) principle [35]. The advantages of the camera sensor include the ability to provide data in real time, low latency in data acquisition and processing, adaptability to extreme lighting conditions (low lighting, bright lighting), accurate estimation of object position and orientation, and low production and implementation costs. The constraints of camera sensors include the need for a direct view of the surrounding objects, susceptibility to unexpected changes in lighting conditions, and the need for greater computer capacity due to the large quantities of data that are constantly generated.
2. Radar sensor generates data based on the reflection duration of radio waves ToF (Time of Flight) when detecting nearby target vehicles [36,37] and uses ML methods to estimate the current and future positions of nearby vehicles [38], respectively, using DL (deep learning) methods to avoid collisions [39]. The benefits of radar sensors include the capacity to provide the location of target vehicles in real time, flexibility to severe weather conditions (rain, snow, fog), and low manufacturing and installation costs. The constraints of radar sensors include the requirement for increased computer capacity due to the massive volumes of data generated on a continuous basis, as well as a reliance on extra hardware systems and software.
3. Lidar sensors provide a system based on generating a point cloud through 2D and 3D laser scanning for real-time localization of static and dynamic objects in proximity [40,41] and applies the YOLO (You Only Look Once) picture segmentation technique [42]. The advantages of lidar sensors include the ability to localize static and moving objects in proximity precisely. The disadvantages of lidar sensors include the need for greater computer power due to the large quantity of data generated continuously, sensitivity to bad weather conditions (rain, snow, fog), and high manufacturing and implementation costs.

Sensor fusion is the process of combining sensor signals [43,44] using CNN (convolutional neural network) neural networks, processing these signals with DL-type AI elements [45,46], detecting nearby objects in real time [47,48], and then making predictions about the evolution of these objects [49–52].

The virtual sensor models presented in this review were created, tested, and calibrated using the CarMaker simulation application from IPG Automotive, which is extensively used in the automotive industry for virtual vehicle model development at all stages. CarMaker is a platform that enables the development of any virtual test scenarios that are connected to other software applications [53].

Yeong et al. [54] classified physical and virtual sensors as smart or non-smart. Smart sensors are directly related to the IoT concept, and they are systems made up of interconnected devices that may collect and transport data remotely without the need for human involvement. A smart sensor is an IoT device that can condition and select incoming signals, process and interpret the generated data, and make decisions without the assistance of a separate processing unit [55].

Virtual sensors can be classed using the following criteria [56–58]:

1. Sensor fidelity could be classified as high-, medium-, or low-income.
2. Method for collecting information from the environment:
    (a) A deterministic strategy based on the simulation application's mathematical apparatus and involving the usage of a vast volume of input parameters to represent the ideal behavior and response of the virtual sensor as accurately as possible;
    (b) A statistical technique based on statistical distribution functions, which include the normal, binomial, Poisson, or exponential distribution;
    (c) An electromagnetic field propagation approach to simulate electromagnetic wave propagation using Maxwell's equations.
3. The objective of using sensors is to develop a vehicle's operating mode based on observed metrics and to perform diagnostics using AI-based maintenance techniques to define the smart maintenance regime.

Virtual systems developed for simulation applications use additional virtual sensors that are intended to replace certain partial functionalities of the main real sensors to reduce the volume of input data, reduce computing power requirements, calibrate the main sensor, and provide an optimized output data stream [29].

CarMaker classifies virtual sensors into three types: ideal sensors, Hi-Fi (high-fidelity) sensors, and RSI (Raw Signal Interface) sensors (Figure 2). These virtual sensor models are intended to maximize the performance of the virtual vehicle model on which they are installed, as well as to assist the command-and-control system in developing and expanding the specific capabilities of each sensor to a higher class of sensors [59,60].

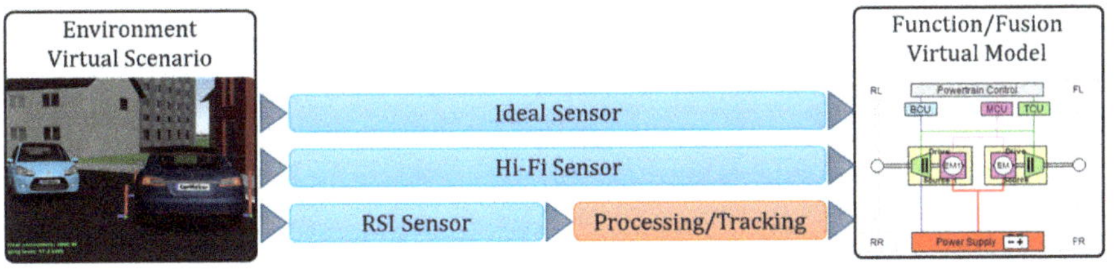

Figure 2. Virtual sensor classification.

The virtual model developed in CarMaker incorporates the following command and control systems for advanced assistance functions: ADAS (Advanced Driver Assistance System): ACC, EBA (Emergency Brake Assist), LDW (Lane Departure Warning), LKA (Lane Keeping Assist), PA (Park Assist), ILA (Intelligent Light Assist), and TSA (Traffic Sign Assist). All these embedded systems evaluate and interpret data about the motor and/or vehicle's operating mode by combining various virtual sensor models [61,62].

2.1.1. Ideal Sensors

The role of ideal sensors in the CarMaker simulation program is to collect information from the simulation environment and transmit it to an embedded system. Ideal sensors are virtual entities developed using software that are independent of technology (Figure 3) and equip the virtual vehicle model with the following: slip angle, inertial, object, free space, traffic sign, line, road, and object-by-line sensor. Physical impacts that occur in the real environment in the case of a model integrated into the HiL (Hardware-in-the-Loop) system have no effect on these ideal sensors that are integrated into the SiL (Software-in-the-Loop) model and do not generate information similar to a real sensor [63].

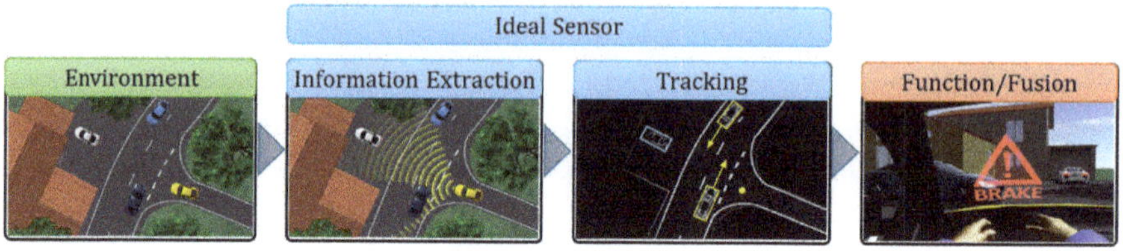

Figure 3. Ideal sensors.

2.1.2. Hi-Fi Sensors

Hi-Fi sensors filter the information supplied to the embedded system and provide data on the physical impacts that occur in the real environment, particularly the detection and classification of static and dynamic objects in the area. The virtual vehicle model is equipped with the following Hi-Fi sensors (Figure 4): camera, global navigation, and radar sensors. Hi-Fi sensors have a role in reducing the impacts of false positives and false negatives that can occur in object perception and identification due to scenarios where part of the objects overlap, or environmental conditions prevent exact identification [64].

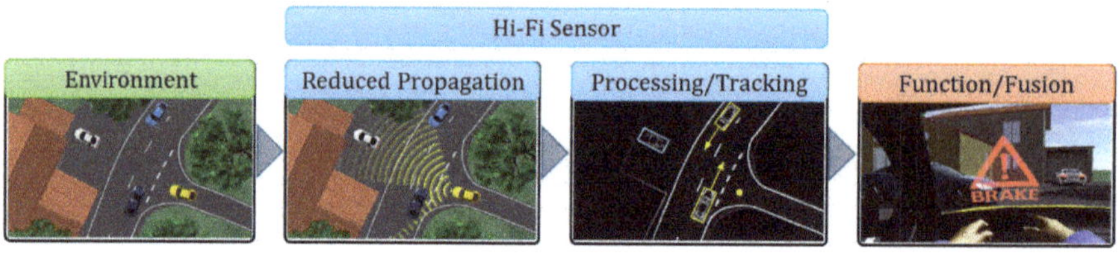

Figure 4. Hi-Fi sensors.

2.1.3. RSI Sensors

RSI sensors provide raw data and function identically to real sensors. The system filters, extracts, and interprets the data sent by the RSI sensors. Processing the information assigned by the RSI sensors necessitates high computational power, particularly for graphics processing provided by the GPU. There are RSI sensor types that conduct post-processing of the input in order to reduce the computing load on the embedded system. In the CarMaker simulation program, the RSI sensors identify objects in traffic and proximity, as well as all 3D surfaces in the surrounding environment. The utility IPGMovie, which is integrated with CarMaker, provides raw information for all these images. The virtual vehicle model is equipped with the following RSI sensors (Figure 5): ultrasonic RSI and lidar RSI [64].

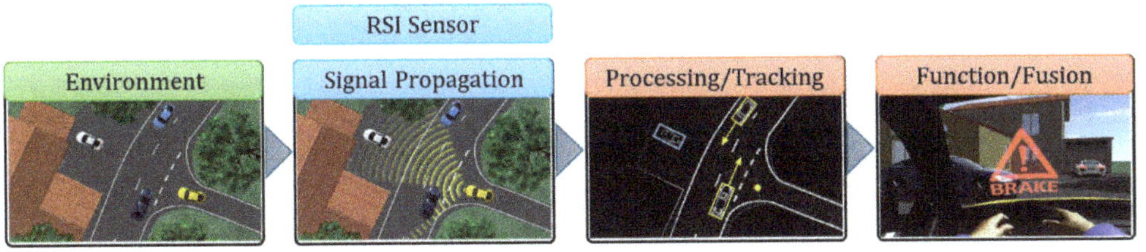

Figure 5. RSI sensors.

The use of RSI sensors in a virtual environment requires modeling the properties of the materials that compose the objects in their vicinity, namely relative electric permittivity for electromagnetic waves and scattering effects. The direction and intensity of the field of waves reflected off 3D surfaces are significantly influenced by the material's characteristic properties [65].

Figures 3–5 illustrate the information cycle, which starts with extracting relevant details from the environment (green) and continues with processing and transmitting these data (blue) to embedded command and control systems (red).

RSI sensors process 3D images and offer real-time output for embedded systems, namely images and videos for the IPGMovie and/or MovieNX simulated scenario rendering system in CarMaker (Figure 6) [66].

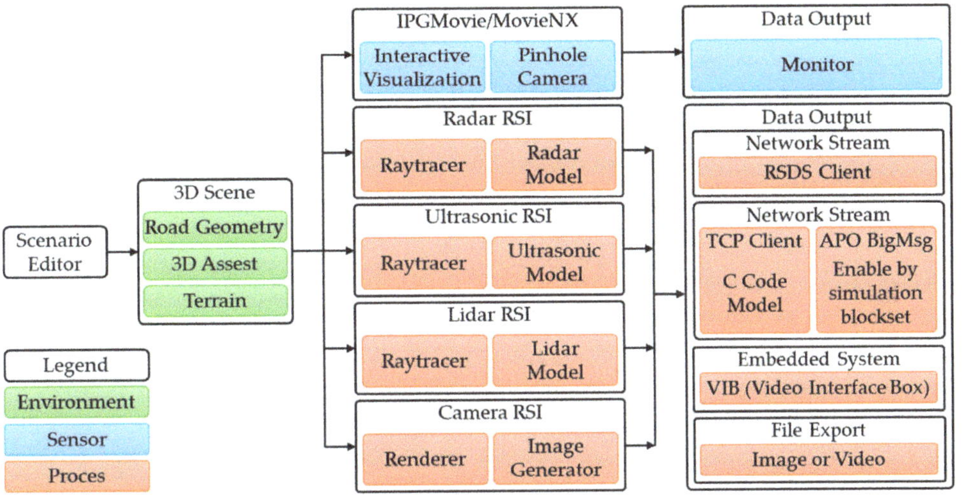

Figure 6. Interfaces and output format for RSI sensors.

2.2. Virtual Vehicle and Environmental Model

A virtual vehicle model is a prototype that precisely replicates the characteristics of the elements and systems of a real model using mathematical and physical models. After validating the model, the simulation method enables the virtual vehicle to run in any user-defined scenario in a short period of time and at a low cost [20,67].

A virtual vehicle model, also referred to as a DTw (Digital Twin), is a digital image of a physical vehicle. Renard et al. [68] define a DTw as an entity made up of a real model in a real space, a virtual model in a virtual space, and the data links that connect the real and virtual models. DTw systems, thanks to bidirectional communication, allow the virtual model to be updated when the real model's state changes, and vice versa [69–72].

Tu et al. [73] argue that autonomous driving technology represents a technological revolution in transportation. It requires the integration of artificial intelligence and smart sensors, which—unlike human drivers—facilitate significantly faster and more intelligent decision-making.

Navya Autonom® Shuttle is an autonomous shuttle bus vehicle designed for public passenger transportation and based on the architecture of a fully electric vehicle. The Navya Autonom® Shuttle was introduced by the French start-up Navya in October 2015, and the main technical requirements (navya.tech) were utilized for developing the virtual model in the CarMaker implementing virtual sensors (Figure 7) [10,27,74–76].

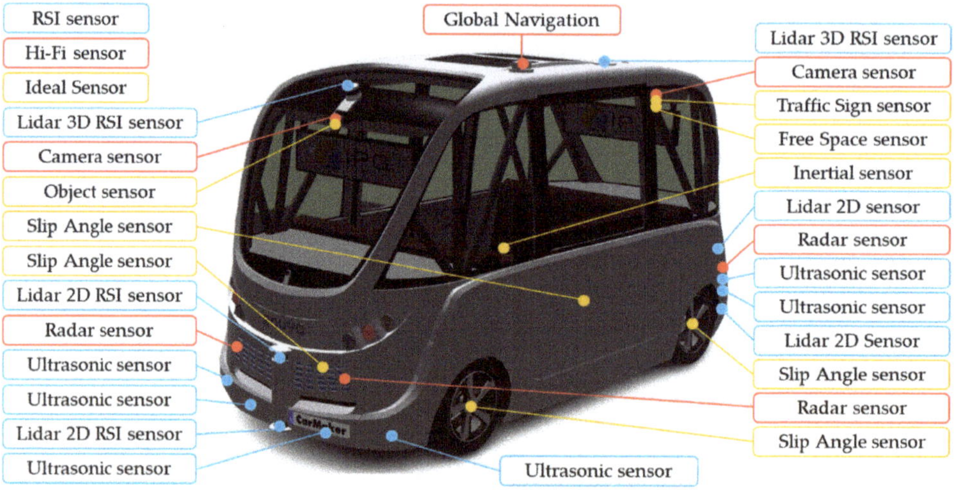

Figure 7. Virtual vehicle model for autonomous shuttle bus equipped with virtual sensors model: ideal sensors (yellow), Hi-Fi sensors (red), and RSI sensors (blue).

The virtual environmental model consists of a virtual road and a virtual environment. The virtual road in which the virtual model of the autonomous vehicle travels was defined by digitizing the real route in Lyon, France, using geographical coordinates (latitude, longitude, and altitude) extracted from Google Earth (Figure 8) [27,77,78]. The digitized route was converted to SRTM (Shuttle Radar Topography Mission) coordinates [78] using the GPSPrune application [79]. The route with the altitude profile was loaded into CarMaker's IPGRoad utility [80], which defined the following parameters in addition to the geographical coordinates (latitude, longitude, altitude): dimensions (length, width), connection angle, curvature, inclination, speed limit, and friction coefficient.

The virtual environment for the computer simulations was created using the CarMaker application's Environment utility, which allowed the following atmospheric conditions to be defined: reference temperature, air density, air pressure, air humidity, cloud model, cloud intensity, fog, visibility, rain rate, wind velocity, and wind angle [62].

The autonomous driving system operating algorithm in CarMaker's Vehicle Control section defines virtual driver behavior by addressing different driving styles corresponding to a human driver's reaction speed and performance criteria under optimal energy consumption conditions [81].

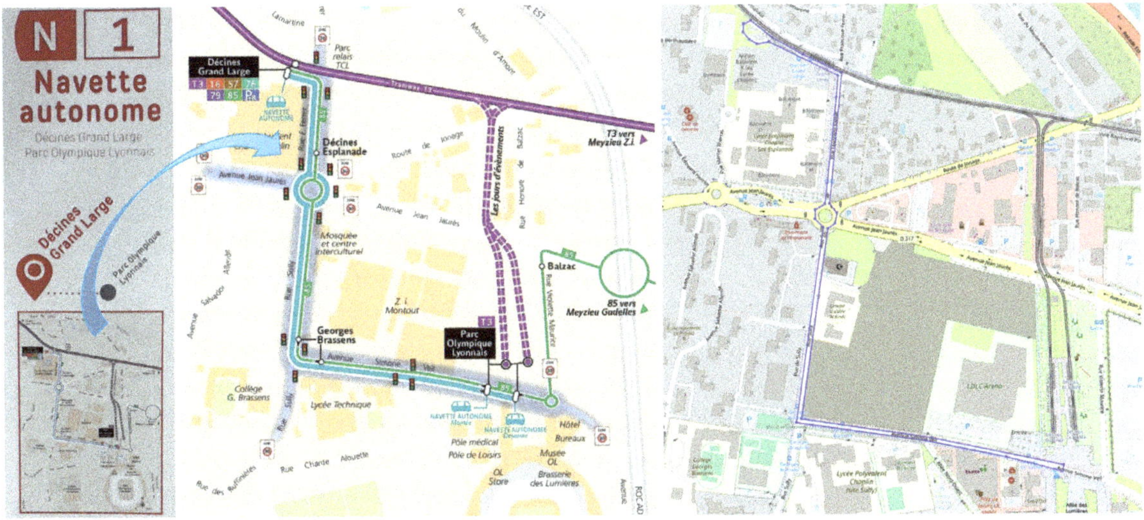

Figure 8. Virtual road vs. real road (TCL Lyon vs. GPSPrune—photo author (C.I.)).

3. Characteristics of the Virtual Sensor

3.1. Characteristics of Ideal Sensor

3.1.1. Slip Angle Sensor

The Pacejka model is essential for realistic and thorough vehicle dynamics simulations, particularly when complex maneuvers and tire behavior are involved. It provides the necessary realism in illustrating tire forces, which are essential for vehicle motion. The Pacejka model is an useful mathematical model for simulating tire behavior in autonomous vehicle simulations in CarMaker because it accurately represents the complicated, nonlinear interaction between the tire and the road surface.

The slip angle sensor monitors the lateral slip angle between the steering wheel angle and the vehicle's direction of motion. The slip angle sensor is located near the vehicle's steering wheel [27,62,66].

The yaw angle is helpful in active safety systems because it controls cornering stability, prevents vehicle rollovers, and avoids lane departure. Controlling the yaw angle is required because a big yaw angle reduces the tires' capacity to create lateral forces and greatly impairs the effectiveness of the vehicle control system. In addition to the yaw angle, the yaw rate is also a required variable for vehicle stability management [82,83].

The Pacejka model [84], a lateral force model, describes the complexity of the interaction between the tire and the road surface during dynamic maneuvers that are specific to autonomous vehicles. The Pacejka model is a semi-empirical mathematical model that represents the behavior of forces and moments created by a tire in contact with the road surface. It additionally provides a nonlinear representation of lateral forces, accounting for both large slip angles and normal forces. The Pacejka model (Figure 9) is commonly used in vehicle simulation and control, particularly for the development of advanced support systems and autonomous vehicles. The Pacejka model includes the following broad shape [83–85].

$$\mathrm{y(x)} = \mathrm{D}\cdot\sin\{\mathrm{C}\cdot\arctan[(1-\mathrm{E})\mathrm{Bx} - \mathrm{E}\cdot\arctan(\mathrm{Bx})]\}, \tag{1}$$

with

$$\mathrm{Y(X)} = \mathrm{y(x)} + \mathrm{S_V}, \tag{2}$$

$$x = X + S_H, \tag{3}$$

where y(x) represents F_x, F_y, M_z; B is the stiffness factor; C is the shape factor; D is the peak value; E is the curvature factor; S_H is the horizontal displacement; and S_V is the vertical displacement.

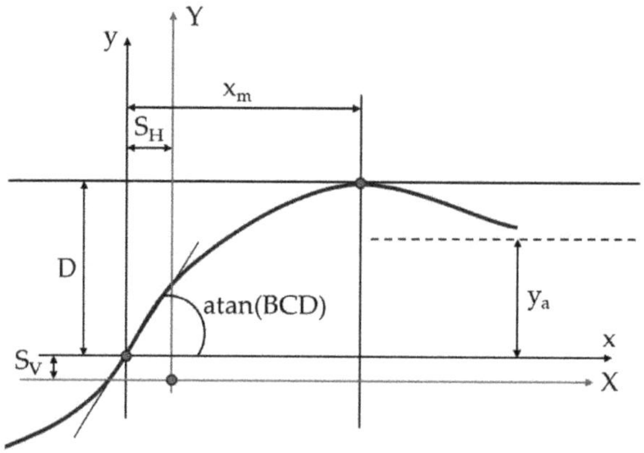

Figure 9. The Pacejka model.

The Pacejka model generates a curve that passes through the origin (x = y = 0), reaches a maximum value, and then tends to a horizontal asymptote. To create a more accurate representation of tire behavior, the Pacejka model allows the curve's position to be adjusted by inserting two translations, S_H and S_V. These translations allow for the adjustment of potential asymmetries in the experimental data, resulting in a better fit between the model and reality. For the specific coefficients B, C, D, and E, the curve exhibits anti-symmetry with respect to the origin. The coefficient D specifies the peak value, whereas the product of B, C, and D determines the curve's beginning slope. The coefficient C changes the operating range limits in the general formula, determining the shape of the curve. The stiffness factor is defined as the coefficient B, which is calculated using the inclination relative to the origin. The coefficient E is introduced to manage the curvature peak while also controlling its horizontal position. The shape factor C can be derived from the peak height and horizontal asymptote using the following formula [83,84,86]:

$$C = 1 \pm \left(1 - \frac{2}{\pi}\arcsin\frac{y_a}{D}\right). \tag{4}$$

The curvature factor E is computed from B and C for the position x_m of the peak value using the following equation (if C > 1) [87]:

$$E = \frac{Bx_m - \tan\left[\frac{\pi}{(2C)}\right]}{Bx_m - \arctan(Bx_m)}. \tag{5}$$

The specific force is expressed as follows in both the longitudinal and transverse directions [88]:

$$\Gamma = D\cdot\sin\{C\cdot\mathrm{arctg}[B\cdot s - E\cdot(B\cdot s - \mathrm{arctg}(B\cdot s))]\}. \tag{6}$$

The tire model parameters B, C, D, and E are all determined by the tire's vertical load, which is defined as follows [89–91]:

$$F_{z1} = l_r \cdot \left(\frac{mg}{2l} + \frac{ma_y h}{2b_f l}\right) - \frac{ma_x h}{2l} F_{z2} = l_r \cdot \left(\frac{mg}{2l} - \frac{ma_y h}{2b_f l}\right) - \frac{ma_x h}{2l} F_{z3} = l_f \cdot \left(\frac{mg}{2l} + \frac{ma_y h}{2b_r l}\right) + \frac{ma_x h}{2l} F_{z4} = l_f \cdot \left(\frac{mg}{2l} - \frac{ma_y h}{2b_r l}\right) + \frac{ma_x h}{2l} \quad (7)$$

where F_{z1}, F_{z2}, F_{z3}, and F_{z4} represent the vertical load of each corresponding tire, h is the height of the vehicle's mass center, g is the acceleration of gravity, lf and lr are the distances from the vehicle's gravity center to the front and rear axles, respectively, and bf and br are the half treads of the front and rear wheels. The sideslip angle of each tire can be defined as

$$a_1 = \delta - \arctan\frac{v_y + l_f\gamma}{v_x + b_f\gamma/2} a_2 = \delta - \arctan\frac{v_y + l_f\gamma}{v_x - b_f\gamma/2} a_3 = -\arctan\frac{v_y - l_r\gamma}{v_x + b_r\gamma/2} a_4 = -\arctan\frac{v_y - l_r\gamma}{v_x - b_r\gamma/2} \quad (8)$$

The tire slip rate can be obtained as follows:

$$s_j = \mathrm{sgn}(n_j r - v_{nj}) \frac{(v_{nj}, n_i r) - (v_{nj}, n_i r)}{(v_{nj}, n_i r)} \quad (9)$$

where s_j is the jth tire's slip rate, n_j is its rotational speed, v_{nj} is its linear velocity, and r is its effective wheel radius.

To run the simulations in CarMaker, the virtual vehicle model dynamics library was utilized, which calculates the wheel slip angle using the force and moment equilibrium equations as the derivative of the virtual vehicle's slip and yaw ratio [92,93]:

$$\dot{B} = -\gamma + \frac{2C_f}{mv_x}\left(\delta - \beta - \frac{l_f\gamma}{v_x}\right) + \frac{2C_r}{mv_x}\left(-\beta + \frac{l_r\gamma}{v_x}\right), \quad (10)$$

$$\dot{\gamma} = \frac{2C_f l_f}{I_z}\left(\delta - \beta - \frac{l_f\gamma}{v_x}\right) + \frac{2C_r l_r}{I_z}\left(\beta + \frac{l_r\gamma}{v_x}\right). \quad (11)$$

For β vehicle slip angle and γ vehicle yaw rate, C_f is the vehicle front tire cornering stiffness, m the vehicle mass and v_x the longitudinal velocity. Vehicle parameters include δ as the front-steering angle, β as the body slip angle, l_f as the distance from center of gravity to the front axle, and v_x as the longitudinal velocity, C_r as the vehicle rear tire cornering stiffness, l_r as the distance between the center of gravity and the rear axle, and I_z as the moment of inertia for the vehicle's yaw axis.

3.1.2. Inertial Sensor

The inertial sensor determines the vehicle's position, speed, and acceleration. It is based on a three-axis accelerometer (x,y,z) that outputs information about the vehicle's translational speed, translational acceleration, and rotational acceleration. The inertial sensor is located in the center of the vehicle [27,62,66].

Inertial sensors, coupled with the slip angle sensor, comprise the inertial positioning system IMU (Inertial Measurement Unit), which also incorporates a three-axis accelerometer. Inertial measurements include linear acceleration, angular velocity, and angular acceleration. The dynamic parameters (roll, pitch, and yaw rate) measured by the inertial sensor are incorporated in the following relationships [94]:

$$\begin{bmatrix} \dot{Q}_0 \\ \dot{q}_1 \\ \dot{q}_2 \\ \dot{q}_3 \end{bmatrix} = \frac{1}{2} \cdot \begin{bmatrix} 0 & -\dot{\phi} & -\dot{\theta} & -\dot{\varphi} \\ \dot{\phi} & 0 & \dot{\varphi} & -\dot{\theta} \\ \dot{\theta} & -\dot{\varphi} & 0 & \dot{\phi} \\ \dot{\varphi} & \dot{\theta} & -\dot{\phi} & 0 \end{bmatrix} \cdot \begin{bmatrix} q_0 \\ q_1 \\ q_2 \\ q_3 \end{bmatrix}. \quad (12)$$

In three-dimensional rotation computations, the quaternion $[q_0 q_1 q_2 q_3]^T = Q$ represents the ϕ roll, θ pitch, and ψ yaw rate. Based on the Taylor series results, the quaternion solution from instant k to moment k + 1 is

$$Q(k+1) = \left(I\left(1 - \frac{\Delta\Theta^2}{8}\right) + \frac{\Delta\Theta}{2} \right) \cdot Q(k), \tag{13}$$

$$\Delta\Theta = \int_k^{k+1} \begin{bmatrix} 0 & -\dot{\phi} & -\dot{\theta} & -\dot{\varphi} \\ \dot{\phi} & 0 & \dot{\varphi} & -\dot{\theta} \\ \dot{\theta} & -\dot{\varphi} & 0 & \dot{\phi} \\ \dot{\varphi} & \dot{\theta} & -\dot{\phi} & 0 \end{bmatrix} dt \approx \begin{bmatrix} 0 & -\Delta\Theta_\phi & -\Delta\Theta_\theta & -\Delta\Theta_\varphi \\ \Delta\Theta_\phi & 0 & \Delta\Theta_\varphi & \Delta\Theta_\theta \\ \Delta\Theta_\theta & -\Delta\Theta_\varphi & 0 & \Delta\Theta_\phi \\ \Delta\Theta_\varphi & \Delta\Theta_\theta & -\Delta\Theta_\phi & 0 \end{bmatrix}. \tag{14}$$

For the dt sampling time, $\Delta\theta = [\Delta\theta_\phi \Delta\theta_\theta \Delta\theta_\varphi]^T$ angular increments include $\Delta\theta_\phi$, $\Delta\theta_\theta$, and $\Delta\theta_\psi$ roll, pitch, and yaw angles.

The inertial sensors IMU and VDM (vehicle dynamics model) can be used to explain the performance and safety of autonomous vehicles by properly determining the slip angle and altitude. VDM is composed of two basic parts: a delayed estimator and a predictor (Figure 10) [95]. The delayed estimator includes two types of estimators: those based on IMU data and those based on vehicle dynamics models. IMU estimators directly estimate variables like speed and attitude, but VDM estimators use mathematical models based on measurements from other sensors, like wheel speed sensors.

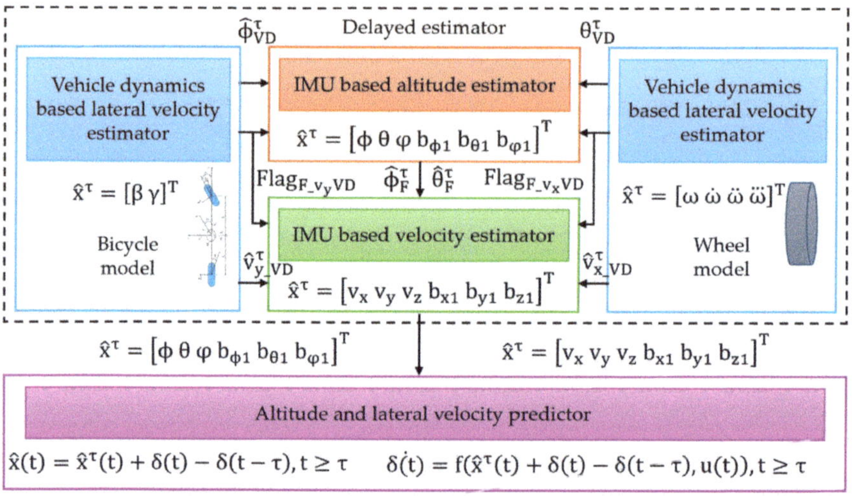

Figure 10. Vehicle dynamics model.

Under normal driving conditions, data from VDM estimators are utilized to correct errors that may develop in IMU estimates, using a Kalman filter to predict the roll and pitch angles [95].

Vehicle dynamics models provide a more in-depth understanding of the vehicle's overall behavior, which enhances estimation accuracy. Dynamic models may become less accurate under harsh driving situations, such as hard breaking or quick bends. In such cases, the IMU estimators are temporarily separated from the VDM estimators to prevent error propagation. To synchronize the input from the two estimators, a delay is added to the VDM-based estimate. To predict the system's current state, the predictor uses delayed estimates as well as information about the vehicle's controls. This enables a more precise estimate of the slip angle and attitude, even in dynamic conditions [96].

3.1.3. Object Sensor

Scanning the environment is an important stage for an autonomous vehicle since it offers information that allows it to perceive and understand its surroundings. This first stage is critical for obtaining a thorough and up-to-date image of the traffic situation, allowing the vehicle to make informed decisions and travel safely. To perform this comprehensive and multidimensional scanning, autonomous vehicles employ a complex network of sensors, each with particular expertise in providing information. The object sensor is a software component that simulates the operation of a real video camera in an autonomous vehicle. It detects objects in traffic and estimates their distance, with the nearest object considered the target. The data are utilized to make decisions in autonomous driving. The object sensor employs image processing techniques and AI to identify and track objects such as vehicles, pedestrians, and bicycles, and the distance between them can be calculated using trigonometric calculations [97].

Bewley et al. in [98] describe a simple and efficient technique for real-time multi-object tracking. This method focuses on the quick connection of items observed in consecutive frames, highlighting the need for accurate detection for quality tracking. Using conventional methods such as the Kalman filter and the Hungarian algorithm, the method achieves an accuracy comparable to complex systems. This method's simplicity and efficiency make it ideal for real-time applications including pedestrian tracking in autonomous driving systems.

To predict the position of an object in the frame, a linear motion model with constant velocity is utilized, which is unaffected by other objects or camera movement.

$$X = \left[u, v, s, r, \dot{u}, \dot{v}, \dot{s}\right]^T, \tag{15}$$

where u and v represent the horizontal and vertical location of the target center in pixels, and s and r indicate the scale (area) and aspect ratio of the target's bounding box, respectively [98,99].

The DPM (Deformable Part Model) algorithm, which detects surrounding vehicles and pedestrians, is an advanced object recognition method. DPM looks for and evaluates the characteristics of target objects in the images collected by the video camera, where objects are defined as a collection of parts organized in a deformable configuration. Each part represents the local attributes of an object's appearance, whereas links between pairs of parts define the deformable configuration. The DPM algorithm learns to identify objects of interest from the background by comparing positive and negative examples. Therefore, the algorithm develops a collection of filters that respond to the object's specific properties, such as edges, corners, and textures. A filter is built using a rectangular template defined by a matrix of dimensional vectors d. The response of the filter F at a position (x,y) on a feature map G is the "local dot product" of the filter and a sub-window of the feature map at (x,y) [100,101].

$$\sum_{x',y'} F\left[x', y'\right] \cdot G\left[x + x', y + y'\right]. \tag{16}$$

A feature pyramid is used to specify an object's size and position in a picture. A pyramid is a series of feature maps with varying resolutions. In practice, pyramids are created by computing a conventional image pyramid, smoothing, and repeatedly downsampling. A feature map is then generated from each point in the picture pyramid, as depicted in Figure 11 [102,103].

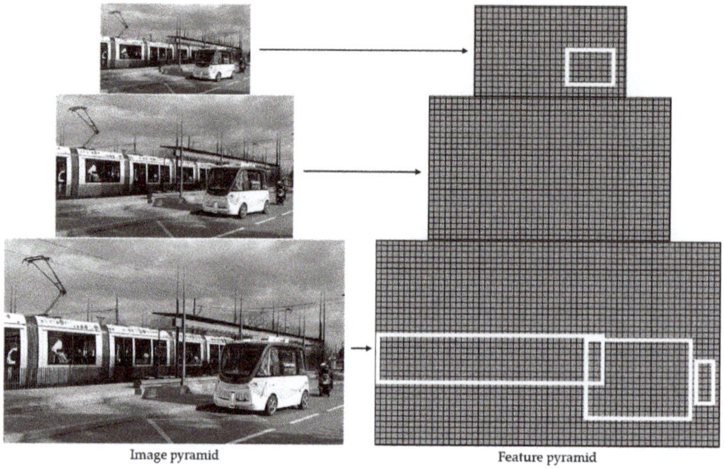

Figure 11. A feature pyramid undergoing instantiation of a person model within it. The part filters are positioned at double the spatial resolution of the root location.

In the CarMaker virtual environment, the object sensor is a virtual sensor that detects objects in traffic and calculates their distance. The most appropriate target object is determined by its proximity to the sensor [27,62]. The object sensor transmits information to the ACC system, which is responsible for automatically adapting the acceleration in the vehicle's movement so that it maintains a consistent speed in comparison to the vehicles in front of it [67].

Figure 12 illustrates the structure of the ACC system, which includes the object sensor. A cluster of two sensors collects information about detected objects, one using an antenna for objects at a distance (long-range antenna) and one using an antenna for objects in close proximity to the vehicle (short-range antenna) and generates a list of intercepted objects. Raw sensor data are used to locate and track detected objects (object-sensor-detected object list), with tracking algorithms performing data fusion to ensure control over the vehicle's cruising speed while maintaining a safe distance from the relevant detected objects [66].

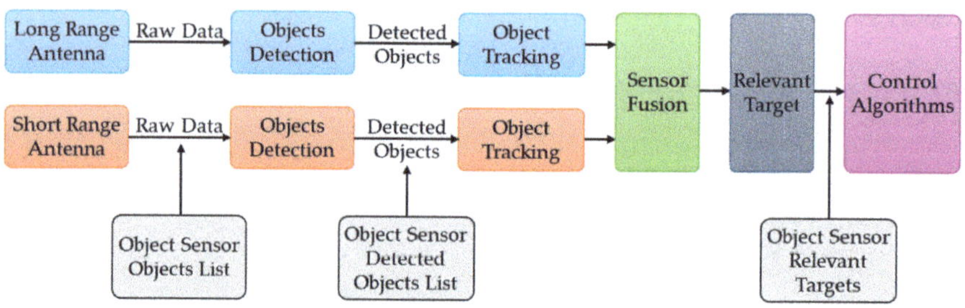

Figure 12. Object sensor integrated into the ACC system.

The object list in the object sensor interface transmits data about the objects in the CarMaker application database that are detected by the sensor's detection field. These objects are identified based on the characteristics of object ID (identifier), object dimensions, object orientation, distance to the object, and object speed, which correspond to the angle of incidence between the sensor beam and the object.

The algorithm for detecting the closest object is based on scanning all objects within range and selecting the relevant target objects for the object sensor by identifying the trajectory and movement lane, respectively, by calculating the movement speed and distance to the object (Figure 13) [66].

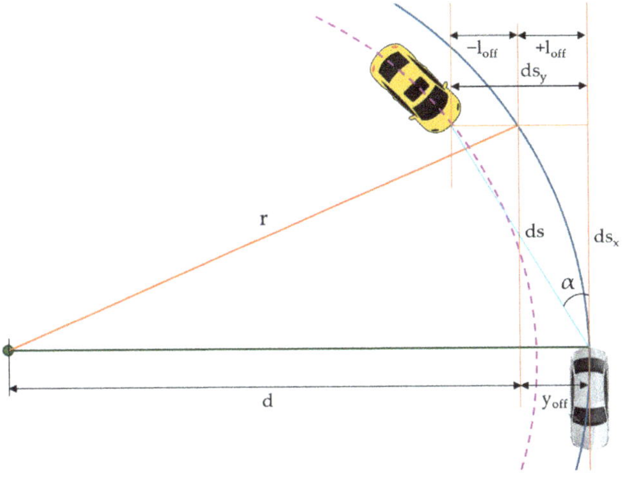

Figure 13. Object trajectory identification algorithm.

In Figure 14, ds represents the projected distance to the target vehicle, α the angle of the target vehicle in sensor frame, ds_x the component in the X direction of the projected distance, dsy the component in Y direction of projected distance in the sensor frame, r the turning radius, y_{off} the imaginary vehicle offset in sensor frame at the target position, l_{off} half of the vehicle lane width, ay the lateral acceleration of vehicle, and v the vehicle speed. Target selection algorithms can configure the sensor response in two modes:

- Nearest object—this is the closest visible object that is considered a relevant target;
- Nearest object in the path—this is the closest object within an interval of an estimated vehicle trajectory.

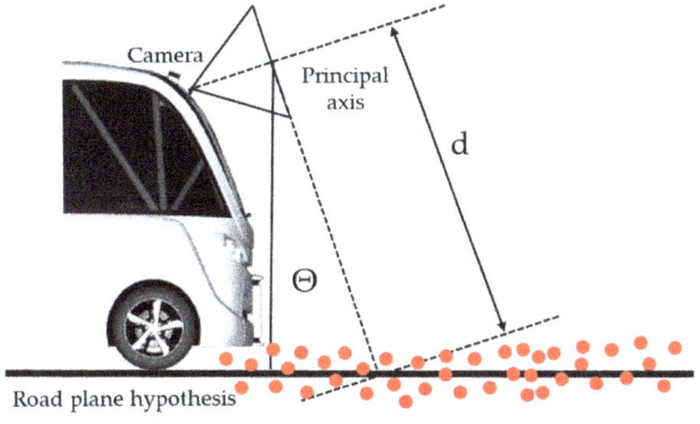

Figure 14. Detecting the road plane from a point cloud.

The following relationships characterize components in object sensors:

$$ds_x = ds \cdot \cos(\alpha), \quad (17)$$

$$Ds_y = ds \cdot \sin(\alpha). \quad (18)$$

The following relationships describe the vehicle's offset in the sensor's perception of the position of adjacent objects, and thus the limits of the vehicle's trajectory.

$$y_{off} = r - d = \left(r - \sqrt{r^2 - ds_x^2}\right) \cdot \text{sign}(a_y), \quad (19)$$

$$L_{off} = \frac{\text{vehicle}_{lane}}{2} + \text{lane}_{offset}, \quad (20)$$

$$\left(y_{off} - l_{off} < ds_y\right) \cap \left(ds_y < y_{off} + l_{off}\right). \quad (21)$$

The CarMaker object sensor module generates an item list for each configured sensor, with quantities for each traffic object in the sensor's view. After scanning the environment, the list of objects that can be recognized around the virtual vehicle model will include the following markers [66]:

- Object ID, a name or a code used to identify an object;
- Path direction (reference and closest point);
- Relative distance and velocity (between the reference and the nearest positions);
- Relative orientation in the axle x-y-z (the reference point);
- Sensor frame's x-y-z distances (between the reference and the nearest point);
- Sensor frame's x-y-z velocity (between the reference and the nearest point);
- Flag object has been identified (in the sensor viewing area);
- Flag object has been identified (in the observation area);
- Incidence angles between the sensor beam and the object being detected from proximity;
- Width, height, length of the object, and height above the ground.

3.1.4. Free Space Sensor

The free space sensor is a software component that detects free space around the vehicle and uses this information to plan routes and avoid obstacles. The sensor creates an accurate map of the surrounding environment by combining data from several sensors, including cameras, lidar, and radar. The sensor data are analyzed to identify barriers and compute their distances. The free space sensor information is utilized to build safe and efficient trajectories, making it an important component in assuring the safety and efficiency of autonomous vehicles.

The free space sensor is an extension of the object sensor, with the sensor beam separated into horizontal and vertical segments. Each segment determines the closest point of the observed objects in traffic, in addition to the vehicle's angle with respect to these objects and their respective speeds. The sensor scans the environment and determines the free and occupied area in the vicinity, guiding the vehicle's progress through it [27,63].

The free space sensor plus is an extension of the sensor that detects all around items using a separate computational approach based on 3D image analysis. Three-dimensional image analysis of objects in proximity uses two filtering methods: the closest point on the object surface (nearest) and the strongest point on the object surface (strongest). The nearest finds the point on the object surface (represented by a pixel in the generated image) that is closest to the sensor position. The strongest determines the point on the object's surface with the least reflection angle relative to the incident vector for each pixel within the sensor's detecting range [66].

Open space identification methods use either 2D models (camera pictures) or 3D models (point clouds obtained from lidar sensors or stereo cameras). Two-dimensional approaches segment the road using low-level cues including color and texture, but they may fail if the road textures are inconsistent. However, 3D algorithms may have difficulties recognizing modest height variations, such as those between the road and the sidewalk.

The hybrid method combines the benefits of 2D and 3D modeling to overcome the limits of each methodology and provide more robust open space identification [104–106]. Therefore, the use of 3D information obtained from the input 3D point cloud renders road plane recognition more efficient. The road plane is determined in a parametric space, which includes the plane distance from the center of the room as well as the angle between the plane normal and the room main axis. The plane in the parametric space is described by the following equation [105]:

$$Z \cdot \sin\theta - y \cdot \cos\theta = d \cdot \cos\theta. \quad (22)$$

This estimate suggests that the camera height and direction to the road remain constant (Figure 14). Encoders mounted on the vehicle wheel are used to correct camera and point cloud translations obtained by simultaneous localization and mapping to a metric space. The distance scale obtained from the encoders is utilized to adjust the camera translation scale, which then automatically scales the point cloud to metric space. This is required since it assists in parametrically altering the plane given a known initialization of d, dependent on the camera height [106].

3.1.5. Traffic Sign Sensor

The traffic sign sensor recognizes pre-selected signs within its defined range and sight region. The sensor determines if the detected traffic sign is directly facing the vehicle and ranks the detected signs in ascending order of distance from it before identifying and classifying them. The information supplied to embedded command and control systems about detected traffic signs aids in the comprehension and interpretation of the traffic rules and conduct specified by them [27,63]. The traffic sign sensor is an ideal camera, equipped with an algorithm for recognizing traffic signs and the colors of traffic lights within its field of vision, which uses the identifiers assigned to traffic signs and traffic lights in close proximity to locate, classify, and interpret their operation [107].

HD (High-Definition) maps can provide insight into the environment in which road traffic evolves. HD maps give precise information about the environment where static road traffic occurs, including details about roads and obstacles, across a radius of more than 200 m, even in locations with no direct vision (in bends). This information, when combined with data from cameras and lidar sensors, allows for exact vehicle localization. Currently, creating HD maps requires a professional technique that includes specialized topography and mapping with an MMS (Mobile Mapping System). These maps are created by integrating road pictures with 3D data extracted from point clouds. Zhang et al. in [108] developed an architecture for real-time HD map production that includes an industrial camera, a cutting-edge GNSS (Global Navigation Satellite System)/IMU, and a high-performance computer platform on-board the vehicle (Figure 15). The semantic SLAM (Simultaneous Localization And Mapping) technology, which is based on an enhanced BiSeNet (Bilateral Segmentation Network), is used to extract semantic data from point clouds, including information about the traffic situation.

SLAM is expensive and takes a long time to create HD maps. Furthermore, HD maps may contain inconsistencies between recorded road signs and real-time local modifications. In addition to supporting drivers, intelligent object identification systems can help with roadside maintenance, including road signs, lane markings, and guardrails. Road sign recognition systems, for example, can check for potential anomalies using autonomous vehicles, as human inspection of a complete road infrastructure is difficult.

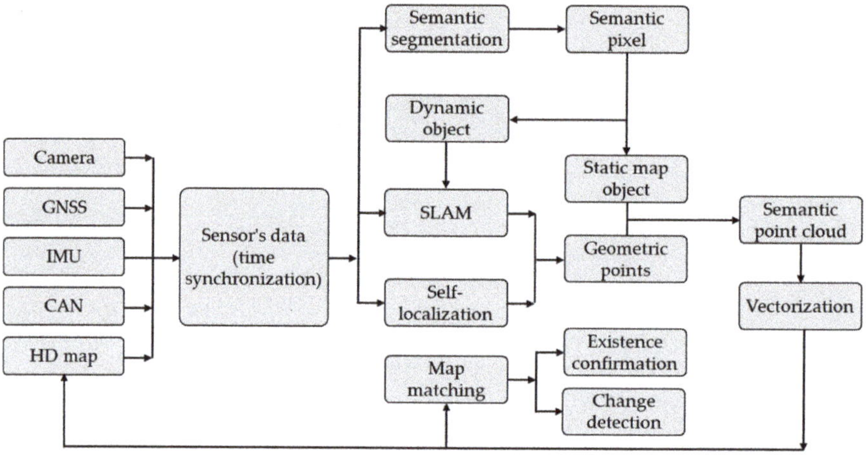

Figure 15. The architecture of the HD map.

As a result, the traffic sign recognition technique is an essential component of both autonomous driving systems and road management systems. Methods for recognizing road signs have centered on researching key aspects such as color and shape [109]. These feature-based approaches are particularly sensitive over long distances and in poor light. The usage of object detection models based on CNN has recently become popular in road sign recognition systems. DL-based object identification algorithms, such as YOLO models, aid in the correct recognition of road signs in traffic. YOLO model-based studies for road sign identification have demonstrated great performance when using publicly available reference data sets [60,110,111]. Figure 16 shows a YOLO model-based arrangement for traffic sign identification [60].

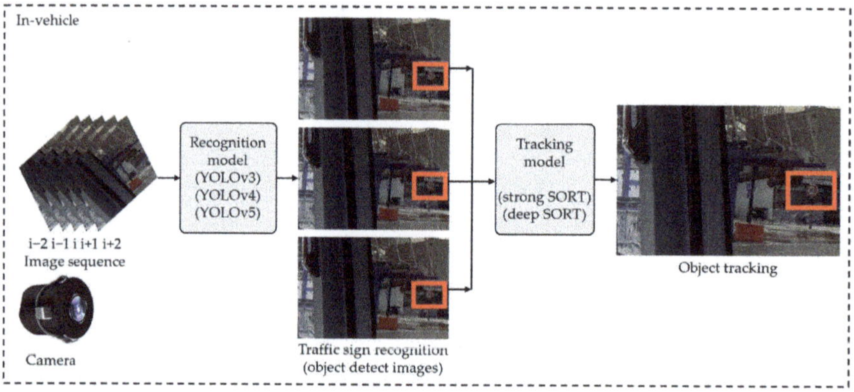

Figure 16. Traffic sign recognition using YOLO models.

3.1.6. Line Sensor

The line sensor detects road markings on the roadway in the direction of driving and organizes them into left and right lines, recording information about each. The recorded data include the lateral distance to the given reference points, the type of lines, their width and height, and the color code. To detect roadway lines, the sensor generates a sequence of planes based on seven points (five vertical planes and three inclined horizontal planes) deployed along the travel direction. Road markings on the roadway are identified by recording the intersection points of the vertical and horizontal planes created by the sensor and the lines on the road surface [66].

The tread detection algorithm, illustrated in Figure 17, starts by capturing an image of the road. The image is processed in multiple phases. Initially, the image is transformed from RGB to grayscale, and then noise is reduced using a symmetric 2D Gaussian filter. The image is then processed to improve contrast in order to recognize road markings. A Sobel operator can be used to detect edges. Finally, a binary image highlighting the tread markings is generated (Figure 18) [112].

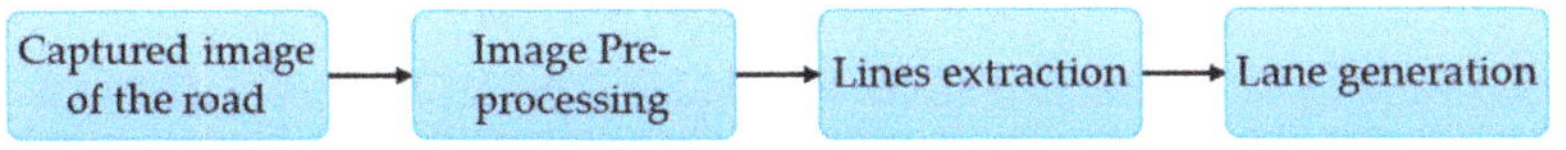

Figure 17. The tread detection algorithm.

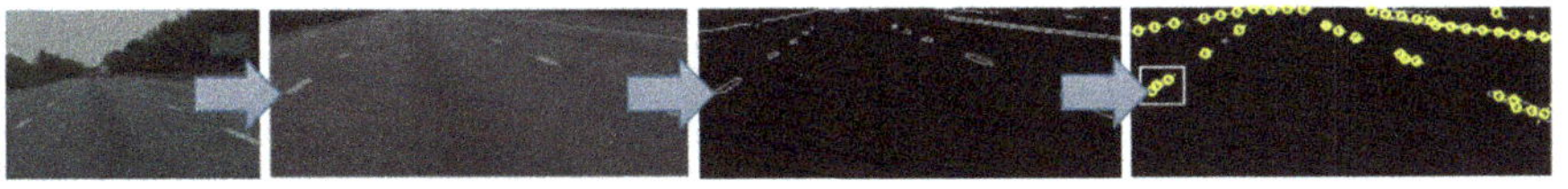

Figure 18. Vector direction of markings on a road.

If a combined laser scanning and video camera system is used for lane marking identification, the method is based on a top-hat transformation, which is preprocessed using the vertical-gradient Prewitt operator to generate a binary image. The binary image is then processed with a PPHT (Progressive Probabilistic Hough Transform) to detect lane markers. Figure 19 displays the lane marking detection algorithm [92].

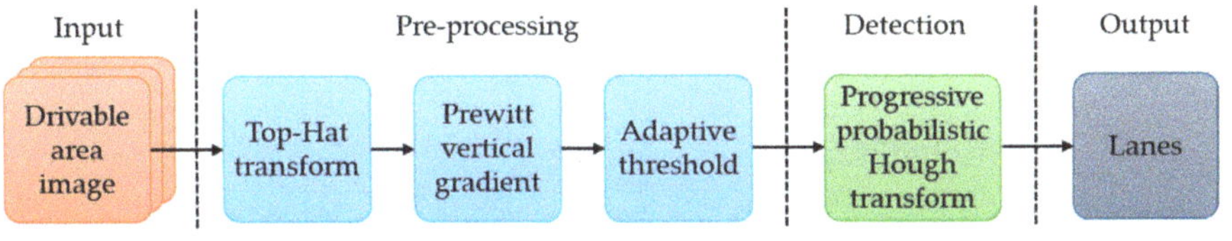

Figure 19. Lane marking detection algorithm.

A Top-Hat transformation is used to increase image contrast while reducing interference from non-tread marks. The Top-Hat transformation is a mathematical model that recovers small-sized bright objects and details from pictures using the following relationship:

$$h = f - (f^{\circ} b), \tag{23}$$

where f is the source image, h is the final image after performing the Top-Hat transformation, and "∘" is an operator that is realized by the Top-Hat transformation and controlled by the choice of the structuring element (b). The size of the structural element b determines how many elements are extracted from the image [92].

The Prewitt vertical-gradient operator uses the following mathematical model to detect vertical edges in an image [113]:

$$\begin{gathered} G(x,y) = |I(x+1,y-1) + I(x+1,y) + I(x+1,y+1) - I(x-1,y-1) - \\ I(x-1,y) - I(x-1,y+1)|, \end{gathered} \tag{24}$$

where I(x,y) represents the pixel intensity at the coordinates (x,y).

PPHT reduces the amount of computing required to correctly detect the markings, utilizing a linear mathematical model:

$$y = mx + b, \tag{25}$$

where m is the slope of the line and b is the intercept at the origin.

3.1.7. Road Sensor

The road sensor determines the following roadway attributes up to a specified distance: road curvature, road marker attributes (speed limits), longitudinal and lateral slope, and the distance and angle of deviation when driving. These data are sent to embedded command and control systems to perform the following functions: LK (Lane Keeping), LDW, AD (Autonomous Driving), SD (Sign Detection), EM (Energy Management), FC (Fuel Consumption), WLD (Wheel Lifting Detection), and PT (Powertrain) (Table 2) [66].

Table 2. Road sensor functions.

Function	LK	LDW	AD	SD	EM	FC	WLD	PT
Road curvature			■					
Longitudinal/lateral slope					■			■
Deviation angle/distance	■	■	■					
Lane information	■	■	■					
Road point position						■	■	
Road marker attributes			■	■	■			■

The background color (gray) indicates which function corresponds (is active) to one of the listed systems (LK (Lane Keeping), LDW, AD (Autonomous Driving), SD (Sign Detection), EM (Energy Management), FC (Fuel Consumption), WLD (Wheel Lifting Detection), and PT (Powertrain)).

The road sensor is located in the middle of the vehicle's front wheels. The sensor's technique for detecting roadway features is based on projecting a point along the route reference line. The deviation represents the lateral offset of the projected point from the route reference line.

3.1.8. Object-by-Line Sensor

Object-by-line sensors detect and transmit information on traffic lanes and traffic objects passing through them by assigning POI (Point Of Interest) points to each of these objects.

The route's number of lanes is divided into LaneScope sections, which include the road axis (LaneScope Center), the left (LaneScope Left), and the right (LaneScope Right).

LaneScopes are used to structure information about objects and traffic lanes (Figure 20), with s_{min}, s_{max}, t_{min}, and t_{max} defining the offsets of the extremities of traffic objects along the route [66].

The LaneScope center is calculated using the POI position. The LaneScope center course is considered to be the basis for establishing the left and right LaneScopes. The center lane course is determined by track lists generated by the computation method. The algorithm determines whether there is a lane where the POI is located. It then creates two lanes (successor/predecessor) beginning with this lane. If no acceptable successor/predecessor route is found, the relevant lane for the object-by-line sensor is terminated.

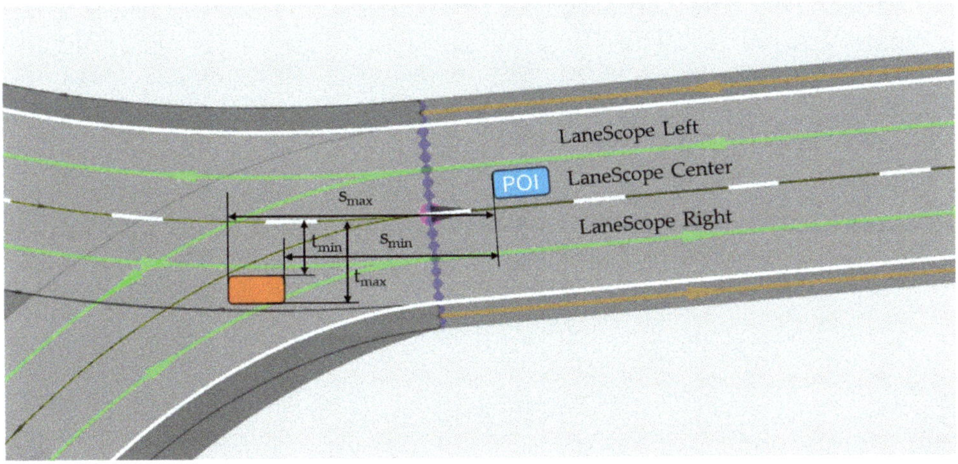

Figure 20. Characteristics of object-by-line sensor.

Lindenmaier et al. in [114] used a GNN (Global Nearest Neighbor) approach to assign detected objects to the reference object category in an interval d_c, respectively, $d_{c,lat}$, resulting in a minimal global association distance. The Mahalanobis distance relation d_{MH}(xi, zj) between the reference object xi and the detection object zj defines the distance matrix $D_{ij} \in R^{+(N,M)}$, which serves as the foundation of the GNN method.

$$D_{ij} = d_{MH}(x_i, z_j) = \left(x_i^{pos} - z_j^{pos}\right)^T \cdot S^{-1} \cdot \left(x_i^{pos} - z_j^{pos}\right), \quad (26)$$

where N and M are the number of reference and detected objects, and x_i^{pos} and z_j^{pos} are the position vectors of the respective objects. The matrix of covariance S is calculated using the cutoff distance ratio $d_r = d_c / d_{c,lat}$ as follows [115]:

$$S = \begin{bmatrix} \cos\alpha & -\sin\alpha \\ \sin\alpha & \cos\alpha \end{bmatrix} \cdot \begin{bmatrix} 1 & 0 \\ 0 & d_r^2 \end{bmatrix} \cdot \begin{bmatrix} \cos\alpha & -\sin\alpha \\ \sin\alpha & \cos\alpha \end{bmatrix}^T, \quad (27)$$

where α represents the angle of the road path at the longitudinal distance of the detected object x_i.

3.2. *Characteristics of Hi-Fi Sensor*

3.2.1. Camera Sensor

The virtual vehicle model is equipped with a sensor camera that is positioned front-to-back to provide a circular image of its surroundings. The sensor camera's objective is to constantly monitor the movement of the virtual vehicle model along the selected route's traffic lane in order to detect and classify static and dynamic objects in the surroundings, as well as recognize traffic signs and traffic light colors. Monocular cameras take 2D images without determining the distances to the monitored objects, whereas stereo cameras may determine the distance by measuring the difference between two images taken from various perspectives [73,116].

Environmental conditions may reduce visibility and affect the identification of nearby objects. The influence of these elements is determined using the following relationship:

$$A_{Env} = \max\left(1.0 - \frac{\text{RainRate}}{\text{RainRate}_{max}}, 0\right) \cdot \min\left(\frac{\text{VisRangeInFog}}{\text{VisRange}_{max}}, 1\right), \quad (28)$$

where RainRate is the rain's intensity, and VisRangeInFog is the direct visibility under foggy conditions.

The camera sensor's maximum error in measuring the distance ($dist_{Err,max}$) to nearby objects will be computed using the following formula:

$$dist_{Err,max} = \frac{dist^2}{f \cdot b} \cdot d_{Err}, \tag{29}$$

where dist is the actual distance to the monitored object, f is the focal length, and b is the baseline. d_{Err} represents the disparity error.

The x and y coordinates of the image of the item acquired by the camera are determined using the following formulas [117,118]:

$$x = \frac{h \cdot \left(x'^2 + f^2 - y' \cdot f \cdot \tan\alpha\right) \cdot \sin\beta}{\sqrt{x'^2 + f^2} \cdot (f \cdot \tan\alpha + y')}, \tag{30}$$

$$Y = \frac{h \cdot \left(x'^2 + f^2 - y' \cdot f \cdot \tan\alpha\right) \cdot \cos\beta}{\sqrt{x'^2 + f^2} \cdot (f \cdot \tan\alpha + y')}, \tag{31}$$

where h is the camera's height from the ground, f is its focal length, and α is the angle between the camera's optical axis and the horizontal line to the target. Breaking down the relationships results [119],

$$\begin{aligned} \tan^{-1}\frac{f \cdot \tan\alpha}{\sqrt{x'^2+f^2}} + \tan^{-1}\frac{y'}{\sqrt{x'^2+f^2}} &= \tan^{-1}\frac{\frac{f \cdot \tan\alpha}{\sqrt{x'^2+f^2}} + \frac{y'}{\sqrt{x'^2+f^2}}}{1 - \frac{f \cdot \tan\alpha}{\sqrt{x'^2+f^2}} \cdot \frac{y'}{\sqrt{x'^2+f^2}}} \\ &= \tan^{-1}\frac{\sqrt{x'^2+f^2} \cdot (f \cdot \tan\alpha + y')}{x'^2+f^2-y' \cdot f \cdot \tan\alpha} = \tan^{-1}\frac{h}{d}, \end{aligned} \tag{32}$$

$$\frac{\sqrt{X'^2 + f^2} \cdot (f \cdot \tan\alpha + y')}{x'^2 + f^2 - y' \cdot f \cdot \tan\alpha} = \frac{h}{d}, \tag{33}$$

$$d = \frac{h\left(x'^2 + f^2 - y' \cdot f \cdot \tan\alpha\right)}{\sqrt{x'^2 + f^2} \cdot (f \cdot \tan\alpha + y')}. \tag{34}$$

Three-dimensional Gaussian Splatting (3DGS) is a technique that significantly enhances real-time, high-quality 3D rendering. The 3DGS method, a volume rendering technique for real-time radiance field rendering, was introduced by Kerbl et al. [120]. The 3DGS technique represents light and density in 3D space using 3D Gaussians, which are volumetric entities with color, size, and transparency. Autonomous driving is strongly reliant on an accurate and efficient representation and understanding of 3D scenes. Three-dimensional Gaussian Splatting provides substantial advantages by allowing for the real-time reconstruction of complex situations.

Important tasks including obstacle identification and path planning depend heavily on spatial information processing. Street Gaussians, for example, model scenes as 3D Gaussian point clouds, capturing both geometry and appearance, and enable the separation of static backgrounds from dynamic foreground elements, which is critical for semantic segmentation in autonomous systems (Figure 21) [121].

Figure 21. Decomposition results of Street Gaussians [121].

Thus, the process is analogous to triangle rasterization in computer graphics, except instead of triangles, it uses 3D Gaussians. This approach enables the representation of complex situations while maintaining a balance between rendering efficiency and visual fidelity [122]. The 3DGS algorithm represents a scene using a collection of 3D Gaussians [121].

Each Gaussian is specified by parameters including transparency, location, and shape, which are tuned to represent the scene from multiple perspectives. To render an image, 3DGS organizes the Gaussians according to their distance from the camera and then mixes them using alpha blending. This approach allows for the production of smooth and realistic visuals in real time. The 3D Gaussian representation can be stated mathematically as follows [121,122]:

$$L_{3DGS(x,y,z,\theta,\phi)} = \sum_i G(x, y, z, \mu_i, \Sigma_i) \cdot c_i(\theta, \phi), \tag{35}$$

where G is the Gaussian function with mean μ_i and covariance Σ_i, and c is the view-dependent color.

Figure 22 illustrates the process from the initial sparse point cloud to the final image creation, with essential steps including initialization, 3D Gaussian representation, projection, rasterization, and adaptive density control [123].

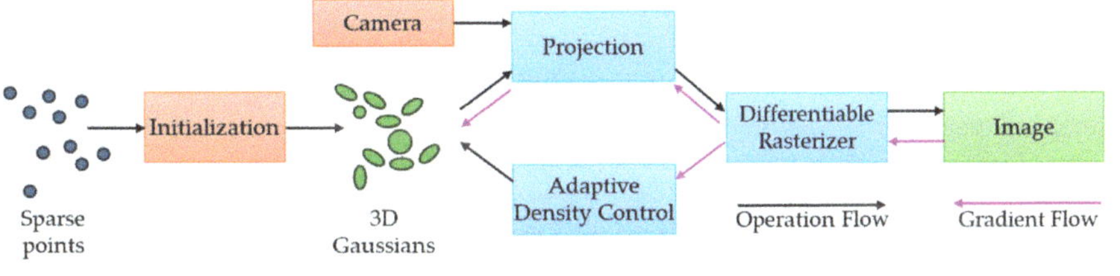

Figure 22. Three-dimensional Gaussian Splatting Process: from sparse point cloud initialization and adaptive density optimization to efficient training via tile-based rendering [123].

The 3DGS technique has various advantages over conventional 3D cameras and rendering systems. It offers a more efficient and scalable rendering mechanism, enabling real-time rendering while preserving great visual quality. The 3DGS method also enables simple scene manipulation and editing, making it a useful tool for a variety of applications.

3.2.2. Global Navigation Sensor

The global navigation sensor locates a vehicle by using the positions of at least four geostationary GPS satellites. The sensor determines real-time positioning in geographic coordinates using information about the transmission time of the extrapolation signal emitted by the satellites and received by the vehicle's receiver (x,y,z,t) [124].

CarMaker can represent any position of the virtual vehicle model in the global road architecture as a geographic point on the Earth's surface. This uses the GCS (Geographic Coordinate System) coordinate system, which consists of latitude, longitude, and altitude.

The origin of the road frame on the Earth's surface is determined using the FlatEarth projection method using GCS reference points (Figure 23). This projection method ignores the Earth's curvature around the GCS reference point when calculating the relative position $\overrightarrow{RefPo}$ of point P in the road frame. The elevation value h at point P is calculated as follows [66]:

$$h = h_R + \overrightarrow{RefPo}[z]_0. \tag{36}$$

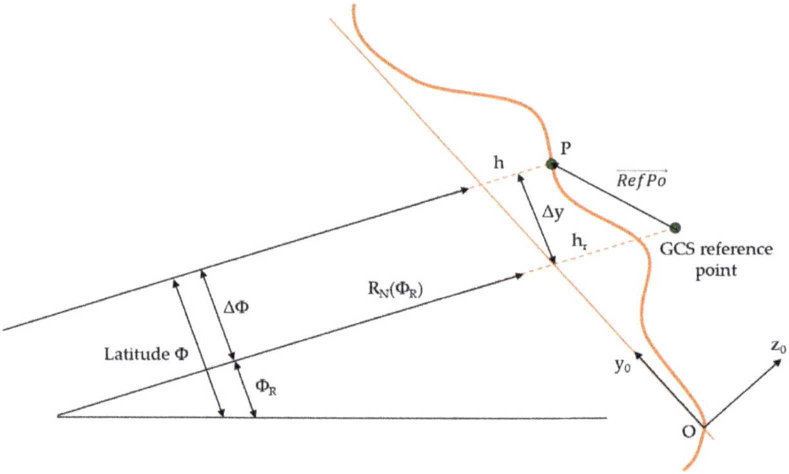

Figure 23. Calculation of the latitude for the global navigation sensor.

The relative latitudc $\Delta\phi$ can be calculated as follows:

$$\Delta\Phi = \text{asin}\left(\frac{\Delta y}{R_N(\phi_R) + h}\right) \approx \frac{\Delta y}{R_N(\phi_R) + h} = \frac{\overrightarrow{RefPo}[y]_0}{R_N(\phi_R) + h}, \tag{37}$$

where $R_N(\phi_R)$ is the radius of the Earth's ellipsoid in the north, which is determined by the latitude of the GCS reference point. The latitude ϕ of point P is transformed into

$$\Phi = \phi_R + \Delta\phi. \tag{38}$$

Similarly, the longitude at point P can be determined using the following formula:

$$\lambda = \lambda_R + \frac{\overrightarrow{\mathrm{RefPo}}[zx]_0}{(R_E(\phi_R)\cdot\cos(\varphi_R))}, \tag{39}$$

where $R_E(\phi_R)$ is the radius of the Earth's ellipsoid in the east, which is determined by the latitude of the GCS reference point. The factor cos(ϕ_R) considers decreasing radius with increasing latitude.

D-GNSS (Differential-Global Navigation Satellite System) with differential RTK (Real-Time Kinematic) correction reduces vehicle positioning errors by applying a differential correction relative to the coordinates of a reference base station, which then uses signals transmitted by satellites. The delivery of D-GNSS differential corrections with RTK from the reference base station to the vehicle takes place via mobile data connections [27].

CarMaker's virtual car model employs Cartesian coordinates (x, y, z), geocentric coordinates ECEF (Earth-Centered, Earth-Fixed), and ellipsoidal coordinates (ϕ latitude, λ longitude, h elevation) for geostationary satellite positioning. GDOP (Geometric Dilution of Precision) refers to the accuracy of the computed position of the vehicle receiver in reference to the geostationary position of visible satellites.

3.2.3. Radar Sensor

The radar sensor detects static and dynamic objects on the virtual vehicle route using the SNR signal. The sensor detects things using cellular units and accounts for the effects of overlapping traffic items. The detected items are identified using particular RCS (Radar Cross-Section) cross-section maps that take into account the angle of incidence and the signal reflected by the traffic objects. Depending on the signal-to-noise ratio, surrounding objects will be recognized, removed, or regarded as false negatives [27].

RCS is defined as the cross-section of a detected object that intercepts the most amount of power transmitted by the radar sensor. It is determined by the following parameters: the size and shape of the object, the antenna orientation angle, the frequency and polarization of the radar waves, the object's electromagnetic properties, and the surface structure. Radar sensor simulation is complex due to the dispersion effect of radar waves in virtual settings, and it is achieved using physically interpretable characteristics such as the distance to nearby objects, the movement speeds of traffic items, and their angular locations.

Elster et al. in [125] used the DVM (Double Validation Metric) methodology to validate radar sensor data virtually. DVM uses the reliability and repeatability of radar sensor readings to effectively quantify deviations between distributions for various types of detected objects. The measurement data (M1, M2) are preprocessed and filtered, and the number of points resulting from the measurements is determined for each data set using the EDF (Empirical Cumulative Distribution Function) methodology.

The deviation of mean values d_{bias} for measurement data M1 is calculated and compared to that of measurement data M2. The shape deviation of the distribution function d_{CAVM} is calculated to highlight the difference in signal scattering between M1 and M2 (Figure 24) [126].

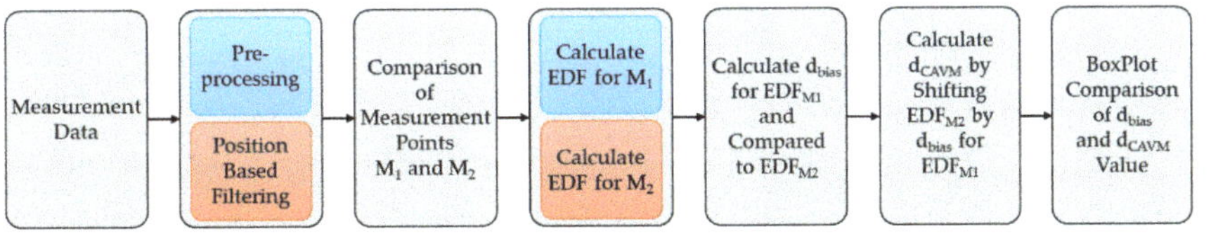

Figure 24. DVM methodology.

The radar equation defines the physical relationships connected to the features of the radar sensor, resulting in the received signal power $\rho_s(r_s,\upsilon_s,\phi_s)$ [66,127,128].

$$|\rho_{sP}(r_s,\upsilon_s,\phi_s)|^2 = \frac{P_{Rx}}{P_{Tx}} = \frac{\lambda_c^2 \cdot G_{Tx}(\upsilon_s,\phi_s) \cdot G_{Rx}(\upsilon_s,\phi_s) \cdot \sigma_s}{(4\pi)^3 \cdot r_s^4} \cdot F(r_s,\upsilon_s,\phi_s)^4, \tag{40}$$

where ρ_s is the impulse response for s point scatterer, positioned at r_s distance at an angular position, for υ_s elevation and ϕ_s azimuth, λ_c is the wavelength of the carrier frequency, G_{Tx} is the gain of the transmitter antenna, G_{Rx} is the gain of the receiver antenna, and σ_s is the reflection coefficient of the point scatterer.

A gain map is used to highlight antenna features, which is characterized by a unidirectional gain factor parameterized by azimuth and elevation in the scanning direction. The following equations describe field strength and antenna gain [129]:

$$f(\theta,\phi) = \frac{\sin\left(\pi\frac{a}{\lambda}\sin\theta\cos\phi\right)}{(\pi v_y)} \frac{\sin\left(\pi\frac{b}{\lambda}\sin\theta\sin\phi\right)}{(\pi v_z)}, \tag{41}$$

The aperture dimensions include θ elevation, ϕ azimuth, and a and b major lobes.

The detection threshold can be determined by the lowest detectable value of SNR_{min} and is calculated using the parameters of the minimal probability of detection P_{Dmin} and the probability of false alarm P_{FA} [66]:

$$SNR_{min} = 2\left(\mathrm{erfc}^{-1}(2P_{FA}) - \mathrm{erfc}^{-1}(2P_{min})\right)^2. \tag{42}$$

The strength S of the received signal is determined by the radar equation:

$$S = \frac{PG^2\lambda^2 RCS}{(4\pi)^3 r^4} \cdot \frac{1}{L_A L_{atm}}, \tag{43}$$

where P is the transmitted power, G is the gain of the antenna, λ is the wavelength, r is the distance from the radar sensor to the object, L_A is the additional system loss, and L_{atm} is the atmospheric loss.

The specific RCS cross-section maps for close objects are determined by radar sensor resolution, object size, direction of incidence, object occlusion, and object merging (Figure 25) [27].

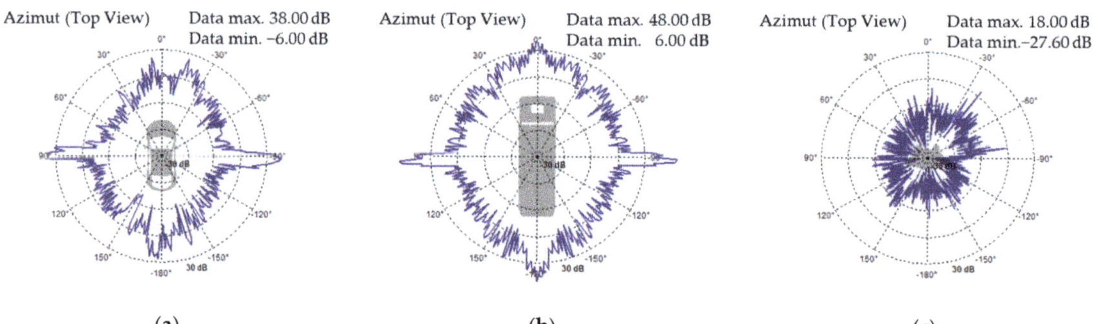

(a) (b) (c)

Figure 25. RCS of the various objects: (**a**) vehicle, (**b**) truck, (**c**) pedestrian.

The transmit gain map adjusts the power of the delivered signal to 3D objects. The transmit gain is calculated using linear interpolation of the parameterized gain map (Figure 26).

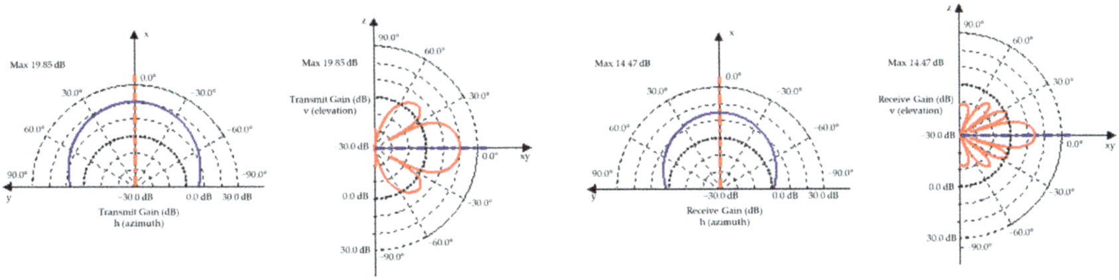

Figure 26. Transmit/receive (azimuth/elevation) gain map.

3.3. Characteristics of RSI Sensor

3.3.1. Lidar RSI Sensor

The lidar (light detection and ranging) sensor operates based on measuring the ToF for a beam of laser pointers with a wavelength of 905 nm that is emitted at objects in the area and computing the reception time of this beam reflected by these items. Lidars, including RSI sensors, generally use the ToF principle to determine distance. This involves generating a laser pulse and measuring its return time after hitting an object [130].

The lidar 2D sensor captures information about nearby objects by sending a single laser beam onto a revolving mirror perpendicular to the axis of rotation. The lidar 3D sensor gathers information about nearby objects by shooting a beam of laser rays through a rotating mechanism, resulting in a point cloud for the contour of these objects and the capacity to build high-precision 3D maps.

Lidar RSI sensors use a rotating equipment to guide a laser beam across their range of vision. This scanning is often accomplished using rotating polygonal mirrors that direct the beam with each facet, resulting in a balance of accuracy and speed. Alternatively, revolving prisms can refract the beam with great precision and stability. A recent technique uses MEMS (Micro-Electro-Mechanical Systems) technology, in which small mirrors direct the beam, allowing for more compact designs but potentially limiting the range and scan angle.

Lidar generates a 3D representation of the viewed reality by measuring the ToF at multiple points within its FoV (field of view). The set of points is referred to as a point cloud. The nth measurement point (p_n) in the lidar reference system {L} can be expressed as follows [131]:

$$\{L\}_{p_n} = \left[\{L\}_{xn}, \{L\}_{y_n}, \{L\}_{zn}\right]^T = \frac{c}{2} t_{ToF,n} \cdot \{L\}_{\hat{s}n}, \tag{44}$$

where c is the speed of light in air, $t_{ToF,n}$ is the ToF measure, and $\hat{s}_n$ is the unitary vector indicating the scanning direction of the lidar in its reference system {L}. Equation (43) shows that accurate and exact point clouds require both the ToF measurement ($t_{ToF,n}$) and the scanning direction ($\hat{s}_n$). The scanning direction of the lidar, designated as $\hat{s}$, is expressed in the laser source's reference system {I}. The mirror's tilt angles (α and β) correspond to horizontal and vertical directions, respectively. This relationship is further developed in the following equations [131]:

$$\{I\}_{\hat{s}} = \{I\}_{\hat{i}} - \left(\gamma(\alpha, \beta, \varphi) - \sqrt{(\gamma(\alpha, \beta, \varphi)^2}\right) \cdot \{I\}_{\hat{n}}, \tag{45}$$

$$\{I\}_{\hat{n}} = \begin{bmatrix} \sin\alpha \cdot \cos\beta \\ \cos\varphi \cdot \sin\beta + \sin\varphi \cdot \cos\alpha \cdot \cos\beta \\ \sin\varphi \cdot \sin\beta - \cos\varphi \cdot \cos\alpha \cdot \cos\beta \end{bmatrix}, \tag{46}$$

$$\gamma(\alpha, \beta, \varphi) = \{I\}_{\hat{n3}} = \sin\varphi \cdot \sin\beta - \cos\varphi \cdot \cos\alpha \cdot \cos\beta. \quad (47)$$

In the laser source reference system {I}, γ represents the third component of the normal vector $\hat{n}$, whereas α and β represent the scanning tilt angles, direction, and mirror surface. Each point in the cloud is described by its (x, y, z) coordinates, which are computed from the measured distance, the horizontal angle (determined by the revolving mirror or prism), and the vertical angle (typically fixed but variable in multi-emitter systems).

For the CarMaker virtual vehicle model, lidar 2D and 3D sensors are defined as lidar RSI sensors that provide information about nearby objects and behave just like genuine sensors. The calibration of the lidar virtual sensor was performed by using simulation data from various surroundings that were correlated with the motion sensor [132].

Lidar 2D sensors are used in pairs of two units to reduce coverage gaps and ensure continuous visibility throughout the entire surface of a horizontal scanning plane (P2D) with a viewing angle of 180° and a field of view centered on the virtual vehicle model's median longitudinal plane. Lidar 3D sensors scan on 16 channels, following a plane (P3D) with successive inclinations along an increasing angular axis with respect to the upper horizontal scanning plane. They have a viewing angle of 30° and an opening of ±15° with respect to the median longitudinal plane of the virtual vehicle model (Figure 27).

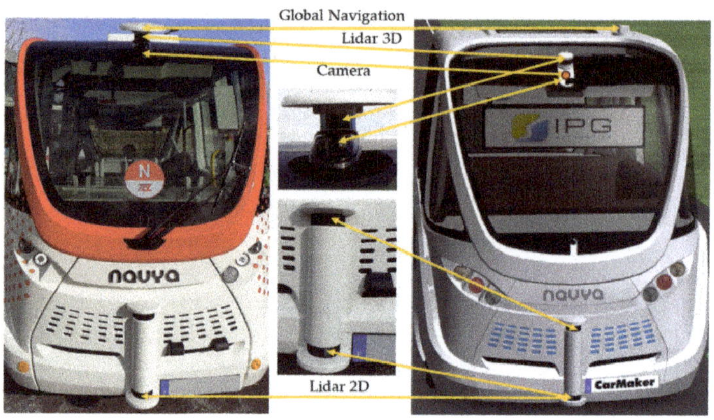

Figure 27. RSI sensor distribution on the virtual vehicle model's body structure.

The interaction modes of the lidar RSI sensor are classified using the following criteria [65]:

1. Uniformly distributed diffuse reflected laser rays, regardless of the direction of the incident ray (Lambertian reflection), with the intensity of the reflected ray decreased by the inverse of the angle between the incident ray and the normal of the reflective surface;
2. Retroreflective, meaning that incident laser rays are reflected back in the direction of the emitter, with the intensity of the reflected ray being reduced based on the reflectance parameters and incident angle;
3. Specular, meaning that the incident and reflected laser rays form identical angles with the normal of the reflective surface, and the incident and reflected laser rays are in the identical plane;
4. Transmissive, meaning that the incident laser photons keep their course but are attenuated by transmission.

3.3.2. Ultrasonic RSI Sensor

CarMaker's ultrasonic RSI model is based on the signal chain of a real ultrasonic sensor. The process begins with a transmitter illuminating the environment and emitting sound waves (Figure 28). This method divides sound waves into a finite number of rays. This provides for a balance between simulation speed and physical accuracy. Environmental circumstances with physical effects on wave propagation are considered, and the Helmholtz equation is used to compute electrical scatter fields on object surfaces utilizing parameterizable material properties to predict reflections. Every suitable ray returns to the receiver [133].

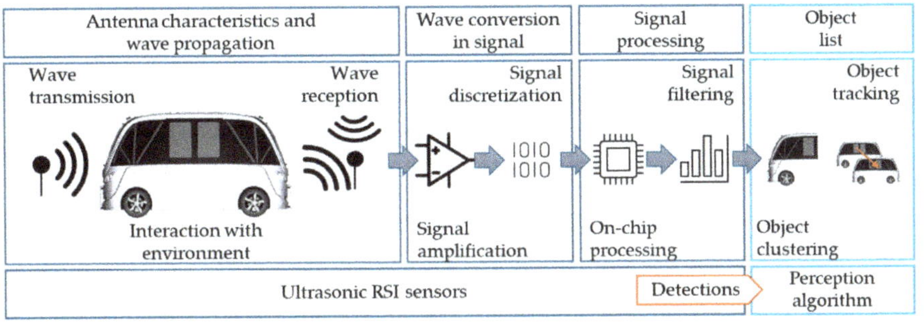

Figure 28. Signal chain of the ultrasonic RSI sensor models.

For all detections, the sensor model returns the sound pressure level as well as the flight time. It accounts for dense packaging and associated effects by optionally sensing cross-echoes between sensors. Interfaces enable the replacement of individual steps with a user-defined code, and clustering on available GPUs improves performance [133], as seen in Figure 29.

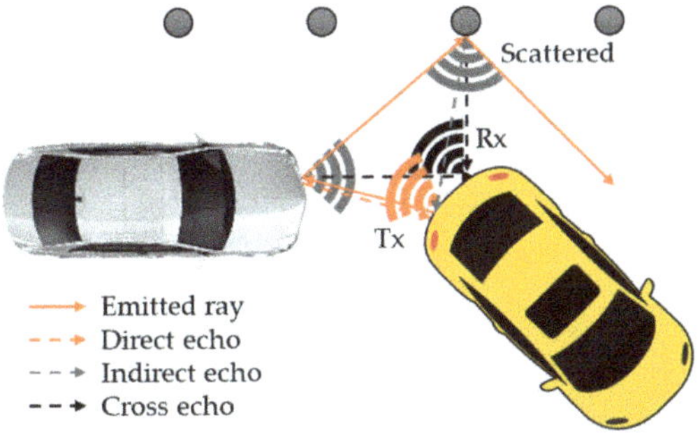

Figure 29. A diagram of the ray tracing algorithm used to simulate a sound wave.

The ultrasonic RSI sensor uses mechanical acoustic pressure waves that are reflected by obstacles in the immediate vicinity of the virtual vehicle model using the ToF principle, and the distance to the respective obstacles is accurately calculated using the SPA (Sound Pressure Amplitude) (Figure 30). The sensor considers the impacts of overlapping objects in the area, the effects of physical propagation, and the classification as false positives or false negatives of the observed objects. The propagation modes of acoustic pressure waves are classified using the following criteria [66]:

1. Direct Echo: The acoustic pressure wave is reflected once by an object in close proximity and received by the emitting sensor.
2. Indirect Echo: The acoustic pressure wave is reflected at least twice by objects or surfaces in the vicinity and received by the emitting sensor.

3. Repeated Echo: The emitting sensor receives the acoustic pressure wave after it has been reflected by nearby objects or surfaces.
4. Cross Echo: The reflected acoustic pressure wave is received by a sensor other than the originating sensor, resulting in a propagation mode known as cross echo.
5. Road Clutter: This occurs when the acoustic pressure wave is reflected by bumps or irregularities in the roadway.

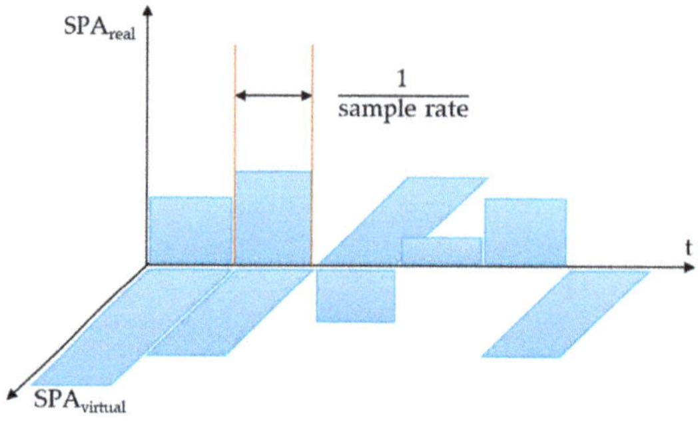

Figure 30. SPA full wave form.

Ultrasonic RSI sensors are installed in the vehicle's front bumper, side panels, and rear bumper.

4. Virtual Sensor Parametrization

4.1. Ideal Sensor Parametrization

The ideal sensors incorporated in the virtual vehicle model are as follows: slip angle, inertial, object, free space, traffic sign, line, road, and object by line.

To highlight the major parameters generated by the indicated ideal sensors, a series of simulations have been performed on the proposed route (see Figure 8), digitized, and implemented in the IPGRoad/CarMaker application.

For the slip angle sensor (Figure 31a), one parameter has been chosen: Sensor.SAngle.SL00.Ang (Figure 31b). The analysis was performed on an area of the route traveled, starting with a straight road section, then a left turn and another straight road section.

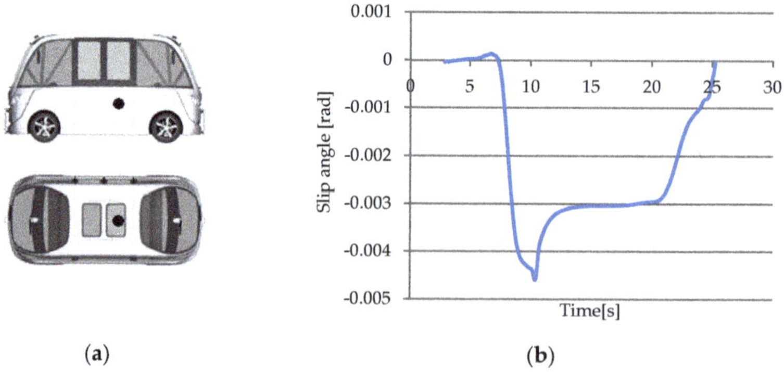

Figure 31. Slip angle sensor parameterization and generated parameter. (**a**) Sensor position (**b**) Sensor parameter.

The graphic illustrates how the slip angle sensor's values change when the vehicle approaches a left curve (second 7) and returns to a straight road segment (second 25).

According to the data displayed on the graph in the image, the maximum slip angle is around 0.0045 radians (0.25 degrees).

For the inertial sensor (Figure 32a), two parameters have been chosen: Sensor.Inertial.YRS.Vel_0.x (red curve) for the speed in the x direction and Sensor.Inertial.YRS. Acc_B.y (blue curve) for the lateral acceleration (Figure 32b).

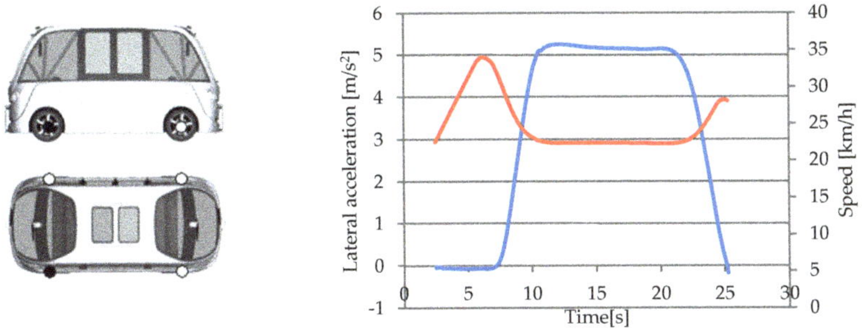

(**a**) (**b**)

Figure 32. Inertial sensor parameterization and generated parameters. (**a**) Sensor position (**b**) Sensor parameters.

The graphic shows that lateral acceleration increases from 0 m/s^2 when moving straight to 5.3 m/s^2 when turning. The vehicle's lateral acceleration reduces as it exits the turn, returning to near-zero m/s^2 levels. The sensor's observed lateral accelerations correspond with typical turns.

For the object sensor (Figure 33a), two parameters have been chosen: Sensor.Object. RadarL.relvTgt.NearPnt.ds.x (red curve) represents the distance (ds) along the specified x-axis, corresponding to the direction of the sensor system axes, from the most relevant target (relvTgt), respectively, the detected object that is most important to track; Sensor.Object.RadarL.relvTgt.NearPnt.ds.y (blue curve) represents the distance (ds) along a specific y-axis parallel to the sensor system axes and the detected object's nearest point (NearPnt) (Figure 33b). This value represents the shortest distance between the sensor and the relevant target (a point on the target), which could be a vehicle, a pedestrian, etc.

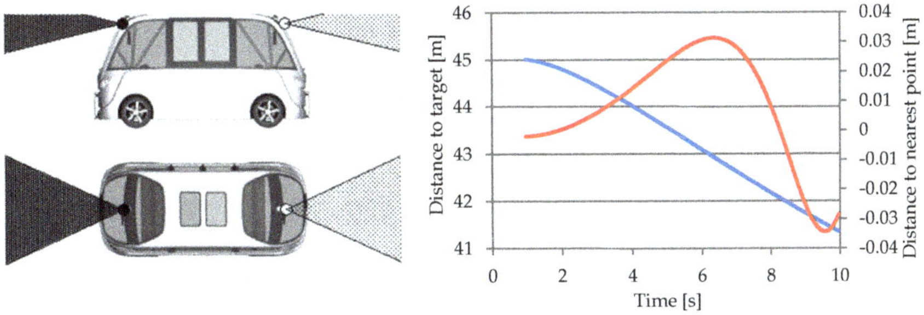

(**a**) (**b**)

Figure 33. Object sensor parameterization and generated parameters. (**a**) Sensor position (**b**) Sensor parameters.

The graphic shows the variations in x and y coordinates over time for the nearest point on an obstacle detected by the sensor. The coordinates corresponding to the z-axis (height) have not been shown since they contain relatively minor fluctuations that have no significant effect on the position of the obstacle.

For the free space sensor (Figure 34a), two parameters have been chosen—Sensor.FSpace.Front.Segm.0.ds.x (red curve), and Sensor.FSpace.Front.Segm.198.ds.x (blue curve)—which represent the segmentation areas, includes the closest point of a detected traffic obstacle. The segmentation area is divided into four quadrants. Figure 34b shows how the various parameters change along the x-axis, with each parameter representing a quadrant of the segmentation area. The graphs illustrate the change in expected parameter values across the segmentation region over the same time frame. Analyzing the graph which corresponds to the parameter Sensor.FSpace.Front.Segm.198.ds.x, one can see no more variations in the specified parameter after second 52, indicating that no obstacle was identified in the associated quadrant.

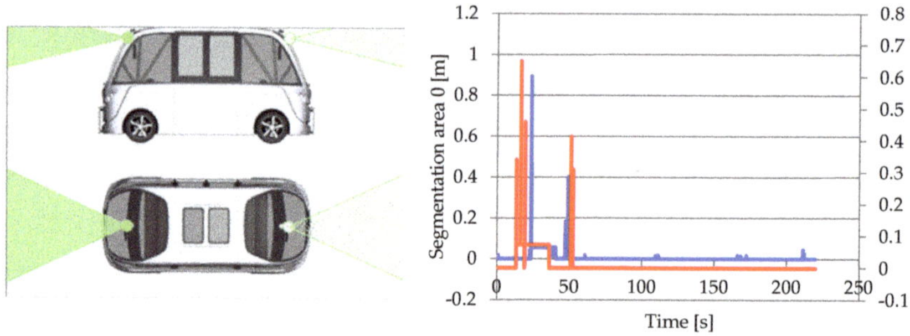

(**a**) (**b**)

Figure 34. Free space sensor parameterization and generated parameters. (**a**) Sensor position (**b**) Sensor parameters.

For the traffic sign sensor (Figure 35a), two parameters have been chosen: Sensor.TSign.FrontCam.SpeedLimit.0.ds.x (blue curve) indicates the distance between the detected indicator (Speed Limit sign) and the direction of travel (x), respectively; Sensor.TSign.FrontCam.SpeedLimit.nSign (red curve) indicates the signal generated by the sensor when it detects a speed limitation sign (Figure 35b). When the distance between the vehicle (sensor) and the road sign is less than 30 m, the signal reaches its maximum value.

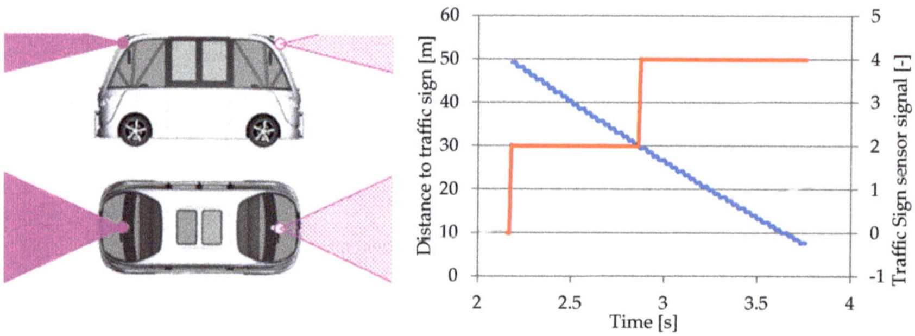

(**a**) (**b**)

Figure 35. Traffic Sign sensor parameterization and generated parameters. (**a**) Sensor position (**b**) Sensor parameters.

For the line sensor (Figure 36a), one parameter has been chosen: Sensor.Line.Front.RLines.1.Type (Figure 36b) indicates the type of longitudinal road marking located on the right side of the road in the virtual vehicle model's direction of travel. The graph shows a variation in the signal of the parameter in the range of 1 to 2. The parameter's value of 1 indicates the presence of a longitudinal road marking made up of a simple dashed line, while the parameter's value of 2 indicates the presence of a longitudinal road marking made up of a single continuous line.

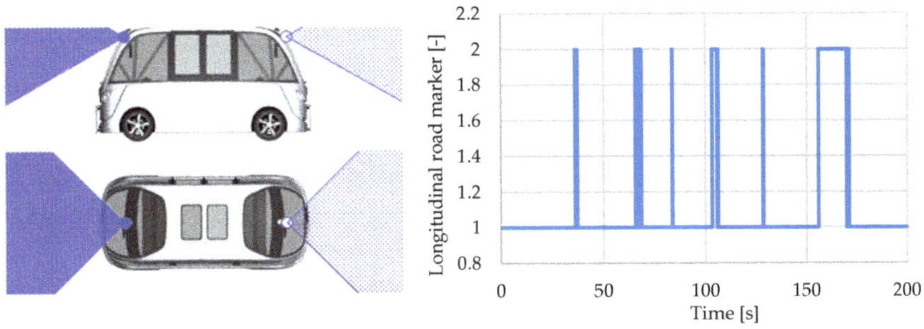

(**a**) (**b**)

Figure 36. Line sensor parameterization and generated parameter. (**a**) Sensor position (**b**) Sensor parameters.

Changing the parameter value from 1 to 2 moves the virtual vehicle model from a road section with a simple dashed line (1) to a road section with a continuous line (2). If the parameter value is constant (1 or 2), the type of longitudinal road marking does not differ; if the values fluctuate frequently, a road section with many changes in longitudinal road marking will result.

Two parameters have been chosen for the road sensor (Figure 37a): Sensor.Road.AlongR.Path.tx (blue curve) represents the vehicle displacement along the x-axis, and Sensor.Road.AlongR.Path.DevDist (red curve) represents the deviation from the planned route (Figure 37b). Over a route length of around 8 m, the deviation ranges between 0 and 7×10^{-9} m.

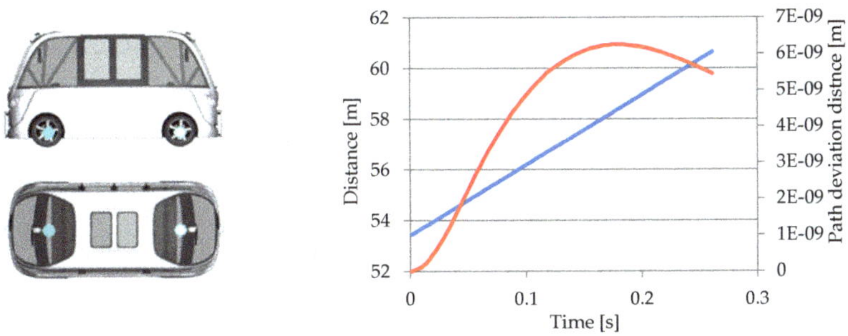

(**a**) (**b**)

Figure 37. Road sensor parameterization and generated parameters. (**a**) Sensor position (**b**) Sensor parameters.

For the object-by-line sensor (Figure 38a), one parameter has been chosen: Sensor.ObjByLine.OBL00.LinesC.0.ObjF.0.sMax (Figure 38b) highlights the variation in maximum distance between the virtual vehicle model and the POI over a 2.5 s period. Figure 21 shows that S_{max} decreases over time, indicating that the vehicle has approached the POI, where LinesC indicates the LaneScope Center section.

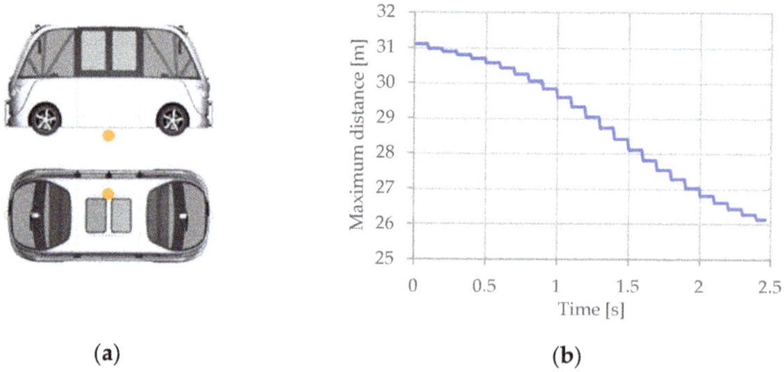

Figure 38. Object-by-line sensor parameterization and generated parameter. (**a**) Sensor position (**b**) Sensor parameter.

4.2. Hi-Fi Sensor Parametrization

The Hi-Fi sensors incorporated in the virtual vehicle model are as follows: (a) camera, (b) global navigation, (c) radar.

To highlight the major parameters generated by the indicated Hi-Fi sensors, a series of simulations have been performed on the proposed route (see Figure 8), digitized, and implemented in the IPGRoad/CarMaker application.

For the camera sensor (Figure 39a), two parameters have been chosen: Sensor.Camera.CA00.Obj.0.nVisPixels (x-axis) indicates the number of visible pixels of the object recognized by the sensor, and Sensor.Camera.CA00.Obj.0.Confidence (y-axis) indicates the confidence degree of object detection (Figure 39b). The graph illustrates the efficacy of object detection based on their size in the image. If an object has many visible pixels but a low confidence level, the recognition algorithm becomes unreliable. The maximum confidence level for recognized items is 1.

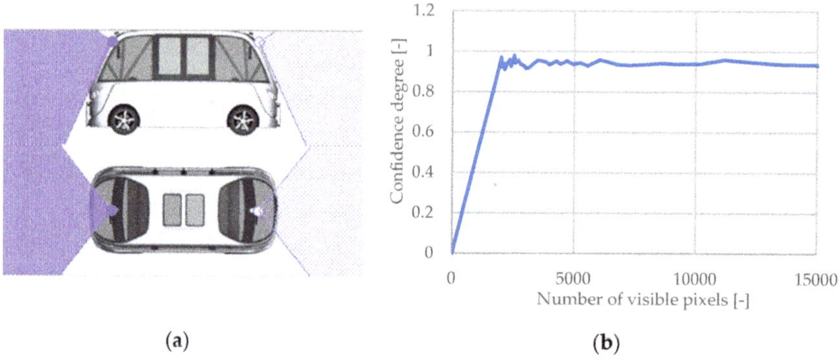

Figure 39. Camera sensor parameterization and generated parameter. (**a**) Sensor position (**b**) Sensor parameter.

For the global navigation sensor (Figure 40a), one parameter has been chosen, Sensor.GNav.Receiver.SatNoDirectView, that highlights the number of satellites that are directly visible to the GNSS receiver in a specific time interval (Figure 40b).

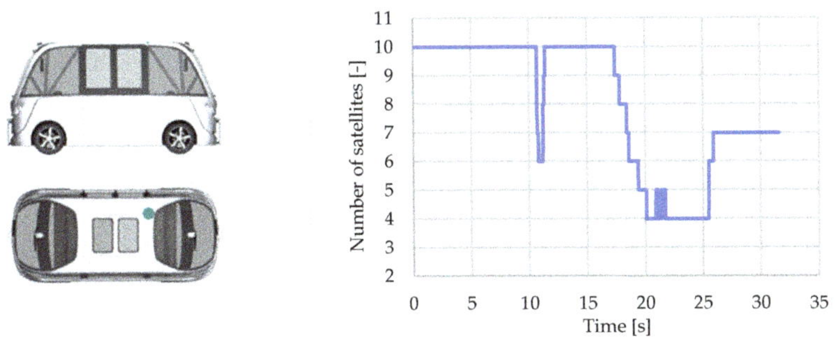

(a) (b)

Figure 40. Global navigation sensor parameterization and generated parameter. (**a**) Sensor position (**b**) Sensor parameter.

The satellites' direct visibility influences the measurement accuracy. The maximum number of visible satellites ranges from 4 to 10 and is influenced by tall buildings or underground passages that the vehicle passes through, and by extreme weather conditions.

For the radar sensor (Figure 41a), two parameters have been chosen: Sensor.Radar.RA00.Obj0.Dist (blue curve) represents the variation in the signal-to-noise ratio (SNR) as a function of the distance from the object, and Sensor.Radar.RA00.Obj0.SNR represents the signal to noise ratio (Figure 41b). The graph illustrates the radar signal's detection efficiency versus background noise as a function of distance from the target object. A higher signal-to-noise ratio suggests better detection. As the object moves away, the signal-to-noise ratio decreases, resulting in poorer detection.

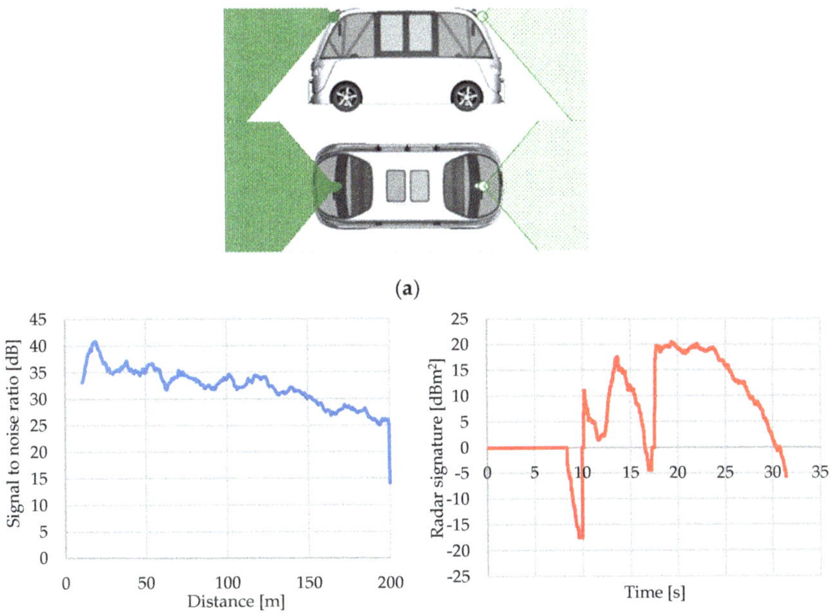

(a)

(b) (c)

Figure 41. Radar sensor parameterization and generated parameters. (**a**) Sensor position (**b**) Sensor parameter signal to noise ratio (**c**) Sensor parameter radar signature.

Another parameter evaluated for the radar sensor is Sensor.Radar.RA00.Obj0.RCS (red curve), which represents a variation in the RCS parameter (target object's radar signature) in the case of detection (Figure 41c). The graph shows significant changes in RCS parameter values, ranging from −18 dBm^2 to 21 dBm^2. The analysis of the RCS parameter indicates that the radar sensor detects obstacles in the virtual vehicle model's travel environment, ranging from small (negative values for poor detection) to large (positive values for good detection).

4.3. RSI Sensor Parametrization

The RSI sensors incorporated in the virtual vehicle model are as follows: (a) lidar RSI, (b) ultrasonic RSI.

To highlight the major parameters generated by the indicated RSI sensors, a series of simulations have been performed on the proposed route (Figure 8), digitized, and implemented in the IPGRoad/CarMaker application.

For the lidar RSI sensor (Figure 42a), one parameter was chosen: Sensor.LidarRSI.LIRS00. nScanPoints indicates the number of points detected in an interval of time (Figure 42b). The graph's parameter variations allow for an investigation of scanning conditions in particular areas of the virtual vehicle model's travel path. A high parameter value results in a congested area with many detected obstacles, whereas a low parameter value results in an area with fewer detected obstacles.

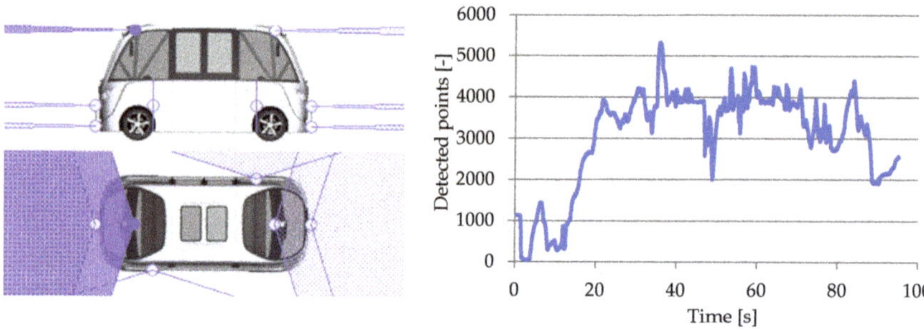

(**a**) (**b**)

Figure 42. Lidar RSI sensor parameterization and generated parameter. (**a**) Sensor position (**b**) Sensor parameter.

The ultrasonic RSI sensor (Figure 43a) shows the variation in a specified parameter for two identical sensors: Sensor.USonicRSI.Tx.USRS00.Rx.USRS00.nDetections (blue curve) for the USRS00 sensor and Sensor.USonicRSI.Tx.USRS01.Rx.USRS01.nDetections (red curve) for the USRS01 sensor. These two parameters define the variation in obstacles identified by two ultrasonic sensors placed in different locations on the virtual vehicle model (Figure 43b). The variation in the vertical curves represents how many times each ultrasonic sensor identified an obstacle, object, or reflection during the time period studied.

The variance of the blue curve corresponds to the signals of the sensor framed in the blue square, whilst the variation in the red curve indicates the oscillations of the sensor's signals framed in the red square.

If an ultrasonic sensor registers an increase in detections, it indicates that an obstacle has been recognized in the neighborhood; if the ultrasonic sensor detects nothing, it indicates that there may be blind zones around the virtual vehicle model.

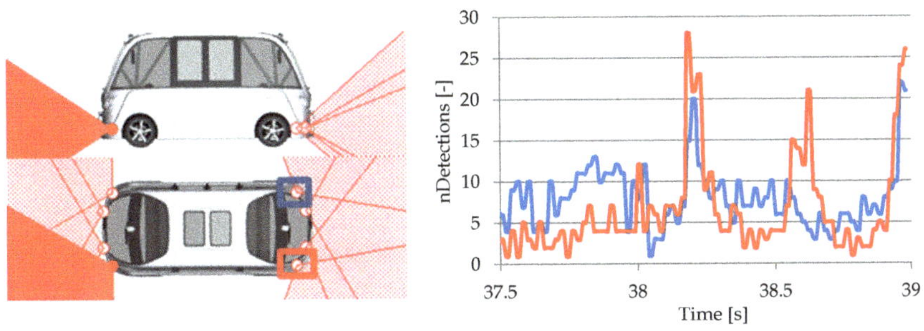

(a) (b)

Figure 43. Ultrasonic RSI sensor parameterization and generated parameters. (**a**) Sensor position (**b**) Sensor parameters.

5. Discussion

Regarding the future evolution of autonomous vehicles, Karp, L., noted in a report to the Stevens Institute of Technology [134] that assigning manufacturer liability for sensors and software algorithms will be a major driver of growth and innovation, given their critical role in ensuring the functional safety of such vehicles.

Replicating the behavior of real-world sensors in a virtual environment is a complex task. This involves capturing elements such as sensor noise, limitations in the field of view, uncertainty in physical parameters, and responses to varying environmental conditions. This can be achieved through advanced data-driven algorithms, AI-based simulation techniques, and innovative virtual sensor designs. Autonomous vehicles utilize a variety of sensors (cameras, radar, lidar, etc.), and combining data from multiple virtual and real sensors through sensor fusion techniques, each with its characteristics and potential errors, is a significant challenge [135]. Their performance can degrade due to changes in nonlinear dynamics, complex physical processes in the environment, and nonlinear interactions between input and output variables [136,137]. High-fidelity sensors can be computationally intensive to simulate real-world sensor data with high precision. This requires robust hardware and efficient algorithms to process the data in real-time [137]. Virtual sensors must operate in real time within the simulation environment, meaning they must generate and process data quickly enough to keep up with the simulation. This can be computationally demanding [138]. AI techniques enable the development of virtual sensors that can estimate physical parameters without traditional sensors, using data-driven models to emulate real-world conditions [11]. The integration of AI introduces potential uncertainties, necessitating validation to maintain trust in these systems [139].

Validating virtual sensor data against real-world data or physical sensor measurements can be difficult. Autonomous vehicles must handle unusual situations, and realistically simulating these circumstances adds another layer of complexity. Virtual sensors in the overall software architecture of autonomous vehicles can be complicated, mainly if the architecture was not designed initially with virtual sensors in mind. The development and validation of virtual sensors can be time-consuming and costly, especially when dealing with complex sensor models or high accuracy requirements.

Future studies could focus on improving the accuracy and robustness of virtual sensor models, particularly in complex and dynamic environments. This can address challenges related to nonlinear dynamics models with complex physical processes, research on environmental factors (such as weather conditions, temperature, and lighting), and detailed real-world behavioral models. Research into advanced AI and ML techniques for sensor fusion and data processing is also essential. Implementing these advancements is likely to result in more accurate and reliable data, thereby enhancing the overall performance of AV driving systems. Additionally, using DL architectures designed explicitly for particular tasks (like object detection or lane keeping) would be highly beneficial. Efforts should also concentrate on developing robust data-driven system models capable of accurately identifying various physical characteristics. Such models can provide a viable and cost-effective alternative to traditional real sensors. Research should include testing virtual sensor performance in rare or unexpected situations, often referred to as edge cases. This requires testing and validating across a wide range of scenarios.

The importance of developing virtual models of autonomous vehicles equipped with intelligent sensors is underscored by several international regulatory and legal frameworks governing automated driving systems:

UN Regulation No. 157 (2021/389) [140] introduces a multi-pillar certification framework for automated driving systems, including physical testing, audit procedures, and validated virtual simulations.

Regulation (EU) 2018/1592 [141] allows simulations to be used for vehicle feature evaluation, provided they accurately replicate real-world behavior.

ISO 26262—Functional Safety of Electrical/Electronic Systems [142] recommends simulation during development to identify and mitigate risks in electrical and electronic systems, ensuring compliance with safety standards.

ISO 34502:2022—Scenario-based Safety Evaluation Framework for Automated Driving Systems [143] presents a structured methodology for assessing safety performance in automated driving systems through standardized scenario libraries and virtual simulations.

ISO/PAS 21448:2022 Road Vehicles—Safety of the Intended Functionality [144] outlines guidelines for mitigating risks from functional limitations in automated systems, particularly in the absence of system failures, and emphasizes the role of scenario-based testing and simulation in evaluating real-world performance.

Validating and calibrating virtual sensors is also important and may involve establishing standardized testing procedures and metrics. Establishing industry standards for development and validation would promote interoperability and facilitate the widespread adoption of this technology. It is also crucial to optimize algorithms and ensure that hardware can meet the computational demands of high-fidelity sensor interfaces, especially in complex scenarios. Exploring hybrid approaches that combine both virtual sensor data and real sensor data could leverage the strengths of each type, thus enhancing overall accuracy and reliability. Also, it is important to examine the necessary levels of simulation fidelity for various autonomous driving tasks to balance the need for accuracy with computational efficiency. Studying how virtual sensor data are presented to human operators in semi-autonomous systems can also be an important step, as it impacts their situational awareness and decision-making processes.

6. Conclusions

This paper has explored the critical role of virtual sensors in the simulation and development of autonomous vehicles. Our study reveals that virtual sensors are not only substitutes for physical sensors but are integral components of advanced driver-assistance systems and autonomous driving functionalities. The classification of virtual sensors into three categories (ideal, Hi-Fi, and RSI) provides a structured approach to understanding their capabilities and limitations, demonstrating a range of fidelity from purely software-based abstractions to signal interfaces mimicking real-world sensor outputs.

Our study demonstrated that virtual sensors can effectively model various sensor functionalities, including slip angle, inertial measurements, object detection, and environmental perception. The examples provided show how these virtual sensors generate data that align with expected real-world behaviors, thus validating their role in simulation environments. The increasing complexity of sensor types, from ideal to RSI, highlights the trade-off between computational efficiency and simulation fidelity, a crucial consideration for real-time applications.

The implications of our research are significant. Accurately simulating a wide range of sensor data enables the development and testing of autonomous driving algorithms in a safe, controlled, and cost-effective virtual environment. This accelerates the development cycle by minimizing the need for physical testing and allowing the exploration of edge cases and rare scenarios that are difficult to replicate in the real world. Additionally, integrating AI and ML techniques into virtual sensor models opens new methods for adaptive and predictive sensing capabilities, enhancing the robustness and reliability of autonomous systems.

The advantages of virtual sensors include the following:

- Cost-effectiveness—virtual sensors are affordable alternatives to real sensors.
- Reduced vehicle mass—replacing physical sensors and associated wiring and connectors helps to reduce vehicle weight, leading to fuel and energy savings and lower emissions.
- Software-based—virtual sensors are developed using software applications and do not require additional hardware.
- Remote updates—firmware upgrades can be performed remotely via the Over-The-Air method, eliminating the need for physical interventions.
- Improved data accuracy and resolution—virtual sensors can improve data accuracy and resolution by merging information from multiple sources and by using data fusion algorithms.
- Expanded coverage—they can provide coverage to locations where physical sensors are unavailable.
- Preprocessing and optimization—when integrated into an embedded system, they can preprocess data, and perform error correction, merging, and optimization of input data.
- Flexibility—they can be simple or sophisticated, depending on the simulated activities and their consequences.
- Data accessibility—they can increase data accessibility from physical sensors, facilitating collaboration and efficient use of resources in interconnected systems like IoT.
- Reduced testing costs and time—using virtual sensors, developers can significantly reduce the cost and time associated with physical testing.
- Exploration of a wider range of scenarios—virtual sensors allow for exploring a wider range of scenarios and edge cases.

However, challenges remain. Replicating the nuanced behavior of real-world sensors, including noise, environmental dependencies, and physical limitations, requires sophisticated modeling techniques.

In conclusion, virtual sensors are indispensable tools in the development of autonomous vehicles. They facilitate comprehensive testing, accelerate development, and enable the exploration of advanced sensing capabilities.

Author Contributions: Conceptualization, I.B. and C.I.; methodology, C.I.; software, C.I.; validation, C.I., H.B. and F.B.S.; formal analysis, I.B. and C.I.; resources, C.I.; data curation, H.B. and F.B.S.; writing—original draft preparation, C.I. and A.M.; writing—review and editing, I.B., C.I., H.B., C.A., A.M. and F.B.S.; visualization, C.I.; supervision, I.B. All authors have read and agreed to the published version of the manuscript.

Funding: This research received no external funding.

Institutional Review Board Statement: Not applicable.

Informed Consent Statement: Not applicable.

Data Availability Statement: Access to the data is available upon request. Access to the data can be requested via e-mail to the corresponding author.

Acknowledgments: The simulations presented in the review were performed using the software CarMaker supported by IPG Automotive GmbH, Karlsruhe, Germany.

Conflicts of Interest: The authors declare no conflicts of interest.

Abbreviations

The following abbreviations are used in this manuscript:

ABS	Antilock Braking System
ADAS	Advanced Driver Assistance System
ACC	Adaptive Cruise Control
AD	Autonomous driving
AEB	Automatic Emergency Braking
AI	Artificial Intelligence
AWD	All-Wheel Drive
BiSeNet	Bilateral Segmentation Network
CNN	Convolutional neural network
D-GNSS	Differential-Global Navigation Satellite System
CPU	Central Processing Unit
DL	Deep Learning
DMS	Driver Monitoring System
DPM	Deformable Part Model
DTw	Digital Twin
DVM	Double Validation Metric
EBA	Emergency Brake Assist
ECEF	Earth-Centered, Earth-Fixed
EDF	Empirical Cumulative Distribution Function
EM	Energy Management
ESC	Electronic Stability Control
FC	Fuel Consumption
FCW	Forward Collision Warning
FET	Field-Effect Transistor
FoV	Field-Of-View
GCS	Geographic Coordinate System
GDOP	Geometric Dilution of Precision
GNN	Global Nearest Neighbor
GNSS	Global Navigation Satellite System
GPS	Global Positioning System
GPU	Graphics Processing Unit
HD	High-Definition
Hi-Fi	High fidelity
HiL	Hardware-in-the-Loop
HV	High voltage
ID	Identifier
ILA	Intelligent Light Assist
IMU	Inertial Measurement Unit
IoT	Internet of Things
LDW	Lane Departure Warning
Lidar	Light detection and ranging
LK	Lane Keeping

LKA	Lane Keeping Assist
LV	Low voltage
MEMS	Micro-Electro-Mechanical Systems
ML	Machine Learning
MMS	Mobile Mapping System
OTA	Over-The-Air
PA	Park Assist
PPHT	Progressive Probabilistic Hough Transform
POI	Point Of Interest
PS	Physical sensors
PT	Powertrain
QTC	Quantum Tunneling Composite
RCS	Radar Cross-Section
RSI	Raw Signal Interface
RTK	Real-Time Kinematic
SAS	Smart Airbag System
SD	Sign Detection
SiL	Software-in-the-Loop
SLAM	Simultaneous localization And mapping
SNR	Signal-to-noise ratio
SPA	Sound Pressure Amplitude
SRTM	Shuttle Radar Topography Mission
ToF	Time of Flight
TSA	Traffic Sign Assist
V2I	Vehicle-to-Infrastructure
V2V	Vehicle-to-Vehicle
VDM	Vehicle dynamics model
VLC	Visible Light Communication
VS	Virtual sensors
WLD	Wheel Lifting Detection
YOLO	You Only Look Once

References

1. Martin, D.; Kühl, N.; Satzger, G. Virtual Sensors. *Bus. Inf. Syst. Eng.* **2021**, *63*, 315–323. [CrossRef]
2. Dahiya, R.; Ozioko, O.; Cheng, G. *Sensory Systems for Robotic Applications*; MIT Press: Cambridge, MA, USA, 2022. [CrossRef]
3. Šabanovič, E.; Kojis, P.; Šukevičius, Š.; Shyrokau, B.; Ivanov, V.; Dhaens, M.; Skrickij, V. Feasibility of a Neural Network-Based Virtual Sensor for Vehicle Unsprung Mass Relative Velocity Estimation. *Sensors* **2021**, *21*, 7139. [CrossRef] [PubMed]
4. Persson, J.A.; Bugeja, J.; Davidsson, P.; Holmberg, J.; Kebande, V.R.; Mihailescu, R.-C.; Sarkheyli-Hägele, A.; Tegen, A. The Concept of Interactive Dynamic Intelligent Virtual Sensors (IDIVS): Bridging the Gap between Sensors, Services, and Users through Machine Learning. *Appl. Sci.* **2023**, *13*, 6516. [CrossRef]
5. Ambarish, P.; Mitradip, B.; Ravinder, D. *Solid-State Sensors*; IEEE Press Series on Sensors; Wiley-IEEE Press: Hoboken, NJ, USA, 2023. [CrossRef]
6. Shin, H.; Kwak, Y. Enhancing digital twin efficiency in indoor environments: Virtual sensor-driven optimization of physical sensor combinations. *Automat. Constr.* **2024**, *161*, 105326. [CrossRef]
7. Stanley, M.; Lee, J. *Sensor Analysis for the Internet of Things*; Morgan & Claypool Publishers: San Rafael, CA, USA; Arizona State University: Tempe, AZ, USA, 2018.
8. Stetter, R. A Fuzzy Virtual Actuator for Automated Guided Vehicles. *Sensors* **2020**, *20*, 4154. [CrossRef]
9. Xie, J.; Yang, R.; Gooi, H.B.; Nguyen, H. PID-based CNN-LSTM for accuracy-boosted virtual sensor in battery thermal management system. *Appl. Energ.* **2023**, *331*, 120424. [CrossRef]
10. Iclodean, C.; Cordos, N.; Varga, B.O. Autonomous Shuttle Bus for Public Transportation: A Review. *Energies* **2020**, *13*, 2917. [CrossRef]
11. Fabiocchi, D.; Giulietti, N.; Carnevale, M.; Giberti, H. AI-Driven Virtual Sensors for Real-Time Dynamic Analysis of Me-chanisms: A Feasibility Study. *Machines* **2024**, *12*, 257. [CrossRef]

12. Kabadayi, S.; Pridgen, A.; Julien, C. Virtual sensors: Abstracting data from physical sensors. In Proceedings of the IEEE International Symposium on a World of Wireless, Mobile and Multimedia Networks, Buffalo-Niagara Falls, NY, USA, 26–29 June 2006. [CrossRef]
13. Compredict. Available online: https://compredict.ai/virtual-sensors-replacing-vehicle-hardware-sensors/ (accessed on 6 February 2025).
14. Prokhorov, D. Virtual Sensors and Their Automotive Applications. In Proceedings of the 2005 International Conference on Intelligent Sensors, Sensor Networks and Information Processing, Melbourne, VIC, Australia, 5–8 December 2005. [CrossRef]
15. Forssell, U.; Ahlqvist, S.; Persson, N.; Gustafsson, F. Virtual Sensors for Vehicle Dynamics Applications. In *Advanced Microsystems for Automotive Applications 2001*; VDI-Buch; Krueger, S., Gessner, W., Eds.; Springer: Berlin/Heidelberg, Germany, 2012. [CrossRef]
16. Hu, X.H.; Cao, L.; Luo, Y.; Chen, A.; Zhang, E.; Zhang, W. A Novel Methodology for Comprehensive Modeling of the Kinetic Behavior of Steerable Catheters. *IEEE/ASME Trans. Mechatron.* **2019**, *24*, 1785–1797. [CrossRef]
17. Cummins. Available online: https://www.cummins.com/news/2024/03/18/virtual-sensors-and-their-role-energy-future (accessed on 6 February 2025).
18. Bucaioni, A.; Pelliccione, P.; Mubeen, S. Modelling centralised automotive E/E software architectures. *Adv. Eng. Inform.* **2024**, *59*, 102289. [CrossRef]
19. Zhang, Q.; Shen, S.; Li, H.; Cao, W.; Tang, W.; Jiang, J.; Deng, M.; Zhang, Y.; Gu, B.; Wu, K.; et al. Digital twin-driven intelligent production line for automotive MEMS pressure sensors. *Adv. Eng. Inform.* **2022**, *54*, 101779. [CrossRef]
20. Ida, N. *Sensors, Actuators, and Their Interfaces: A multidisciplinary Introduction*, 2nd ed.; The Institution of Engineering and Technology: London, UK, 2020. [CrossRef]
21. Masti, D.; Bernardini, D.; Bemporad, A. A machine-learning approach to synthesize virtual sensors for parameter-varying systems. *Eur. J. Control.* **2021**, *61*, 40–49. [CrossRef]
22. Paepae, T.; Bokoro, P.N.; Kyamakya, K. From fully physical to virtual sensing for water quality assessment: A comprehensive review of the relevant state-of-the-art. *Sensors* **2021**, *21*, 6971. [CrossRef] [PubMed]
23. Tihanyi, V.; Tettamanti, T.; Csonthó, M.; Eichberger, A.; Ficzere, D.; Gangel, K.; Hörmann, L.B.; Klaffenböck, M.A.; Knauder, C.; Luley, P.; et al. Motorway Measurement Campaign to Support R&D Activities in the Field of Automated Driving Technologies. *Sensors* **2021**, *21*, 2169. [CrossRef]
24. Tactile Mobility. Available online: https://tactilemobility.com/ (accessed on 6 February 2025).
25. Compredict-Virtual Sensor Platform. Available online: https://compredict.ai/virtual-sensor-platform/ (accessed on 6 February 2025).
26. Mordor Intellingence. Available online: https://www.mordorintelligence.com/industry-reports/virtual-sensors-market (accessed on 6 February 2025).
27. Iclodean, C.; Varga, B.O.; Cordoș, N. Autonomous Driving Technical Characteristics. In *Autonomous Vehicles for Public Transportation, Green Energy and Technology*; Springer: Berlin/Heidelberg, Germany, 2022; pp. 15–68. [CrossRef]
28. SAE. Available online: https://www.sae.org/standards/content/j3016_202104/ (accessed on 6 February 2025).
29. Muhovič, J.; Perš, J. Correcting Decalibration of Stereo Cameras in Self-Driving Vehicles. *Sensors* **2020**, *20*, 3241. [CrossRef] [PubMed]
30. Hamidaoui, M.; Talhaoui, M.Z.; Li, M.; Midoun, M.A.; Haouassi, S.; Mekkaoui, D.E.; Smaili, A.; Cherraf, A.; Benyoub, F.Z. Survey of Autonomous Vehicles' Collision Avoidance Algorithms. *Sensors* **2025**, *25*, 395. [CrossRef]
31. Cabon, Y.; Murray, N.; Humenberger, M. Virtual KITTI 2. *arXiv* **2020**, arXiv:2001.10773. [CrossRef]
32. Mallik, A.; Gaopande, M.L.; Singh, G.; Ravindran, A.; Iqbal, Z.; Chao, S.; Revalla, H.; Nagasamy, V. Real-time Detection and Avoidance of Obstacles in the Path of Autonomous Vehicles Using Monocular RGB Camera. *SAE Int. J. Adv. Curr. Pract. Mobil.* **2022**, *5*, 622–632. [CrossRef]
33. Zhe, T.; Huang, L.; Wu, Q.; Zhang, J.; Pei, C.; Li, L. Inter-Vehicle Distance Estimation Method Based on Monocular Vision Using 3D Detection. *IEEE Trans. Veh. Technol.* **2020**, *69*, 4907–4919. [CrossRef]
34. Rill, R.A.; Faragó, K.B. Collision avoidance using deep learning-based monocular vision. *SN Comput. Sci.* **2021**, *2*, 375. [CrossRef]
35. He, J.; Tang, K.; He, J.; Shi, J. Effective vehicle-to-vehicle positioning method using monocular camera based on VLC. *Opt. Express* **2020**, *28*, 4433–4443. [CrossRef] [PubMed]
36. Choi, W.Y.; Yang, J.H.; Chung, C.C. Data-Driven Object Vehicle Estimation by Radar Accuracy Modeling with Weighted Interpolation. *Sensors* **2021**, *21*, 2317. [CrossRef] [PubMed]
37. Muckenhuber, S.; Museljic, E.; Stettinger, G. Performance evaluation of a state-of-the-art automotive radar and corres-ponding modeling approaches based on a large labeled dataset. *J. Intell. Transport. Syst.* **2022**, *26*, 655–674. [CrossRef]
38. Sohail, M.; Khan, A.U.; Sandhu, M.; Shoukat, I.A.; Jafri, M.; Shin, H. Radar sensor based Machine Learning approach for precise vehicle position estimation. *Sci. Rep.* **2023**, *13*, 13837. [CrossRef]
39. Srivastav, A.; Mandal, S. Radars for autonomous driving: A review of deep learning methods and challenges. *IEEE Access* **2023**, *11*, 97147–97168. [CrossRef]

40. Poulose, A.; Baek, M.; Han, D.S. Point cloud map generation and localization for autonomous vehicles using 3D lidar scans. In Proceedings of the 2022 27th Asia Pacific Conference on Communications (APCC), Jeju, Republic of Korea, 19–21 October 2022; pp. 336–341. [CrossRef]
41. Saha, A.; Dhara, B.C. 3D LiDAR-based obstacle detection and tracking for autonomous navigation in dynamic environments. *Int. J. Intell. Robot. Appl.* **2024**, *8*, 39–60. [CrossRef]
42. Dazlee, N.M.A.A.; Khalil, S.A.; Rahman, S.A.; Mutalib, S. Object detection for autonomous vehicles with sensor-based technology using YOLO. *Int. J. Intell. Syst. Appl. Eng.* **2022**, *10*, 129–134. [CrossRef]
43. Guan, L.; Chen, Y.; Wang, G.; Lei, X. Real-time vehicle detection framework based on the fusion of LiDAR and camera. *Electronics* **2020**, *9*, 451. [CrossRef]
44. Kotur, M.; Lukić, N.; Krunić, M.; Lukač, Ž. Camera and LiDAR sensor fusion for 3d object tracking in a collision avoidance system. In Proceedings of the 2021 Zooming Innovation in Consumer Technologies Conference (ZINC), Novi Sad, Serbia, 26–27 May 2021; pp. 198–202. [CrossRef]
45. Choi, W.Y.; Kang, C.M.; Lee, S.H.; Chung, C.C. Radar accuracy modeling and its application to object vehicle tracking. *Int. J. Control. Autom. Syst.* **2020**, *18*, 3146–3158. [CrossRef]
46. Simcenter. Available online: https://blogs.sw.siemens.com/simcenter/the-sense-of-virtual-sensors/ (accessed on 6 February 2025).
47. Kim, J.; Kim, Y.; Kum, D. Low-level sensor fusion network for 3D vehicle detection using radar range-azimuth heatmap and monocular image. In Proceedings of the Asian Conference on Computer Vision, Kyoto, Japan, 30 November–4 December 2020. [CrossRef]
48. Lim, S.; Jung, J.; Lee, B.H.; Choi, J.; Kim, S.C. Radar sensor-based estimation of vehicle orientation for autonomous driving. *IEEE Sensors J.* **2022**, *22*, 21924–21932. [CrossRef]
49. Caesar, H.; Bankiti, V.; Lang, A.H.; Vora, S.; Liong, V.E.; Xu, Q.; Krishnan, A.; Pan, Y.; Baldan, G.; Beijbom, O. nuScenes: A multimodal dataset for autonomous driving. In Proceedings of the IEEE/CVF Conference on Computer Vision and Pattern Recognition, Seattle, WA, USA, 13–19 June 2020; pp. 11621–11631. [CrossRef]
50. Robsrud, D.N.; Øvsthus, Ø.; Muggerud, L.; Amendola, J.; Cenkeramaddi, L.R.; Tyapin, I.; Jha, A. Lidar-mmW Radar Fusion for Safer UGV Autonomous Navigation with Collision Avoidance. In Proceedings of the 2023 11th International Conference on Control, Mechatronics and Automation (ICCMA), Grimstad, Norway, 1–3 November 2023; pp. 189–194. [CrossRef]
51. Wang, Y.; Jiang, Z.; Gao, X.; Hwang, J.N.; Xing, G.; Liu, H. RODnet: Radar object detection using cross-modal supervision. In Proceedings of the IEEE/CVF Winter Conference on Applications of Computer Vision, Virtual, 5–9 January 2021; pp. 504–513. [CrossRef]
52. Rövid, A.; Remeli, V.; Paufler, N.; Lengyel, H.; Zöldy, M.; Szalay, Z. Towards Reliable Multisensory Perception and Its Automotive Applications. *Period. Polytech. Transp. Eng.* **2020**, *48*, 334–340. [CrossRef]
53. IPG, CarMaker. Available online: https://www.ipg-automotive.com/en/products-solutions/software/carmaker/ (accessed on 6 February 2025).
54. Yeong, D.J.; Velasco-Hernandez, G.; Barry, J.; Walsh, J. Sensor and Sensor Fusion Technology in Autonomous Vehicles: A Review. *Sensors* **2021**, *21*, 2140. [CrossRef] [PubMed]
55. Liu, X.; Baiocchi, O. A comparison of the definitions for smart sensors, smart objects and Things in IoT. In Proceedings of the 2016 IEEE 7th Annual Information Technology, Electronics and Mobile Communication Conference (IEMCON), Vancouver, BC, Canada, 13–15 October 2016; pp. 1–4. [CrossRef]
56. Peinado-Asensi, I.; Montés, N.; García, E. Virtual Sensor of Gravity Centres for Real-Time Condition Monitoring of an Industrial Stamping Press in the Automotive Industry. *Sensors* **2023**, *23*, 6569. [CrossRef] [PubMed]
57. Stetter, R.; Witczak, M.; Pazera, M. Virtual Diagnostic Sensors Design for an Automated Guided Vehicle. *Appl. Sci.* **2018**, *8*, 702. [CrossRef]
58. Lengyel, H.; Maral, S.; Kerebekov, S.; Szalay, Z.; Török, Á. Modelling and simulating automated vehicular functions in critical situations—Application of a novel accident reconstruction concept. *Vehicles* **2023**, *5*, 266–285. [CrossRef]
59. Dörr, D. *Using Virtualization to Accelerate the Development of ADAS & Automated Driving Functions*; IPG Automotive, GTC Europe München: München, Germany, 28 September 2017.
60. Kim, J.; Park, S.; Kim, J.; Yoo, J. A Deep Reinforcement Learning Strategy for Surrounding Vehicles-Based Lane-Keeping Control. *Sensors* **2023**, *23*, 9843. [CrossRef]
61. Pannagger, P.; Nilac, D.; Orucevic, F.; Eichberger, A.; Rogic, B. Advanced Lane Detection Model for the Virtual Development of Highly Automated Functions. *arXiv* **2021**, arXiv:2104.07481. [CrossRef]
62. IPG Automotive. *IPG Guide-User's Guide Version 12.0.1 CarMaker*; IPG Automotive: München, Germany, 2023.
63. Iclodean, C.; Varga, B.O.; Cordoș, N. Virtual Model. In *Autonomous Vehicles for Public Transportation, Green Energy and Technology*; Springer: Berlin/Heidelberg, Germany, 2022; pp. 195–335. [CrossRef]

64. Schäferle, S. Choosing the Correct Sensor Model for Your Application. IPG Automotive. 2019. Available online: https://www.ipg-automotive.com/uploads/tx_pbfaqtickets/files/98/SensorModelLevels.pdf (accessed on 6 February 2025).
65. Magosi, Z.F.; Wellershaus, C.; Tihanyi, V.R.; Luley, P.; Eichberger, A. Evaluation Methodology for Physical Radar Perception Sensor Models Based on On-Road Measurements for the Testing and Validation of Automated Driving. *Energies* **2022**, *15*, 2545. [CrossRef]
66. IPG Automotive. *Reference Manual Version 12.0.1 CarMaker*; IPG Automotive: München, Germany, 2023.
67. Iclodean, C. *Introducere în Sistemele Autovehiculelor*; Risoprint: Cluj-Napoca, Romania, 2023.
68. Renard, D.; Saddem, R.; Annebicque, D.; Riera, B. From Sensors to Digital Twins toward an Iterative Approach for Existing Manufacturing Systems. *Sensors* **2024**, *24*, 1434. [CrossRef]
69. Brucherseifer, E.; Winter, H.; Mentges, A.; Mühlhäuser, M.; Hellmann, M. Digital Twin conceptual framework for improving critical infrastructure resilience. *at-Automatisierungstechnik* **2021**, *69*, 1062–1080. [CrossRef]
70. Grieves, M.; Vickers, J. Digital twin: Mitigating unpredictable, undesirable emergent behavior in complex systems. In *Transdisciplinary Perspectives on Complex Systems: New Findings and Approaches*; Springer: Berlin/Heidelberg, Germany, 2016; pp. 85–113. [CrossRef]
71. Kritzinger, W.; Karner, M.; Traar, G.; Henjes, J.; Sihn, W. Digital Twin in manufacturing: A categorical literature review and classification. *Ifac-PapersOnline* **2018**, *51*, 1016–1022. [CrossRef]
72. Shoukat, M.U.; Yan, L.; Yan, Y.; Zhang, F.; Zhai, Y.; Han, P.; Nawaz, S.A.; Raza, M.A.; Akbar, M.W.; Hussain, A. Autonomous driving test system under hybrid reality: The role of digital twin technology. *Internet Things* **2024**, *27*, 101301. [CrossRef]
73. Tu, L.; Xu, M. An Analysis of the Use of Autonomous Vehicles in the Shared Mobility Market: Opportunities and Challenges. *Sustainability* **2024**, *16*, 6795. [CrossRef]
74. Navya. Brochure-Autonom-Shuttle-Evo. Available online: https://navya.tech/wp-content/uploads/documents/Brochure-Autonom-Shuttle-Evo-EN.pdf (accessed on 6 February 2025).
75. Navya. Self-Driving Shuttle for Passenger Transportation. Available online: https://www.navya.tech/en/solutions/moving-people/self-driving-shuttle-for-passenger-transportation/ (accessed on 6 February 2025).
76. Patentimage. Available online: https://patentimages.storage.googleapis.com/12/0f/d1/33f8d2096f49f6/US20180095473A1.pdf (accessed on 6 February 2025).
77. AVENUE Autonomous Vehicles to Evolve to a New Urban Experience Report. Available online: https://h2020-avenue.eu/wp-content/uploads/2023/03/Keolis-LyonH2020-AVENUE_Deliverable_7.6_V2-not-approved.pdf (accessed on 6 February 2025).
78. EarthData Search. Available online: https://search.earthdata.nasa.gov/search?q=SRTM (accessed on 6 February 2025).
79. GpsPrune. Available online: https://activityworkshop.net/software/gpsprune/download.html (accessed on 6 February 2025).
80. IPG Automotive. *InfoFile Description Version 12.0.1 IPGRoad*; IPG Automotive: München, Germany, 2023.
81. IPG Automotive. *User Manual Version 12.0.1 IPGDriver*; IPG Automotive: München, Germany, 2023.
82. Piyabongkarn, D.N.; Rajamani, R.; Grogg, J.A.; Lew, J.Y. Development and Experimental Evaluation of a Slip Angle Estimator for Vehicle Stability Control. *IEEE Trans. Control. Syst. Technol.* **2009**, *17*, 78–88. [CrossRef]
83. IPG Automotive. *CarMaker Reference Manual 12.0.2 CarMaker*; IPG Automotive: München, Germany, 2023.
84. Pacejka, H.B. *Tyre and Vehicle Dynamics*, 2nd ed.; Elsevier's Science and Technology: Oxford, UK, 2006.
85. Salminen, H. Parametrizing Tyre Wear Using a Brush Tyre Model. Master's Thesis, Royal Institute of Technology, Stockholm, Sweden, 15 December 2014. Available online: https://kth.diva-portal.org/smash/get/diva2:802101/FULLTEXT01.pdf (accessed on 12 February 2025).
86. Pacjka, H.B.; Besselink, I.J.M. Magic Formula Tyre Model with Transient Properties. *Veh. Syst. Dyn.* **1997**, *27*, 234–249. [CrossRef]
87. Pacejka, H.B. Chapter 4—Semi-Empirical Tire Models. In *Tire and Vehicle Dynamics*, 3rd ed.; Pacejka, H.B., Ed.; Butterworth-Heinemann: Oxford, UK, 2012; pp. 149–209. [CrossRef]
88. Guo, Q.; Xu, Z.; Wu, Q.; Duan, J. The Application of in-the-Loop Design Method for Controller. In Proceedings of the 2nd IEEE Conference on Industrial Electronics and Applications, Harbin, China, 23–25 May 2007; pp. 78–81. [CrossRef]
89. Chen, T.; Chen, L.; Xu, X.; Cai, Y.; Jiang, H.; Sun, X. Sideslip Angle Fusion Estimation Method of an Autonomous Electric Vehicle Based on Robust Cubature Kalman Filter with Redundant Measurement Information. *World Electr. Veh. J.* **2019**, *10*, 34. [CrossRef]
90. Jin, L.; Xie, X.; Shen, C.; Wang, F.; Wang, F.; Ji, S.; Guan, X.; Xu, J. Study on electronic stability program control strategy based on the fuzzy logical and genetic optimization method. *Adv. Mech. Eng.* **2017**, *9*, 168781401769935. [CrossRef]
91. Zhao, Z.; Chen, H.; Yang, J.; Wu, X.; Yu, Z. Estimation of the vehicle speed in the driving mode for a hybrid electric car based on an unscented Kalman filter. *Proc. Inst. Mech. Eng. Part D J. Automob. Eng.* **2014**, *229*, 437–456. [CrossRef]
92. Li, Q.; Chen, L.; Li, M.; Shaw, S.-L.; Nuchter, A. A Sensor-Fusion Drivable-Region and Lane-Detection System for Auto-nomous Vehicle Navigation in Challenging Road Scenarios. *IEEE Trans. Veh. Technol.* **2013**, *63*, 540–555. [CrossRef]
93. Rana, M.M. Attack Resilient Wireless Sensor Networks for Smart Electric Vehicles. *IEEE Sens. Lett.* **2017**, *1*, 5500204. [CrossRef]

94. Xia, X.; Xiong, L.; Huang, Y.; Lu, Y.; Gao, L.; Xu, N.; Yu, Z. Estimation on IMU yaw misalignment by fusing information of automotive onboard sensors. *Mech. Syst. Signal Process.* **2022**, *162*, 107993. [CrossRef]
95. Sieberg, P.M.; Schramm, D. Ensuring the Reliability of Virtual Sensors Based on Artificial Intelligence within Vehicle Dynamics Control Systems. *Sensors* **2022**, *22*, 3513. [CrossRef] [PubMed]
96. Xiong, L.; Xia, X.; Lu, Y.; Liu, W.; Gao, L.; Song, S.; Han, Y.; Yu, Z. IMU-Based Automated Vehicle Slip Angle and Attitude Estimation Aided by Vehicle Dynamics. *Sensors* **2019**, *19*, 1930. [CrossRef] [PubMed]
97. Ess, A.; Schindler, K.; Leibe, B.; Van Gool, L. Object detection and tracking for autonomous navigation in dynamic environments. *Int. J. Robot. Res.* **2010**, *29*, 1707–1725. [CrossRef]
98. Bewley, A.; Ge, Z.; Ott, L.; Ramos, F.; Upcroft, B. Simple online and realtime tracking. In Proceedings of the 2016 IEEE International Conference on Image Processing (ICIP), Phoenix, AZ, USA, 25–28 September 2016. [CrossRef]
99. Banerjee, S.; Serra, J.G.; Chopp, H.H.; Cossairt, O.; Katsaggelos, A.K. An Adaptive Video Acquisition Scheme for Object Tracking. In Proceedings of the 27th European Signal Processing Conference (EUSIPCO), A Coruna, Spain, 2–6 September 2019. [CrossRef]
100. Chen, N.; Li, M.; Yuan, H.; Su, X.; Li, Y. Survey of pedestrian detection with occlusion. *Complex Intell. Syst.* **2021**, *7*, 577–587. [CrossRef]
101. Liu, Z.; Chen, W.; Wu, X. Salient region detection using high level feature. In Proceedings of the 13th International Conference on Control Automation Robotics & Vision (ICARCV), Singapore, 10–12 December 2014. [CrossRef]
102. Felzenszwalb, P.; Girshick, R.; McAllester, D.; Ramanan, D. Visual object detection with deformable part models. *Commun. ACM* **2013**, *56*, 97–105. [CrossRef]
103. Kato, S.; Takeuchi, E.; Ishiguro, Y.; Ninomiya, Y.; Takeda, K.; Hamada, T. An Open Approach to Autonomous Vehicles. *IEEE Micro* **2015**, *35*, 60–68. [CrossRef]
104. Broggi, A.; Cattani, S.; Patander, M.; Sabbatelli, M.; Zani, P. A full-3D voxel-based dynamic obstacle detection for urban scenario using stereo vision. In Proceedings of the 16th International IEEE Conference on Intelligent Transportation Systems (ITSC 2013), The Hague, The Netherlands, 6–9 October 2013; pp. 71–76. [CrossRef]
105. Patra, S.; Maheshwari, P.; Yadav, S.; Arora, C.; Banerjee, S. A Joint 3D-2D based Method for Free Space Detection on Roads. *arXiv* **2018**, arXiv:1711.02144. [CrossRef]
106. Vitor, G.B.; Lima, D.A.; Victorino, A.C.; Ferreira, J.V. A 2D/3D vision based approach applied to road detection in urban environments. In Proceedings of the IEEE Intelligent Vehicles Symposium (IV 2013), Gold Coast, QLD, Australia, 23–26 June 2013; pp. 952–957.
107. Heinz, L. *CarMaker Tips & Tricks No. 3-011 Detect Traffic Lights*; IPG Automotive: München, Germany, 2019; Available online: https://www.ipg-automotive.com/uploads/tx_pbfaqtickets/files/100/DetectTrafficLights.pdf (accessed on 6 February 2025).
108. Zhang, P.; Zhang, M.; Liu, J. Real-time HD map change detection for crowdsourcing update based on mid-to-high-end sensors. *Sensors* **2021**, *21*, 2477. [CrossRef]
109. Bahlmann, C.; Zhu, Y.; Ramesh, V.; Pellkofer, M.; Koehler, T. A System for Traffic Sign Detection, Tracking, and Recognition Using Color, Shape, and Motion Information. In Proceedings of the IEEE Proceedings of Intelligent Vehicles Symposium, Las Vegas, NV, USA, 6–8 June 2005; pp. 255–260. [CrossRef]
110. Fazekas, Z.; Gerencsér, L.; Gáspár, P. Detecting Change between Urban Road Environments along a Route Based on Static Road Object Occurrences. *Appl. Sci.* **2021**, *11*, 3666. [CrossRef]
111. Liu, C.; Tao, Y.; Liang, J.; Li, K.; Chen, Y. Object detection based on YOLO network. In Proceedings of the 2018 IEEE 4th Information Technology and Mechatronics Engineering Conference (ITOEC), Chongqing, China, 14–16 December 2018; pp. 799–803. [CrossRef]
112. Nuthong, C.; Charoenpong, T. Lane Detection using Smoothing Spline. In Proceedings of the 3rd International Congress on Image and Signal Processing, Yantai, China, 16–18 October 2010; pp. 989–993. [CrossRef]
113. Dou, J.; Li, J. Robust object detection based on deformable part model and improved scale invariant feature transform. *Optik* **2013**, *124*, 6485–6492. [CrossRef]
114. Lindenmaier, L.; Aradi, S.; Bécsi, T.; Törő, O.; Gáspár, P. Object-Level Data-Driven Sensor Simulation for Automotive Environment Perception. *IEEE Trans. Intell. Veh.* **2023**, *8*, 4341–4356. [CrossRef]
115. Bird, J.; Bird, J. *Higher Engineering Mathematics*, 5th ed.; Routledge: London, UK, 2006. [CrossRef]
116. Ainsalu, J.; Arffman, V.; Bellone, M.; Ellner, M.; Haapamäki, T.; Haavisto, N.; Josefson, E.; Ismailogullari, A.; Lee, B.; Ma-dland, O.; et al. State of the Art of Automated Buses. *Sustainability* **2018**, *10*, 3118. [CrossRef]
117. Lian, H.; Li, M.; Li, T.; Zhang, Y.; Shi, Y.; Fan, Y.; Yang, W.; Jiang, H.; Zhou, P.; Wu, H. Vehicle speed measurement method using monocular cameras. *Sci. Rep.* **2025**, *15*, 2755. [CrossRef]
118. Vivacqua, R.; Vassallo, R.; Martins, F. A Low Cost Sensors Approach for Accurate Vehicle Localization and Autonomous Driving Application. *Sensors* **2017**, *17*, 2359. [CrossRef] [PubMed]

119. Xue, L.; Li, M.; Fan, L.; Sun, A.; Gao, T. Monocular Vision Ranging and Camera Focal Length Calibration. *Sci. Program.* **2021**, *2021*, 979111. [CrossRef]
120. Kerbl, B.; Kopanas, G.; Leimkuhler, T.; Drettakis, G. 3D Gaussian Splatting for Real-Time Radiance Field Rendering. *ACM Trans. Graph.* **2023**, *42*, 139. [CrossRef]
121. Santosh Reddy, P.; Abhiram, H.; Archish, K.S. A Survey of 3D Gaussian Splatting: Optimization Techniques, Applications, and AI-Driven Advancements. In Proceedings of the 2025 International Conference on Intelligent and Innovative Technologies in Computing, Electrical and Electronics (IITCEE), Bangalore, India, 16–17 January 2025. [CrossRef]
122. Qiu, S.; Xie, B.; Liu, Q.; Heng, P.-A. Creating Virtual Environments with 3D Gaussian Splatting: A Comparative Study. In Proceedings of the 2025 IEEE Conference on Virtual Reality and 3D User Interfaces Abstracts and Workshops (VRW), Saint Malo, France, 8–12 March 2025. [CrossRef]
123. Hornáček, M.; Rozinaj, G. Exploring 3D Gaussian Splatting: An Algorithmic Perspective. In Proceedings of the 2024 International Symposium ELMAR, Zadar, Croatia, 16–18 September 2024. [CrossRef]
124. Rosique, F.; Navarro, P.J.; Fernández, C.; Padilla, A. A Systematic Review of Perception System and Si-mulators for Autonomous Vehicles Research. *Sensors* **2019**, *19*, 648. [CrossRef]
125. Elster, L.; Staab, J.P.; Peters, S. Making Automotive Radar Sensor Validation Measurements Comparable. *Appl. Sci.* **2023**, *13*, 11405. [CrossRef]
126. Roy, C.J.; Balch, M.S. A Holistic Approach to Uncertainty Quantification with Application to Supersonic Nozzle Thrust. *Int. J. Uncertain. Quantif.* **2021**, *2*, 363–381. [CrossRef]
127. Magosi, Z.F.; Eichberger, A. A Novel Approach for Simulation of Automotive Radar Sensors Designed for Systematic Support of Vehicle Development. *Sensors* **2023**, *23*, 3227. [CrossRef] [PubMed]
128. Maier, M.; Makkapati, V.P.; Horn, M. Adapting Phong into a Simulation for Stimulation of Automotive Radar Sensors. In Proceedings of the 2018 IEEE MTT-S International Conference on Microwaves for Intelligent Mobility (ICMIM), Munich, Germany, 15–17 April 2018; pp. 1–4. [CrossRef]
129. Minin, I.V.; Minin, O.V. Lens Candidates to Antenna Array. In *Basic Principles of Fresnel Antenna Arrays*; Lecture Notes Electrical Engineering; Springer: Berlin/Heidelberg, Germany, 2008; Volume 19, pp. 71–127. [CrossRef]
130. Sensors Partners. LiDAR Laser: What Is LiDAR and How Does It Work? | Sensor Partners. Available online: https://sensorpartners.com/en/knowledge-base/how-a-lidar-laser-works/ (accessed on 6 March 2025).
131. García-Gómez, P.; Royo, S.; Rodrigo, N.; Casas, J.R. Geometric Model and Calibration Method for a Solid-State LiDAR. *Sensors* **2020**, *20*, 2898. [CrossRef] [PubMed]
132. Kim, G. Performance Index for Extrinsic Calibration of LiDAR and Motion Sensor for Mapping and Localization. *Sensors* **2022**, *22*, 106. [CrossRef] [PubMed]
133. Schmoll, L.; Kemper, H.; Hagenmüller, S.; Brown, C.L. Validation of an Ultrasonic Sensor Model for Application in a Simulation Platform. *ATZelectronics Worldw.* **2024**, *19*, 8–13. [CrossRef]
134. Stevens Institude of Thechnology 1870. Available online: https://www.stevens.edu/news/autonomous-vehicles-will-add-us81-billion-new-premiums-auto-insurers-2025-according-accenture-report (accessed on 18 May 2025).
135. Sen, S.; Husom, E.J.; Goknil, A.; Tverdal, S.; Nguyen, P. Uncertainty-Aware Virtual Sensors for Cyber-Physical Systems. *IEEE Softw.* **2024**, *41*, 77–87. [CrossRef]
136. Ying, Z.; Wang, Y.; He, Y.; Wang, J. Virtual Sensing Techniques for Nonlinear Dynamic Processes Using Weighted Proba-bility Dynamic Dual-Latent Variable Model and Its Industrial Applications. *Knowl.-Based Syst.* **2022**, *235*, 107642. [CrossRef]
137. Yuan, X.; Rao, J.; Wang, Y.; Ye, L.; Wang, K. Virtual Sensor Modeling for Nonlinear Dynamic Processes Based on Local Weighted PSFA. *IEEE Sens. J.* **2022**, *22*, 20655–20664. [CrossRef]
138. Zheng, T. Algorithmic Sensing: A Joint Sensing and Learning Perspective. In Proceedings of the 21st Annual International Conference on Mobile Systems, Applications and Services, Helsinki, Finland, 18–22 June 2023; Association for Computing Machinery: New York, NY, USA, 2023; pp. 624–626. [CrossRef]
139. Es-haghi, M.S.; Anitescu, C.; Rabczuk, T. Methods for Enabling Real-Time Analysis in Digital Twins: A Literature Review. *Comput. Struct.* **2024**, *297*, 107342. [CrossRef]
140. EUR-Lex, UN Regulation No 157—Uniform Provisions Concerning the Approval of Vehicles with Regards to Automated Lane Keeping Systems [2021/389]. Available online: https://eur-lex.europa.eu/eli/reg/2021/389/oj/eng (accessed on 18 May 2025).
141. EUR-Lex, Regulation No 140 of the Economic Commission for Europe of the United Nations (UN/ECE)—Uniform Provisions Concerning the Approval of Passenger Cars with Regard to Electronic Stability Control (ESC) Systems [2018/1592]. Available online: https://eur-lex.europa.eu/eli/reg/2018/1592/oj/eng (accessed on 18 May 2025).
142. International Organization for Standardization. ISO 26262-1:2018(en) Road Vehicles—Functional Safety—Part 1: Vocabulary. Available online: https://www.iso.org/obp/ui/en/#iso:std:iso:26262:-1:ed-2:v1:en (accessed on 19 May 2025).

143. International Organization for Standardization. ISO 34502:2022 Road Vehicles—Test Scenarios for Automated Driving Systems—Scenario Based Safety Evaluation Framework. Available online: https://www.iso.org/standard/78951.html (accessed on 19 May 2025).
144. International Organization for Standardization. ISO 21448:2022(en); Road Vehicles—Safety of the Intended Functionality. Available online: https://www.iso.org/obp/ui/en/#iso:std:iso:21448:ed-1:v1:en (accessed on 19 May 2025).

Disclaimer/Publisher's Note: The statements, opinions and data contained in all publications are solely those of the individual author(s) and contributor(s) and not of MDPI and/or the editor(s). MDPI and/or the editor(s) disclaim responsibility for any injury to people or property resulting from any ideas, methods, instructions or products referred to in the content.

sensors

MDPI

Review

An Overview of Autonomous Parking Systems: Strategies, Challenges, and Future Directions

Javier Santiago Olmos Medina [1], Jessica Gissella Maradey Lázaro [1,*], Anton Rassõlkin [2,*] and Hernán González Acuña [1]

[1] Program of Mechatronic Engineering, Universidad Autónoma de Bucaramanga, Bucaramanga 680003, Colombia; jolmos@unab.edu.co (J.S.O.M.); hgonzalez3@unab.edu.co (H.G.A.)

[2] Department of Electrical Power Engineering and Mechatronics, Tallinn University of Technology, 19086 Tallinn, Estonia

* Correspondence: jmaradey@unab.edu.co (J.G.M.L.); anton.rassolkin@taltech.ee (A.R.); Tel.: +57-3164633309 (J.G.M.L.)

Abstract: Autonomous Parking Systems (APSs) are rapidly evolving, promising enhanced convenience, safety, and efficiency. This review critically examines the current strategies in perception, path planning, and vehicle control, alongside system-level aspects like integration, validation, and security. While significant progress has been made, particularly with the advent of deep learning and sophisticated sensor fusion, formidable challenges persist. This paper delves into the inherent trade-offs, such as balancing computational cost with real-time performance demands; unresolved foundational issues, including the verification of non-deterministic AI components; and the profound difficulty of ensuring robust real-world deployment across diverse and unpredictable conditions, ranging from cluttered urban canyons to poorly lit, ambiguously marked parking structures. We also explore the limitations of current technologies, the complexities of safety assurance in dynamic environments, the pervasive impact of cost considerations on system capabilities, and the critical, often underestimated, need for genuine user trust. Future research must address not only these technological gaps with innovative solutions but also the intricate socio-technical dimensions to realize the full potential of APS.

Keywords: autonomous parking system; path planning; intelligent systems; environment recognition; end-to-end model-based system

1. Introduction

1.1. Context and Motivation

Autonomous Parking Systems (APSs), encompassing functionalities ranging from Autonomous Parking Assist (APA) where the system handles steering while the driver controls speed, to fully Autonomous Valet Parking (AVP) where the vehicle parks itself without a driver inside, represent a significant area of advancement within automotive technology. These systems aim to alleviate the common stress and difficulty associated with parking maneuvers, particularly in constrained urban environments where precision is paramount and margins for error are minimal. The potential benefits are manifold, including enhanced driver convenience by automating a tedious task, improved spatial utilization in increasingly crowded parking areas through more precise and consistent parking, a reduction in minor collisions (dings, scrapes) often occurring during low-speed parking maneuvers, and contributing valuable technological building blocks to the broader development of fully autonomous vehicles (AV) [1]. The integration of Artificial Intelligence (AI), especially in perception and decision-making, has notably accelerated the development of these features, enabling more adaptive and seemingly intelligent behaviors [2]. Market forecasts reflect this potential, with the automotive AI market expected to grow significantly, driven partly by the demand for Advanced Driver Assistance Systems (ADASs) like APS.

However, despite considerable progress and the availability of various systems on the market, deploying APSs that can operate reliably and robustly across the full spectrum of real-world complexity remains a formidable challenge [3]. The transition from controlled demonstrations in well-defined test tracks to ubiquitous, robust real-world deployment in everyday parking scenarios is hampered not only by sheer environmental complexity, such as varying weather conditions, diverse parking lot layouts, and inconsistent human behaviors, but also by fundamental trade-offs. These include the delicate balance between system performance (e.g., speed and smoothness of parking) and the computational resources available on automotive-grade hardware, the rigorous demands of safety assurance versus the desire for rapid feature deployment, and the overarching influence of cost constraints on sensor suites and processing power. Furthermore, the persistent challenge of validating systems against the unpredictability of human behavior (e.g., pedestrians suddenly appearing, other drivers making unexpected maneuvers) and diverse, often poorly maintained, operational conditions adds layers of difficulty [4,5]. This review will critically examine these underlying issues, aiming to provide a nuanced understanding of the current state and future trajectory of APSs. Figure 1 illustrates the general architecture of such a system, outlining the primary components and the flow of information from sensor input to vehicle control output.

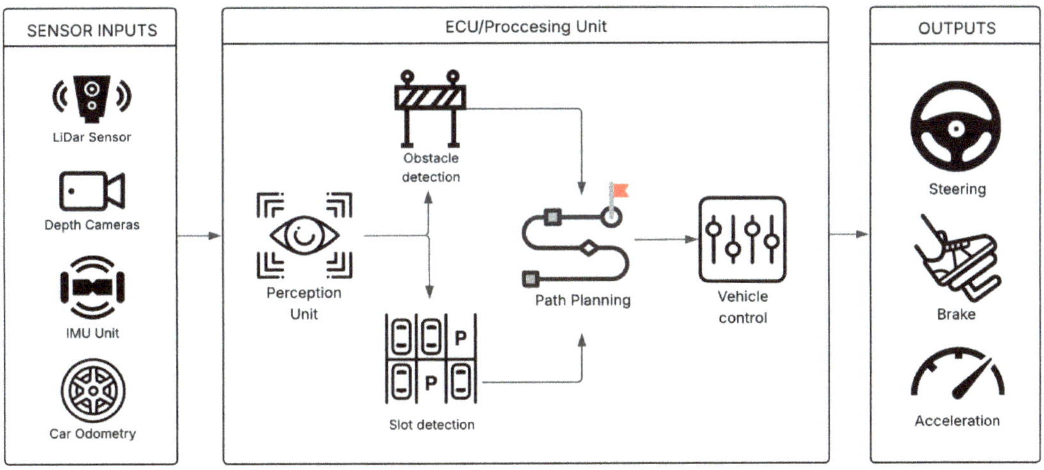

Figure 1. General architecture of a modular Autonomous Parking System (APS) illustrating core components and data flow.

1.2. Recent Drivers of Progress

Recent years have witnessed continued progress in APS research, fueled by several key technological drivers. Advances in sensor technologies, particularly the increasing resolution, field-of-view, and performance of automotive radar (e.g., 77 GHz and 79 GHz systems offering finer angular separation) [6] and LiDAR (light detection and ranging) providing denser point clouds for detailed 3D mapping [7,8], provide richer and more detailed environmental data. Simultaneously, more advanced fusion algorithms are emerging to seamlessly combine inputs from diverse sensor types—ultrasonic, camera, radar, and LiDAR, enhancing the system's overall perception capabilities [9,10]. This fusion aims to overcome the limitations of individual sensors—for example, the poor performance of cameras in low light can be compensated for by the all-weather capabilities of radar, thereby improving overall perception accuracy and robustness against environmental uncertainties. Deep learning (DL) has become the dominant paradigm for perception tasks such as parking slot detection, free-space estimation, and obstacle identification [3,8,11], offering superior performance compared to traditional computer vision (CV) methods, especially in handling vast variations in lighting, weather, and object appearances where rule-based systems falter. Furthermore, research continues to yield more advanced path planning algorithms capable of generating smooth, kinematically feasible trajectories in cluttered and confined spaces, taking into account vehicle dynamics and passenger comfort [12–14]. Moreover, these refined vehicle control strategies enable precise and comfortable maneuver execution, minimizing jerky movements and ensuring accurate path tracking [15,16]. Therefore, these advancements collectively push the capabilities of APSs towards greater autonomy, reliability, and a more seamless user experience. These systems are generally designed using one of two primary architectural paradigms, the traditional modular pipeline or the more recent end-to-end approach, as compared in Figure 2.

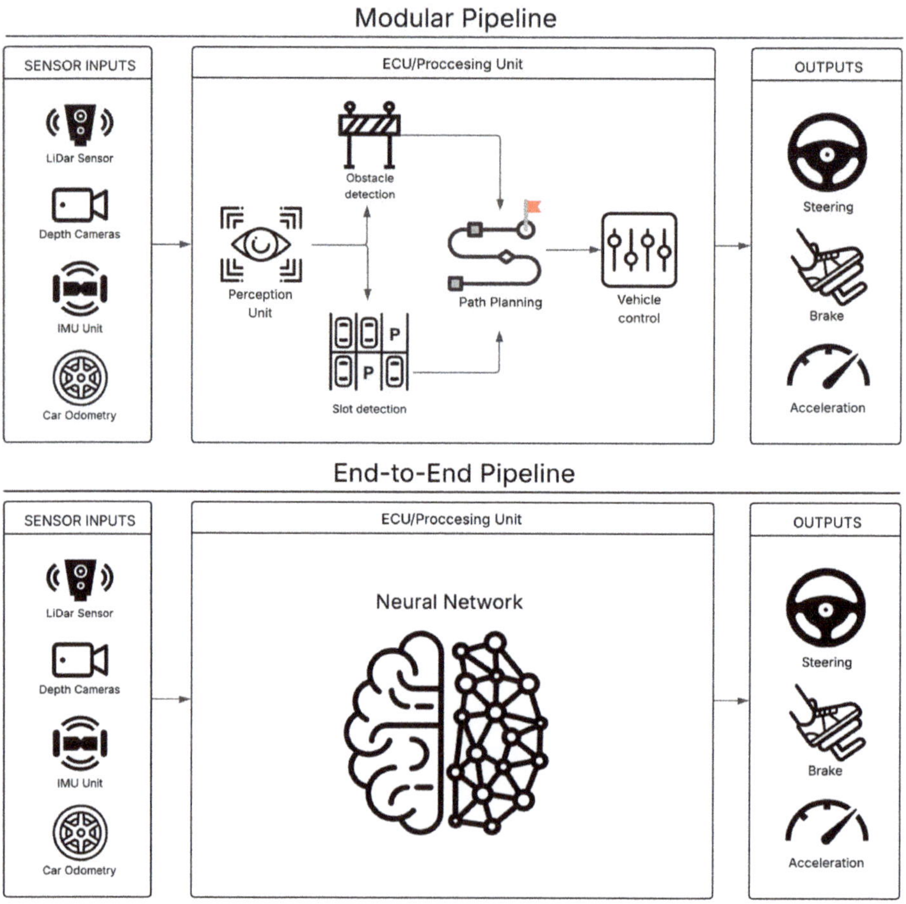

Figure 2. Comparison of modular pipeline and end-to-end (E2E) pipeline architectural paradigms for Autonomous Parking Systems.

2. Perception Methods for Parking Environments

Accurate and robust perception of the surrounding environment is fundamental for any APS. Research conducted in recent years has focused on enhancing sensor capabilities, refining fusion techniques, leveraging deep learning for detection tasks, and critically, improving performance under the challenging and diverse conditions encountered in real-world parking scenarios [3,17].

2.1. Sensor Technologies and Fusion

This section examines the specific sensor technologies utilized in Advanced Parking Systems (APSs), as well as the data fusion strategies applied to integrate their outputs and achieve a comprehensive understanding of the parking environment.

2.1.1. Critical Analysis of Sensor Technologies

Table 1 provides a critical comparative analysis of key individual sensor technologies relevant to the development and deployment of Autonomous Parking Systems (APSs). Indeed, it aims to synthesize current knowledge, highlighting not only reported strengths but also inherent weaknesses, practical limitations, specific fusion challenges, and crucial cost–performance trade-offs.

2.1.2. Critical Analysis of Sensor Fusion Approaches

Table 2 reports a critical comparative analysis of key sensor fusion approaches relevant to APS development and highlights strengths, weaknesses, limitations, fusion challenges, and cost–performance trade-offs.

Table 1. Critical Analysis of Sensor Technologies in Autonomous Parking Systems (APS) [1].

Sensor Technology	Key Reported Strengths	Inherent Weaknesses/Practical Limitations	Specific Fusion Challenges	Cost vs. Performance Considerations
Standard Cameras (RGB, Fisheye, Rectilinear)	- Rich semantic info (color, texture) for classification, recognition, identification [18]. - Fisheye: Wide FoV for near-vehicle sensing [19]. - Cost-effective; foundational ADAS/APS component [18]. - Note: Performance limits are a system challenge. - Advanced AI/DL support; mature algorithms [18]. - Passive sensing; avoids interference [18].	- Illumination dependent: Degrades in low light, glare, shadows [18]. Critical for APSs in garages/night. - Susceptible to adverse weather (rain, fog, snow) [20]. - Limited direct depth (monocular); stereo adds cost/complexity [18]. - Lens occlusion/soiling: Critical reliability issue [18,21]. - Fisheye distortion: Requires correction, adds overhead [22].	- Calibration (intrinsic/extrinsic) with 3D sensors critical [20]. - Temporal synchronization with other sensors essential [20]. - Data association: Matching features/objects across sensors. - Resolving conflicting data from different sensor types. - Handling sensor degradation (soiling, occlusion) and adapting and fusion [21,23].	- Low unit cost: Enables multi-camera 360° APS [18]. - High processing cost: Rich data needs powerful ECUs (Electronic Control Units)/SoCs [22]. - Data deluge can be a bottleneck. - Central to basic APSs (with ultrasonics); limited in diverse conditions [22]. - Insufficient alone for advanced APS safety/ODDs (Operational Design Domains); needs fusion [21,22].
Ultrasonic Sensors	- Extremely low cost: Ubiquitous for basic parking aid [24]. - Reliable short-range detection (<5–10 m): Crucial for final maneuvers, low obstacles [18]. - Acts as "last-centimeter" guardian. - Independent of lighting conditions [18]. - Detects various material types (large, hard surfaces) [18].	- Limited range (very short distances) [18]. - Poor object classification/identification; only distance [18]. - Sparse data, hard to fuse intelligently. - Narrow FoV per sensor; multiple units needed, may leave gaps [25]. - Susceptible to environment (wind, temp, heavy rain/snow) [18]. - Struggles with soft, curved, thin, small, sound-absorbing objects [18]. - Contamination/damage: Susceptible to dirt, ice; silent failures possible [26].	- Complementary fusion only: Provides short-range safety bubble [19]. - Data association: Linking sparse pings to rich camera/LiDAR data. - Handling false positives/negatives: Robust filtering needed. - Sensor degradation detection (obstruction/damage) important [26,27]. - Self-check or cross-validation needed.	- Extremely low cost: Main driver for adoption [24]. - Essential for low-speed maneuver safety [28]. - Not standalone for advanced APS; part of larger suite [19].

Table 1. *Cont.*

Sensor Technology	Key Reported Strengths	Inherent Weaknesses/Practical Limitations	Specific Fusion Challenges	Cost vs. Performance Considerations
Solid-State LiDAR (SSL) (e.g., MEMS, OPA, Flash)	- Improved reliability/durability: No/fewer moving parts [29]. - Compact size and lower weight: Easier vehicle integration [29]. - Potential for lower cost (mass production) [30]. - "Democratization" of LiDAR is key. - High data rate/resolution: Detailed 3D mapping/detection [29]. - Fast scanning (some types); Flash LiDAR illuminates entire scene [30]. - Good performance in various lighting (active sensor) [18].	- Limited FoV per unit: Multiple units often needed for 360°, offsetting cost benefits [31]. - Shorter range (historically/certain types), though improving [29]. - Adverse weather performance degraded (rain, snow, fog) [18]. - Thermal management challenges: Heat build-up affects performance/durability [31]. - Near-field detection issues/blind spots (sub-meter to few meters) for some types [32]. - Irregular scan patterns (some MEMS): Can complicate processing [33].	- Calibration of multiple SSL units: Crucial and complex [20]. - Data synchronization: Multiple SSLs and other sensors. - Point cloud registration/stitching: Merging data from multiple SSLs. - High data bandwidth and processing demands [20]. - Handling sensor degradation (soiling, thermal effects).	- Current cost still a factor, especially for multiple units [29]. - Trade-off: FoV/Range vs. Cost (lower-cost SSLs may be limited). - Performance benefits for APS: High resolution/accuracy for small obstacles, mapping, localization [30]. - Mass-market viability target: Achieving performance at suitable price [34].
4D Imaging Radar	- Adverse weather robustness: Excellent in rain, fog, snow [18]. - Crucial baseline sensor for APS continuity. - Direct velocity measurement (Doppler) [18]. - Elevation information (4th D): Better 3D object detection/classification [35]. - Long detection range [36]. - Can "see through" some non-metallic obstructions [35]. - Improved angular resolution (vs. traditional radar) [36].	- Lower resolution than LiDAR/Cameras: Difficult for detailed classification/boundaries [18]. - Point cloud sparsity (vs. LiDAR) [37]. - Noisy data/clutter/multipath: Especially in dense metallic environments (garages) [18]. - Limited material differentiation (no color/texture) [18]. - Challenges with stationary objects (differentiation) [35]. - Frequency regulation hurdles for higher resolution [35]. - Short-range blind spots (ground reflection/DC noise) [35].	- Fusing sparse radar with dense camera/LiDAR: Sophisticated techniques needed [38]. - Calibration and Synchronization: Precise spatio-temporal alignment crucial [20]. - Resolving conflicting detections. - Computational load for raw 4D radar tensor processing (e.g., RadarOcc [37]). - Handling sensor degradation.	- Higher cost for high performance (vs. traditional radar/cameras); projected cheaper than LiDAR [35]. - Production costs a barrier [39]. - Single-chip solutions aim to reduce cost for mass-market [40]. - Value for APS: All-weather, velocity, elevation data for reliability [35]. - Trade-off: Resolution vs. Cost (higher-res is more expensive) [35].

Table 1. *Cont.*

Sensor Technology	Key Reported Strengths	Inherent Weaknesses/Practical Limitations	Specific Fusion Challenges	Cost vs. Performance Considerations
Event-Based Cameras (DVS, DAVIS)	- High temporal resolution (μs-scale): Captures fast dynamics (e.g., sudden pedestrian) [38]. - Low latency (μs to sub-ms): Critical for reactive APSs [38]. - High Dynamic Range (HDR >120 dB): Adapts to extreme lighting (garage entry/exit) [38]. - Reduced motion blur [38]. - Potential low power consumption (data only on change) [38]. - Data sparsity: Efficiently represents dynamic scenes [41].	- No/Limited static scene info: Major limitation for APSs (stationary obstacles/lines) [38]. - Positioned as "dynamic specialists." - Grayscale info: Typically event polarity, not absolute intensity/color (DAVIS adds frames) [38]. - Noisy events/data interpretation: Requires specialized algorithms [38]. - Lack of inherent semantic info [41]. - High data volume (highly dynamic scenes): Can negate low-bandwidth advantage [41]. - The "data paradox." - Maturity and Cost and: Emerging tech, cost may be higher, algorithms less mature [38].	- Fusing asynchronous data with synchronous frame-based data: Significant challenge [19]. - Requires new algorithmic paradigms (change-driven). - Event representation for fusion: Converting sparse events can lose info/add cost. - Calibration: Precise spatio-temporal needed. - Complementary fusion for static scenes: Heavily rely on frame-based cameras/LiDAR [42].	- Higher unit cost (currently) vs. standard CMOS cameras [38]. - Potential system-level savings (low power/data in some scenarios)—depends on processing. - Niche performance benefits: Unmatched HDR/high-speed. Useful for sudden intrusions/extreme lighting in APS. - Not a standalone APS solution: Due to static scene issues; more for specialized dynamic threat detection.
Thermal Cameras (LWIR)	- Low/No visible light detection: Operates in complete darkness (e.g., underground garages) [18]. - "Lifesaver" for detecting pedestrians/animals when visible cameras fail. - Robustness to visual obscurants (smoke, haze, light fog) [43]. - Good for detecting animate objects (pedestrians/animals via heat) [43]. - Reduced glare issues (sunlight/headlights) [28]. - High contrast imaging (for warm objects) [43].	- Lower resolution vs. visible cameras [44]. - Lack of color and fine and texture: Grayscale (temperature-based); no detailed visual ID [43]. - "What, not who" limitation. - Cannot see through glass/water [43]. - Temperature-dependent contrast: Image quality affected by ambient/object temp difference [43]. - Difficulty with isothermal scenes/cold obstacles [43]. - Higher cost than standard cameras [43].	- Fusing low-res thermal with high-res visible: Requires careful registration/scaling. - Different data modalities: Fusing heat-based with reflection-based (visible, LiDAR) images. - Calibration and Synchronization and: Precise spatio-temporal essential [45]. - Complementary role: Often supplements visible cameras (low-light pedestrian detection) [45]. - Fusion logic needs to weigh inputs based on conditions.	- Higher cost component: Increases sensor suite cost [46]. - Cost–benefit often positions for premium APS/specific ODDs. - Significant performance boost in specific APS scenarios (dark lots, underground, missed pedestrians/animals) [46]. - Niche capability vs. Cost trade-off: Depends on target ODD/safety requirements. - Deep Learning for Thermal (e.g., YOLO-Thermal [44]) aims to improve performance.

Table 1. *Cont.*

Sensor Technology	Key Reported Strengths	Inherent Weaknesses/Practical Limitations	Specific Fusion Challenges	Cost vs. Performance Considerations
Multispectral Cameras (RGB + NIR, Red Edge, etc.)	- Improved object/material discrimination: Captures multiple spectral bands (visible, NIR, SWIR) for detailed material analysis [47]. - Potential "surface condition specialist" (black ice, oil slicks with SWIR). - Enhanced performance in specific conditions (e.g., NIR for light haze/fog penetration) [47]. - Simultaneous multiband imaging (snapshot cameras) for dynamic scenes [48].	- Higher cost and complexity and vs. standard RGB cameras [47]. - Data volume and processing and: Multiple bands increase data/computational load [48]. - Data richness can be "curse of dimensionality." - Limited bands vs. hyperspectral: Less finesse in spectral analysis [47]. - Application-specific band selection: General APSs might not leverage all specialized bands. - Illumination dependent: Still reliant on external/active illumination in specific bands.	- Registration of multiple bands: Ensuring perfect alignment crucial (snapshot cameras mitigate) [48]. - Fusion with other sensor modalities: Integrating multiband with LiDAR, radar, thermal. - Feature extraction and selection: Identifying relevant spectral features for APSs (e.g., wet vs. dry pavement). - Real-time processing: Can be computationally demanding [49].	- High cost for automotive integration: Likely prohibitive for mass APS [47]. - Niche APS benefits: E.g., road surface condition detection. - Justifying performance gain vs. Cost: Incremental gain must be significant. - Visible–Infrared focus: Current research often on visible–IR fusion for object detection [49]. - Could be cost–performance step if key bands identified for critical APS issues.

[1] This table presents a critical analysis of sensor technologies in Autonomous Parking Systems (APSs), synthesized from the information provided in the correspondingly cited references listed within the table (in the 'Critical References' column). For definitions of abbreviations used, please refer to the 'Abbreviations' section at the end of the manuscript. Italicized text or sub-bullets within table cells provide supplementary commentary or specific details.

Table 2. Critical Analysis of Sensor Fusion Approaches in Autonomous Parking Systems (APS) [2].

Sensor Fusion Approach	Key Reported Strengths	Inherent Weaknesses/Practical Limitations	Specific Fusion Challenges	Cost vs. Performance Considerations
Tightly-Coupled LiDAR-IMU Fusion (e.g., LIO-SAM, FAST-LIO, GF-LIO)	- Improved SLAM (Simultaneous Localization and Mapping)/Odometry: Higher accuracy/robustness in state estimation/mapping, especially GPS-denied (garages) [50]. - De facto standard for robust localization in critical APS ODDs. - Motion distortion correction: IMU de-skews LiDAR clouds [50]. - Enhanced state estimation (degenerate scenarios): IMU aids when LiDAR data sparse [50]. - Improved attitude estimation (roll, pitch, yaw) [51]. - Real-time capability with modern algorithms [50].	- Computational complexity: Joint optimization is intensive [50]. - Balancing accuracy vs. real-time on embedded hardware. - Sensitivity to initialization and calibration: Small errors amplified [50]. - IMU noise and bias drift: Needs accurate online estimation [52]. - Dependence on accurate sensor models [53].	- Precise spatio-temporal calibration (LiDAR-IMU) [20]. - Complexity of IMU preintegration [50]. - Factor graph optimization management (size/complexity) [50]. - Loop closure integration with LIO factors [50]. - Handling reflective/symmetric structures (LiDAR-specific garage challenges) [51,54].	- Adds IMU cost (higher-grade IMUs improve performance but cost more). - Significant performance gain for APS localization/mapping in GPS-denied areas [54]. - Enables higher automation levels (prerequisite for precise, continuous localization). - Computational cost implication: May need more powerful ECUs.
BEV (Bird's Eye View) Fusion (Cameras, LiDAR, Radar)	- Unified spatial representation: Common top-down view for fusing heterogeneous data [55]. - BEV as "lingua franca" for sensor modalities. - Facilitates multimodal fusion: Combines features from different perspectives/structures [55]. - Improved situational awareness: 360° view for navigation/decision-making [56]. - Potential for dense fusion: Preserves more contextual info [57]. - Directly applicable to planning: BEV maps usable by motion planners.	- Information loss in view transformation: Projecting to BEV can cause distortion, ambiguity (distant/occluded objects) [7]. - Lack of depth-aware vision-to-BEV can cause gaps. - Standard 2D BEV loses height info; drives research to 3D/pseudo-3D BEV. - Handling occlusions: Difficult, especially with camera-only systems [55]. - Computational cost: Generating/processing BEV from multiple high-res sensors can be intensive [56]. - Dependence on accurate calibration and synchronization [20]. - Fixed grid resolution: Limits accuracy vs. computational load.	- Camera-to-BEV transformation: Accurate projection, often needs depth estimation or learned transformers (e.g., LSS) [7]. - LiDAR/Radar-to-BEV representation: Efficiently projecting sparse/dense data. - Cross-modal feature alignment and fusion: Camera appearance, LiDAR geometry, radar velocity [55]. - Temporal fusion in BEV: Incorporating history to improve consistency/handle occlusions [7]. - Synchronization delays for multi-camera BEV (RT-BEV aims to mitigate) [56]. - Handling sensor degradation: Affects fused BEV quality.	- Sensor suite cost (multi-camera, LiDAR, radar). - Vision-centric BEV aims for lower cost but faces performance limits [56]. - Computational hardware: Requires powerful ECUs/SoCs with GPUs [56]. - Performance gains for APS: Holistic understanding for complex maneuvers, spot finding, clutter navigation [19]. - Scalability: Sensor number/resolution impacts performance/cost.

Table 2. *Cont.*

Sensor Fusion Approach	Key Reported Strengths	Inherent Weaknesses/Practical Limitations	Specific Fusion Challenges	Cost vs. Performance Considerations
Probabilistic Fusion Frameworks (e.g., Kalman Filters, Particle Filters, Bayesian Networks, Occupancy Grids)	- Uncertainty management: Explicitly model/manage sensor noise, environmental variability, model inaccuracies [20]. - Crucial for safety-critical APS decisions. - Conflict resolution: Principled ways to fuse conflicting info by weighting data based on reliability/uncertainty [20]. - Improved SNR and fault tolerance (Raw Data Fusion) [58]. - State estimation and tracking: Kalman/Particle filters widely used [20]. - Semantic mapping: Probabilistic generation of semantic maps (road, curb) with confidence levels [59]. - Occupancy Grids: Represent free space/obstacles probabilistically for APS path planning [58].	- Model dependence: Performance relies on accuracy of system/sensor models [20]. - Computational complexity: Some methods (Particle Filters for high-dim, full Bayesian) very expensive for real-time APS [20]. - Assumptions (e.g., Gaussian noise for LKF/EKF): May not hold in real-world APSs [60]. - Data association challenges: Associating measurements to tracks/map features in clutter. - Scalability: Maintaining real-time performance with increasing objects/map size.	- Data alignment and synchronization: Critical prerequisite [20]. - Handling heterogeneous data: Integrating diverse sensor types (point clouds, images, radar) with different noise/resolutions [20]. - Dynamic noise covariance estimation: Adaptive estimation important but challenging [28]. - Non-linearities and Non-Gaussianities: Standard KFs struggle; need EKF, UKF, Particle Filters (with own trade-offs) [61]. - Computational latency for high-bandwidth data.	- Trade-off: Robustness vs. Computational Cost (sophisticated models cost more). - Enabling safer decisions: Quantifying uncertainty allows more informed APS actions. - Development cost: Implementing/validating complex probabilistic fusion requires expertise/testing. - Use of lower-cost sensors (with Raw Data Fusion): May enable by improving SNR/overcoming individual failures.

[2] This table presents a critical analysis of sensor fusion approaches in Autonomous Parking Systems (APSs), synthesized from the information provided in the correspondingly cited references listed within the table (in the 'Critical References' column). For definitions of abbreviations used, please refer to the 'Abbreviations' section at the end of the manuscript. Italicized text or sub-bullets within table cells provide supplementary commentary or specific details.

Modern APSs continue to rely on a suite of sensors, typically including ultrasonic sensors, cameras, radar, and increasingly, LiDAR, to capture comprehensive information about the vehicle's surroundings. Ultrasonic sensors remain prevalent for short-range distance measurements, crucial for final parking maneuvers and detecting low-lying curbs. Cameras provide rich visual information, which is essential for detecting parking lines, reading textual information on signs, and classifying objects [9]. Radar offers robustness in adverse weather conditions (rain, snow, fog) and provides direct velocity measurements of moving objects [6]. LiDAR provides precise 3D point clouds, enabling detailed environmental mapping, accurate object dimensioning, and improved localization [7,51]. A notable trend is the push towards higher-frequency radar systems (e.g., 77 GHz and 79 GHz), which offer improved resolution and object separation capabilities compared to traditional automotive radars, enabling finer environmental details to be captured, such as the delineation between closely parked vehicles [6,62], enabling finer environmental details to be captured, such as the delineation between closely parked vehicles. Indeed, the maturation of 4D imaging radar technology is evidenced by commercially available systems that provide dense point clouds detailing range, azimuth, elevation, and velocity, offering robust perception data essential for advanced APS functions [63].

However, the inherent limitations of individual sensors introduce critical challenges that may propagate throughout the entire APS workflow. For instance, while cameras provide rich visual data, their sensitivity to illumination changes (e.g., sudden glare from the sun, deep shadows in underground garages, rapid transitions from bright outdoor to dim indoor lighting) and adverse weather (e.g., rain droplets or mud obscuring lenses, heavy fog reducing visibility to near zero) can severely degrade performance [9,17,64]. So, a camera might fail to detect faded parking lines on a wet surface or misinterpret reflections as obstacles. Current mitigation strategies, such as High Dynamic Range (HDR) imaging and infrared (IR) illumination, may prove insufficient under all-weather and variable lighting conditions, particularly in the context of cost-sensitive mass-market vehicles, where the integration of high-end, expensive sensors is often not feasible. The practical consequences of sensor performance degradation can vary widely, ranging from incorrect parking space detection and failed parking attempts to the inability to identify small obstacles, such as animals or debris, thereby directly impacting both safety and system reliability.

Sensor fusion is indispensable for creating a perception system that is more reliable and accurate than the sum of its parts, mitigating these inherent weaknesses by leveraging the complementary strengths of different sensor types [9,10,17,64]. Some authors explore various fusion strategies, ranging from geometry-based approaches that analyze patterns from fused multi-sensor data [10] to more complex, tightly coupled fusions of LiDAR and Inertial Measurement Unit (IMU) data for improved pose estimation and robust point cloud mapping, even during vehicle motion [8,65], as illustrated in Figure 3. Bird's Eye View (BEV) fusion frameworks, which project sensor data onto a common top-down plane, are also gaining traction for their intuitive representation and ability to integrate information from multiple cameras and other sensors. Moreover, sensor fusion is intended to enhance the robustness of perception systems. However, the inherent complexity and current limitations of fusion algorithms introduce significant, often under-acknowledged, challenges [1,17]. A critical question is the following: What are the inherent failure modes of current fusion algorithms? For example, how are inconsistencies between sensor modalities resolved, particularly in scenarios where radar detects an object that a camera fails to register due to occlusion or variations in material reflectivity? Similar challenges arise when a sensor provides clearly erroneous data as a result of malfunction, signal interference (e.g., radar jamming in dense environments), or specific environmental conditions, such as ultrasonic sensors generating 'ghost' readings from highly reflective surfaces. Moreover, does the reliance on sensor fusion introduce additional points of failure, including errors from miscalibration or temporal desynchronization, which could potentially lead to misaligned datasets? In conclusion, to what extent does the computational overhead incurred by integrating high-bandwidth sensor data streams induce latency levels that compromise the timeliness and reliability of decision-making processes in real-time, safety-critical applications? Furthermore, maintaining accurate calibration and temporal synchronization between diverse sensors in harsh automotive environments (subject to constant vibrations, wide temperature extremes, and potential physical damage) is a persistent operational challenge. Sensor drift over time, if not continuously monitored and corrected, can significantly degrade fusion reliability, leading to a gradual or sudden decline in perception performance.

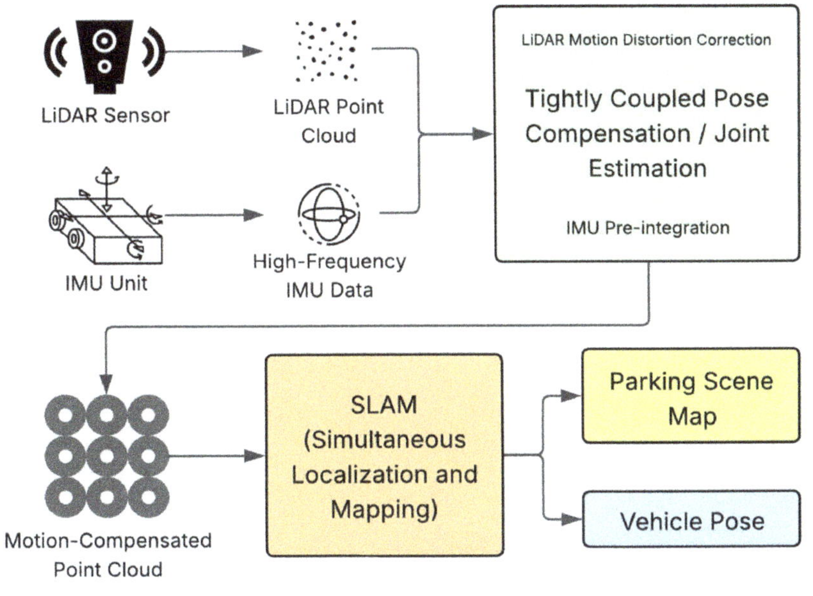

Figure 3. Architecture for tightly coupled LiDAR-IMU fusion, demonstrating the process from sensor input to motion-compensated point cloud generation and subsequent SLAM for map creation and vehicle pose estimation.

Contemporary research embodies a pragmatic paradigm that meticulously reconciles technological innovation with prevailing market imperatives and rigorous cost-efficiency criteria. While advanced techniques, such as tightly coupled LiDAR-IMU fusion [8,65], push performance boundaries for high-end systems, significant effort is also directed towards enhancing the capabilities and robustness of lower-cost sensor configurations, often relying heavily on vision and ultrasonic sensors. This leads to a critical analysis point: while optimizing lower-cost sensor configurations is essential for increasing accessibility and market adoption of APS technology, it is critical to assess the quantifiable performance limits and safety trade-offs of these budget-conscious systems compared to high-end alternatives, especially under challenging conditions such as heavy snowfall or in complex environments like densely packed or irregularly shaped parking spaces. Is there a definable threshold in sensor quality or configuration below which reliable and safe APS operation becomes unfeasible, regardless of advances in algorithmic processing or fusion strategies? This trade-off between cost and performance, especially concerning safety-critical functions where failure can have severe consequences, requires more transparent evaluation and clear communication of system limitations to end-users.

2.2. Deep Learning for Parking Space and Obstacle Detection

DL, particularly using Convolutional Neural Networks (CNNs), has become the standard approach for parking space detection, largely superseding traditional computer vision techniques that relied on handcrafted features like edge detection or Hough transforms [3,9]. DL methods demonstrate superior performance, especially in handling vast variations in lighting, weather conditions, diverse parking space markings (or lack thereof), and varied appearances of obstacles, where traditional methods often falter due to their rigidity [11,66,67]. Architectures like YOLO (You Only Look Once) and its variants are frequently adapted for their balance of speed and accuracy in detecting parking slots as bounding boxes [3,68], with enhancements such as oriented bounding boxes and attention mechanisms being developed to improve performance in specific parking scenarios (Figure 4). Recent work includes improvements that focus not only on accuracy (e.g., mean Average Precision, mAP) but also on inference speed (Frames Per Second, FPS) and reducing model size for deployment on resource-constrained embedded automotive hardware [3]. Segmentation-based models (e.g., Mask R-CNN) [69], which classify each pixel, are also employed for more precise delineation of parking spaces and drivable areas. Hybrid DL-computer vision approaches, which might use DL for initial region proposal and traditional techniques for refinement [11], are also explored to balance performance and computational load.

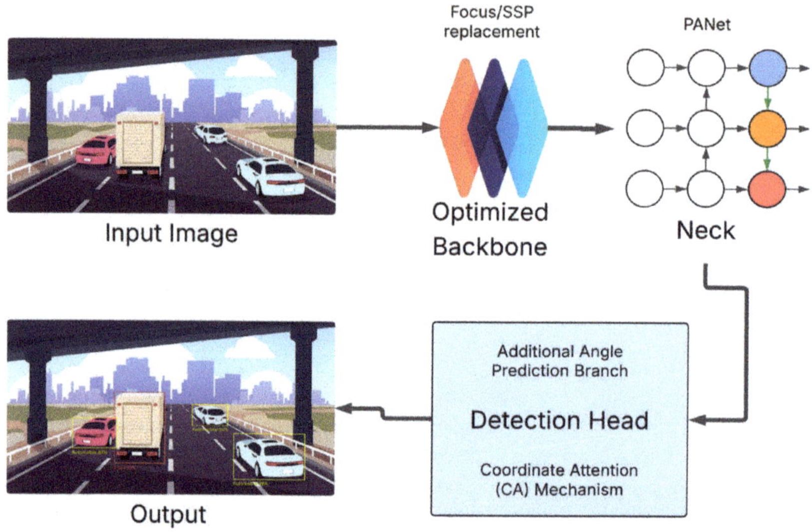

Figure 4. Enhanced YOLOv5-OBB model architecture for oriented bounding box detection in parking scenarios, featuring an optimized backbone, PANet neck, and a detection head with an additional angle prediction branch and Coordinate Attention (CA) mechanism.

While metrics like mAP and FPS demonstrate progress in academic benchmarks and controlled tests [3,11], a critical question is the following: How well do these correlate with real-world robustness and safety, especially concerning rare but critical 'edge cases' (e.g., an unusually shaped obstacle, a parking spot partially obscured by overhanging foliage, or ambiguous markings during construction) not well-represented in training datasets [3]? And, do current benchmarks, often composed of curated and relatively clean data, adequately capture a model's ability to generalize to entirely novel parking environments, different geographical regions with unique parking conventions, or unforeseen obstacle types? The focus on benchmark performance might not fully translate to reliability in the unpredictable and often messy real world, potentially leading to a gap between perceived and actual system capability.

The superior performance of DL methods often comes at the cost of interpretability [2,66]. This "black box" nature, where the internal reasoning of the network is opaque even to its developers, poses a significant challenge for safety validation and debugging. How can we ensure the reliability and predictability of DL-based perception when the decision-making process is not transparent, and what are the implications for certifying such systems as safe according to stringent automotive safety standards like ISO 26262 [70]? To mitigate the risks associated with this lack of transparency, the research community is actively pursuing several key strategies aimed at rendering AI-based systems more scrutable and trustworthy. Methodologies such as:

Explainable AI (XAI) seeks to provide insights into the internal reasoning of a network, for instance, by generating visual attribution maps that highlight which input features most influenced a particular decision. Another prominent approach is the development of hybrid models, which strategically combine the high-performance capabilities of deep learning for perception tasks with the formal verifiability of traditional, model-based algorithms for safety-critical logic and decision-making. Furthermore, uncertainty estimation is a critical technique wherein the model outputs not only a prediction but also a corresponding confidence level. This allows the system to recognize when it is operating in a low-confidence or out-of-distribution scenario, enabling it to trigger fallback maneuvers or request human intervention, thereby providing a crucial safety margin.

Furthermore, DL's heavy reliance on large datasets introduces concerns about data dependency and bias [3,66]. Given DL's heavy reliance on large datasets, how do biases present in these datasets (e.g., overrepresentation of certain parking types like perpendicular slots in well-lit conditions, specific weather conditions common to the data collection region, or particular geographic locations with uniform parking infrastructure) affect the fairness and robustness of deployed APS? What are the risks if a system performs exceptionally well in one region or under specific conditions but poorly in another due to dataset bias, potentially leading to inequitable performance or unexpected failures when encountering underrepresented scenarios? This can lead to systems that are not equitable or universally reliable, for instance, performing worse in regions with non-standard parking markings or in weather conditions rare in the training data.

To systematically address these challenges of data dependency, bias, and the critical need to cover rare but consequential 'edge cases,' the field employs a multi-pronged data strategy. A foundational technique is **data augmentation**, where existing data is algorithmically manipulated by altering lighting, applying simulated weather effects, or adding noise to create a wider variety of training examples without new data collection. For scenarios that are too dangerous, costly, or infrequent to capture in reality, **synthetic data generation** using high-fidelity automotive simulators like CARLA has become indispensable; it allows for the creation of vast, perfectly annotated datasets encompassing a nearly infinite variety of environmental conditions and event sequences. Additionally, more advanced paradigms are employed: **active learning** optimizes the expensive process of human annotation by having the model itself flag uncertain or novel scenarios that it cannot confidently interpret, thereby focusing expert review on the most valuable learning opportunities. To overcome privacy barriers and geographic bias, **federated learning** allows a central model to learn from the real-world experiences of a globally distributed fleet of vehicles without the raw sensor data ever leaving the individual car. Finally, these methods are complemented by **targeted real-world sampling**, a strategic approach where data collection fleets are specifically deployed to underrepresented domains such as regions with non-standard infrastructure or during specific adverse weather events to consciously fill known coverage gaps and mitigate inherent biases in the training data.

A significant practical constraint is computational efficiency for automotive-grade embedded systems, which have limited processing power and strict thermal envelopes [2,3]. The drive for computational efficiency often leads to model compression techniques (e.g., pruning, quantization) or the use of shallower, less complex neural networks. Critically, what is the quantifiable impact of such optimizations on the model's ability to detect subtle features (e.g., a faintly visible parking line, a small, low-contrast obstacle), handle occlusions effectively, or maintain performance in challenging, noisy conditions? Is there a point where efficiency gains, driven by cost or power limitations, lead to an unacceptable degradation in safety-critical perception tasks, potentially missing a critical obstacle or misjudging a parking space boundary? This trade-off requires careful, systematic balancing and rigorous testing to ensure that safety is not unduly compromised for the sake of speed or cost.

2.3. *Achieving Robustness in Diverse Conditions*

A persistent challenge for APS perception systems is maintaining reliable performance across the wide range of conditions encountered in real-world parking environments [3,17,64]. These include variations in illumination (bright sunlight causing lens flare, deep shadows obscuring details, rapidly changing light at dusk/dawn, various types of artificial indoor lighting with different spectral properties), adverse weather (rain, snow, fog, sleet, hail) [64,71], occlusions (by other vehicles, pedestrians, shopping carts, structural pillars), and inconsistencies or degradation in parking space markings (faded lines, non-standard colors or types, temporary markings, or complete absence of markings in informal parking areas).

Training DL models on large, diverse datasets that encompass a wide variety of these conditions is a common and essential strategy [11]. Sensor fusion inherently contributes to robustness by providing complementary information; for instance, radar's performance is less affected by poor lighting or precipitation compared to cameras [6,62], while LiDAR can penetrate some level of fog or rain better than cameras.

The scarcity of comprehensive public datasets is a recurring theme and a significant impediment to progress [7,72]. However, it is crucial to analyze why creating such datasets is profoundly difficult. Beyond sheer volume and the cost of collection and annotation, the challenge lies in capturing the long tail of rare events—those infrequent but potentially critical scenarios that a system must handle safely. This includes unusual obstacle types, peculiar parking geometries, complex interactions with other road users, and diverse sensor noise profiles under extreme conditions. Furthermore, capturing the subtle environmental cues that human drivers intuitively process (e.g., slight changes in road texture indicating a parking boundary, or the behavior of other drivers suggesting an imminent maneuver) is exceptionally hard to codify and include in datasets. Are current data collection (e.g., fleet vehicles) and annotation methods (manual and semi-automated) scalable or sophisticated enough to address this fundamental bottleneck effectively and economically? This data scarcity significantly hampers the development and rigorous evaluation of robust perception algorithms, making it difficult to benchmark different approaches fairly [7,72], consistently reproduce research findings, and train models that generalize well to unseen environments.

While training on diverse datasets [11,71] and employing sensor fusion are common strategies, to what extent do these approaches guarantee robustness against truly 'out-of-distribution' scenarios (i.e., situations fundamentally different from anything in the training data) or compound failures (e.g., a primary sensor malfunction occurring simultaneously with adverse weather conditions)? Current strategies often represent incremental improvements that may still leave significant performance gaps in extreme or unforeseen conditions, rather than providing a comprehensive, verifiable solution to the robustness problem. They may make the system more robust, but not necessarily sufficiently robust for all conceivable real-world encounters.

Table 3 provides a comparative summary of several recent deep learning-based methods, highlighting their architectural innovations, the datasets used for evaluation, and key performance metrics.

Table 3. Comparison of recent DL-based parking slot detection Methods [3].

Method Category	Specific Method/Key Innovation	Dataset(s) Used	Key Performance Metrics Reported	Robustness Aspects Addressed/Target Platform	Reference(s)
Object Detection (YOLO)	Improved YOLOv5-OBB (Oriented BBox, Backbone opt., CA mechanism)	Homemade	mAP +8.4%, FPS +2.87, Size −1 M	Lighting variations; low-compute embedded	[3]
Object Detection (YOLO)	YOLOv5 (Fine-tuned)	PKLot, Custom	Valid. Acc: 92.9%	Real time (PARKTag system)	[68]
Hybrid (DL + CV)	Two-Stage: YoloV11 (Key points) + CV (Special kernel for rotation)	ps2.0 (Public)	Acc: 98.24%, Inf. Time: 12.3 ms (Desktop), 16.8 ms (Laptop)	Speed; varied cond. (ps2.0)	[11]
Segmentation	Mask R-CNN	Tongji Parking-slot DS	Precision: 94.5%, Recall: 94.5%	Lighting variability; occlusions	[69]
Segmentation	Novel Convolutions (Directional, Large Field) for low-level features	Public Remote Sensing	Improved perf. vs. baseline	Potential for parking	[73]
End-to-End Learning	LSS-based Transformer/BEVFusion (Camera -> Control)	CARLA (Simulated)	Success: 85.16%, Pos Err: 0.24 m, Orient Err: 0.34 deg	End-to-end pipeline, dataset creation	[7]

[3] This table provides a comparison of recent deep learning (DL)-based parking slot detection methods, synthesized from the information provided in the correspondingly cited references listed within the table (in the 'Reference(s)' column). For definitions of abbreviations used, please refer to the 'Abbreviations' section at the end of the manuscript. Italicized text or sub-bullets within table cells, if any, provide supplementary commentary or specific details.

3. Path Planning Algorithms for APS

Generating a safe, feasible, and efficient path from the vehicle's current position to the target parking spot is a core function of APSs, requiring a delicate balance between geometric constraints, vehicle dynamics, safety margins, and passenger comfort [74]. Research in recent years has advanced various planning paradigms, often combining techniques to leverage their respective strengths, particularly for navigating complex and constrained parking environments.

3.1. Optimization and Control-Based Planning

Formulating parking trajectory generation as an Optimal Control Problem (OCP) is a powerful approach that allows for the direct minimization of objectives like parking time, path length, or control effort (e.g., minimizing steering changes) while simultaneously satisfying vehicle dynamic constraints (e.g., maximum curvature, velocity limits) and collision avoidance criteria [12]. The general form of an OCP can be expressed as:
minimize the cost functional J:

$$J = \phi(x(t_f), t_f) + \int_{t_0}^{t_f} L(x(t), u(t), t)dt, \tag{1}$$

subject to system dynamics:

$$\dot{x}(t) = f(x(t), u(t), t), \tag{2}$$

initial conditions:

$$x(t_0) = x_0, \tag{3}$$

and path/terminal constraints:

$$\psi(x(t), u(t)) \leq 0, \tag{4}$$

$$\chi(x(t_f)) = 0. \tag{5}$$

where J is the performance index, $\phi(x(t_f), t_f)$ is the terminal cost, $L(x(t), u(t), t)$ is the instantaneous cost (Lagrangian), $x(t)$ is the state vector, $u(t)$ is the control input vector, $f(x(t), u(t), t)$ represents the vehicle dynamics, $\psi(x(t)$ and $u(t))$ are path constraints (e.g., actuator limits, obstacle avoidance), and $\chi(x(t_f))$ are terminal constraints (e.g., reaching the desired parking pose).

Recent work has demonstrated the use of indirect optimal control methods, solved efficiently using specialized numerical tools, which can achieve computation times suitable for near real-time application, even with fine discretization needed for navigating close to obstacles [12]. Numerical optimization techniques, such as Sequential Quadratic Programming (SQP), are also employed, often in conjunction with other methods. For instance, SQP can optimize the control points of Bézier curves used for path generation, minimizing a cost function that includes terms for path smoothness (curvature, jerk), safety (based on an Artificial Potential Field representing obstacles), and proximity to the target pose [16]. Model Predictive Control (MPC) is another relevant technique, used either directly for planning over a receding horizon [75] or, more commonly in the recent literature, for trajectory optimization and tracking [13]. These methods can produce high-quality, smooth, and dynamically feasible trajectories, but they often face computational challenges, especially when dealing with complex environments featuring many obstacles and non-convex constraints. Finding a globally optimal solution can be time-consuming, and the performance heavily depends on the problem formulation and the availability of a good initial guess or starting point for the optimization process, which if poor, can lead to local minima or slow convergence [12,74]. This dependency often leads researchers to adopt hierarchical or hybrid approaches [13,14,76].

3.2. Search and Sampling-Based Planning

Graph search and random sampling algorithms remain popular for exploring the configuration space (the space of all possible vehicle positions and orientations) and finding kinematically feasible paths, especially in complex or initially unknown environments [74]. The A* algorithm, a cornerstone of graph search, evaluates nodes using the function:

$$f(n) = g(n) + h(n), \tag{6}$$

where $f(n)$ is the estimated total cost of the path from the start node to the goal node through the node n, $g(n)$ is the actual cost from the start node to the node n, and $h(n)$ is the heuristic estimate of the cost from the node n to the goal.

The Hybrid A* algorithm, which combines the grid-based search of A* with continuous state expansion respecting vehicle kinematics (e.g., Reeds–Shepp curves), is particularly prevalent [13,14]. Recent enhancements focus on improving its efficiency and the quality of the generated path, for example, through adaptive node expansion strategies that adjust the search step length and angle based on environmental complexity and proximity to the goal [14]. Heuristic functions are also being improved by incorporating information from simpler, faster global planners to better guide the Hybrid A* search.

Rapidly exploring Random Tree (RRT) algorithms and their variants (RRT*, Bi-RRT*, RRT-Connect) are also widely used due to their ability to quickly explore high-dimensional spaces and find paths without explicit free-space modeling [77,78]. Improvements aim to address the inherent randomness and sometimes suboptimal (e.g., overly long or jerky) paths generated by basic RRT. While effective at finding an initial, collision-free, and kinematically feasible path, the resulting path often lacks the smoothness (e.g., continuous curvature) and optimality required for comfortable and precise execution by a real vehicle [14,74].

The prevalence of hybrid approaches, which combine a search/sampling-based algorithm for generating an initial path with an optimization-based method for subsequent refinement (as shown in Figure 5), strongly suggests that neither paradigm is sufficient in isolation for the multifaceted complexities of APSs [13,14,76]. Critically, does this reliance on multi-stage planning introduce cumulative errors, where small inaccuracies in the initial path are magnified during refinement? Does it increase computational overheads due to sequential processing, or create new challenges in ensuring global optimality or even consistent, predictable behavior across different scenarios? For example, a suboptimal initial path might lead the refiner into a local minimum from which it cannot escape to find a better overall solution.

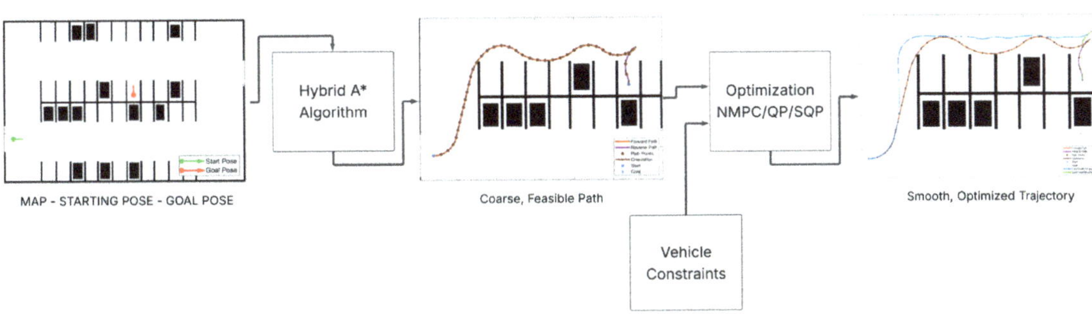

Figure 5. Hierarchical path planning framework for APSs, showing an initial path generated by the Hybrid A* algorithm refined by an Optimization stage (e.g., NMPC (Nonlinear Model Predictive Control)/QP/SQP) considering vehicle constraints.

Delving into the trade-offs, optimization-based methods [12,16], while capable of producing smooth, dynamically feasible paths, often struggle with non-convex environments leading to local minima (becoming stuck in a suboptimal solution), and their computational cost can be prohibitive for real-time replanning in dynamic scenarios where obstacles might move or new information becomes available. The complexity of formulating all relevant constraints accurately can also be a hurdle. Conversely, while sampling-based planners like RRT* [77] can navigate complex spaces and are probabilistically complete (meaning they will find a path if one exists, given enough time), the paths generated are often suboptimal in terms of length or smoothness, requiring significant post-smoothing. This post-smoothing process may itself re-introduce constraint violations (e.g., by smoothing a corner too tightly, making it kinematically infeasible) or cause the path to deviate significantly from the initial 'safe' path found by the sampler.

What are the fundamental limits of each approach such as the curse of dimensionality for grid-based search or the difficulty of incorporating complex cost functions into sampling that necessitate these hybrid solutions, and what compromises are being made in terms of path quality, computational time, guarantee of finding a solution, and predictability of the final trajectory?

Moreover, although many planners effectively manage static obstacles with well-defined geometries, the real-time, robust, and efficient navigation around unpredictable dynamic entities such as pedestrians abruptly crossing, vehicles unexpectedly reversing, or freely moving shopping carts within the constrained spaces and limited visibility typical of parking lots, continues to represent a significant and largely unresolved challenge [16,79]. How do current approaches balance proactive avoidance (maintaining safe distances) with maintaining progress towards the goal without resorting to overly conservative (e.g., stopping frequently) or jerky maneuvers that could be disconcerting to passengers or other road users? The ability to predict the intent and future motion of these dynamic actors is crucial yet exceedingly difficult.

3.3. Addressing Constraints and Complexities

A major focus of recent APS path planning research is tackling challenging real-world scenarios. This includes planning maneuvers in narrow or tightly constrained parking spots where clearance is minimal [12,13,76], operating in unstructured environments without clear markings or predefined layouts (e.g., gravel lots, informal street parking), and safely navigating environments with multiple static or dynamic obstacles [79,80]. Non-holonomic kinematic constraints (minimum turning radius, steering angle limits, vehicle dimensions) are explicitly considered during node expansion in algorithms like Hybrid A* [13,14] or are formulated as hard constraints within OCP or MPC frameworks [12]. A common representation used for vehicle kinematics is the bicycle model:

$$\dot{x}_v = v\cos(\psi), \tag{7}$$

$$\dot{y}_v = v\sin(\psi), \tag{8}$$

$$\dot{\psi} = \frac{v}{L}\tan(\delta), \tag{9}$$

with constraints such as:

$$|\delta| \leq \delta_{max}, \tag{10}$$

$$\left|\dot{\delta}\right| \leq \dot{\delta}_{max}, \tag{11}$$

$$R_{min} = \frac{L}{|\tan(\delta_{max})|}, \tag{12}$$

where $(\dot{x}_v, \dot{y}_v)$ are the vehicle's longitudinal and lateral velocities, v It's speed, ψ is the orientation (yaw) angle, δ is the steering angle of the front wheels, L is the wheelbase, δ_{max} and $\dot{\delta}_{max}$ are the maximum steering angle and rate, and R_{min} is the minimum turning radius.

Collision avoidance is handled through various means: smooth, differentiable penalty functions added to OCP cost functions that penalize proximity to obstacles [12]; repulsive forces generated by obstacles within an Artificial Potential Field (APF) framework [16,77]; or geometric constructions like the Improved Safe Travel Corridor (I-STC), illustrated conceptually in Figure 6, that define a collision-free corridor around an initial path, simplifying the constraints for subsequent optimization stages [76]. In APF methods, the total potential $U(q)$ at a configuration q is often a sum of an attractive potential towards the goal $U_{att}(q)$ and a repulsive potential from obstacles $U_{rep}(q)$:

$$U(q) = U_{att}(q) + U_{rep}(q), \tag{13}$$

an attractive potential can be:

$$U_{att}(q) = \frac{1}{2}k_{att}\|q - q_{goal}\|^2, \tag{14}$$

and a repulsive potential for an obstacle i is:

$$U_{rep,i}(q) = \begin{cases} \frac{1}{2}k_{rep}\left(\frac{1}{\rho(q,obs_i)} - \frac{1}{\rho_0}\right)^2 & \text{if } \rho(q,obs_i) \leq \rho_0 \\ 0 & \text{if } \rho(q,obs_i) > \rho_0 \end{cases}, \tag{15}$$

where k_{att} and k_{rep} are gain coefficients, q_{goal} is the goal configuration, $\rho(q, obs_i)$ is the minimum distance to obstacle i, and ρ_0 is its distance of influence. The vehicle then follows the negative gradient of this field: $F(q) = -\nabla U(q)$.

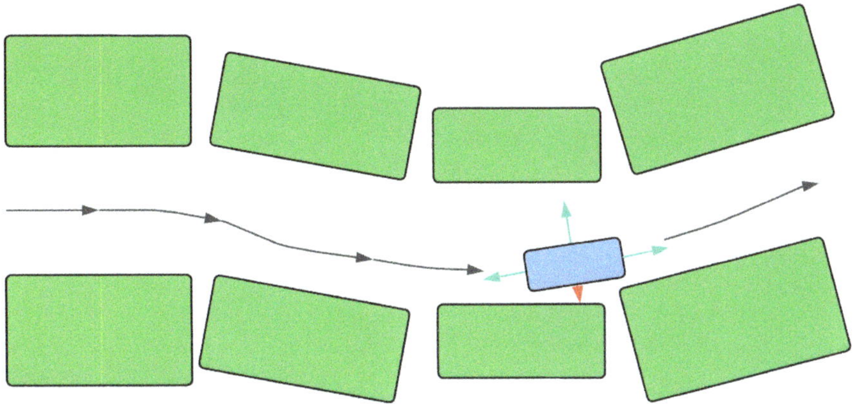

Figure 6. Conceptual illustration of the Improved Safe Travel Corridor (I-STC), which defines a collision-free rectangular region around the vehicle, oriented along its current heading, by expanding until it hits obstacles or a maximum distance. The green arrows represent the calculated safe clearance distances, while the red arrow indicates a dangerously close proximity.

Planners are also being developed to handle complex maneuvers involving multiple gear shifts (forward and reverse segments), often required for parking in tight parallel or perpendicular spots [12,81].

3.4. Path Smoothing and Refinement

Generating paths that are not only feasible and collision-free but also smooth and comfortable for passengers is crucial for practical APS. Raw paths from search or sampling algorithms often contain sharp turns, discontinuities in curvature, or unnecessary oscillations [14,74]. Therefore, post-processing or integrated smoothing techniques are commonly applied. Geometric curves known for their smoothness properties, such as Bézier curves (offering flexibility through control points) [16] and Clothoid curves (Euler spirals, whose curvature varies linearly with arc length, naturally representing steering transitions), are frequently used.

A cubic Bézier curve, for instance, is defined by four control points (P_0, P_1, P_2, P_3) as:

$$B(t) = P_0(1-t)^3 + 3P_1t(1-t)^2 + 3P_2t^2(1-t) + P_3t^3, t \in [0,1], \quad (16)$$

Clothoid curves are characterized by their curvature κ varying linearly with arc length s:

$$\kappa(s) = \kappa_0 + \sigma s, \quad (17)$$

or, if starting with zero curvature, $\kappa(s) = s/A^2$, where A is the clothoid parameter and σ is the sharpness.

Optimization-based smoothing, utilizing techniques such as Quadratic Programming (QP) [14] or Sequential Quadratic Programming (SQP) [16], adjusts path points or curve parameters to minimize objectives related to curvature, jerk, and deviation from the original path.

The necessity of incorporating explicit path smoothing stages reveals a fundamental shortcoming in many initial path generation algorithms. These algorithms typically focus on quickly identifying any feasible path that satisfies basic constraints, often at the expense of path quality metrics such as smoothness, continuity, and minimal curvature. This prioritization results in trajectories that may be suboptimal for practical execution, necessitating subsequent refinement processes to enhance drivability, passenger comfort, and overall system performance. Does this post-processing step guarantee the maintenance of safety margins established by the initial planner, or could smoothing inadvertently move the path closer to obstacles, potentially reducing clearance below acceptable levels in an attempt to achieve a more aesthetically pleasing or comfortable trajectory? For example, smoothing a sharp turn might slightly cut a corner, bringing the vehicle closer to a parked car than the initial planner intended. Moreover, how can the trade-off between path smoothness—critical for passenger comfort and precise trackability by the vehicle's control system—and path length or maneuver time—key factors for operational efficiency and user satisfaction—be rigorously quantified and systematically optimized, rather than relying on heuristic adjustments driven by developer intuition or limited empirical validation? The absence of standardized, objective metrics to evaluate and balance these competing criteria often results in inconsistent system performance and complicates comparative assessments of various path smoothing methodologies.

Table 4 provides a comparative overview of the diverse path planning strategies discussed, summarizing their key features, how they handle vehicle constraints, and their validation methods.

Table 4. Overview of recent APS path planning algorithms [4].

Category	Specific Method/ Combination	Key Features	Constraint Handling	Smoothing Method	Validation	Reference(s)
Search + Opt.	Hybrid A* + NMPC	- Hierarchical - NMPC optimizes coarse path	- Narrow spaces - Kinematics	Implicit (NMPC)	Sim, Real	[13]
Search + Opt.	Hybrid A* + QP Smoothing + S-Curve Speed Planning	- Adaptive search - Improved heuristic - QP smoothing - S-curve speed	- Constrained env. - Kinematics	- QP - S-Curve	Sim, Real	[14]

Table 4. *Cont.*

Category	Specific Method/ Combination	Key Features	Constraint Handling	Smoothing Method	Validation	Reference(s)
Search + Opt.	Graph Search (Hybrid A*) + Numerical Opt. + I-STC	- Hierarchical - Warm start - I-STC simplifies collision constraints	- Parallel parking - Narrow spaces	Numerical Opt.	Sim	[81]
Search + Opt.	Hybrid A* + GA (Genetic Algorithm) Opt. + Geometric Curves (Bézier and Clothoid)	- Hierarchical - GA local opt. - Curve smoothing	- Tight spaces - AVP	- Bézier - Clothoid	Sim	[76]
Search + APF	RRT* + Improved APF + Variable Probability Strategy	- Guided sampling - APF avoids local minima - Adaptive sampling	Obstacle avoidance	Enhanced APF	Sim	[77]
Optimal Control	Indirect OCP (Minimum Time) + pNN Controller	- Efficient OCP solver (PINS) - Smooth 3D penalty functions - Complex maneuvers	- Narrow spaces - Unstructured - Kinematics	Implicit (OCP)	Sim, Real	[12]
Prediction + Opt.	Prediction + APF + Bézier Curve + SQP Optimization	- Integrates prediction - APF target selection - SQP optimizes Bézier CPs	- Dynamic obstacles - Collision avoidance	Bézier Curve + SQP	Sim	[16]
Reinforcement Learning	Soft Actor-Critic (SAC) + Segmented Training Framework	- Optimizes safety, comfort, efficiency, accuracy - Handles neighbor deviation	- Non-ideal scenarios - Kinematics	Implicit (Learned Policy)	Sim	[15]
Reinforcement Learning	Hybrid RL (Rule-based RS + Learning-based Planner)	- Combines rule-based feasibility with learned adjustments	Complex environments	Implicit (Learned Policy)	Sim, Real	[82]
Multi-Agent	Improved Conflict-Based Search (ICBS) + Swarm Opt. (IACA-IA) + Adaptive A*	- Multi-UGV coordination - Conflict resolution - Slot allocation	- High-density - Multi-vehicle	Adaptive A*	Sim	[79]

[4] This table provides an overview of recent Autonomous Parking Systems (APSs) path planning algorithms, synthesized from the information provided in the correspondingly cited references listed within the table (in the 'Reference(s)' column). For definitions of abbreviations used, please refer to the 'Abbreviations' section at the end of the manuscript. Italicized text or sub-bullets within table cells, if any, provide supplementary commentary or specific details.

4. Vehicle Control Strategies

Once a suitable path is planned, the vehicle's control system is responsible for executing the maneuver by accurately tracking the trajectory and managing the vehicle's actuators (steering, throttle, brakes). Research in this area focuses on developing controllers that ensure precision, stability, smoothness, and robustness while respecting physical limitations and passenger comfort [83].

4.1. Advanced Trajectory Tracking Controllers

Several advanced control strategies have been investigated and applied to the APS problem:

- Model Predictive Control (MPC) remains a prominent technique due to its inherent ability to handle constraints (on states and inputs) explicitly and optimize control actions over a future prediction horizon [13,75]. It can anticipate future path requirements and adjust current inputs accordingly, leading to smoother control [4]. A common discrete-time cost function for MPC in trajectory tracking is:

$$J(x_k, U_k) = \sum_{i=0}^{N_p-1} \left(\|x_{k+i|k} - x_{ref,k+i}\|_Q^2 + \|u_{k+i|k} - u_{ref,k+i}\|_R^2 \right) + \|x_{k+N_p|k} - x_{ref,k+N_p}\|_P^2 \quad (18)$$

Subject to the discretized system dynamics $x_{k+i+1|k} = f_d\left(x_{k+i|k}, u_{k+i|k}\right)$ and constraints on states x and inputs u. Here, N_p is the prediction horizon, x_{ref} and u_{ref} are reference states and inputs, and Q, R, P are weighting matrices.

- Reinforcement Learning (RL) techniques, which are fundamentally based on agents learning optimal behavior through trial-and-error interactions with an environment to maximize a cumulative reward signal, are gaining traction for learning control policies directly from these interactions (or, more commonly, a simulation) [84,85]. Algorithms like Soft Actor-Critic (SAC), whose architecture is detailed in Figure 7, have been used to train parking strategies that explicitly balance multiple objectives, including safety, comfort (e.g., minimizing jerk), efficiency (e.g., minimizing time), and accuracy (e.g., final pose error) [15,82]. SAC maximizes an entropy-regularized objective:

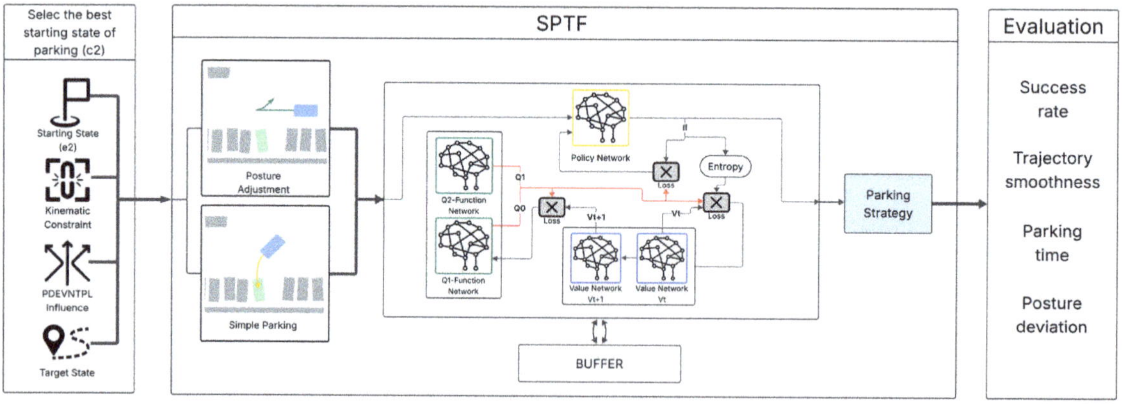

Figure 7. Architecture of the Segmented Parking Training Framework (SPTF) employing a Soft Actor–Critic (SAC) agent, illustrating inputs, training stages, network components, and evaluation metrics [15].

$$J(\pi) = \sum_{t=0}^{T} \mathbb{E}_{(s_t,a_t)\sim\rho_\pi}[r(s_t, a_t) + \alpha\mathcal{H}(\pi(\cdot|s_t))] \quad (19)$$

where $r(s_t, a_t)$ is the reward, α is a temperature parameter, and $\mathcal{H}(\pi(\cdot|s_t))$ is the policy entropy, encouraging exploration. Hybrid RL approaches, combining rule-based planners with learned components [82,86], are also being explored to leverage the strengths of both paradigms.

The increasing application of Deep Reinforcement Learning (DRL), particularly model-free methods like Soft Actor–Critic, represents a significant trend in vehicle control for autonomous parking. Unlike traditional controllers that rigidly follow pre-computed geometric paths, DRL agents can learn complex, adaptive control policies directly from interaction. This approach is particularly well-suited to the non-linear dynamics and multi-objective nature of parking, where the system must continuously balance safety margins, passenger comfort (e.g., minimizing jerk), maneuver efficiency, and final placement accuracy. By defining these objectives within a reward function, DRL can discover novel, high-performance parking strategies that are difficult to hand-engineer. However, as noted in the critical evaluation of these strategies, this power comes with significant challenges in formal verification, sim-to-real transfer, and ensuring robust performance against out-of-distribution scenarios, which remains a primary focus of ongoing research.

- Neural Network (NN)-Based Controllers; Beyond standard RL algorithms, researchers are developing bespoke NN-based controllers. One example is a pseudo-neural network (PNN) steering controller designed with a physics-driven structure, using feedforward components and feedback terms, trained via supervised learning on data from high-fidelity simulations [12].

Reinforcement Learning (RL), particularly SAC [15,82], shows promise for optimizing complex objectives and adapting to nuanced situations not easily captured by traditional models. However, its application in safety-critical APS faces significant hurdles: the 'black box' nature of learned policies makes formal verification of their safety and stability exceptionally difficult [2]; ensuring safety during the trial-and-error learning phase (especially if conducted in real-world interactions, which is rarely feasible) is problematic and typically relies on carefully designed simulators and reward functions; and the sim-to-real gap can lead to unpredictable or degraded real-world performance if the simulation doesn't perfectly capture all aspects of vehicle dynamics and environmental interactions [15,82,86]. How are researchers addressing these fundamental safety and reliability concerns beyond demonstrating high reward scores in simulated environments? Are current validation methods for RL policies, often based on empirical testing over a limited set of scenarios, sufficient for safety certification in a domain where unexpected failures can have severe consequences?

MPC's effectiveness is intrinsically linked to the fidelity of the vehicle model it uses for prediction [4,83]. In real-world parking scenarios, unmodelled dynamics (e.g., tire slip on varied surfaces like ice, wet leaves, or loose gravel; suspension changes due to passenger load or uneven weight distribution; aerodynamic effects at higher speeds, though less relevant for parking; or even subtle changes in tire pressure) or sudden external disturbances (e.g., wind gusts in exposed parking lots) can degrade MPC performance, leading to tracking errors or suboptimal control [13]. How robust are current MPC strategies to such model-reality mismatches, and what are the adaptation mechanisms in place (e.g., adaptive MPC, robust MPC formulations)? Over-reliance on an imperfect model can lead to suboptimal or even unsafe control actions if the controller's predictions diverge significantly from reality.

Similar to DL in perception, NN-based controllers [12], whether learned through RL or supervised methods, raise profound questions about robustness to out-of-distribution states (i.e., situations or sensor inputs significantly different from those encountered during training) and the formal verification of their stability and safety envelopes across the entire operational domain. Without rigorous verification methods that can provide mathematical guarantees of safe behavior, their deployment in safety-critical systems like APSs remains contentious and reliant on extensive, but inevitably incomplete, empirical validation.

4.2. Ensuring Precision, Stability, and Handling Constraints

The primary goal of the trajectory tracking controller is to minimize the error between the vehicle's actual position and orientation and the desired path. High precision is critical, especially in tight parking spots where millimeters can make the difference between a successful park and a collision [87]. Stability during the maneuver is paramount; controllers must ensure the vehicle does not exhibit undesirable oscillations or unstable behavior [83]. Furthermore, they must operate within the physical limitations of the vehicle's actuators, respecting constraints on steering angle, steering rate, acceleration, and braking capabilities. Passenger comfort, often related to minimizing jerk (the rate of change of acceleration) [15,16], is another crucial consideration, as overly aggressive or jerky maneuvers can be unsettling.

Achieving high precision is often framed as a primary goal, and some research reports impressively low tracking errors [87]. However, is there a point where striving for sub-centimeter precision yields diminishing returns in practical parking success, especially if it comes at the cost of increased maneuver time (due to slower, more cautious movements), higher computational load (for more complex controllers), or overly sensitive control actions that might reduce passenger comfort or perceived safety (e.g., frequent, tiny steering adjustments)? How is 'sufficient precision' defined and justified in the context of overall system performance, user acceptance, and varying environmental conditions (e.g., is sub-centimeter precision truly necessary or even achievable on a highly uneven surface)? An overemphasis on precision in one aspect might neglect other important operational qualities or lead to an overly complex and sensitive system.

4.3. Critical Evaluation of Vehicle Control Strategies in Autonomous Parking Systems (APS)

Introduction to the Comparative Table

The following table, "Table 5: Critical Evaluation of Vehicle Control Strategies in Autonomous Parking Systems (APSs)," provides a synthesized and critical overview of prominent vehicle control strategies employed in Autonomous Parking Systems.

Table 5. Critical evaluation of vehicle control strategies in Autonomous Parking Systems (APSs) [5].

Control Strategy	Operating Principles	Reported Advantages for APS	Verification/Safety and Robustness Challenges	Handling Precision vs. Comfort/Efficiency Trade-Offs
Model Predictive Control (MPC)—Standard	- Explicit vehicle model (e.g., bicycle [88]) predicts future states (Np) [89]. - Optimizes control sequence (Nc) by minimizing cost function (tracking error, effort, constraints) [88]. - Applies first control input; repeats (receding horizon) [90]. - Explicitly handles system constraints (actuator/state limits, obstacle avoidance) [89,90].	- High precision/accuracy for tight spaces [90]. - Optimal performance and constraint handling (vehicle limits, obstacles) [89]. - Improved comfort (smoother motion vs. simpler controllers) [91]. - Enhanced robustness (vs. some traditional methods) [91]. - Predictive capability: Proactive control [89]. - Solves constrained motion problems optimally [46].	- Model fidelity: Performance/safety depend on model accuracy; mismatches problematic [89]. - Obtaining perfect models for all conditions is challenging. - Computational complexity: Real-time optimization can be demanding [89]. - Recursive feasibility: Ensuring a solution exists at every step is a major challenge [92]. - Tuning complexity: Many parameters (weights, horizons) need careful tuning [46]. - Uncertainty handling: Standard MPC assumes deterministic models; robust variants increase complexity/conservatism [92].	- Adjusting weights in cost function (Q, R, P) [88]. - Precision: ↑ weight on tracking error. - Comfort: Penalize aggressive inputs (steering rate, acceleration). - Efficiency: Penalize path length/time. - Hard constraints for precision; soft for comfort/feasibility. - Prediction/control horizons (N_p, N_c) influence behavior.
Learning-Based MPC (e.g., residual models, Koopman, GP, NN-adaptive)	- Integrates ML to improve prediction model or compensate for deficiencies [89]. - Residual Model Learning: Learns unknown dynamics (GP, NN) to augment nominal model [89]. - Koopman Operator Theory: Lifts nonlinear dynamics to linear space for prediction [93]. - NN-Adaptive LPV-MPC: NN adapts LPV-MPC parameters online (e.g., tire stiffness) [44].	- Enhanced model accuracy and control performance [89]. - Adaptability to varying conditions (road friction, tires) [46]. - Reduced dimensionality for learning (residual models) [89].	- Safety of learned components (GP, NN): Formal verification difficult [93]. - Data requirements and generalization: Poor OOD performance risks safety [89]. - Verification of hybrid system (MPC + learned parts). - Computational overhead of online learning/adaptation. - Defining valid feature space for learned component to avoid unsafe extrapolation [89].	- MPC cost function still key for balance. - Learned component might implicitly balance if trained on such data. - Economic MPC (EMPC): Cost function directly optimizes efficiency (e.g., energy [94]) alongside performance. - Learning helps achieve trade-offs if model is more accurate.

Table 5. *Cont.*

Control Strategy	Operating Principles	Reported Advantages for APS	Verification/Safety and Robustness Challenges	Handling Precision vs. Comfort/Efficiency Trade-Offs
Reinforcement Learning (RL)—General	- Agent learns actions (steering, accel.) by interacting with environment (parking scenario) [95]. - Learns policy (state -> action) to maximize cumulative reward [95]. - Trial-and-error learning, often in simulation [95]. - Model-free (direct policy) or model-based (learns model) [49]. - End-to-end RL: Sensor inputs -> control commands [96].	- Adaptability to complex/dynamic environments [95,97]. - Learning complex maneuvers for tight spaces [98]. - Optimization through interaction; potential for novel solutions [95]. - Reduced reliance on hand-engineered rules [99].	- 'Black box' nature: Formal safety verification very difficult [95]. - Safety during learning (exploration): Mostly simulated, leading to sim-to-real gap [95]. - Sim-to-real transfer gap: Policies often perform poorly in real world [82]. - Bridging gap (domain randomization, OGM (Occupancy Grid Map) [82]) is critical. - Robustness to OOD states: Catastrophic failure possible [99]. - Reward hacking and specification: Defining correct reward function is hard [99]. - Stability and convergence issues [96]. - Data bias and sufficiency [95]. - Computational demands for training DRL [95].	- Primarily via Reward Function Design [99]: - Precision: Reward proximity/alignment; penalize collisions/deviations. - Comfort: Penalize jerk, high accel/decel, large steering rates. - Efficiency: Reward speed, short paths, low energy; penalize maneuvers. - Multi-Objective RL (MORL): Explicitly multiple rewards, balancing complex. - Constraint-driven safety RL (e.g., CMDP): Safety as hard constraint [99]. - Hierarchical RL: Different rewards per level (strategic vs. maneuver) [100]. - Reward changes can lead to unpredictable policy changes.
Reinforcement Learning (RL)—Soft Actor-Critic (SAC)	- Off-policy, model-free, actor-critic DRL for continuous actions [99]. - Actor (policy) decides actions; Critic (value/Q-net) evaluates [99,101]. - Entropy regularization: Maximizes reward + policy entropy (encourages exploration) [101]. - Off-policy learning from replay buffer (sample efficiency) [99,101].	- High parking success rates (simulated) [97]. - Reduced maneuver times (vs. traditional/other DRL) [101]. - Robust handling of dynamic obstacles [101]. - Fine-grained vehicle control (continuous actions) [101]. - Improved sample efficiency (with HER for sparse rewards) [97].	- Same general RL challenges: Black box, verification, safety, sim-to-real, OOD, reward, stability, data, computation [95]. - SAC properties may mitigate some training issues (sample efficiency, exploration) but fundamental safety challenges persist.	- Similar to general RL (reward function design). - Entropy regularization might naturally lead to smoother policies (comfort) [101]. - Explicit control over trade-offs still heavily relies on reward component formulation.

Table 5. *Cont.*

Control Strategy	Operating Principles	Reported Advantages for APS	Verification/Safety and Robustness Challenges	Handling Precision vs. Comfort/Efficiency Trade-Offs
Neural Network (NN) Based Controllers—Direct NN Controllers (End-to-End)	- NN (often deep) maps sensor inputs (LiDAR [102], camera, sonar [103]) or states to control commands [102]. - Learns mapping from data (imitation learning [102]) or RL. - Internal layers learn hierarchical features. - NNs with ReLU: Complex piecewise linear functions [102]. - Sigmoid/tanh: Smooth non-linearities [104].	- Learning complex kinematics/dynamics from data [103]. - Automatic generation of control commands [103]. - Adaptability to different situations (if trained on diverse data) [103]. - Reduced parking time; system-level, connected CAVs (Connected Automated Vehicles) [105]. - Handling high-dimensional inputs (e.g., LiDAR images) [102].	- 'Black box' and lack of interpretability: Hard to debug, provide safety guarantees [102]. - Formal verification difficulties: Computationally challenging for deep NNs [102]. - Techniques exist (hybrid systems, SMC, reachability) but have limits. - System-level safety (closed-loop) very complex. - Robustness to OOD states/adversarial attacks: NNs can be brittle [106]. - Robust OOD detection essential [107]. - Data dependency and generalization issues [106]. - Sim-to-real gap. - Lack of precise mathematical specifications for "correct" behavior [102].	- Implicitly learned from training data (imitation or RL reward). - If trained on human data, mimics that balance. - Hard to predictably adjust behavior post-deployment without retraining. - Neural abstractions trade abstraction precision vs. verification time [108].
Neural Network (NN) Based Controllers—Jordan Neural Networks	- Recurrent NN: Output fed back as input via "context units" [103]. - Inputs: Sensor readings, odometer, current maneuver state. - Outputs: Control commands (speed, steering), next maneuver state [103]. - Learns from examples of successful maneuvers (supervised learning, e.g., RPROP) [103].	- Automatic knowledge acquisition from examples [103]. - Adaptability and robustness: Potential generalization to new situations (vs. rigid rules) [103]. - Simplified development: Focus on collecting examples [103].	- Similar to direct NN: Black box, verification, data dependency, generalization, sim-to-real. - Recurrent structure adds complexity to formal analysis. - Ensuring stability/convergence of recurrent dynamics can be challenging.	- Trade-offs primarily learned implicitly from training examples. - Quality of demonstrated maneuvers dictates learned policy balance. - Explicit tuning post-training difficult without new data.

Table 5. *Cont.*

Control Strategy	Operating Principles	Reported Advantages for APS	Verification/Safety and Robustness Challenges	Handling Precision vs. Comfort/Efficiency Trade-Offs
Hybrid RL Approaches—RL + Rule-Based Planners (e.g., RL with Reeds-Shepp (RS) curves, A*)	- Combines stability/guarantees of traditional planners (RS, A*) with RL's adaptability [82]. - Rule-based planner: Initial reference/candidate paths [109]. - RL agent: Refines trajectory, selects candidates, or makes high-level decisions [82]. - E.g., RS path + RL speed/steering adjustments from LiDAR OGM [82].	- Improved generalizability/adaptability [82]. - Higher planning success rates (vs. pure rule-based/RL) [110]. - Enhanced training efficiency for RL (rule-based guidance) [110]. - Bridging sim-to-real gap (e.g., LiDAR OGM for consistent representation) [82].	- Complexity of verifying hybrid systems: Interaction between learning and rule-based parts [111]. - Safety of the RL component (black box, OOD). - Interface consistency/robustness: Misinterpretations cause failures. - Balancing control authority between RL and traditional parts. - "Weakest link" problem: Safety depends on both components and interaction.	- Traditional planner: Focus on feasible/efficient paths (precision, efficiency) [109]. - RL agent: Refines for comfort, adapts to dynamics, optimizes further [100]. - Modular assignment simplifies design/tuning.
Hybrid RL Approaches—RL + Model Predictive Control (MPC)	- Integrates RL with MPC. - RL: High-level decisions (strategy, MPC objectives/constraints) [110]. - MPC: Low-level trajectory optimization/control [110]. - Alt: RL learns model for MPC or tunes MPC parameters.	- Optimal low-level control: MPC for constraint-aware execution [110]. - Strategic high-level learning: RL adapts MPC for different scenarios. - Improved trajectory quality and planning time (NN hierarchical DRL + opt. layer) [112].	- Verification complexity of interacting learning/model-based parts [111]. - Safety of RL decision-making component. - Ensuring MPC recursive feasibility given RL goals. - Potential negative interference if RL gives unsuitable objectives to MPC. - "Weakest link" problem applies.	- Hierarchical task decomposition. - RL: Higher-level strategic goals (efficiency, context adaptation). - MPC: Manages precision/comfort for low-level execution (cost function).
Hybrid RL Approaches—DRL + Kinematic-Based Co-pilot	- DRL agent learns primary driving policy [113]. - Kinematic co-pilot: Guidance/constraints for DRL during training (efficiency) and operation (safety/decision support) [113]. - May include rule-based system to assess/mediate final actions for safety [113].	- Enhanced training efficiency for DRL (kinematic guidance) [113]. - Flexible decision-making guidance from co-pilot [113]. - Improved safety/reliability: Rule-based system as safety net [113].	- Verification of interaction (DRL, co-pilot, rule-supervisor) is complex. - Ensuring co-pilot guidance is always safe/beneficial. - Determining override logic for rules (not too conservative/missing DRL insights). - "Weakest link" principle applies.	- Kinematic co-pilot/rules enforce comfort/safety (precision) constraints, guiding DRL [113]. - DRL reward function still primary driver for efficiency/other aspects within safety envelope.

[5] This table presents a critical evaluation of vehicle control strategies in Autonomous Parking Systems (APSs), synthesized from the information provided in the correspondingly cited references listed within the table (in the 'Key References' column). For definitions of abbreviations used, please refer to the 'Abbreviations' section at the end of the manuscript. Italicized text or sub-bullets within table cells provide supplementary commentary or specific details.

5. Comparative Analysis of Recent APS Research and System Developments

5.1. Introduction to Comparative Framework

To synthesize the rapid advancements in Autonomous Parking Systems, this section provides a comparative analysis of pivotal research studies and system developments. The focus is primarily on contributions from 2023 to 2025, supplemented by earlier works that represent critical foundational advances directly influencing current APS technology.

Grounding the advanced research presented here in practical reality, it is essential to analyze the state of Autonomous Parking Systems (APSs) currently available in the commercial market. These systems are not monolithic; rather, they exist on a spectrum of automation, a distinction this review makes between Automatic Parking Assist (APA) and fully Autonomous Valet Parking (AVP). The most common systems deployed by major manufacturers function at the APA level, providing steering control while the human driver remains responsible for managing the vehicle's speed and monitoring the environment.

Moving toward greater autonomy, well-known systems like Tesla's Autopark automate the entire steering and speed control sequence for parallel or perpendicular parking while the driver is inside. Tesla's more advanced Smart Summon feature exemplifies an AVP-like capability, allowing a user to call the vehicle from its parking spot to their location remotely. Similarly, Hyundai's Remote Smart Parking Assist (RSPA) provides remote control for maneuvering into tight spaces. While impressive, the real-world performance of these systems is often constrained; they function most reliably in well-marked lots with ample space and can exhibit hesitant or unpredictable behavior in crowded, dynamic environments. This performance envelope is a direct consequence of their reliance on a sensor suite composed primarily of cameras and ultrasonic sensors, which can struggle with the edge cases and adverse conditions detailed throughout this review. This highlights the significant gap that advanced research, focusing on robust multi-modal fusion with technologies like LiDAR and 4D radar, aims to close.

5.2. Comparative Table of Pivotal APS Studies and Systems

Table 6 summarizes selected recent and critical studies and system developments in APSs.

Table 6. Comparative table of pivotal APSs studies and systems [6].

Study/System	Core Technological Innovation	Sensor Suite Utilized	AI/ML Methodologies Applied	Key Performance Metrics/Findings	Specific Relevance/Contribution to APS Advancement
RL-OGM-Parking [82]	Hybrid RL (Reeds-Shepp + SAC) planner using LiDAR OGM.	LiDAR	- DRL (SAC) - Rule-based (Reeds-Shepp)	- High PSR (Parking Success Rate) - Reduced ANGS (Number of Gear Shifts) and PL (Path Length) - Outperforms pure rule/learning.	Addresses sim-to-real for learned planners; stable/adaptive maneuvers.
SAC-based DRL for Parking [15]	DRL (SAC) for continuous vehicle control.	LiDAR, Camera	DRL (SAC)	- High PSR - Reduced maneuver times - Robust to dynamic obstacles - Outperforms traditional and other DRL.	Fine-grained control, efficient path gen in dynamic scenarios.
U-Park [67]	User-centric smart parking recommendation for e-micromobility.	Implies Camera (CNN for space detection)	CNN (parking space detection, hazy/foggy)	Tailored recommendations (user pref., conditions).	Extends smart parking to micromobility; user-centric, robust perception.
Diffusion Model Planning [114]	Diffusion models for diverse, feasible motion trajectories.	General AV sensors	Diffusion Models	SOTA PDM score (94.85) on NAVSIM.	Potential for diverse, smooth, context-aware parking trajectories.
4D Imaging Radar for Automotive [62]	4D mmWave radar for high-res point clouds (range, azimuth, elevation, velocity).	4D mmWave Radar	Doppler/angle resolution algos, dynamic CFAR (Constant False Alarm Rate)	Demonstrated 4D high-res imagery in parking lots.	Enhances all-weather perception, velocity measurement for APS.
Monocular VIO (Visual-Inertial Odometry) with Planar Regularities [114]	Monocular VIO regularized by planar features using MSCKF (Multi-State Constraint Kalman Filter).	Monocular Camera, IMU	- MSCKF - Custom Plane Detection	Improved ATE (Absolute Trajectory Error) (1–3 cm accuracy, structured env.).	Precise, cost-effective ego-motion for structured parking (garages).
E2E Parking Dataset and Model) [7]	Open-source dataset (10k+ scenarios) and benchmark for E2E parking.	Multi-camera, Vehicle sensors	- E2E Learning (Transformers) - BEV representation	Baseline: 85.16% success, 0.24 m pos err, 0.34 deg orient err.	Facilitates reproducible research, standardized E2E model benchmarking.

Table 6. *Cont.*

Study/System	Core Technological Innovation	Sensor Suite Utilized	AI/ML Methodologies Applied	Key Performance Metrics/Findings	Specific Relevance/Contribution to APS Advancement
REDFormer [115]	Transformer-based camera-radar fusion for 3D object detection (low-visibility).	Camera, Radar (nuScenes)	- Transformers - BEV Fusion	Improved perf. in rain/night on nuScenes.	Enhances APS perception robustness in adverse conditions.
Improved CILQR (Constrained Iterative Linear Quadratic Regulator) [116]	Enhanced Constrained Iterative LQR for stable, human-like, efficient trajectory planning.	Assumed std. AV sensors	- CILQR - Hybrid A*	Improved human-like driving, traffic efficiency, real-time capability.	Potential for higher quality, smoother, natural parking trajectories.
AVP HMI (Human–Machine Interface) Explanations [117]	Scenario-based XAI framework for AVP HMI.	N/A (HMI study)	XAI principles	Improved driver trust and UX; reduced mental workload; better user perf.	Improves user understanding, trust, acceptance of AVP via transparency.
Hyundai Mobis Parking System [118]	Commercial Advanced Automated Parking System (AAPS).	Ultrasonic, Surround-view cameras	Proprietary map gen and autonomous parking algos	Seamless autonomous parking (single touch).	Real-world deployment, commercialization of in-vehicle APS tech.

[6] This table presents a comparative overview of pivotal Autonomous Parking Systems (APSs) studies and systems, synthesized from the information provided in the correspondingly cited references listed within the table (in the 'Study/System column). For definitions of abbreviations used, please refer to the 'Abbreviations' section at the end of the manuscript. Italicized text or sub-bullets within table cells, if any, provide supplementary commentary or specific details.

5.3. Analysis of Overarching Trends and Innovations

The comparative analysis presented in the previous section reveals several dominant trends and significant innovations in the APS landscape between 2023 and 2025. There is a clear and accelerating convergence towards AI-centric solutions, particularly Deep Reinforcement Learning (DRL) and end-to-end learning models, for tackling the core planning and control logic of autonomous parking. This indicates a strategic shift away from purely geometric or rigidly rule-based systems, especially for navigating complex parking maneuvers in cluttered and dynamic real-world environments. The inherent ability of these AI paradigms to learn from data and manage high-dimensional state-action spaces makes them increasingly suitable for the nuanced demands of modern APS.

At the same time, the performance and reliability of advanced AI methodologies are intrinsically contingent upon concurrent developments in sensor suite architectures, as the availability of high-resolution, multimodal data streams serves as a foundational enabler for robust perception, context-aware reasoning, and real-time decision-making in autonomous systems. Also, Table 6 shows the growing adoption of Solid-State LiDAR (SSL) for its improved reliability and cost-effectiveness [64], 4D imaging radar for its enhanced environmental understanding (including velocity data) [62], and specialized camera systems (thermal, event-based, multi-spectral) designed for robust perception in challenging conditions [64]. The rich, high-density data provided by these advanced sensors facilitates more detailed and reliable state representations, which are crucial for training effective DRL agents and other data-driven perception models. This co-evolution of sensor technology and AI algorithms is a critical enabler for progress in APS.

5.4. Identified Gaps and Future Research Pointers from Comparative Analysis

In fact, Table 6 provides impressive performance metrics and innovative solutions, a deeper examination suggests areas where current approaches may have limitations and where future research is needed. Notably, although many state-of-the-art AI-based planning systems report high success rates and notable efficiency improvements, primary research publications often place less emphasis on formal safety verification and comprehensive adherence to SOTIF (Safety of the Intended Functionality) standards. Then, it may indicate a common research trajectory where achieving peak functional performance with complex AI is an initial priority, with the development and integration of rigorous safety assurance methodologies for these newer, often black-box, techniques still an evolving frontier. This potential gap highlights the need for ongoing research into verifiable and interpretable AI for safety-critical applications, such as autonomous parking. Ensuring robust performance across a truly exhaustive range of Out-of-Distribution (OOD) scenarios and developing universally accepted, comprehensive validation frameworks for highly adaptive learning-based APS remains a significant challenge.

6. System Aspects and Emerging Challenges

Beyond the core algorithms for perception, planning, and control, the successful deployment of APSs involves addressing complex system-level integration, rigorous and comprehensive testing and validation methodologies, and tackling emerging challenges related to cybersecurity, user trust, and operational complexities in diverse real-world settings.

6.1. System Integration, Simulation, and Validation

Integrating the various software components (perception, planning, control, HMI) with the vehicle's hardware platform (sensors, actuators, ECUs) is a complex engineering task. Research also explores hardware aspects, such as the utilization of powerful System-on-Chip (SoC) platforms that integrate Field-Programmable Gate Arrays (FPGAs) for parallel processing and ARM processor cores to handle demanding real-time computations [10]. Methodologies like Model-Based Systems Engineering (MBSE) and digital engineering workflows are being explored to manage the complexity of designing, developing, and validating these integrated autonomous systems systematically [4].

Simulation plays an indispensable role in the development lifecycle of APSs. Various simulation environments are employed, ranging from robotics simulators like Gazebo to specialized automotive simulators like CARLA and high-fidelity vehicle dynamics simulators like CarSim, often co-simulated with sensor simulators (e.g., for radar or LiDAR) and control design tools like MATLAB/Simulink [87,119]. Simulation enables rapid prototyping, cost-effective testing of algorithms under a wide range of reproducible conditions (including scenarios that cannot be safely tested in the real world), and systematic evaluation of system performance.

While simulation is crucial, validating APS performance in the real world is essential for ensuring safety and reliability [4]. A significant trend observed in recent literature is the increased emphasis on real-vehicle testing [12–14,80]. Researchers are deploying and evaluating their algorithms on actual vehicle platforms, often instrumented with research-grade sensors and drive-by-wire capabilities, in controlled test environments or real parking lots. Standardized test scenarios and performance evaluation frameworks are being developed to allow for more systematic and comparable assessment of different APS technologies [119].

The increasing emphasis on real-vehicle testing is crucial, yet the 'reality gap'—the often significant difference between performance in simulation and performance in the physical world—persists and poses a significant challenge. Critically analyze why this gap is so challenging to close. Are current simulation tools [7,87], despite their increasing sophistication, capable of faithfully reproducing the full spectrum of sensor noise (e.g., thermal noise, interference, multipath reflections for radar), complex material reflectivity variations (e.g., how different paints, wet surfaces, or dirty vehicles affect LiDAR returns), intricate lighting interactions (e.g., dappled sunlight through trees creating complex patterns, reflections from wet pavement), and the nuanced, often unpredictable, and sometimes irrational behavior of other road users (pedestrians, cyclists, other drivers) encountered in real parking lots? If not, what are the limits of what can be reliably validated in simulation versus what must be validated extensively and expensively in the real world? And how can the efficiency and coverage of real-world testing be maximized to address the combinatorial explosion of potential scenarios without incurring prohibitive costs or timelines?

While efforts towards standardized test scenarios (e.g., predefined parking maneuvers, obstacle configurations) are noted and valuable for benchmarking [119], how comprehensive can these be in covering the combinatorial explosion of real-world variations in environmental conditions, parking lot layouts, vehicle types, and dynamic actor behaviors? Is there a risk of 'teaching to the test,' where systems are optimized to perform well on these specific standardized scenarios but lack generalized robustness when faced with novel situations not included in the test suite? Over-reliance on a limited set of standardized tests might create a false sense of security regarding a system's overall real-world capabilities and safety.

6.2. Security Vulnerabilities in APS

As vehicles become more connected and automated, relying on complex software and communication networks, cybersecurity emerges as a critical concern [120]. APSs, especially those involving connectivity, such as AVP systems (which may communicate with infrastructure, remote operators, or user smartphones), are potential targets for malicious attacks. In-vehicle networks, such as the Controller Area Network (CAN) bus, are known to have vulnerabilities (e.g., lack of authentication or encryption) that could be exploited through attacks like message injection or denial-of-service, potentially compromising vehicle control by manipulating steering, braking, or acceleration [120]. Robust security measures, including Intrusion Detection Systems (IDSs)—potentially leveraging machine learning techniques to identify anomalous network traffic or sensor data—and secure software development practices are needed. Blockchain technology has also been proposed to enhance security and transparency in related areas, such as smart parking management systems that interact with APS, particularly for functions like automated fee calculation or access control [68].

So, in the case of fully autonomous valet parking (AVP) systems, which inherently depend on broader connectivity, including cloud services, parking infrastructure, remote human oversight, and user-facing mobile applications, the attack surface extends well beyond the confines of in-vehicle networks [120]. What are the unique, critical vulnerabilities introduced by these external communication links? Examples could include spoofing infrastructure signals (e.g., falsely indicating a parking spot is free or occupied), intercepting or replaying remote commands for vehicle movement, denial-of-service attacks on parking facility management systems or the AVP service itself (rendering vehicles unable to park or retrieve), or exploiting vulnerabilities in user authentication mechanisms to gain unauthorized access. What are the potential cascading failures if an AVP system's security is compromised—could it lead to widespread vehicle immobilization within a garage, unauthorized vehicle movement causing damage or theft, large-scale data breaches of user information, or even the creation of "botnets" of compromised vehicles? Are current automotive cybersecurity paradigms, often focused on securing the vehicle itself, adequately addressing these distributed, networked threats that span multiple entities and communication channels? The security focus must extend beyond the individual vehicle to encompass the entire AVP ecosystem, including backend servers, communication protocols, and infrastructure components.

6.3. User Interaction, Acceptance, and Trust

Technological capability alone does not guarantee the success of APS; user acceptance and trust are crucial factors influencing adoption and effective use [5,117,121]. Public perception can be negatively impacted by reports of AV accidents, even if unrelated to parking, and concerns about safety, reliability, and loss of control remain significant barriers [5]. Factors influencing acceptance include perceived ease of use, usefulness, safety, risk, and the perceived competence of the system. A specific concern identified is Human-Manipulated Risk Perception (HMRP)—the perceived risk associated with scenarios requiring human intervention or oversight in partially automated systems, which can negatively impact user attitudes and their sense of control [5].

Building user trust requires the careful design of the Human–Machine Interface (HMI) and the leveraging of Explainable AI (XAI) to make the system's behavior transparent and understandable [117,118]. Recent research has investigated scenario-based explanation frameworks for AVP systems. Studies show that providing clear, timely, and relevant explanations for the AV's decisions and actions (e.g., "waiting for pedestrian to cross," "adjusting path for narrow space") via the HMI can significantly improve driver trust, enhance the overall user experience (UX), reduce the mental workload associated with monitoring the system, and even improve objective performance by enabling users to anticipate system actions [117,118]. Importantly, the effectiveness of explanations can vary depending on the user's familiarity with the technology and their cognitive style. Customizing explanations—providing more detailed guidance and transparency for new or anxious users, while focusing on efficiency and critical safety information for frequent, confident users—has been found to be significantly more effective and satisfying than generic, one-size-fits-all explanations [117,118].

While XAI aims to enhance transparency and thereby foster trust, what are the limitations of current explanation methods, especially when attempting to explain the complex, often emergent behavior of deep learning systems or the nuanced decisions resulting from multi-objective optimizers in path planning? Can an explanation that is itself an approximation, a simplification, or a post hoc rationalization of an opaque process truly foster deep, resilient trust, particularly if the system still makes occasional, inexplicable errors that contradict the provided explanations? Superficial explanations might even erode trust if they are perceived as unhelpful or misleading. Beyond HMI design and XAI, what are the deeper socio-technical factors (e.g., perceived loss of control and agency, ethical concerns about algorithmic decision-making in potential conflict scenarios even in low-speed parking, negative media narratives shaping public opinion, lack of clear liability frameworks in case of malfunction, or anxieties about data privacy) that create persistent barriers to widespread acceptance [5]? How can these be addressed systemically—through public education, transparent development practices, robust regulatory frameworks, and ethical guidelines—rather than just at the interface level? True trust requires more than just a well-designed dashboard explanation; it involves a broader societal adaptation, clear accountability, and consistent, reliable system performance over extended periods.

6.4. Operational Challenges

Several operational challenges continue to impede the development and deployment of fully robust and universally applicable APSs. Operating reliably in diverse and unstructured environments, including those with poor or non-existent markings, unusual geometries (e.g., angled parking on steep slopes), or during adverse weather conditions that degrade sensor performance, remains profoundly difficult. A significant contributing factor, as previously highlighted, is the lack of large-scale, diverse, and publicly available datasets for training and benchmarking perception and planning algorithms [7,72]. This is particularly acute for end-to-end learning approaches, which are exceptionally data-hungry, and for rigorous robustness testing under a wide array of adverse conditions. Additionally, coordinating multiple autonomous vehicles within shared parking facilities, such as in multi-UGV (Unmanned Ground Vehicle) or AVP scenarios in large garages, presents significant challenges in path planning, real-time conflict resolution, efficient resource allocation (e.g., assigning parking spots or charging stations), and communication protocols [79].

7. Future Directions

Based on this critical review, future research should intensify its focus on the following key directions:

- End-to-End Learning with Verifiability and Safety Guarantees. A key emerging research direction is the development of end-to-end learning approaches that map sensor inputs directly to control actions. While still a nascent area for production systems, this paradigm holds significant potential for simplifying the traditional, modular APS pipeline and discovering novel, holistic solutions [7,72]. However, this approach intensifies challenges related to data dependency, interpretability, and safety verification. Future work must integrate mechanisms for robust interpretability, comprehensive uncertainty quantification, and, where possible, formal verification or runtime monitoring with safety fallbacks into these architectures.
- Robust Multi-Modal Perception and Advanced Fusion. Research into sensor fusion techniques that can gracefully handle sensor degradation or complete failure of one or more modalities, resolve conflicting information with high reliability using

uncertainty-aware methods, and adapt dynamically to extreme environmental conditions is crucial [6–10,17,62,64,65]. This includes developing better domain adaptation methods and creating self-assessment capabilities within perception systems to flag low-confidence situations.

- Safety Verification and Validation (V&V) for AI-based Systems. Developing more rigorous, scalable, and widely accepted V&V methodologies specifically tailored for AI-driven systems is paramount. This includes advancing formal methods applicable to neural networks, investing in large-scale realistic simulation platforms with a strong focus on automated edge case generation, and establishing standardized safety metrics and benchmarks aligned with automotive safety standards like ISO 26,262 and SOTIF (ISO 21448) [4,119,122–124].
- Human-Centric APS Design and Trust Calibration. Deeper investigation into XAI techniques that provide causal, contrastive, and actionable explanations is needed [117,118]. Research should focus on adaptive HMIs that manage user expectations and reduce cognitive load. Longitudinal studies on trust dynamics—how trust is built, lost, and potentially repaired over time—are needed to ensure users neither dangerously over-trust an imperfect system nor under-utilize its capabilities [5].
- Cooperative and Multi-Agent Systems with Scalable Coordination. As vehicle connectivity (Vehicle-to-Everything (V2X)) increases, research into decentralized, robust, and scalable algorithms for coordinating multiple autonomous vehicles in shared parking environments (e.g., efficient allocation of spots, collision-free maneuvering, negotiation for shared resources) will become increasingly important [79,120,125].
- This includes not only vehicle-to-vehicle (V2V) coordination but also vehicle-to-infrastructure (V2I) cooperation, which presents significant opportunities. For instance, future work could explore smart parking lots that communicate directly with vehicles to guide them to available spots, or cloud services that offload some of the complex computational logic for path planning, reducing the burden on the vehicle's onboard systems.
- Proactive Cybersecurity for Connected APS Architectures. Dedicated research is urgently needed to identify specific vulnerabilities in distributed APS architectures—spanning sensors, ECUs, V2X links, and backend cloud infrastructure—and to develop tailored, adaptive, and resilient intrusion detection and prevention mechanisms [68,120,125].
- Addressing the Data Bottleneck Systematically and Collaboratively. Concerted efforts towards creating large-scale, high-quality, diverse, and well-annotated public datasets are essential [7,65,71,72]. This may involve exploring novel data collection strategies, advancing synthetic data generation techniques, and investigating federated learning approaches.

Addressing these research directions with a critical, holistic, and collaborative perspective will be key to overcoming the remaining hurdles and realizing the full societal and economic benefits of Autonomous Parking Systems in future mobility solutions, ultimately leading to safer, more efficient, and less stressful parking experiences

8. Conclusions

This review has charted the significant evolution of Autonomous Parking Systems (APS), highlighting a clear trend towards AI-centric solutions. Deep learning now forms the foundation of perception, while hybrid algorithms that pair search-based methods with optimization techniques dominate path planning. In vehicle control, there is a notable shift toward learning-based strategies like reinforcement learning to enhance adaptability. The field is also maturing beyond pure algorithms, with an increasing emphasis on real-world vehicle testing and a growing recognition of the criticality of user trust, as evidenced by research into explainable AI and human–machine interfaces.

Despite this progress, formidable challenges persist. The primary obstacle remains achieving robust system performance across the full, unpredictable spectrum of real-world conditions, from adverse weather to poorly maintained infrastructure and erratic human behavior. Guaranteeing the safety and security of opaque AI components, closing the persistent gap between simulation and reality, and overcoming public skepticism are critical hurdles. The scarcity of comprehensive public datasets continues to be a fundamental bottleneck, hindering the development of truly generalizable AI. Ultimately, while the technological building blocks of APSs have advanced considerably, the path to deploying fully reliable, safe, and trusted systems requires a continued, concerted effort to solve these deep-seated challenges.

Author Contributions: Conceptualization, J.S.O.M., J.G.M.L., A.R. and H.G.A.; methodology, J.G.M.L. and J.S.O.M.; formal analysis, J.S.O.M. and H.G.A.; investigation, J.S.O.M.; resources, J.G.M.L.; writing—original draft preparation, J.S.O.M.; writing—review and editing, J.G.M.L. and A.R.; All authors have read and agreed to the published version of the manuscript.

Funding: This work has been supported by the Research Office of the Universidad Autónoma de Bucaramanga, through a research grant that supports bachelor's students in Mechatronics Engineering in automation and digital transformation, and, research investment plan for researchers professors. Additionally, the research work of Anton Rassõlkin has been supported by the Estonian Research Council under grant PRG2532 "Advanced Digital Tools to Accelerate the Development of Software-Defined Electric Vehicles".

Conflicts of Interest: The authors declare no conflicts of interest.

Abbreviations

The following abbreviations are used in this manuscript:

AAPS	Advanced Automated Parking System
ADAS	Advanced Driver Assistance Systems
AI	Artificial Intelligence
ANGS	Number of Gear Shifts
APA	Autonomous Parking Assist
APF	Artificial Potential Field
APS	Autonomous Parking Systems
ATE	Absolute Trajectory Error
AV	Autonomous Vehicle
AVP	Autonomous Valet Parking
BEV	Bird's Eye View
Bi-RRT*	Bidirectional Rapidly exploring Random Tree*
CAN	Controller Area Network
CAV	Connected Automated Vehicles
CFAR	Constant False Alarm Rate
CILQR	Constrained Iterative Linear Quadratic Regulator
CMDP	Constraint-driven Markov Decision Process
CMOS	Complementary Metal-Oxide Semiconductor
CNN	Convolutional Neural Networks
CP	Control Points
CV	Computer Vision
DAVIS	Dynamic and Active Pixel Vision Sensor
DL	Deep Learning

DRL	Deep Reinforcement Learning
DVS	Dynamic Vision Sensor
ECU	Electronic Control Unit
EKF	Extended Kalman Filter
EMPC	Economic Model Predictive Control
E2E	End-to-End
FAST-LIO	Fast LiDAR-Inertial Odometry
FoV	Field-of-View
FPGA	Field-Programmable Gate Array
FPS	Frames Per Second
GA	Genetic Algorithm
GF-LIO	Global Factor-based LiDAR-Inertial Odometry
GP	Gaussian Process
GPS	Global Positioning System
GPU	Graphics Processing Unit
HDR	High Dynamic Range
HER	Hindsight Experience Replay
HMI	Human–Machine Interface
HMRP	Human-Manipulated Risk Perception
ICBS	Improved Conflict-Based Search
IDS	Intrusion Detection Systems
IMU	Inertial Measurement Unit
IR	Infrared
I-STC	Improved Safe Travel Corridor
LIO-SAM	LiDAR-Inertial Odometry via Smoothing and Mapping
LKF	Linear Kalman Filter
LPV	Linear Parameter Varying
LSS	Lift Scene Splatting
LWIR	Long-Wave Infrared
mAP	mean Average Precision
MBSE	Model-Based Systems Engineering
MEMS	Micro-Electro-Mechanical Systems
ML	Machine Learning
MORL	Multi-Objective Reinforcement Learning
MPC	Model Predictive Control
MPS	Mobis Parking System
MSCKF	Multi-State Constraint Kalman Filter
MTBF	Mean Time Between Failures
NIR	Near-Infrared
NMPC	Nonlinear Model Predictive Control
NN	Neural Network
OCP	Optimal Control Problem
ODD	Operational Design Domain
OGM	Occupancy Grid Map
OPA	Optical Phased Array
PDM	Probabilistic Diffusion Model
PID	Proportional Integral Derivative
PINS	Pontryagin's Indirect Method Solver
PL	Path Lenght
PNN	Pseudo-Neural Network
PSR	Parking Success Rate
QP	Quadratic Programming
ReLU	Rectified Linear Unit
RGB	Red

RL	Reinforcement Learning
RPROP	Resilient Propagation
RS	Reeds-Shepp
RRT	Rapidly exploring Random Tree
RRT*	Rapidly exploring Random Tree Star
RRT-Connect	Rapidly exploring Random Tree Connect
SAC	Soft Actor-Critic
SLAM	Simultaneous Localization and Mapping
SMC	Satisfiability Modulo Convex
SNR	Signal-to-Noise Ratio
SoC	System-on-Chip
SOTA	State-Of-The-Art
SOTIF	Safety Of The Intended Functionality
SQP	Sequential Quadratic Programming
SSL	Solid-State LiDAR
SWIR	Short-Wave Infrared
T-ITS	IEEE Transactions on Intelligent Transportation Systems
T-IV	IEEE Transactions on Intelligent Vehicles
UGV	Unmanned Ground Vehicle
UKF	Unscented Kalman Filter
UX	User Experience
V&V	Verification and Validation
V2X	Vehicle-to-Everything
V2I	Vehicle-to-infrastructure
VIO	Visual-Inertial Odometry
XAI	Explainable AI
YOLO	You Only Look Once

References

1. Wang, W.; Song, Y.; Zhang, J.; Deng, H. Automatic parking of vehicles: A review of literatures. *Int. J Automot. Technol.* **2014**, *15*, 967–978. [CrossRef]
2. Hossain, M.; Rahim, M.; Rahman, M.; Ramasamy, D. Artificial Intelligence Revolutionising the Automotive Sector: A Comprehensive Review of Current Insights, Challenges, and Future Scope. *Comput. Mater. Contin.* **2025**, *82*, 3643–3692. [CrossRef]
3. Yu, B.; Lin, L.; Chen, J. (Eds.) *Recent Advance in Intelligent Vehicle*; MDPI-Multidisciplinary Digital Publishing Institute: Basel, Switzerland, 2024; ISBN 978-3-7258-1469-5.
4. Samak, T.V.; Samak, C.V.; Brault, J.; Harber, C.; McCane, K.; Smereka, J.; Brudnak, M.; Gorsich, D.; Krovi, V. A Systematic Digital Engineering Approach to Verification & Validation of Autonomous Ground Vehicles in Off-Road Environments. *arXiv* **2025**. [CrossRef]
5. Deng, M.; Guo, Y.; Guo, Y.; Wang, C. The Role of Technical Safety Riskiness and Behavioral Interventions in the Public Acceptance of Autonomous Vehicles in China. *J. Transp. Eng. Part A Syst.* **2023**, *149*, 04022122. [CrossRef]
6. Feng, Z. Application and Development of Radar Sensors in Autonomous Driving Technology. *Appl. Comput. Eng.* **2025**, *140*, 48–52. [CrossRef]
7. Gao, K.; Zhou, L.; Liu, M.; Knoll, A. E2E Parking Dataset: An Open Benchmark for End-to-End Autonomous Parking. *arXiv* **2025**. [CrossRef]
8. Chen, J.; Li, F.; Liu, X.; Yuan, Y. Robust Parking Space Recognition Approach Based on Tightly Coupled Polarized Lidar and Pre-Integration IMU. *Appl. Sci.* **2024**, *14*, 9181. [CrossRef]
9. Wang, X.; Miao, H.; Liang, J.; Li, K.; Tan, J.; Luo, R.; Jiang, Y. Multi-Dimensional Research and Progress in Parking Space Detection Techniques. *Electronics* **2025**, *14*, 748. [CrossRef]
10. Chinnaiah, M.C.; Vani, G.D.; Karumuri, S.R.; Srikanthan, T.; Lam, S.-K.; Narambhatla, J.; Krishna, D.H.; Dubey, S. Geometry-Based Parking Assistance Using Sensor Fusion for Robots With Hardware Schemes. *IEEE Sens. J.* **2024**, *24*, 8821–8834. [CrossRef]
11. Jiang, J.; Tang, R.; Kang, W.; Xu, Z.; Qian, C. Two-Stage Efficient Parking Space Detection Method Based on Deep Learning and Computer Vision. *Appl. Sci.* **2025**, *15*, 1004. [CrossRef]
12. Pagot, E.; Piccinini, M.; Bertolazzi, E.; Biral, F. Fast Planning and Tracking of Complex Autonomous Parking Maneuvers With Optimal Control and Pseudo-Neural Networks. *IEEE Access* **2023**, *11*, 124163–124180. [CrossRef]

13. Zhang, P.; Zhou, S.; Hu, J.; Zhao, W.; Zheng, J.; Zhang, Z.; Gao, C. Automatic parking trajectory planning in narrow spaces based on Hybrid A* and NMPC. *Sci. Rep.* **2025**, *15*, 1384. [CrossRef]
14. Li, Y.; Li, G.; Wang, X. Research on Trajectory Planning of Autonomous Vehicles in Constrained Spaces. *Sensors* **2024**, *24*, 5746. [CrossRef] [PubMed]
15. Tang, X.; Yang, Y.; Liu, T.; Lin, X.; Yang, K.; Li, S. Path Planning and Tracking Control for Parking via Soft Actor-Critic Under Non-Ideal Scenarios. *IEEE/CAA J. Autom. Sin.* **2024**, *11*, 181–195. [CrossRef]
16. Ahn, S.; Oh, T.; Yoo, J. Collision Avoidance Path Planning for Automated Vehicles Using Prediction Information and Artificial Potential Field. *Sensors* **2024**, *24*, 7292. [CrossRef]
17. Nahata, D.; Othman, K.; Nahata, D.; Othman, K. Exploring the challenges and opportunities of image processing and sensor fusion in autonomous vehicles: A comprehensive review. *AIMS Electron. Electr. Eng.* **2023**, *7*, 271–321. [CrossRef]
18. Hasanujjaman, M.; Chowdhury, M.Z.; Jang, Y.M. Sensor Fusion in Autonomous Vehicle with Traffic Surveillance Camera System: Detection, Localization, and AI Networking. *Sensors* **2023**, *23*, 3335. [CrossRef] [PubMed]
19. Automated Parking | Develop APS Efficiently. Available online: https://www.appliedintuition.com/use-cases/automated-parking (accessed on 19 May 2025).
20. Mehta, M.D. Sensor Fusion Techniques in Autonomous Systems: A Review of Methods and Applications. *Int. Res. J. Eng.* **2025**, *12*, 1902–1908.
21. Beránek, F.; Diviš, V.; Gruber, I. Soiling Detection for Advanced Driver Assistance Systems. In Proceedings of the Seventeenth International Conference on Machine Vision (ICMV 2024), Edinburgh, UK, 10–13 October 2024; SPIE: Bellingham, WA, USA, 2024; Volume 13517, pp. 174–182.
22. Heimberger, M.; Horgan, J.; Hughes, C.; McDonald, J.; Yogamani, S. Computer vision in automated parking systems: Design, implementation and challenges. *Image Vis. Comput.* **2017**, *68*, 88–101. [CrossRef]
23. Kumar, S.; Truong, H.; Sharma, S.; Sistu, G.; Scanlan, T.; Grua, E.; Eising, C. Minimizing Occlusion Effect on Multi-View Camera Perception in BEV with Multi-Sensor Fusion. *arXiv* **2025**. [CrossRef]
24. The Rise of Ultrasonic Sensor Technology in Park Assist Systems Market to Grow to USD 3,277.4 Million, with a CAGR of 14.4% | Future Market Insights, Inc. Morningstar, Inc. 2025. Available online: https://www.accessnewswire.com/newsroom/en/automotive/the-rise-of-ultrasonic-sensor-technology-in-park-assist-systems-market-to-grow-to-usd-3-2-987599 (accessed on 19 May 2025).
25. Automotive Applications. Available online: https://www.monolithicpower.com/en/learning/mpscholar/sensors/real-world-applications/automotive-applications (accessed on 19 May 2025).
26. Lim, B.S.; Keoh, S.L.; Thing, V.L.L. Autonomous vehicle ultrasonic sensor vulnerability and impact assessment. In Proceedings of the 2018 IEEE 4th World Forum on Internet of Things (WF-IoT), Singapore, 5–8 February 2018; pp. 231–236.
27. Acharya, R. Sensor blockage in autonomous vehicles: AI-driven detection and mitigation strategies. *World J. Adv. Eng. Technol. Sci.* **2025**, *15*, 321–331. [CrossRef]
28. Yeong, D.J.; Panduru, K.; Walsh, J. Exploring the Unseen: A Survey of Multi-Sensor Fusion and the Role of Explainable AI (XAI) in Autonomous Vehicles. *Sensors* **2025**, *25*, 856. [CrossRef] [PubMed]
29. Lopac, N.; Jurdana, I.; Brnelić, A.; Krljan, T. Application of Laser Systems for Detection and Ranging in the Modern Road Transportation and Maritime Sector. *Sensors* **2022**, *22*, 5946. [CrossRef]
30. Mak, H. Explore the Different Types of Lidar Technology: From Mechanical to Solid-State Lidar. Available online: https://globalgpssystems.com/lidar/explore-the-different-types-of-lidar-technology-from-mechanical-to-solid-state-lidar/ (accessed on 19 May 2025).
31. Solid State Lidar. Neuvition | Solid-State Lidar, Lidar Sensor Suppliers, Lidar Technology, Lidar Sensor. Available online: https://www.neuvition.com/solid-state-lidar-neuvition (accessed on 19 May 2025).
32. Wei, W.; Shirinzadeh, B.; Nowell, R.; Ghafarian, M.; Ammar, M.M.A.; Shen, T. Enhancing Solid State LiDAR Mapping with a 2D Spinning LiDAR in Urban Scenario SLAM on Ground Vehicles. *Sensors* **2021**, *21*, 1773. [CrossRef] [PubMed]
33. Royo, S.; Ballesta-Garcia, M. An Overview of Lidar Imaging Systems for Autonomous Vehicles. *Appl. Sci.* **2019**, *9*, 4093. [CrossRef]
34. Emilio, M.D.P. The Future of LiDAR Lies in ADAS. EE Times. 2023. Available online: https://www.eetimes.com/the-future-of-lidar-lies-in-adas/ (accessed on 19 May 2025).
35. Han, Z.; Wang, J.; Xu, Z.; Yang, S.; He, L.; Xu, S.; Wang, J.; Li, K. 4D Millimeter-Wave Radar in Autonomous Driving: A Survey. *arXiv* **2024**. [CrossRef]
36. Lee, J.-E. 4D Radar Reimagining Next-Gen Mobility. Geospatial World. 2023. Available online: https://geospatialworld.net/prime/business-and-industry-trends/4d-radar-reimagining-next-gen-mobility/ (accessed on 19 May 2025).
37. Ding, F.; Wen, X.; Zhu, Y.; Li, Y.; Lu, C.X. RadarOcc: Robust 3D Occupancy Prediction with 4D Imaging Radar. *arXiv* **2024**. [CrossRef]
38. Gallego, G.; Delbruck, T.; Orchard, G.; Bartolozzi, C.; Taba, B.; Censi, A.; Leutenegger, S.; Davison, A.J.; Conradt, J.; Daniilidis, K.; et al. Event-Based Vision: A Survey. *IEEE Trans. Pattern Anal. Mach. Intell.* **2022**, *44*, 154–180. [CrossRef]

39. Huang, J.-T. Indoor Localization and Mapping with 4D mmWave Imaging Radar. Master's Thesis, Carnegie Mellon University, Pittsburgh, PA, USA, 2024. Available online: https://www.ri.cmu.edu/publications/indoor-localization-and-mapping-with-4d-mmwave-imaging-radar/ (accessed on 19 May 2025).
40. 4D Single-Chip Waveguide Radar & Multi-Chip Cascade 4D Imaging Radar Report Analysis-Electronics Headlines-EEWORLD. Available online: https://en.eeworld.com.cn/mp/ICVIS/a387462.jspx (accessed on 19 May 2025).
41. Wang, H.; Guo, R.; Ma, P.; Ruan, C.; Luo, X.; Ding, W.; Zhong, T.; Xu, J.; Liu, Y.; Chen, X. Towards Mobile Sensing with Event Cameras on High-agility Resource-constrained Devices: A Survey. *arXiv* **2025**. [CrossRef]
42. Chen, Y.; Gilitshenski, M.I.; Amini, A. *Real World Application of Event-Based End to End Autonomous Driving*; MIT Department of Mathematics: Cambridge, MA, USA, 2020.
43. Vollmer, M.; Möllmann, K. *Infrared Thermal Imaging: Fundamentals, Research and Applications*, 1st ed.; Wiley: Hoboken, NJ, USA, 2017; ISBN 978-3-527-41351-5.
44. Kebbati, Y.; Puig, V.; Ait-Oufroukh, N.; Vigneron, V.; Ichalal, D. Optimized adaptive MPC for lateral control of autonomous vehicles. In Proceedings of the 2021 9th International Conference on Control, Mechatronics and Automation (ICCMA), Belval, Luxembourg, 11 November 2021; IEEE: Belval, Luxembourg, 2021; pp. 95–103.
45. Mirlach, J.; Wan, L.; Wiedholz, A.; Keen, H.E.; Eich, A. R-LiViT: A LiDAR-Visual-Thermal Dataset Enabling Vulnerable Road User Focused Roadside Perception. *arXiv* **2025**. [CrossRef]
46. Ni, X.; Kuehnel, C.; Jiang, X. Thermal Detection of People with Mobility Restrictions for Barrier Reduction at Traffic Lights Controlled Intersections. *arXiv* **2025**. [CrossRef]
47. Multispectral Drones & Cameras. Available online: https://advexure.com/collections/multispectral-drones-sensors (accessed on 19 May 2025).
48. Snapshot Multispectral Cameras. Spectral Devices. Available online: https://spectraldevices.com/collections/snapshot-multispectral-cameras (accessed on 19 May 2025).
49. Sun, J.; Yin, M.; Wang, Z.; Xie, T.; Bei, S. Multispectral Object Detection Based on Multilevel Feature Fusion and Dual Feature Modulation. *Electronics* **2024**, *13*, 443. [CrossRef]
50. Zhang, Y.; Shi, P.; Li, J. LiDAR-Based Place Recognition For Autonomous Driving: A Survey. *ACM Comput. Surv.* **2025**, *57*, 1–36. [CrossRef]
51. Kuang, Y.; Hu, T.; Ouyang, M.; Yang, Y.; Zhang, X. Tightly Coupled LIDAR/IMU/UWB Fusion via Resilient Factor Graph for Quadruped Robot Positioning. *Remote Sens.* **2024**, *16*, 4171. [CrossRef]
52. Ibrahim, Q.; Ali, Z. A Comprehensive Review of Autonomous Vehicle Architecture, Sensor Integration, and Communication Networks: Challenges and Performance Evaluation. *Preprint* **2025**. [CrossRef]
53. Yang, L.; Tao, Y.; Li, M.; Zhou, J.; Jiao, K.; Li, Z. Lidar-Inertial SLAM Method Integrated with Visual QR Codes for Indoor Mobile Robots. *Res. Sq.* **2025**. [CrossRef]
54. Huang, C.; Wang, Y.; Sun, X.; Yang, S. Research on Digital Terrain Construction Based on IMU and LiDAR Fusion Perception. *Sensors* **2024**, *25*, 15. [CrossRef]
55. Sahoo, L.K.; Varadarajan, V. Deep learning for autonomous driving systems: Technological innovations, strategic implementations, and business implications—A comprehensive review. *Complex Eng. Syst.* **2025**, *5*, 83. [CrossRef]
56. Liu, L.; Lee, J.; Shin, K.G. RT-BEV: Enhancing Real-Time BEV Perception for Autonomous Vehicles. In Proceedings of the 2024 IEEE Real-Time Systems Symposium (RTSS), York, UK, 10 December 2024; IEEE: New York, NY, USA, 2024; pp. 267–279.
57. Shi, K.; He, S.; Shi, Z.; Chen, A.; Xiong, Z.; Chen, J.; Luo, J. Radar and Camera Fusion for Object Detection and Tracking: A Comprehensive Survey. *arXiv* **2024**. [CrossRef]
58. Yang, B.; Li, J.; Zeng, T. A Review of Environmental Perception Technology Based on Multi-Sensor Information Fusion in Autonomous Driving. *World Electr. Veh. J.* **2025**, *16*, 20. [CrossRef]
59. Paz, D.; Zhang, H.; Li, Q.; Xiang, H.; Christensen, H. Probabilistic Semantic Mapping for Urban Autonomous Driving Applications. *arXiv* **2020**. [CrossRef]
60. Wang, S.; Ahmad, N.S. A Comprehensive Review on Sensor Fusion Techniques for Localization of a Dynamic Target in GPS-Denied Environments. *IEEE Access* **2025**, *13*, 2252–2285. [CrossRef]
61. Zaim, H.Ç.; Yolaçan, E.N. Taxonomy of sensor fusion techniques for various application areas: A review. *NOHU J. Eng. Sci.* **2025**, *14*, 392–411. [CrossRef]
62. Jiang, M.; Xu, G.; Pei, H.; Feng, Z.; Ma, S.; Zhang, H.; Hong, W. 4D High-Resolution Imagery of Point Clouds for Automotive mmWave Radar. *IEEE Trans. Intell. Transport. Syst.* **2024**, *25*, 998–1012. [CrossRef]
63. Edwards, A.; Giacobbe, M.; Abate, A. On the Trade-off Between Efficiency and Precision of Neural Abstraction. *arXiv* **2023**. [CrossRef]
64. Marti, E.; de Miguel, M.A.; Garcia, F.; Perez, J. A Review of Sensor Technologies for Perception in Automated Driving. *IEEE Intell. Transp. Syst. Mag.* **2019**, *11*, 94–108. [CrossRef]

65. Fan, Z.; Zhang, L.; Wang, X.; Shen, Y.; Deng, F. LiDAR, IMU, and camera fusion for simultaneous localization and mapping: A systematic review. *Artif. Intell. Rev.* **2025**, *58*, 174. [CrossRef]
66. Ganesh, N.; Shankar, R.; Mahdal, M.; Murugan, J.S.; Chohan, J.S.; Kalita, K. Exploring Deep Learning Methods for Computer Vision Applications across Multiple Sectors: Challenges and Future Trends. *Comput. Model. Eng. Sci.* **2024**, *139*, 103–141. [CrossRef]
67. Yan, S.; O'Connor, N.E.; Liu, M. U-Park: A User-Centric Smart Parking Recommendation System for Electric Shared Micromobility Services. *IEEE Trans. Artif. Intell.* **2024**, *5*, 5179–5193. [CrossRef]
68. Kalbhor, A.; Nair, R.S.; Phansalkar, S.; Sonkamble, R.; Sharma, A.; Mohan, H.; Wong, C.H.; Lim, W.H. PARKTag: An AI–Blockchain Integrated Solution for an Efficient, Trusted, and Scalable Parking Management System. *Technologies* **2024**, *12*, 155. [CrossRef]
69. Ajeenkya Dy Patil School of Engineering, P. International Conference on International Conference on Multidisciplinary Research in Engineering & Technology (ICMRET- 2025). *Int. J. Multidiscip. Res. Sci. Eng. Technol.* **2025**, *3*, 1–375.
70. ISO 26262-1:2018. Available online: https://www.iso.org/standard/68383.html (accessed on 6 July 2025).
71. Zhang, R.; Ma, Y.; Li, T.; Lin, Z.; Wu, Y.; Chen, J.; Zhang, L.; Hu, J.; Qiu, T.Z.; Guo, K. A Robust Real-Time Lane Detection Method with Fog-Enhanced Feature Fusion for Foggy Conditions. *arXiv* **2025**. [CrossRef]
72. Wang, Y.; Xing, S.; Can, C.; Li, R.; Hua, H.; Tian, K.; Mo, Z.; Gao, X.; Wu, K.; Zhou, S.; et al. Generative AI for Autonomous Driving: Frontiers and Opportunities. *arXiv* **2025**. [CrossRef]
73. Xiao, R.; Zhong, C.; Zeng, W.; Cheng, M.; Wang, C. Novel Convolutions for Semantic Segmentation of Remote Sensing Images. *IEEE Trans. Geosci. Remote Sens.* **2023**, *61*, 1–13. [CrossRef]
74. Han, Z.; Sun, H.; Huang, J.; Xu, J.; Tang, Y.; Liu, X. Path Planning Algorithms for Smart Parking: Review and Prospects. *World Electr. Veh. J.* **2024**, *15*, 322. [CrossRef]
75. Hsu, T.-H.; Liu, J.-F.; Yu, P.-N.; Lee, W.-S.; Hsu, J.-S. Development of an automatic parking system for vehicle. In Proceedings of the 2008 IEEE Vehicle Power and Propulsion Conference, Harbin, China, 3–5 September 2008; pp. 1–6.
76. Cui, G.; Yin, Y.; Xu, Q.; Song, C.; Li, G.; Li, S. Efficient Path Planning for Automated Valet Parking: Integrating Hybrid A* Search with Geometric Curves. *Int.J Automot. Technol.* **2025**, *26*, 243–253. [CrossRef]
77. Tao, F.; Ding, Z.; Fu, Z.; Li, M.; Ji, B. Efficient path planning for autonomous vehicles based on RRT* with variable probability strategy and artificial potential field approach. *Sci. Rep.* **2024**, *14*, 24698. [CrossRef]
78. Li, J.; Huang, C.; Pan, M. Path-planning algorithms for self-driving vehicles based on improved RRT-Connect. *Transp. Saf. Environ.* **2023**, *5*, tdac061. [CrossRef]
79. Zeng, D.; Chen, H.; Yu, Y.; Hu, Y.; Deng, Z.; Leng, B.; Xiong, L.; Sun, Z. UGV Parking Planning Based on Swarm Optimization and Improved CBS in High-Density Scenarios for Innovative Urban Mobility. *Drones* **2023**, *7*, 295. [CrossRef]
80. Hu, Z.; Chen, X.; Yang, Z.; Yu, M.G.; Qin, H.; Gao, M. Path Planning with Multiple Obstacle-Avoidance Modes for Intelligent Vehicles. *Automob. Eng.* **2025**, *47*, 402. [CrossRef]
81. Jin, X.; Tao, Y.; Opinat Ikiela, N.V. Trajectory Planning Design for Parallel Parking of Autonomous Ground Vehicles with Improved Safe Travel Corridor. *Symmetry* **2024**, *16*, 1129. [CrossRef]
82. Wang, Z.; Chen, Z.; Jiang, M.; Qin, T.; Yang, M. RL-OGM-Parking: Lidar OGM-Based Hybrid Reinforcement Learning Planner for Autonomous Parking. *arXiv* **2025**. [CrossRef]
83. Aryan Rezaie, A. Development of Path Tracking Control Strategies for Autonomous Vehicles and Validation Using a High-Fidelity Driving Simulator. Master's Thesis, Politecnico di Torino, Torino, Italia, 2025.
84. Sutton, R.S.; Barto, A. Reinforcement learning: An introduction. In *Adaptive Computation and Machine Learning*, 2nd ed.; The MIT Press: Cambridge, MA, USA; London, UK, 2020; ISBN 978-0-262-03924-6.
85. Zhang, Z.; Luo, Y.; Chen, Y.; Zhao, H.; Ma, Z.; Liu, H. Automated Parking Trajectory Generation Using Deep Reinforcement Learning. *arXiv* **2025**. [CrossRef]
86. Zhang, C.; Zhou, R.; Lei, L.; Yang, X. Research on Automatic Parking System Strategy. *World Electr. Veh. J.* **2021**, *12*, 200. [CrossRef]
87. Wang, J.; Li, Q.; Ma, Q. Research on Active Avoidance Control of Intelligent Vehicles Based on Layered Control Method. *World Electr. Veh. J.* **2025**, *16*, 211. [CrossRef]
88. Nuhel, A.K.; Al Amin, M.; Paul, D.; Bhatia, D.; Paul, R.; Sazid, M.M. Model Predictive Control (MPC) and Proportional Integral Derivative Control (PID) for Autonomous Lane Keeping Maneuvers: A Comparative Study of Their Efficacy and Stability. In *Cognitive Computing and Cyber Physical Systems*; Pareek, P., Gupta, N., Reis, M.J.C.S., Eds.; Lecture Notes of the Institute for Computer Sciences, Social Informatics and Telecommunications Engineering; Springer Nature: Cham, Switzerland, 2024; Volume 537, pp. 107–121. ISBN 978-3-031-48890-0.
89. Li, Y.; Huang, C.; Yang, D.; Liu, W.; Li, J. Learning Based MPC for Autonomous Driving Using a Low Dimensional Residual Model. *arXiv* **2024**. [CrossRef]
90. Zhao, Y. Automatic parking planning control method based on improved A* algorithm. *arXiv* **2024**. [CrossRef]
91. Yu, L.; Wang, X.; Hou, Z. *Path Tracking for Driverless Vehicle Under Parallel Parking Based on Model Predictive Control*; SAE International: Warrendale, PA, USA, 2021.

92. Batkovic, I. *Enabling Safe Autonomous Driving in Uncertain Environments: Based on a Model Predivtive Control Approach*; Chalmers University of Technology: Göteborg, Sweden, 2022; ISBN 978-91-7905-623-0.
93. Liang, K.; Yang, G.; Cai, M.; Vasile, C.-I. Safe Navigation in Dynamic Environments Using Data-Driven Koopman Operators and Conformal Prediction. *arXiv* **2025**. [CrossRef]
94. Ammaturo, P. Energy-Efficient Adaptive Cruise Control: An Economic MPC Framework Based on Constant Time Gap. Master's Thesis, Politecnico di Torino: Torino TO, Italia, 2025.
95. Kiran, B.R.; Sobh, I.; Talpaert, V.; Mannion, P.; Sallab, A.A.A.; Yogamani, S.; Pérez, P. Deep Reinforcement Learning for Autonomous Driving: A Survey. *arXiv* **2021**. [CrossRef]
96. Zhang, P.; Xiong, L.; Yu, Z.; Fang, P.; Yan, S.; Yao, J.; Zhou, Y. Reinforcement Learning-Based End-to-End Parking for Automatic Parking System. *Sensors* **2019**, *19*, 3996. [CrossRef]
97. Kim, T.; Kang, T.; Son, S.; Ko, K.W.; Har, D. Goal-Conditioned Reinforcement Learning Approach for Autonomous Parking in Complex Environments. In Proceedings of the 2025 International Conference on Artificial Intelligence in Information and Communication (ICAIIC), Fukuoka, Japan, 18 February 2025; pp. 465–470.
98. pg3328 Autonomous-Parking-Using-Reinforcement-Learning. Available online: https://github.com/pg3328/Autonomous-Parking-Using-Reinforcement-Learning (accessed on 19 May 2025).
99. Gao, F.; Wang, X.; Fan, Y.; Gao, Z.; Zhao, R. Constraints Driven Safe Reinforcement Learning for Autonomous Driving Decision-Making. *IEEE Access* **2024**, *12*, 128007–128023. [CrossRef]
100. Li, Z.; Jin, G.; Yu, R.; Chen, Z.; Li, N.; Han, W.; Xiong, L.; Leng, B.; Hu, J.; Kolmanovsky, I.; et al. A Survey of Reinforcement Learning-Based Motion Planning for Autonomous Driving: Lessons Learned from a Driving Task Perspective. *arXiv* **2025**. [CrossRef]
101. Haarnoja, T.; Zhou, A.; Abbeel, P.; Levine, S. Soft Actor-Critic: Off-Policy Maximum Entropy Deep Reinforce-ment Learning with a Stochastic Actor. In Proceedings of the 35th International Conference on Machine Learning, Stockholm, Sweden, 10–15 July 2018; Dy, J., Krause, A., Eds.; Volume 80, pp. 1861–1870. Available online: https://proceedings.mlr.press/v80/haarnoja18b.html (accessed on 19 May 2025).
102. Sun, X.; Khedr, H.; Shoukry, Y. Formal Verification of Neural Network Controlled Autonomous Systems. *arXiv* **2018**. [CrossRef]
103. Heinen, M.R.; Osorio, F.S.; Heinen, F.; Kelber, C. SEVA3D: Using Artificial Neural Networks to Autonomous Vehicle Parking Control. In Proceedings of the 2006 IEEE International Joint Conference on Neural Network Proceedings, Vancouver, BC, Canadá, 16–21 July 2006; pp. 4704–4711.
104. Ivanov, R.; Carpenter, T.J.; Weimer, J.; Alur, R.; Pappas, G.J.; Lee, I. Verifying the Safety of Autonomous Systems with Neural Network Controllers. *ACM Trans. Embed. Comput. Syst.* **2021**, *20*, 1–26. [CrossRef]
105. Shen, X.; Choi, Y.; Wong, A.; Borrelli, F.; Moura, S.; Woo, S. Parking of Connected Automated Vehicles: Vehicle Control, Parking Assignment, and Multi-agent Simulation. *arXiv* **2024**. [CrossRef]
106. Yuan, Y.; Wang, S.; Su, Z. Precise and Generalized Robustness Certification for Neural Networks. *arXiv* **2023**. [CrossRef]
107. Chen, J.; Li, Y.; Wu, X.; Liang, Y.; Jha, S. Robust Out-of-distribution Detection for Neural Networks. *arXiv* **2021**. [CrossRef]
108. Jiang, M.; Li, Y.; Zhang, S.; Chen, S.; Wang, C.; Yang, M. HOPE: A Reinforcement Learning-Based Hybrid Policy Path Planner for Diverse Parking Scenarios. *IEEE Trans. Intell. Transport. Syst.* **2025**, *26*, 6130–6141. [CrossRef]
109. Xu, G.; Chen, L.; Zhao, X.; Liu, W.; Yu, Y.; Huang, F.; Wang, Y.; Chen, Y. Dual-Layer Path Planning Model for Autonomous Vehicles in Urban Road Networks Using an Improved Deep Q-Network Algorithm with Proportional–Integral–Derivative Control. *Electronics* **2025**, *14*, 116. [CrossRef]
110. Fulton, N.; Platzer, A. Safe AI for CPS (Invited Paper). In Proceedings of the 2018 IEEE International Test Conference (ITC), Phoenix, AZ, USA, 29 October–1 November 2018; pp. 1–7.
111. Yuan, Z.; Wang, Z.; Li, X.; Li, L.; Zhang, L. Hierarchical Trajectory Planning for Narrow-Space Automated Parking with Deep Reinforcement Learning: A Federated Learning Scheme. *Sensors* **2023**, *23*, 4087. [CrossRef]
112. Lu, Y.; Ma, H.; Smart, E.; Yu, H. Enhancing Autonomous Driving Decision: A Hybrid Deep Reinforcement Learning-Kinematic-Based Autopilot Framework for Complex Motorway Scenes. *IEEE Trans. Intell. Transport. Syst.* **2025**, *26*, 3198–3209. [CrossRef]
113. Hamidaoui, M.; Talhaoui, M.Z.; Li, M.; Midoun, M.A.; Haouassi, S.; Mekkaoui, D.E.; Smaili, A.; Cherraf, A.; Benyoub, F.Z. Survey of Autonomous Vehicles' Collision Avoidance Algorithms. *Sensors* **2025**, *25*, 395. [CrossRef] [PubMed]
114. Chen, C.; Geneva, P.; Peng, Y.; Lee, W.; Huang, G. Monocular Visual-Inertial Odometry with Planar Regularities. In Proceedings of the 2023 IEEE International Conference on Robotics and Automation (ICRA), London, UK, 29 May 2023; pp. 6224–6231.
115. Cui, C.; Ma, Y.; Lu, J.; Wang, Z. REDFormer: Radar Enlightens the Darkness of Camera Perception With Transformers. *IEEE Trans. Intell. Veh.* **2024**, *9*, 1358–1368. [CrossRef]
116. Li, Q.; He, H.; Hu, M.; Wang, Y. Spatio-Temporal Joint Trajectory Planning for Autonomous Vehicles Based on Improved Constrained Iterative LQR. *Sensors* **2025**, *25*, 512. [CrossRef]
117. Ma, J.; Feng, X. Analysing the Effects of Scenario-Based Explanations on Automated Vehicle HMIs from Objective and Subjective Perspectives. *Sustainability* **2024**, *16*, 63. [CrossRef]

118. Markets, R. Automated Parking System Market Report 2025, with Profiles of Key Players Including Westfalia Technologies, Unitronics Systems, Klaus Multiparking Systems, Robotic Parking Systems & City Lift Parking. Available online: https://www.globenewswire.com/news-release/2025/01/28/3016636/28124/en/Automated-Parking-System-Market-Report-2025-with-Profiles-of-Key-Players-including-Westfalia-Technologies-Unitronics-Systems-Klaus-Multiparking-Systems-Robotic-Parking-Systems-City.html (accessed on 19 May 2025).
119. Liang, J.; Li, Y.; Yin, G.; Xu, L.; Lu, Y.; Feng, J.; Shen, T.; Cai, G. A MAS-Based Hierarchical Architecture for the Cooperation Control of Connected and Automated Vehicles. *IEEE Trans. Veh. Technol.* **2023**, *72*, 1559–1573. [CrossRef]
120. Fonzone, A.; Fountas, G.; Downey, L. Automated bus services—To whom are they appealing in their early stages? *Travel Behav. Soc.* **2024**, *34*, 100647. [CrossRef]
121. Patel, M.; Jung, R.; Khatun, M. A Systematic Literature Review on Safety of the Intended Functionality for Automated Driving Systems. *arXiv* **2025**, arXiv:2503.02498.
122. Song, Q.; Tan, K.; Runeson, P.; Persson, S. Critical scenario identification for realistic testing of autonomous driving systems. *Softw. Qual. J.* **2023**, *31*, 441–469. [CrossRef]
123. Zhao, X.; Yan, Y. A Deep Reinforcement Learning and Graph Convolution Approach to On-Street Parking Search Navigation. *Sensors* **2025**, *25*, 2389. [CrossRef]
124. Aledhari, M.; Razzak, R.; Rahouti, M.; Yazdinejad, A.; Parizi, R.M.; Qolomany, B.; Guizani, M.; Qadir, J.; Al-Fuqaha, A. Safeguarding connected autonomous vehicle communication: Protocols, intra- and inter-vehicular attacks and defenses. *Comput. Secur.* **2025**, *151*, 104352. [CrossRef]
125. Chen, Z.; Xu, H.; Zhao, J.; Liu, H. Curbside Parking Monitoring With Roadside LiDAR. *Transp. Res. Rec. J. Transp. Res. Board* **2023**, *2677*, 824–838. [CrossRef]

Disclaimer/Publisher's Note: The statements, opinions and data contained in all publications are solely those of the individual author(s) and contributor(s) and not of MDPI and/or the editor(s). MDPI and/or the editor(s) disclaim responsibility for any injury to people or property resulting from any ideas, methods, instructions or products referred to in the content.

sensors

MDPI

Review

Recent Developments on Drivable Area Estimation: A Survey and a Functional Analysis

Juan Luis Hortelano *, Jorge Villagrá *, Jorge Godoy and Víctor Jiménez

Centro de Automática y Robótica, CSIC—Universidad Politécnica de Madrid, Ctra. Campo Real, Km 0.200, Arganda del Rey, 28500 Madrid, Spain; jorge.godoy@csic.es (J.G.); victor.jimenez@csic.es (V.J.)
* Correspondence: juanluis.hortelano@csic.es (J.L.H.); jorge.villagra@csic.es (J.V.)

Abstract: Most advanced autonomous driving systems (ADS) today rely on the prior creation of high-definition maps (HD maps). This process is expensive and needs to be performed frequently to keep up with the changing conditions of the road environment. Creating accurate navigation maps online is an alternative to reduce the cost and broaden the current operational design domains (ODD) of modern ADS. This paper offers a snapshot of the state of the art in drivable area estimation, which is an essential technology to deploy ADS in ODDs where HD maps are limited or unavailable. The proposed review introduces a novel architecture breakdown that fits learning-based and non-learning-based techniques and allows the analysis of a set of impactful and recent drivable area algorithms. In addition to that, complimentary information for practitioners is provided: (i) an assessment of the influence of modern sensing technologies on the task under study and (ii) a selection of relevant datasets for evaluation and benchmarking purposes.

Keywords: autonomous vehicles; drivable area estimation; road estimation; perception; computer vision

1. Introduction

The last decade has seen autonomous vehicles take their first steps out of academia into the commercial world with some companies starting to offer completely unmanned taxi services in selected American cities. These services are usually bound to the prior generation of HD maps. These digital maps are highly precise representations of the driving scene containing centimeter accurate descriptions of the road shape, driving lanes, traffic markings and traffic signs. The creation and maintenance of HD maps stands as a barrier for the commercial viability and general deployment of autonomous vehicles because it is a time-consuming and expensive operation [1]. HD maps are usually captured with an extensive array of sensors, such as GPS, 3D LiDAR, HD RGB cameras, radar and aerial imaging, that then need to be processed offline and stitched to existing sections of the map. The output of this process is a static map that is sensitive to the everyday variability of the drivable area [2]. An alternative to this process is to build accurate online maps in the autonomous vehicles themselves while they are out operating.

Building navigation maps online requires several highly complex tasks performed at once: high precision localization, online identification of traffic markings and signs and an accurate and fast drivable area estimation [3]. This publication focuses on the latter by drawing a picture of the state of the art in drivable area estimation in regards to algorithms, sensors and datasets. The main contributions of this survey are:

- An analysis of the state of the art of the last eight years in regards to drivable area estimation
- An architecture breakdown and a taxonomy that fit learning-based and non-learning-based algorithms

- A study of the existing relevant datasets to assess the performance of these algorithms and the influence of modern sensing technologies in their performance.
- A proposal for future research directions for the field

The rest of the paper is structured as follows: Section 1.1 focuses the scope of the paper in relation to its adjacent research fields and Section 1.2 presents other relevant surveys. Section 2 presents the challenges of drivable area perception with a brief description on especially adverse situations for sensors. Section 3 describes the datasets that are useful for the research and development of algorithms in the field. Section 4 presents the description and analysis of the chosen algorithms plus the proposed architecture breakdown and taxonomy. Section 5 tries to synthesize the findings into a proposal for future research directions, and finally, Section 6 draws some concluding remarks.

1.1. Scope

In this work, drivable area is defined as the area that is physically accessible and designed to be traversed by vehicles. The publication main objective is to study the techniques and the literature works that allow an autonomous vehicle to differentiate road from sidewalks, curbs and other non-navigable ground areas. Publications that, on top of estimating the drivable area, offer additional information such as lane-lines are also considered. The estimation of traffic rules from traffic signs and markings is considered out of scope for this publication.

Drivable area estimation can also be considered a subset of larger segmentation tasks such as semantic, instance or panoptic segmentation. Semantic segmentation tries to label every pixel of an image or point in a pointcloud into the clusters that belong to the same class. Instance segmentation produces a specific label for each occurrence of the same object in the scene but leaves out the background if it cannot be assigned to an object. Panoptic segmentation combines semantic and instance segmentation in order to produce a differentiated label for everything that can be seen in one view of the scene.

Detecting the drivable area has specific problems and characteristics that makes it different from segmenting other classical segmentation labels such as cars, pedestrians or traffic signs. The drivable area is usually locally static, and it is located underneath the vehicle and surrounding it. It needs to be detected constantly since the absence of it should cause an immediate emergency stop. Its appearance varies in color, texture and shape with no standard form. Furthermore, opposed to some other labels, the drivable area can be merged with offline registries such as in low fidelity maps.

Studying this problem in isolation allows for the creation of an architecture breakdown of the problem that sheds light on how each stage of the strategies affects the detection of the drivable area. This analysis is significantly different than using holistic segmentation techniques, which may hinder the understanding of how each stage affects each label. Even though some of the datasets will be shared by the study of the different disciplines, the proposed algorithms and functional analysis are distinct enough to warrant a stand-alone survey.

1.2. Related Surveys

Drivable area estimation has been studied in other high impact surveys that treat the problem with varying degrees of specificity. In [4], the authors offer a thorough review of the state of the art of the 2000s, an exhaustive and fitting architecture breakdown for non-learning-based techniques and the identification of gaps in the research. The future research directions they suggest have proven very inspired, as all of them were explored in the following decade: the creation of public benchmarks, the application of machine learning techniques and the implementation of sensor fusion. In [5], the author presents a survey of terrain traversability for unmanned ground vehicles that offers an interesting perspective on the ground variability that a vehicle could encounter, which helps broaden the focus from standard roads. The authors in [6] published a modern survey that presents learning-based lane estimation methods with a focus on ego-lane detection. Finally, the authors in [7] present a very comprehensive survey of learning-based 3D LiDAR semantic segmentation that is useful for the subtask of drivable area estimation because it specifically provides a thorough review of the current datasets and discusses how those data are currently generated and consumed.

The present survey collects publications without discriminating between learning-based and non-learning-based methods with a focus on drivable area estimation plus an analysis on the sensors and datasets needed to develop the state-of-the-art research in the field.

2. Drivable Area Perception

Drivable area estimation demands a specific performance that the sensors achieve or miss depending on the situation. It is the norm to mount an arrangement of different sensors in the car since it is understood that one single sensor type would not adequately adapt to the myriad different scenarios in real-world driving. A brief performance analysis of each sensor relative to drivable area estimation is presented below.

2.1. Rgb

RGB cameras are a constant bet by autonomous driving researchers and industrial developers due to its low cost, size and technology maturity. They are also the closest sensing technology to human sight, which explains why most of the traffic language is visual and, therefore, adapted to cameras. Figure 1 shows an example RGB view of a roundabout entrance using the camera of a Google Pixel 4a with a resolution of 4032 × 3024.

(a) (b)

Figure 1. RGB view of a roundabout entrance. (**a**) Day view in dry conditions. (**b**) Night view in wet conditions.

The challenge of using RGB cameras is that the exact same visual traffic information can have very different sensor values depending on external factors. Figure 1a,b show how going from day to night and from dry to wet conditions changes the values for color, intensity and reflectance of the image. These differences can even appear on the same image due to shadow and illumination changes.

2.2. 3D LiDAR

3D LiDAR is the technology that has seen the biggest improvement in the past decade. The progress has come as an increase in the number of layers, point density, reflective sensitivity and output frequency. Figure 2 shows an example road view of three state of the art 128-Layer 3D LiDARs. These sensors display a dense representation of the surroundings that allows for a clear road perception. It is interesting to see how those three top-of-their-line sensors differ in their reflective sensitivity; in Figure 2b, the sensor is able to give very high reflectivity values for the lane lines and the vegetation, making the road more apparent. The sensor in Figure 2c is able to pick up on the lane markings but struggles with vegetation. In Figure 2d, the sensor is able to discern between road, vegetation and lane markings in reflectivity alone but misses less reflective road points that are further away from it.

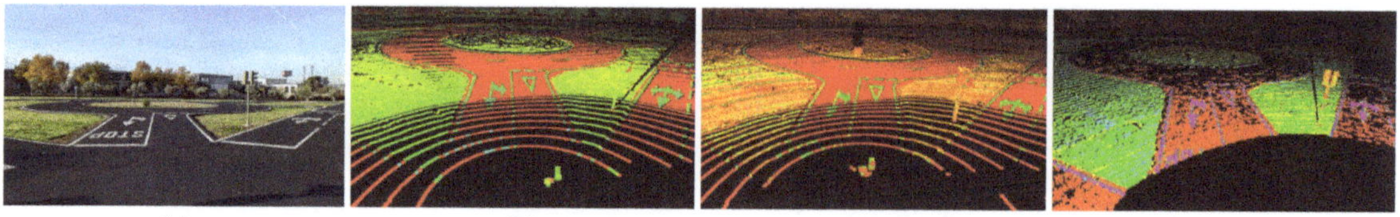

(a) (b) (c) (d)

Figure 2. 3D LiDAR view of a roundabout entrance color-coded by reflectivity (red is lower, green is higher). (**a**) Reference picture. (**b**) Robosense Ruby Plus. (**c**) Velodyne Alpha Prime. (**d**) Ouster 0S2.

Even though the technology has taken a considerable step forward, it still has its drawbacks when tackling the drivable area estimation task. Some of these shortcomings can be seen in Figure 3. Figure 3a,b show how a high-end 3D LiDAR can suffer data loss from poorly reflective surfaces such as black vehicles. Figure 3c,d illustrate how it can be hard to discern between pathways and drivable area in 3D LiDAR pointclouds.

(a) (b) (c) (d)

Figure 3. 3D LiDAR drawbacks. (**a**) Black car picture. (**b**) Black car in pointcloud. (**c**) Road to pathway picture. (**d**) Road to pathway in pointcloud.

2.3. Open Source Maps

Another common way of acquiring information about autonomous vehicle surroundings is through the use of open source maps. The most widely adopted provider is Open Street Maps (OSM) [8], which is a crowd-sourced repository of map features that contains geolocalized road topology, signals, traffic rules and buildings. This source of information can be useful to tackle the drivable area estimation task since it contains geospacial limits to the road, several works already make use of it [9,10]. However it cannot be a complete and standalone solution due to the potential errors introduced in the map caused by manual data acquisition and entry and due to the fact that the rate at which the environment changes is higher than the rate at which users update the open database. Figure 4a shows a section of a roundabout stored in the OSM database. Figure 4b shows the OSM data from Figure 4a transformed into the Lanelet2 [11] format in order to process each individual lane by using the Commonroads open-source OSM-Lanelet2 converter [12].

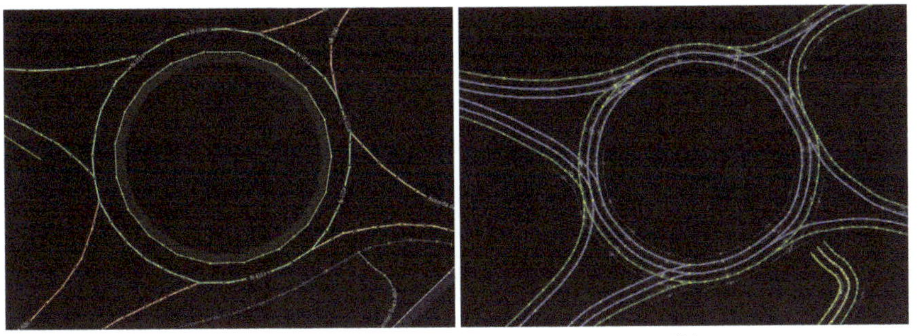

(a) (b)

Figure 4. Open source maps. (**a**) Raw OSM. (**b**) Lanelets from OSM.

Figure 5 shows some of the errors and drawbacks of open-sourced maps. In Figure 5a the difficulty in aligning the lanes derived from OSM with satellite imagery in some specific scenarios can be seen. These issues are similar to the ones that may arise when aligning the images with the GPS position from an autonomous vehicle. Figure 5b shows that the number of lanes stored in the OSM database does not always match the current state of the road.

(a) (b)

Figure 5. Lanelets from OSM drawbacks. (**a**) Misalignment. (**b**) Wrong number of lanes.

3. Datasets

As can be seen in previous surveys [4], one of the biggest gaps in the state of the art was the lack of dedicated datasets that would help to initiate, validate and compare research in the drivable area estimation task. This section presents a set of datasets that have emerged in the last ten years and have proven to be especially useful to tackle the problem at hand. Table 1 presents a summarized view of those datasets. The exclusive criteria used to choose them was modernity, public availability and applicability to the problem, which is characterized by presenting urban scenes with a wide choice of sensor configurations. The non-exclusive criteria was the presence of benchmark mechanisms (BM), which can be leaderboards or competitions that allow for public comparison, the inclusion of geolocalization to potentially merge them with other sources of information, sequentiality, separate testing and training datasets and the relevance of the publications that make use of them. Below, there is a description of the selected datasets based on their applicability to the problem.

Table 1. Dataset comparison.

Dataset	Year	Sensor Type	Sensors	Frames	Map	BM
Waymo Perception [13]	2023	3D LiDAR RGB	5 × Proprietary LiDAR 5 × 1920 × 1280 RGB	390,000	✓	✓
Argoverse 2 [14]	2023	3D LiDAR RGB	2× VLP-32C 7 × 2048 × 1550 RGB	6,000,000	✓	✓
KITTI ROAD [15]	2015	3D LiDAR RGB	Velodyne HDL-64E 1242 × 375 RGB	579	✗	✓
nuScenes [16]	2020	3D LiDAR RGB RADAR	Velodyne HDL32E 1600 × 900 RGB ARS 408-21	40,000	✓	✓
BDD100K [17]	2020	RGB	1280 × 720 RGB	100,000	✗	✓
CityScapes [18]	2016	Stereo RGB	2048 × 1024 RGB	24,998	✗	✓
KITTI-360 [19]	2021	3D LiDAR Stereo RGB 360° RGB	Velodyne HDL-64E SICK LMS 200 4 × 1408 × 376 RGB	83,000	✗	✓
Semantic KITTI [20]	2019	3D LiDAR	Velodyne HDL-64E	43,552	✗	✓
DIODE [21]	2019	RGB-D	FARO Focus S350	27,858	✗	✗
3DHD CityScenes [22]	2022	3D LiDAR	Trimble Mx8	-	✓	✗
OpenLane V2 [23]	2023	Built on: nuScenes and Argoverse			✓	✓
OMC Benchmark [24]	2021	Built on: nuScenes			✓	✗

Waymo Perception as a part of the Waymo Open Dataset is one of the most significant sensor datasets in terms of the sheer quantity and variety of annotation types. On top of its sensor suite, it offers labels for bounding box, key point, 2D panoptic segmentation and 3D semantic segmentation. It shows urban and residential scenes with diverse weather in both daytime and nighttime environments. It was updated in 2023 with an HD Map, and it also includes a python devkit to streamline development.

Argoverse 2 builds its value proposal on top of the long range (200 m) of its LiDAR and annotations, the six different US cities it collected its data in and the amount of different labels it offers (30). It also offers a vector map with lane-level geometry and a ground height raster map to ease the filtering of ground LiDAR returns.

nuScenes is a dataset that implements a full sensor array with RGB cameras spanning a 360° view, a 32-layer 3D LiDAR and a RADAR array. It includes a high number of annotations in inner-city traffic scenes with changing weather and heavy traffic. It has a leaderboard that differentiates between LiDAR only and any sensor modality for semantic segmentation and panoptic segmentation.

CityScapes is a Stereo RGB dataset of complex urban scenes that offers fine and coarse semantic annotations. It depicts busy inner-city driving with high levels of traffic and pedestrian interaction and has separate annotations for road and sidewalk.

KITTI is one of the oldest and most popular autonomous driving perception datasets available. The specific subset for drivable area estimation is named KITTI ROAD, and it consists of a non-sequential set of images and pointclouds taken in non-busy suburban environments with and without lane markings in single- and multi-lane roads. It has become the standard benchmark for the task. and one of its main strengths is the dedicated leaderboard, which acts as a powerful surveying tool of the current state of the art. Its main drawbacks are that it has no weather or illumination diversity and the relative sparsity of its traffic scenes.

Semantic KITTI spawned as a complementary dataset from KITTI's popularity and it expands on its features by adding semantic annotations for all sequences of the odometry subset.

BDD100K is an extensive RGB-only dataset that offers 100.000 sequential images of driving in four different American cities. This dataset offers lane marking and drivable area annotations, making a distinction between directly drivable and alternatively drivable areas. They use (i) the directly drivable tag for road areas in which the ego-vehicle is currently driving on and has the priority, and (ii) the alternative drivable tag for road areas that the ego-vehicle is not currently driving on but could through a lane change. This dataset also offers a dedicated leaderboard based on yearly competitions.

KITTI-360 calls itself the successor of KITTI and expands on the latter by offering richer sensor modalities, semantic instance annotations and a more accurate localization in suburban scenes with moderate traffic. It expands on the sensor suite of the original KITTI by adding an additional LiDAR, a pair of front-facing RGB cameras to produce disparity maps and two lateral fisheye cameras to complete their 360° scene perception. The main value proposition of this dataset is that it is the only segmentation-oriented dataset with all sensor modalities simultaneously available.

DIODE is a combined indoor/outdoor dataset that offers RGB, depth and normal information using the same sensing and imaging setup. It achieves this by using the FARO Focus S350 sensor, which is an actuated phase-shift laser scanner that creates RGB and depth scenes with very high accuracy, resolution and FOV. The downside of using this dataset for drivable area estimation is that it has no sequential scenes as all of them are static due to the nature of the sensor.

3DHD CityScenes is a dataset that combines high-definition maps with high-density synchronized and georeferenced pointclouds taken by a high-end spatial imaging sensor.

OpenLane V2 is a dataset built on top of nuScenes and Argoverse 2 that focuses on scene structure perception and reasoning by offering a dynamic map that takes into account traffic elements such as ground markings and traffic lights and signals. One of its highlights is the fact that they offer 3D-annotated lanes in their map, as opposed to the 2D lanes present in most datasets.

Online HD Map Construction Benchmark offers a set of vectorized and rasterized maps from camera images that is built on top of the nuScenes dataset.

4. Drivable Area Estimation

This section presents a review of the state of the art in drivable area estimation methods and the architecture breakdown, plus the taxonomy that can be inferred from them. The architecture breakdown and taxonomy are presented first in an attempt to provide a framework to analyze the algorithms that can be seen in the latter part of the section.

4.1. Architecture

A thorough study of the literature unveils a set of common stages for most of the drivable area estimation algorithms. Figure 6 presents an architecture breakdown based on the similarities found in the shared modules of the algorithms. It is important to note that this diagram depicts a generic architecture and not all works employ every stage or necessarily follow the same order. Nonetheless, it is possible to map most of the research to a subset of this architecture breakdown. In the following, there is a description of each stage.

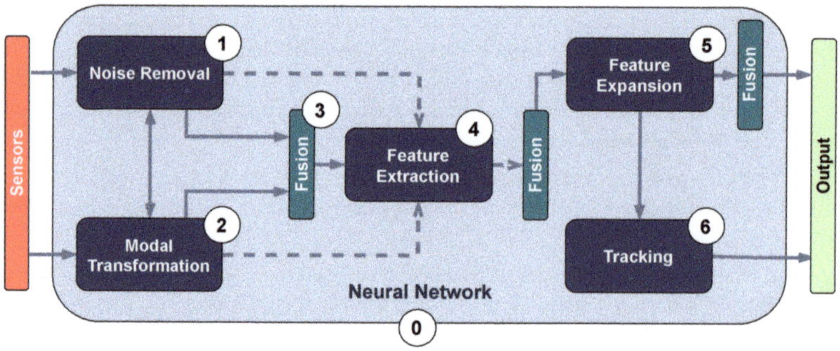

Figure 6. Drivable area estimation architecture breakdown.

Noise Removal is the process of identifying and correcting sensor data with the purpose of improving algorithmic or computational performance. Two types of noise are identified: static and motion noise. Static noise encompasses data that are irrelevant or harmful to the algorithm, such as outliers or out-of-range information, and are inherent to the sensor. The removal process of this type of noise can range from simple and fast through thresholding processes, such as min–maxing points that exceed a set height value, to more complex and costly, such as plane fitting through least-squares or RANSAC. Motion noise causes deformations in relevant information to the algorithm and is created by the vehicle's movement. Example correction processes for this issue are pointcloud deskewing or attitude alignment.

Modal Transformation: As a previous step to fusion or in order to apply cross-field techniques, a modal transformation to sensor data can be performed. Modal transformations usually affect data dimensionality by means of projecting (e.g., 3D LiDAR pointclouds onto image coordinates) or deriving (e.g., creating a map of normals from an image or a pointcloud).

Fusion is the process of combining data coming from different sensors or stages to increase the accuracy that would be achieved in isolation. It is a stage that can be performed at different times during an algorithm execution as it can be applied to sensor information or to already generated features. Common fusion techniques are grid maps and Kalman filtering. Fusion inputs can be diverse: raw sensor data, processed sensor data, low-level features and high-level features.

Feature Extraction: The translation from high-dimensional rich sensor data into basic features takes place in feature extraction. Features useful for drivable area estimation can be low level, such as image color or texture, or high level, such as detected lane lines in RGB images or height differentials extracted from pointclouds.

Feature Expansion formulates hypotheses of the drivable area and fits them to the available extracted features. It generates drivable area proposals and uncertainty estimations. Applicable techniques in this stage are image upscaling, graph search or model fitting.

Tracking: Using an estimation of vehicle displacement and geospacial information, this stage matches the produced estimations through time in order to improve the output. Tracking can be achieved by applying Bayesian fusion or Kalman filtering.

Neural Network deep learning technologies have proven useful to act as a backend in any of the drivable area estimation stages. Neural networks can be trained to perform any stage in isolation and also to perform several tasks at once or even the complete algorithm from start to finish. Network training can be performed at different points in the algorithm's timeline while taking a variety of different inputs (e.g., raw sensor data, data after modal transformation, data after fusion, features, etc).

4.2. Taxonomy

Drivable area estimation is now a mature field that has spawned a varied set of techniques that can be applied to tackle the problem. The analysis of this field enabled the synthesis of a taxonomy that can be seen in Figure 7. It is important to note that a complex problem such as drivable area estimation usually requires the application of several techniques at once, and therefore, those sets are not exclusive, and it is common that one set of techniques feeds off another. A description of each family of techniques is provided below.

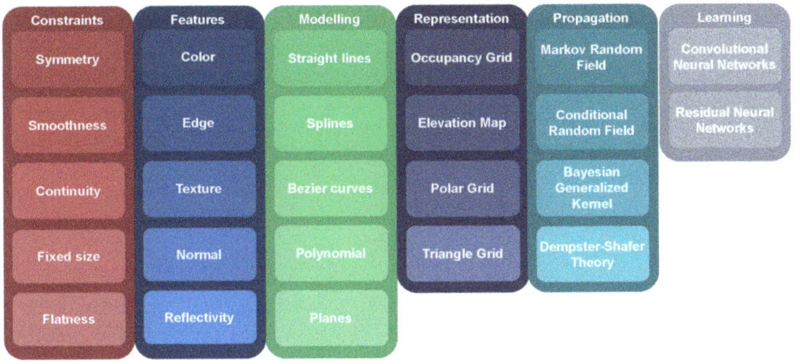

Figure 7. Drivable area estimation taxonomy.

Constraints. Drivable area estimation deals with the detection of man-made, industrially designed artifacts such as roads, curbs, sidewalks, etc. As such, it is useful to limit the search space using carefully chosen constraints. Symmetry constraints are commonly both finding features and fitting models, as most roads are defined by left and right boundaries that are also parallel. Finding the twin equivalent of a feature or constraining a model to remove outliers is a sound geometric foundation for road detection. Smoothness constraints. Roads are made for non-flexible, bound-to-the-ground and several hundred kilos heavy vehicles that cannot make abrupt direction changes and, therefore, require gradual evolution to be traversable. Smoothness constraints help find drivable area estimation candidates and predict its evolution. Continuity constraints are useful to find matches between consecutive sensor reads since drivable areas do not usually abruptly end or change. Fixed size constraints. Most roads maintain a fixed width that can be bound between a maximum and a minimum value that is useful to weed out outliers and find equivalents when also applying the symmetry constraint. Flatness constraints are useful because surfaces need to be locally flat when underneath a vehicle, or otherwise, at least one of the wheels would not be in contact with the ground.

Features are acquired by processing raw data coming from the sensors and creating data points that host new information. Color features use pixel properties in RGB images to differentiate the drivable area from other vehicles, pedestrians or non-drivable surfaces. They are sensible to lighting changes, adverse weather or similarly colored bodies that do not belong to the same category. Edge features try to identify the drivable area by detecting hard gradients in color, intensity or geometry, as those are commonly found in the road limits with sidewalks or ditches. Texture features extract information from how color is arranged spatially in an RGB image and help produce drivable area candidates on the assumption that texture patterns remain consistent within the same road objects. Normal features are produced by calculating the normal vector of a group of geometric points and clustering them based on their angle. Usually, road normals point in the same direction or change gradually. Reflectivity features can be acquired using modern LiDAR sensors and give information about how any surface is able to reflect or absorb light. It is useful to detect road markings and signs as they are usually highly reflective. It could even be used to detect the road itself as a poorly reflective area if the sensor is sensitive enough.

Modeling fits the sensor data to mathematical models that define lines or planes. Straight lines are a common model that is cheap computationally and can be directly applied to simple roads or as a collection of segments to model complex roads. Splines are smooth piecewise functions defined by polynomials. They have variable complexity depending on their order and, therefore, are useful to model curves. Bezier curves are smooth global functions defined by polynomials. They are attractive as a model because they offer curvature continuity at every point and are not computationally expensive to compute in their closed-form expressions. Polynomials of the second or third degree are useful to model curves under some constraints (p.e flatness) as they are simple to understand and compute. Their drawbacks are that modifying any points affects the complete curve or undesirable effects at the boundaries. Planes are especially useful in LiDAR data processing as planes are a very common feature in LiDAR pointclouds that appear in roads, sidewalks or buildings.

All of these model proposals need to be fitted to the data using a mathematical approach. A common method to estimate the parameters of a model is random sample consensus (RANSAC), which consists of iteratively checking the fitness of a random sample of the data against a previously set model. It is useful to separate inliers from outliers but is a non-deterministic model, which means that its accuracy depends on the amount of iterations it has run. Another technique is principal components analysis (PCA), which is a method for dimensionality reduction that tries to keep most information in the dataset. It linearly transforms the data into a new coordinate system where the majority of their variance can be described using less dimensions than the original data. The challenge in applying PCA in this field is usually selecting the data subset in which the model needs to be fitted. Finally, least median squares (LMS) tries to minimize the sum of the squared difference between an observed data point and the fitted value provided by a model. It is a simple method to apply and understand but is sensitive to outliers and can overfit the model.

Representation methods offer alternative representations for sensor and feature information that reduce dimensionality and ease the computational load. They are also widely used for data and feature fusion. Occupancy Grids represent the world with evenly spaced cells that host a binary variable (occupied or free) that is estimated and assigned a probability. These grids are used to make drivable area assumptions over the free area. Elevation Map models the world by keeping height information that can be used to derive the drivable area by detecting abrupt height gradients. Polar Grids host the information in polar coordinates instead of in Cartesian coordinates. They usually result in smaller and, therefore, faster to compute grids, have higher resolution near the vehicle and are adequate representations of roads because they both grow radially from the vehicle. Triangle Grids are common world representations in fields such as videogames because they can accurately model height and complex shapes. Their main characteristics are planarity, which means that every point of a triangle can be at different heights and still belong to the same plane, and simplicity, since they have the lowest amount of vertices of any polygon.

Propagation techniques create relational systems between data points to mitigate the issue of data gaps caused by occlusions or at great distances from the sensor. Markov random field is an undirected and cyclic graph technique that sets each data point as a node and tries to assign labels and their probability to them by performing inference techniques on the graph, such as belief propagation. Conditional random field is a special case of Markov random field in which the graph takes into account the influence between neighboring nodes by modeling the dependencies between nodes. Bayesian Generalized Kernel propagates information by assuming continuity between adjacent data points and inferring the missing information by applying a kernel to the neighboring observed points. Dempster–Shafer theory is a general mathematical framework to reason with uncertainty and deal with information gaps. It is designed to work with sets of different labels combining evidence from different sources and producing a degree of belief.

Learning-Based methods are ubiquitous due to their flexibility that allows them to morph into several different techniques. They can have sensor data, features or grid representations as inputs and can be used to perform model fitting, feature extraction or generate drivable area candidates directly. Their main drawback is the need for large annotated datasets for training. Convolutional Neural Networks are very popular learning-based approaches in the field. They are usually applied directly to RGB images from the road or to images derived from LiDAR pointclouds. They work by training image kernels that learn to identify specific features in an image and then passing those features to a fully-connected layer that recognizes larger elements in the scene. Residual Neural Networks are a special case of convolutional neural networks that, at the same time, try to reduce the number of layers of the network and tackle the vanishing gradient problem that occurs in deep networks. They work by providing residual connections between skipping layers that avoid the activation functions of previous layers that would reduce their derivatives. This results in shallower networks that are more efficient to train.

4.3. Algorithms

This section presents the set of algorithms that describe the current state of the art, their functional analyses and their contributions and limitations. The algorithms and their analyses are collected in Table 2. The criteria for choosing this specific set of algorithms was: (i) *scientific relevance*, the impact that this set of algorithms has in the state of the art and the novelty of their contributions; (ii) *ensuring a broad representation of techniques*, making sure that this survey presented a set of algorithms that covered every type of sensor combination and showed a plural application of the available theory; and (iii) *benchmark appearance*, including the highest performing algorithms today in terms of drivable area estimation quality measures and computation time.

The diagrams appearing in the architecture column depict the functional analysis of each algorithm in a simplified schematic manner by fitting them into the stages appearing in Figure 6. For reference, ⓪ is neural network, ① is noise removal, ② is modal transformation, ③ is fusion, ④ is feature extraction, ⑤ is feature expansion and ⑥ is tracking. Algorithms that show several branches describe works that perform stages in parallel.

The neural network stage depicts the moment of training the network in the algorithm and the inputs it receives to do so. Once this stage appears in an algorithm, all subsequent stages are then performed by neural network inference. For example ②→⓪→③→④→⑤ implements modal transformation and then uses this information to train a neural network that performs fusion with raw data, feature extraction and feature expansion.

Table 2. Drivable area segmentation algorithms.

Name	Year	Input	Architecture	Methodology	Output	Contributions	Limitations
BGK [25]	2021	3D LiDAR	1 → 4 → 5 → 6	Min–maxing, Bilateral Filtering, BGK, Normal Estimation Multi-frame tracking	Segmented Grid	Generates information from unobserved areas without LiDAR hits	Computational and algorithmic performance tied to grid resolution
SNE-RoadSeg [26]	2020	RGB-D	2 → 0 → 3 → 4 → 5	SNE, CNN	Surface Normals Segmented Image	SNE module that can be plugged into other CNNs with proven improvement	Surface normal information might missclassify sidewalks as road
PLARD [27]	2019	3D LiDAR RGB	2 → 0 → 3 → 4 → 5	ADT, CNN, FSA	Altitude Difference Image Segmented Image	Leverages LiDAR to make RGB data robust to shadows	Requires high-end GPU to perform
USNet [28]	2022	RGB-D	2 → 0 → 4 → 5 → 3 2 → 0 → 4 → 5 → 3	SNE, MEC, Uncertainty Map	Uncertainty Image Probabilistic Segmented Image	Good tradeoff between computational and algorithmic performance Uncertainty map could be used by other modules	Requires high-end GPU to perform
Unsupervised RD [29]	2017	3D LiDAR RGB	2 → 3 → 4 → 3 → 5 2 → 3 → 4 → 3	Superpixel, Delaunay triangulation, MRF, Belief propagation	Superpixel Probabilistic Segmented Image Probabilistic Segmented Pointcloud	Robust to illumination changes	Detection of 3D features dictates whole algorithm performance
Map-Supervised RD [9]	2016	Map RGB	3 → 0 → 4 → 5	OSM, CNN	Segmented Image	Able to work with or without map on inference	Reliant on extrinsic camera calibration and GPS quality
RBANet [30]	2019	RGB	0 → 4 → 5	SegNet, BA, RA, CNN	Probabilistic Segmented Image	Residual stages makes algorithm inspectable at different stages	Requires high-end GPU to perform
HID-LS [31]	2019	3D LiDAR RGB	2 → 3 → 4 → 5	Superpixel, Depth Map, Height Map, Histograms	Superpixel Segmented Image	Transforms spatially discrete LiDAR pointclouds into a continuous and organized structure	Reliant on LiDAR resolution and parametrization
Curb Detection [32]	2018	3D LiDAR	1 → 4 → 5	RANSAC Sliding beam segmentation	Curb Points	Good performance on CPU	Relies on curbs to detect drivable area
LiDAR-Histogram [33]	2017	3D LiDAR	2 → 4 → 5	RANSAC, Histogram	LiDAR Imagery Segmented Pointcloud	Detects positive/negative obstacles and estimates road drivability degree	Makes assumptions on voids in LiDAR data which could missclassify poorly reflective obstacles
CyberMELD [10]	2020	Map 3D LiDAR RGB	1 → 2 → 3 → 4 → 5	Delaunay triangulation, IPM	Segmented Image	Leverages OSM to deal with missing features caused by shadows or occlusions	Only validated on single lane two-way roads Sensitive to OSM errors
RoadNet3 [34]	2019	RGB	0 → 4 → 5 → 3 0 → 4 → 5 → 3	CNN, LSTM	Segmented Image	Reduces feature map resolution with a CNN to achieve high performance	Assumes continuity of drivable area
Double Projection [35]	2021	3D LiDAR	2 → 4 → 5 → 3 → 6 2 → 4 → 5 → 3	Range Map, Elevation Map Forward flood fill	Ground Model Segmented Pointcloud	Able to deal with rough terrain and offroad situations	Only considers positive obstacles
Pseudo-LiDAR [36]	2022	RGB	0 → 2 → 4 → 3 → 5 2 → 4 → 3	RNN	Pseudo-LiDAR Segmented Image	Takes advantage of 3D features but only requires an RGB camera	Requires additional depth estimating network to produce pseudo-LiDAR

Table 2. *Cont.*

Name	Year	Input	Architecture	Methodology	Output	Contributions	Limitations
CLCFNet [37]	2021	3D LiDAR RGB	2 → 0 → 4 → 5 → 3 2 → 0 → 4 → 5 → 3 3 → 0 → 4 → 5 → 3	CNN	LiDAR Imagery Segmented Pointcloud Segmented Image	Runs on LiDAR only in case RGB is negated	LiDAR imagery over raw pointclouds makes algorithm sensitive to occlusions
Multi-Cue [38]	2016	RGB-D	4 → 5	SVR	Curb Points Road Boundary Model Segmented Image	Highest ranking stereo vision algorithm in the KITTI dataset	Uses surface normals to find road boundaries which can missclassify sidewalks as road
TRAVEL [39]	2022	3D LiDAR	1 → 4 → 5	Breadth-first graph search, Tri-grid field	Segmented Pointcloud	Manages to perform on sloped surfaces and rough terrain One of few non-NN with open-source code	Only focus on traversable does not differentiate pathways from road
Road Markings [40]	2022	3D LiDAR	1 → 4 → 5	Otsu thresholding, Reflectivity, Line fitting	Lane Boundary Model	Takes advantage of a rarely used data type	Only tested in traffic-free environments sensitive to parameter tuning
Line Fitting [41]	2019	3D LiDAR	1 → 4 → 5	Channel-based segmentation B-spline curve fitting	Road Boundary Model Segmented Pointcloud	Can deal with occlusions Extracted lines can be fused with OSM data	Relies on curb to detect drivable area
YOLOP [42]	2021	RGB	0 → 4 → 5	CSPDarknet, SPP, FPN, CNN	Multi-label Segmented Image	Detects opposite lane as non-drivable area Produces drivable area plus lane estimations	Can missclassify gaps in the drivable area as lane-lines
HybridNets [43]	2022	RGB	0 → 4 → 5	EfficientNet-B3, FPN, CNN	Multi-label Segmented Image	Same as YOLOP while improving computational and algorithmic performance	Limited by camera FOV need to be very close to area to detect it as drivable
Rangenet++ [44]	2019	3D LiDAR	0 → 4 → 5	kNN, CNN	Multi-label Segmented Pointcloud	Offers open-source trained pointclouds tackles over-segmentation in postprocessing	Requires high-end GPU to perform Hard-linked to LiDAR specifics in training
SpatioTemporal CRF [45]	2017	3D LiDAR	4 → 5 → 6	CRF	Segmented Pointcloud	Robust to changes in slope and temporarily obstructed areas	No quantitative results Implementation details bare in the publication
Urban Road Filter [46]	2021	3D LiDAR	4 → 5	Channel-based segmentation	Curb Points Road Boundary Model	Works in real-time on CPU	To detect road it requires the curb to be visible and the sidewalk to be smooth
Evidential Grids [47]	2015	Map 3D LiDAR	3 → 4 → 5	DST	Evidential Grid	Deals very well with noisy input data	Requires previously segmented data Assumes that the map used is high-fidelity
RoadSLAM [48]	2019	Map 3D LiDAR	4 → 5 → 3	OSM, B-Spline, UKF, Channel-based segmentation, GraphSLAM	Road Boundary Model	Is able to produce road estimations in areas out of sensor reach	Does not work on complex road geometries such as junctions
SemanticDepth [49]	2019	RGB	2 → 3 → 4 → 5	CNN	Pseudo-LiDAR Segmented Image	Robust to occlusions Specifically designed to work without lane-lines	Computationally expensive

Following, there is a description of each selected publication's stages and methodology:

BGK [25] first performs coarse ground segmentation through a min–max height difference approach and projects it into a grid. Then, they estimate the missing height information in the grid through Bayesian generalized kernel inference. From the generated dense height grid, they estimate the normal vectors and compute their angle difference to produce navigable candidates through angle thresholding. Afterwards they apply bilateral filtering to preserve edges, and finally, they apply multi-frame tracking on consecutive grids.

SNE-RoadSeg [26] estimates surface normals from the depth information coming from a RGB-D sensor using a surface normal estimation (SNE) module. Then, it feeds the estimated surface normal data and the raw RGB to a convolutional neural network (CNN) with a parallel encoder–decoder architecture with densely-connected skip connections.

PLARD [27] performs an altitude difference-based transformation (ADT) to produce an altitude difference image from the LiDAR data. Then it inputs the RGB data plus the produced altitude difference image into a CNN that fuses the generated features through feature space adaption (FSA) in its intermediate stages to finally produce a road estimation candidate.

USNet [28] performs surface normal estimation from the depth information coming from the RGB-D sensor. It then feeds the estimated data and the raw RGB data into two separate CNNs that generate feature maps. Finally, those feature maps are fed to a module called multi-scale evidence collection (MEC) that generates separate prediction and uncertainty maps.

Unsupervised RD [29] fuses RGB and LiDAR data to create the superpixel data structure. Then, it performs Delaunay triangulation to assign spatial surface to the superpixels. From the estimated surfaces, it computes its normals and uses them to estimate obstacle points through flatness edge thresholding. Then, it creates a ray map by casting rays from the origin of the sensor to the detected obstacles, which produces a rough road estimation. Finally it uses Markov random fields(MRF) and belief propagation to perform feature fusion with the objective of increased robustness.

Map-Supervised RD [9] generates training annotations using OSM and geo-referenced images from the KITTI dataset. It then refines the generated labels by clustering pixels with similar color. Finally, it then uses the automatically generated labeled data to train a CNN.

RBANet [30] uses SegNet [50] to create residual feature maps and then apply reversed attention (RA) and boundary attention (BA) units to them to generate road estimations.

HID-LS [31] fuses RGB and LiDAR data to create superpixels. Then it performs Delaunay triangulation to interpolate the missing gaps in the 3D information. In the generated data, it computes inverse depth and height maps. It then uses the inverse depth map to compute horizontal and vertical histograms and derive first road region estimation. Using row and column scanning in the height map, it derives the second road region estimation. Finally, it fuses both estimations to produce a final road candidate.

Curb Detection [32] first applies RANSAC to extract the ground plane with sidewalk, curb and road points. Then, it performs sliding-beam segmentation to divide the extracted ground areas in regions of interest. Finally, it extracts features through angle thresholding to detect curbs.

LiDAR-Histogram [33] creates LiDAR imagery by displaying the points in a 2D plane organized by pitch and rotation angle. Then, it computes histograms from the LiDAR imagery. Afterwards, it performs RANSAC line fitting on the histograms to estimate road lines. Finally, it segments the pointcloud based on the estimated road lines.

CyberMELD [10] computes vertical and horizontal slope maps and sums them to, afterwards, grow and detect a general region of interest (ROI) road area. It then performs Delaunay triangulation to map ROI 3D Points onto the RGB image. In the generated data, it applies inverse perspective mapping (IPM) and finds lane lines on it using gradients. Finally, it fuses the lane lines with OSM to generate an ego-lane estimation.

RoadNet3 [34] uses cascaded CNNs to reduce the resolution of the feature map followed by a long–short-term memory (LSTM) network to find the contour of the drivable area. It applies the same architecture in parallel in the full image and in a smaller region in the center of the image and fuses them to produce the road estimation.

Double Projection [35] creates a 2.5D elevation map with sets of points falling under the same cell through a min–maxing process. It also creates a 2.5D range map by projecting LiDAR points to a virtual cylinder plane. Then, the elevation map is used to detect obstacles through thresholding, and a reachable area is estimated through forward flood fill. The range map produces a road area using a smoothness constraint. Finally, the reachable area and the road area are fused through Bayesian decision theory to extract the drivable area.

Pseudo-LiDAR [36] uses a big to small [51] neural network to estimate depth from RGB. It then transforms depth into attitude space to generate a Pseudo-LiDAR [52] pointcloud. It then applies a residual neural network (RNN) to process features in parallel for RGB and Pseudo-LiDAR and fuse them at different points of the pipeline to obtain the final result.

CLCFNet [37] transforms a LiDAR pointcloud into LiDAR imagery. Then, it performs a perspective transformation to put the LIDAR imagery in the camera view. Then, it inputs the LiDAR imagery, LiDAR pointcloud and RGB into three cascaded CNNs to extract and fuse the road features. It works with LiDAR only or with both LiDAR–RGB data depending on light conditions.

Multi-Cue [38] computes normal vectors on the disparity maps from stereo images. It then finds interest boundary areas from highly diverging normals. Curve fitting on the boundary pixels by using support vector regression(SVR) [53].

TRAVEL [39] first corrects pointcloud skew and attitude caused by ego-motion. Then, it models the terrain by grouping subsets of pointcloud points into tri-grid field nodes. Afterwards, it uses breadth-first traversable graph search to classify traversable nodes by measuring acceptable concavity and convexity. Finally, it applies model fitting to match the traversable nodes to pointcloud points, assigning them a label.

Road Markings [40] implement coarse ground segmentation through RANSAC plane fitting, along with regional grow-clustering to weed out points belonging to the curb. Then, they apply adaptive thresholding based on Otsu's method [54] on the reflectivity information coming from the sensor. Finally, they produce lane boundary proposals by line model fitting.

Line Fitting [41] first implements coarse ground segmentation through the channel-based clustering of points in a 2.5D polar grid map, along with height thresholding. Then, it identifies boundary points by checking the angle, distance and height difference between adjacent points. Finally, it performs B-spline curve fitting to produce the road boundary candidate.

YOLOP [42] uses CSPDarknet [55] as a backbone to extract feature maps. Then, it implements spatial pyramid pooling (SPP) [56] to fuse features of different scales and a feature pyramid network (FPN) [57] to fuse features of different semantic levels. Then, it uses an upsampling process to restore the original size of the image from the feature map and generate two separate segmented images of the drivable area and lane lines.

HybridNets [43] is a neural network that uses EfficientNet-B3 [58] as a backbone to extract feature maps and an FPN as neck to fuse features across feature maps with different resolutions. Then, it upsamples feature maps up to half the original resolution and feeds it to its proprietary segmentation head to to produce the final multi-label segmented image.

Rangenet++ [44] performs a spherical projection of the LiDAR pointcloud into a range image and then feeds it into a fully convolutional neural network with an encoder–decoder hour-glass-shaped architecture. Finally it performs a k-nearest neighbor (kNN) search to reduce noise and shadow-like artifacts in the produced multi-labeled output.

Urban Road Filter [46] tries to detect the road by applying three different techniques to LiDAR pointclouds. The three different techniques differ in the way they divide the pointcloud, by channels, by beams and by sliding windows. It then applies different heuristics to the clusters to detect anomalies in height or angle differences to produce curb candidates. Finally, those curb candidates are used to produce a drivable area polygon.

Evidential Grids [47] use the Dempster–Shafer theory to efficiently fuse information sources that provide partial information of the environment. It offers a framework to fuse occupancy grids created from LiDAR information with meta-knowledge obtained from a high-fidelity map. It outputs drivable and non-drivable space and can differentiate from stationary and moving objects through temporal fusion.

RoadSLAM [48] separates the pointcloud into ground and non-ground through coarse segmentation and clusters it in sets of free areas. Then, a robust weighted least-squares curve fit is applied to each side of the selected free area in order to find the instance that maximizes the likelihood given the current route of the vehicle. An unscented Kalman filter (UKF) is used to produce a prediction of the control points of the B-Splines representing the road boundaries with the free area boundaries as input. Finally, all information is accumulated and fused using GraphSLAM [59] with OSM as prior information.

SemanticDepth [49] creates two information sources from a single RGB input. First, through monocular-depth estimation, they generate a disparity map from a single RGB image and then they transform it to a full pointcloud. Then, they perform semantic segmentation on the original RGB image using a CNN. Then, they overlay the segmented image as masks on the generated pointcloud to calculate the road width ahead of the ego-vehicle.

Table 3 presents the taxonomical classification of the algorithms. The categories in the column are based on the study presented in Section 4.2.

The acronyms used in the table are the following: For the constraints group: symmetry (Sym), smoothness (Smo), continuity (Con), fixed size (FS) and flatness (Flat). For the Features group: color (C), edge (E), texture (T), normal (N) and reflectivity (R). For the model group: straight line (SL), splines (Sp), polynomial (P) and planes (Pl). For the representation group: occupancy grid (OG), elevation map (EM), polar grid (PG) and triangle grid (TG). For the propagation group: Markov random fields (MRF), conditional random fields (CRF), Bayesian generalized kernel (BGK) and Dempster—Shafer Theory(DST). Finally, for the learning group: convolutional neural network (CNN) and residual neural network (RNN). The Bezier category present in Section 4.2 was left out due to no selected algorithm making use of it.

Table 4 shows the algorithms performance as is reported in their publications or in the leaderboard where they appear.

Table 3. Algorithms taxonomy.

Name			Constraints					Features				Model				Representation				Propagation			Learning	
	Sym	Smo	Con	FS	Flat	C	E	T	N	R	SL	Sp	P	Pl	OG	EM	PG	TG	MRF	CRF	BGK	DST	CNN	RNN
BGK [25]	–	–	–	–	–	–	✓	–	✓	–	–	–	–	–	–	✓	–	–	–	–	✓	–	–	–
SNE-RoadSeg [26]	–	–	✓	–	–	✓	–	–	✓	–	–	–	–	–	–	–	–	–	–	–	–	–	✓	–
PLARD [27]	–	–	✓	–	–	✓	–	✓	–	–	–	–	–	–	–	–	–	–	–	–	–	–	✓	–
USNet [28]	–	–	✓	–	–	✓	–	✓	–	–	–	–	–	–	–	–	–	–	–	–	–	–	✓	–
Unsupervised RD [29]	–	–	–	–	✓	✓	✓	✓	✓	–	–	–	–	–	–	–	–	–	✓	–	–	–	–	–
Map-Supervised RD [9]	–	–	✓	–	–	✓	–	✓	–	–	–	–	–	–	–	–	–	–	–	–	–	–	✓	–
RBANet [30]	–	–	✓	–	–	✓	✓	✓	–	–	–	–	–	–	–	–	–	–	–	–	–	–	–	✓
HID-LS [31]	–	–	–	–	✓	✓	✓	–	✓	–	–	–	–	–	–	✓	–	–	–	–	–	–	–	–
Curb Detection [32]	✓	✓	–	✓	–	–	✓	–	–	–	–	–	✓	–	–	–	–	–	–	–	–	–	–	–
LiDAR-Histogram [33]	–	–	–	–	✓	–	✓	–	–	–	✓	–	–	–	–	–	–	–	–	–	–	–	–	–
CyberMELD [10]	–	–	✓	–	✓	–	✓	–	–	–	✓	–	–	–	–	–	–	–	–	–	–	–	–	–
RoadNet3 [34]	–	–	✓	–	–	✓	–	–	–	–	–	–	–	–	–	–	–	–	–	–	–	–	✓	–
Double Projection [35]	–	✓	✓	–	–	–	✓	–	–	–	–	–	–	–	✓	✓	–	–	–	–	–	–	–	–
Pseudo-LiDAR [36]	–	–	✓	–	–	✓	✓	✓	–	–	–	–	–	–	–	–	–	–	–	–	–	–	–	✓
CLCFNet [37]	–	–	✓	–	–	✓	–	✓	–	–	–	–	–	–	–	–	–	–	–	–	–	–	✓	–
Multi-Cue [38]	–	–	✓	–	–	✓	✓	–	✓	–	–	–	–	✓	–	–	–	–	–	–	–	–	–	–
TRAVEL [39]	–	–	✓	–	✓	–	✓	–	✓	–	–	–	–	–	–	–	–	✓	–	–	–	–	–	–
Road Markings [40]	✓	–	–	–	–	–	–	–	–	✓	✓	–	–	–	–	–	–	–	–	–	–	–	–	–
Line Fitting [41]	–	–	✓	–	–	–	✓	–	–	–	–	✓	–	–	–	–	✓	–	–	–	–	–	–	–
YOLOP [42]	–	–	✓	–	–	✓	–	✓	–	–	–	–	–	–	–	–	–	–	–	–	–	–	✓	–
HybridNets [43]	–	–	✓	–	–	✓	–	✓	–	–	–	–	–	–	–	–	–	–	–	–	–	–	✓	–
Rangenet++ [44]	–	✓	✓	–	–	–	–	–	–	–	–	–	–	–	–	–	–	–	–	–	–	–	✓	–
SpatioTemporal CRF [45]	–	–	–	–	–	–	–	–	–	–	–	–	–	–	–	✓	–	–	–	✓	–	–	–	–
Urban Road Filter [46]	–	✓	✓	–	✓	–	✓	–	–	–	–	–	–	–	–	–	–	–	–	–	–	–	–	–
Evidential Grids [47]	–	–	–	–	–	–	–	–	–	–	–	–	–	–	✓	–	–	–	–	–	–	✓	–	–
RoadSLAM [48]	–	✓	✓	–	–	–	–	–	✓	–	–	✓	–	–	–	–	–	–	–	–	–	–	–	–
SemanticDepth [49]	–	–	–	–	–	–	–	–	–	–	–	–	–	–	–	–	–	–	–	–	–	✓	–	–

Table 4. Algorithm reported performance.

Name	Time [ms]	Pre.	Rec.	F1	GPU	Dataset	Code
BGK [25]	45	98.66	72.72	83.73	✗	SemKITTI	✗
SNE-RoadSeg [26]	80	96.90	96.61	96.75	✓	KITTI	✓
PLARD [27]	160	96.79	96.86	96.83	✓	KITTI	✓
USNet [28]	22	96.51	97.27	96.89	✓	KITTI	✓
Unsupervised RD [29]	-	83.97	91.83	87.72	✗	KITTI	✗
Map-Supervised RD [9]	280	86.01	89.66	87.80	✓	KITTI	✗
RBANet [30]	160	95.14	97.50	96.30	✓	KITTI	✗
HID-LS [31]	250	92.52	93.71	93.11	✗	KITTI	✗
Curb Detection [32]	12	87.64	89.28	86.98	✗	Proprietary	✗
LiDAR-Histogram [33]	100	93.06	88.41	90.67	✗	KITTI	✗
CyberMELD [10]	50	95.94	91.30	93.56	✗	KITTI	✓
RoadNet3 [34]	16	88.12	90.06	89.08	✓	KITTI	✗
Double Projection [35]	77	95.91	99.28	95.00	✗	SemKITTI	✗
Pseudo-LiDAR [36]	460	97.30	97.54	97.42	✓	KITTI	✗
CLCFNet [37]	23	96.38	96.39	96.38	✓	KITTI	✗
Multi-Cue [38]	2500	84.95	88.55	86.71	✗	KITTI	✗
TRAVEL [39]	19	90.00	96.70	93.10	✗	SemKITTI	✓
Road Markings [40]	-	97.04	94.03	95.51	✗	Proprietary	✗
Line Fitting [41]	36	94.95	94.95	94.95	✗	Proprietary	✗
RoadSLAM [48]	-	87.00	92.00	89.43	✗	Proprietary	✗
YOLOP [42]	43		mIoU: 91.5		✓	BDD100K	✓
HybridNets [43]	37		mIoU: 90.5		✓	BDD100K	✓
Rangenet++ [44]	76		mIoU: 91.8		✓	SemKITTI	✓
SpatioTemporal CRF [45]	147		only qualitative		✓	KITTI	✗
Urban Road Filter [46]	15		only qualitative		✗	Proprietary	✓
Evidential Grids [47]	-		only qualitative		✗	Proprietary	✗
SemanticDepth [49]	637		MAE: 0.48 m		✗	Proprietary	✗

The computational performance in the time column is shown in milliseconds per frame, representing the time it takes to produce one instance of drivable area estimation. The algorithmic performance is represented by the reported measurements of precision, recall, F-score, mean intersection over union (mIoU) and mean absolute error (MAE), with the exception of the three algorithms that did only include qualitative experiments in the publication. The GPU column denotes that the algorithm requires a dedicated parallel processor to reach the reported performance. The dataset column indicates which of their reported performance numbers were chosen to be included on the table. The code column indicates whether those publications offer access to their code. The criteria to choose between datasets whenever more than one set of performance metrics was available was the frequency of appearance in the other publications in order to facilitate fair comparison. Most of the publications included performance measured at least on the KITTI dataset, which explains its prevalence on the table.

Finally, some interesting remarks that can be inferred from studying Tables 2–4 are:

- There is, overall, high reported performance in all the publications. Every algorithm reports at least an F1 score of 83, with one of the methods even reaching 97.
- Execution time and performance are not clearly correlated as the highest performer has a runtime of 460 ms but a very close second only needs 22 ms.
- Learning-based methods usually offer the higher performance [26–28,30,36]. However, there are non-learning-based methods that reach similar performance levels [35,40,41].
- In terms of sensing capabilities, it seems that fusing semantic information plus depth (p.e. LiDAR or RGB-D + RGB) pays off in terms of performance. Nevertheless it is not a must as publications with a single sensor input also perform reasonably well [30,36,40].
- Neural networks offer consistently high performance at a relatively low computational cost but they all require a GPU to perform. The best tradeoff between computation time and algorithm performance is offered by USNet.

- With roughly 65% of the works under 100 ms in execution time and 47% under 50 ms, it is possible to see that a majority of the works today can handle sensor inputs at 10 Hz in real time and almost half can handle 20 Hz, which are two standard 3D LiDAR frequencies.

5. Discussion—Future Research

Drivable area estimation has taken a leap forward in the last decade with the advancements in sensor technology, computational power and algorithmic innovation. The maturity of learning-based methods has broken into this field and raised the bar in terms of computational time and performance measures. Three-dimensional LiDAR technology has seen an increase in data quality, resolution and frequency while lowering its cost, finally making it a viable option for researchers and industrial players. Dedicated computational platforms with integrated GPUs have appeared in the market and are used as the processing core of autonomous vehicles. All of this paints a thriving image of the field today, which, nonetheless, still has some threads to pull until it becomes a solved problem:

- **Fusion variability:** Sensor technology advances have brought the importance of sensor fusion to the foreground. Different sensors fill gaps in the drawbacks of others, and having a diverse array has become vital to reach a robust solution. Most of the methods currently tackle this by optimizing an algorithm for a sensor type and then modal transforming the sensor data from a different type into the optimized one. A potential research direction that would increase performance is delaying modal transformation and fusion to later stages of the algorithm and to optimize separate parallel algorithms for each sensor type. The authors in [37] propose a study with results from different fusion architectures that follows this direction.
- **Three-dimensional LiDAR reflectivity:** Modern developments in sensor technology have brought to the market 3D LiDARs with more sensitive receptors to light reflection intensity. The technology has come to a point where road shape can be understood from reflectivity alone. The drivable area estimation algorithms reliant on reflectivity features could pose a new breakthrough in the state of the art. The authors in [40] introduce a technique for lane marking estimation based on lidar reflectivity that could be adapted for drivable area estimation.
- **Low-fidelity map fusion:** Open-source low-fidelity maps offer valuable context information for drivable area estimation algorithms that is currently underrepresented in the state of the art. Some of the map providers see their contributor base increasing each year [60], while new players backed by big tech companies are appearing now [61]. A research direction for the field could be algorithms that take advantage of those databases as a core part of their proposal. A publication that is already working in that direction is [10].
- **Streamlined data processing:** Technological advancements in the sensor industry have brought sensors that produce high-density and high-quality outputs. Learning-based methods have their performance tied to having access to a significant amount of data. Therefore, the field would greatly benefit from new data storage and processing systems that could streamline the flow of data.
- **RGB strongholds:** RGB-based drivable area estimation is being spearheaded by learning-based methods while LiDAR does not have a clear technique leader yet. A potential gap in the current research is novel learning-based methods that only use 3D LiDAR.
- **Algorithm output:** The algorithms express their road estimation differently in terms of data type and quantity. Depending on the desired application, some outputs could be more interesting than others. Some relevant criteria to study the algorithms could be: (i) if the algorithm outputs is single or multi-label, (ii) if the algorithm outputs an expression of probability or uncertainty together with the estimation, (iii) if the output type is an image, a map, a pointcloud or a road model. All of these can significantly influence the choice in research direction. One example could be the need for an

uncertainty measure helping to deal with potential sensor errors. Another could be the fact that a segmented image is likely to need an extra processing step to obtain a general road representation while a road model can be directly used by the next module.

- **Multi-frame estimation:** Algorithms that fuse estimations from different timesteps are not very common. The few ones that combine estimations carry them out in tracking or recursive neural networks. A potential research direction is to apply multi-frame estimation to produce reliable and robust intermittent-noise drivable area candidates.

6. Conclusions

In this publication, a thorough study of the drivable area estimation literature is conducted in order to take a snapshot of the state of the art as it stands today. This investigation has produced an analysis of the modern sensors used to address the problem and their drawbacks, a summary of the most adequate datasets for research validation, a novel architecture breakdown that helps understand and analyze drivable area estimation algorithms, an outline of the most relevant algorithms in the field with a performance comparison and a suggestion for future research directions. Altogether, this publication aims to offer a snapshot of the state of the field at this point in time and an aid to find a path forward.

Author Contributions: Conceptualization, J.L.H., J.V. and J.G.; data curation, J.L.H., J.V., J.G. and V.J.; formal analysis, J.L.H., J.V. and J.G.; funding acquisition, J.V.; investigation, J.L.H., J.V., J.G. and V.J.; methodology, J.L.H., J.V. and J.G.; project administration, J.V.; resources, J.L.H., J.V. and J.G.; supervision, J.V.; validation, J.L.H., J.V., J.G. and V.J.; writing—original draft, J.L.H.; writing—review and editing, J.L.H., J.V., J.G. and V.J. All authors have read and agreed to the published version of the manuscript.

Funding: This work was partially funded by MCIN/AEI/10.13039/501100011033 and by ERDF A way of making Europe with the project DISCERN (PID2021- 125850OB-I00), and by the Community of Madrid through the SEGVAUTO 4.0-CM Programme (S2018-EMT-4362).

Institutional Review Board Statement: Not applicable.

Informed Consent Statement: Not applicable.

Data Availability Statement: Not applicable.

Conflicts of Interest: The authors declare no conflict of interest.

References

1. Pauls, J.H.; Strauss, T.; Hasberg, C.; Lauer, M.; Stiller, C. Can we trust our maps? An evaluation of road changes and a dataset for map validation. In Proceedings of the 2018 21st International Conference on Intelligent Transportation Systems (ITSC), Maui, HI, USA, 4–7 November 2018; pp. 2639–2644.
2. Wong, K.; Gu, Y.; Kamijo, S. Mapping for autonomous driving: Opportunities and challenges. *IEEE Intell. Transp. Syst. Mag.* **2020**, *13*, 91–106.
3. Gwon, G.P.; Hur, W.S.; Kim, S.W.; Seo, S.W. Generation of a precise and efficient lane-level road map for intelligent vehicle systems. *IEEE Trans. Veh. Technol.* **2016**, *66*, 4517–4533.
4. Bar Hillel, A.; Lerner, R.; Levi, D.; Raz, G. Recent progress in road and lane detection: A survey. *Mach. Vis. Appl.* **2014**, *25*, 727–745.
5. Papadakis, P. Terrain traversability analysis methods for unmanned ground vehicles: A survey. *Eng. Appl. Artif. Intell.* **2013**, *26*, 1373–1385.
6. Liang, D.; Guo, Y.C.; Zhang, S.K.; Mu, T.J.; Huang, X. Lane detection: A survey with new results. *J. Comput. Sci. Technol.* **2020**, *35*, 493–505.
7. Gao, B.; Pan, Y.; Li, C.; Geng, S.; Zhao, H. Are we hungry for 3D LiDAR data for semantic segmentation? a survey of datasets and methods. *arXiv* **2021**, arXiv:2006.04307

8. OpenStreetMap Contributors. 2017. Available online: https://www.openstreetmap.org (accessed on 9 April 2023).
9. Laddha, A.; Kocamaz, M.K.; Navarro-Serment, L.E.; Hebert, M. Map-supervised road detection. In Proceedings of the 2016 IEEE Intelligent Vehicles Symposium (IV), Gothenburg, Sweden, 19–22 June 2016; pp. 118–123.
10. Wang, X.; Qian, Y.; Wang, C.; Yang, M. Map-enhanced ego-lane detection in the missing feature scenarios. *IEEE Access* **2020**, *8*, 107958–107968.
11. Poggenhans, F.; Pauls, J.H.; Janosovits, J.; Orf, S.; Naumann, M.; Kuhnt, F.; Mayr, M. Lanelet2: A High-Definition Map Framework for the Future of Automated Driving. In Proceedings of the 2018 21st International Conference on Intelligent Transportation Systems (ITSC), Maui, HI, USA, 4–7 November 2018.
12. Maierhofer, S.; Klischat, M.; Althoff, M. Commonroad scenario designer: An open-source toolbox for map conversion and scenario creation for autonomous vehicles. In Proceedings of the 2021 IEEE Intelligent Transportation Systems Conference (ITSC), Indianapolis, IN, USA, 19–22 September 2021; pp. 3176–3182.
13. Sun, P.; Kretzschmar, H.; Dotiwalla, X.; Chouard, A.; Patnaik, V.; Tsui, P.; Guo, J.; Zhou, Y.; Chai, Y.; Caine, B.; et al. Scalability in perception for autonomous driving: Waymo open dataset. In Proceedings of the IEEE/CVF Conference on Computer Vision and Pattern Recognition, Seattle, WA, USA, 13–19 June 2020; pp. 2446–2454.
14. Wilson, B.; Qi, W.; Agarwal, T.; Lambert, J.; Singh, J.; Khandelwal, S.; Pan, B.; Kumar, R.; Hartnett, A.; Pontes, J.K.; et al. Argoverse 2: Next generation datasets for self-driving perception and forecasting. *arXiv* **2023**, arXiv:2301.00493.
15. Fritsch, J.; Kuehnl, T.; Geiger, A. A New Performance Measure and Evaluation Benchmark for Road Detection Algorithms. In Proceedings of the International Conference on Intelligent Transportation Systems (ITSC), The Hague, The Netherlands, 6–9 October 2013.
16. Caesar, H.; Bankiti, V.; Lang, A.H.; Vora, S.; Liong, V.E.; Xu, Q.; Krishnan, A.; Pan, Y.; Baldan, G.; Beijbom, O. nuScenes: A multimodal dataset for autonomous driving. In Proceedings of the 2020 IEEE/CVF Conference on Computer Vision and Pattern Recognition (CVPR), Virtual Conference, Seattle, WA, USA, 13–19 June 2020.
17. Yu, F.; Chen, H.; Wang, X.; Xian, W.; Chen, Y.; Liu, F.; Madhavan, V.; Darrell, T. BDD100K: A Diverse Driving Dataset for Heterogeneous Multitask Learning. In Proceedings of the IEEE/CVF Conference on Computer Vision and Pattern Recognition, Nashville, TN, USA, 20–25 June 2020.
18. Cordts, M.; Omran, M.; Ramos, S.; Rehfeld, T.; Enzweiler, M.; Benenson, R.; Franke, U.; Roth, S.; Schiele, B. The Cityscapes Dataset for Semantic Urban Scene Understanding. In Proceedings of the IEEE Conference on Computer Vision and Pattern Recognition (CVPR), Las Vegas, NV, USA, 27–30 June 2016.
19. Liao, Y.; Xie, J.; Geiger, A. KITTI-360: A Novel Dataset and Benchmarks for Urban Scene Understanding in 2D and 3D. *arXiv* **2021**, arXiv:2109.13410.
20. Behley, J.; Garbade, M.; Milioto, A.; Quenzel, J.; Behnke, S.; Stachniss, C.; Gall, J. SemanticKITTI: A Dataset for Semantic Scene Understanding of LiDAR Sequences. In Proceedings of the IEEE/CVF Conference on Computer Vision (ICCV), Seoul, Republic of Korea, 27 October–2 November 2019.
21. Vasiljevic, I.; Kolkin, N.; Zhang, S.; Luo, R.; Wang, H.; Dai, F.Z.; Daniele, A.F.; Mostajabi, M.; Basart, S.; Walter, M.R.; et al. DIODE: A Dense Indoor and Outdoor DEpth Dataset. *arXiv* **2019**, arXiv:1908.00463.
22. Plachetka, C.; Sertolli, B.; Fricke, J.; Klingner, M.; Fingscheidt, T. 3DHD CityScenes: High-Definition Maps in High-Density Point Clouds. In Proceedings of the 2022 IEEE 25th International Conference on Intelligent Transportation Systems (ITSC), Macau, China, 8–12 October 2022; pp. 627–634.
23. Wang, H.; Liu, Z.; Li, Y.; Li, T.; Chen, L.; Sima, C.; Wang, Y.; Jiang, S.; Wen, F.; Xu, H.; et al. Road Genome: A Topology Reasoning Benchmark for Scene Understanding in Autonomous Driving. *arXiv* **2023**, arXiv:2304.10440.
24. Li, Q.; Wang, Y.; Wang, Y.; Zhao, H. HDMapNet: An Online HD Map Construction and Evaluation Framework. *arXiv* **2021**, arXiv:2107.06307.
25. Xue, H.; Fu, H.; Ren, R.; Zhang, J.; Liu, B.; Fan, Y.; Dai, B. LiDAR-based Drivable Region Detection for Autonomous Driving. In Proceedings of the 2021 IEEE/RSJ International Conference on Intelligent Robots and Systems (IROS), Prague, Czech Republic, 27 September–1 October 2021; pp. 1110–1116.
26. Fan, R.; Wang, H.; Cai, P.; Liu, M. Sne-roadseg: Incorporating surface normal information into semantic segmentation for accurate freespace detection. In Proceedings of the European Conference on Computer Vision, Glasgow, UK, 23–28 August 2020; pp. 340–356.
27. Chen, Z.; Zhang, J.; Tao, D. Progressive lidar adaptation for road detection. *IEEE/CAA J. Autom. Sin.* **2019**, *6*, 693–702. [CrossRef]
28. Chang, Y.; Xue, F.; Sheng, F.; Liang, W.; Ming, A. Fast Road Segmentation via Uncertainty-aware Symmetric Network. *arXiv* **2022**, arXiv:2203.04537.
29. Liu, Z.; Yu, S.; Wang, X.; Zheng, N. Detecting drivable area for self-driving cars: An unsupervised approach. *arXiv* **2017**, arXiv:1705.00451.
30. Sun, J.Y.; Kim, S.W.; Lee, S.W.; Kim, Y.W.; Ko, S.J. Reverse and boundary attention network for road segmentation. In Proceedings of the Proceedings of the IEEE/CVF Conference on Computer Vision Workshops, Seoul, Republic of Korea, 27–28 October 2019.
31. Gu, S.; Zhang, Y.; Yuan, X.; Yang, J.; Wu, T.; Kong, H. Histograms of the normalized inverse depth and line scanning for urban road detection. *IEEE Trans. Intell. Transp. Syst.* **2018**, *20*, 3070–3080. [CrossRef]
32. Zhang, Y.; Wang, J.; Wang, X.; Dolan, J.M. Road-segmentation-based curb detection method for self-driving via a 3D-LiDAR sensor. *IEEE Trans. Intell. Transp. Syst.* **2018**, *19*, 3981–3991. [CrossRef]

33. Chen, L.; Yang, J.; Kong, H. Lidar-histogram for fast road and obstacle detection. In Proceedings of the 2017 IEEE International Conference on Robotics and Automation (ICRA), Singapore, 29 May–3 June 2017; pp. 1343–1348.
34. Lyu, Y.; Bai, L.; Huang, X. Road segmentation using cnn and distributed lstm. In Proceedings of the 2019 IEEE International Symposium on Circuits and Systems (ISCAS), Sapporo, Japan, 26–29 May 2019; pp. 1–5.
35. Xu, F.; Liang, H.; Wang, Z.; Lin, L. A Framework for Drivable Area Detection Via Point Cloud Double Projection on Rough Roads. *J. Intell. Robot. Syst.* **2021**, *102*, 1–19.
36. Sun, L.; Zhang, H.; Yin, W. Pseudo-LiDAR-Based Road Detection. *IEEE Trans. Circuits Syst. Video Technol.* **2022**, *32*, 5386–5398. [CrossRef]
37. Gu, S.; Yang, J.; Kong, H. A cascaded lidar-camera fusion network for road detection. In Proceedings of the 2021 IEEE International Conference on Robotics and Automation (ICRA), Xi'an, China, 30 May 2021–5 June 2021; pp. 13308–13314.
38. Wang, L.; Wu, T.; Xiao, Z.; Xiao, L.; Zhao, D.; Han, J. Multi-cue road boundary detection using stereo vision. In Proceedings of the 2016 IEEE International Conference on Vehicular Electronics and Safety (ICVES), Beijing, China, 10–12 July 2016; pp. 1–6.
39. Oh, M.; Jung, E.; Lim, H.; Song, W.; Hu, S.; Lee, E.M.; Park, J.; Kim, J.; Lee, J.; Myung, H. TRAVEL: Traversable Ground and Above-Ground Object Segmentation Using Graph Representation of 3D LiDAR Scans. *arXiv* **2022**, arXiv:2206.03190.
40. Certad, N.; Morales-Alvarez, W.; Olaverri-Monreal, C. Road Markings Segmentation from LIDAR Point Clouds using Reflectivity Information. In Proceedings of the 2022 IEEE International Conference on Vehicular Electronics and Safety (ICVES), Bogota, Colombia, 14–16 November 2022; pp. 1–6.
41. Sun, P.; Zhao, X.; Xu, Z.; Wang, R.; Min, H. A 3D LiDAR data-based dedicated road boundary detection algorithm for autonomous vehicles. *IEEE Access* **2019**, *7*, 29623–29638. [CrossRef]
42. Wu, D.; Liao, M.W.; Zhang, W.T.; Wang, X.G.; Bai, X.; Cheng, W.Q.; Liu, W.Y. Yolop: You only look once for panoptic driving perception. *Mach. Intell. Res.* **2022**, *19*, 550–562. [CrossRef]
43. Vu, D.; Ngo, B.; Phan, H. HybridNets: End-to-End Perception Network. *arXiv* **2022**, arXiv:2203.09035.
44. Milioto, A.; Vizzo, I.; Behley, J.; Stachniss, C. Rangenet++: Fast and accurate lidar semantic segmentation. In Proceedings of the 2019 IEEE/RSJ International Conference on Intelligent Robots and Systems (IROS), Macau, China, 3–8 November 2019; pp. 4213–4220.
45. Rummelhard, L.; Paigwar, A.; Nègre, A.; Laugier, C. Ground estimation and point cloud segmentation using spatiotemporal conditional random field. In Proceedings of the 2017 IEEE Intelligent Vehicles Symposium (IV), Los Angeles, CA, USA, 11–14 June 2017; pp. 1105–1110.
46. Horváth, E.; Pozna, C.; Unger, M. Real-time LiDAR-based urban road and sidewalk detection for autonomous vehicles. *Sensors* **2021**, *22*, 194. [CrossRef] [PubMed]
47. Kurdej, M.; Moras, J.; Cherfaoui, V.; Bonnifait, P. Map-aided evidential grids for driving scene understanding. *IEEE Intell. Transp. Syst. Mag.* **2015**, *7*, 30–41. [CrossRef]
48. Burger, P.; Naujoks, B.; Wuensche, H.J. Unstructured road slam using map predictive road tracking. In Proceedings of the 2019 IEEE Intelligent Transportation Systems Conference (ITSC), Auckland, New Zealand, 27–30 October 2019; pp. 1276–1282.
49. Palafox, P.R.; Betz, J.; Nobis, F.; Riedl, K.; Lienkamp, M. Semanticdepth: Fusing semantic segmentation and monocular depth estimation for enabling autonomous driving in roads without lane lines. *Sensors* **2019**, *19*, 3224. [CrossRef]
50. Badrinarayanan, V.; Kendall, A.; Cipolla, R. Segnet: A deep convolutional encoder-decoder architecture for image segmentation. *IEEE Trans. Pattern Anal. Mach. Intell.* **2017**, *39*, 2481–2495. [CrossRef] [PubMed]
51. Lee, J.H.; Han, M.K.; Ko, D.W.; Suh, I.H. From big to small: Multi-scale local planar guidance for monocular depth estimation. *arXiv* **2019**, arXiv:1907.10326.
52. Wang, Y.; Chao, W.L.; Garg, D.; Hariharan, B.; Campbell, M.; Weinberger, K.Q. Pseudo-lidar from visual depth estimation: Bridging the gap in 3d object detection for autonomous driving. In Proceedings of the IEEE/CVF Conference on Computer Vision and Pattern Recognition, Long Beach, CA, USA, 15–20 June 2019; pp. 8445–8453.
53. Drucker, H.; Burges, C.J.; Kaufman, L.; Smola, A.; Vapnik, V. Support vector regression machines. *Adv. Neural Inf. Process. Syst.* **1996**, *9*, 1–7.
54. Otsu, N. A threshold selection method from gray-level histograms. *IEEE Trans. Syst. Man, Cybern.* **1979**, *9*, 62–66. [CrossRef]
55. Wang, C.Y.; Bochkovskiy, A.; Liao, H.Y.M. Scaled-yolov4: Scaling cross stage partial network. In Proceedings of the IEEE/cvf Conference on Computer Vision and Pattern Recognition, Nashville, TN, USA, 20–25 June 2021; pp. 13029–13038.
56. He, K.; Zhang, X.; Ren, S.; Sun, J. Spatial pyramid pooling in deep convolutional networks for visual recognition. *IEEE Trans. Pattern Anal. Mach. Intell.* **2015**, *37*, 1904–1916. [CrossRef] [PubMed]
57. Lin, T.Y.; Dollár, P.; Girshick, R.; He, K.; Hariharan, B.; Belongie, S. Feature pyramid networks for object detection. In Proceedings of the IEEE Conference on Computer Vision and Pattern Recognition, Honolulu, HI, USA, 21–26 July 2017; pp. 2117–2125.
58. Tan, M.; Le, Q. Efficientnet: Rethinking model scaling for convolutional neural networks. In Proceedings of the International Conference on Machine Learning (PMLR), Long Beach, CA, USA, 9–15 June 2019; pp. 6105–6114.
59. Thrun, S.; Montemerlo, M. The graph SLAM algorithm with applications to large-scale mapping of urban structures. *Int. J. Robot. Res.* **2006**, *25*, 403–429. [CrossRef]

60. OSM Maps User Stats. Available online: https://osmstats.neis-one.org/?item=members (accessed on 21 July 2023)
61. Overture Maps. Available online: https://overturemaps.org/(accessed on 21 July 2023).

Disclaimer/Publisher's Note: The statements, opinions and data contained in all publications are solely those of the individual author(s) and contributor(s) and not of MDPI and/or the editor(s). MDPI and/or the editor(s) disclaim responsibility for any injury to people or property resulting from any ideas, methods, instructions or products referred to in the content.

sensors

MDPI

Article

How Do We Calibrate a Battery Electric Vehicle Model Based on Controller Area Network Bus Data?

Dávid Tollner [1,2,*], Ádám Nyerges [1], Mahmoud Said Jneid [1,2], Attila Geleta [2] and Máté Zöldy [1,*]

1 Department of Automotive Technologies, Budapest University of Technology and Economics, 1111 Budapest, Hungary; nyerges.adam@kjk.bme.hu (Á.N.); mahmoud.jneid@kjk.bme.hu (M.S.J.)
2 Robert Bosch Kft., 1103 Budapest, Hungary; attila.geleta@hu.bosch.com
* Correspondence: tollner.david@kjk.bme.hu (D.T.); zoldy.mate@kjk.bme.hu (M.Z.)

Abstract: Transforming an up-to-date vehicle into a measurement system is a rewarding task due to the large number of different sensors in the onboard control and diagnostic systems. These procedures are not performed by a single control unit; it is necessary to share the signal values over a communication network, to which an external device can be connected to record the real traffic. The paper aims to use these recorded data for 1 DOF longitudinal vehicle and powertrain model validation. For repeatability, three city routes are selected: plain road, smaller road grade, and higher road grade in both directions. Therefore, the drivetrain system is tested in a high load range, even with long-term recuperation. The altitude changes are recorded with a DGPS system. By the recorded measurements, the vehicle and the drivetrain model can be calibrated, such as the air drag parameters, the rolling resistances, and the efficiencies of the drivetrain. The validation criteria are defined for speed tracking, and the relative tolerance of the cumulated energy should be below 10%. At the end of the day, a developed model is ready for energetic analysis or control strategy design. The energy balance of the applied cycles is also presented to prove that.

Keywords: data logging; simulation; electrical consumption; electric vehicles

1. Introduction

In recent years, sustainability has strongly influenced the automotive industry and mobility in design and manufacturing processes. Tighter emission standards are supporting the development of alternative fuel powertrains [1]. The electric powertrain control module had to process a large number of sensor values simultaneously, such as the voltage levels of the cells in the high-voltage (HV) battery and the inverter and motor temperatures, to ensure that they were within the appropriate range before damage or overheating occurred. These can be prevented by reducing power. Measuring stator temperature is typically challenging, as the sensor only measures the temperature at one point. The temperature can be determined based on a model for the rest of the surface. Determining the temperature of the rotor is an even more complicated problem because of its rotating motion. Furthermore, the actual State of Charge (SoC) and State of Health (SoH) of the HV battery is also a complex task, usually based on a battery model [2–5].

The inputs to these models are quantities measured by different sensors. An electric car powertrain needs 1 or 2 speed and position sensors, 15–20 current sensors, and 20–30 temperature sensors, usually NTCs [6]. The values of these sensors are required by several electronic control units (ECUs); therefore, they can be found on the vehicle's communication network. Based on these, it was obvious for us to log these data from the car for the simulation instead of installing new sensors in the vehicle.

Modeling and simulating the vehicle and drivetrain operation helps check the measured signals, estimate unmeasurable signals, and diagnose faults in the systems. It is also useful to improve the sampled and measured signals for better resolution.

Implementing battery electric vehicle (BEV) models is simpler than conventional drivetrains [7]. The lack of clutch and gear selection strategies makes the modeling process more accurate; it is easier to reproduce real-world driving [8].

Modern BEVs appeared less than two decades ago, and their development is a very focused area even today. The "optimal" BEV layout is not yet formed; there will be significant developments in efficiency, range, dynamics, and thermal management [9].

The paper will present a model that can be used for energetic analysis and control strategy development. The energy transformation is deeply understandable through model verification and validation, and the key development areas can be found.

2. Measurement Strategy

The powertrain of an electric car has significantly fewer components than an internal combustion version. However, much more information is available on the EVs' in-vehicle communication network. There are several reasons for this: First, the number of ECUs in the vehicle continuously increases due to more advanced driver assistance systems and comfort functions [10]. Another reason is that in an electric car, the drive-related control systems must monitor many more parameters, from battery output current through inverter temperature to rotor position. It is also essential to maximize efficiency and lifetime, for example, by constantly monitoring the battery cell voltages to ensure that none of the cells are discharged below a threshold value, as this would permanently damage them. The development of ECUs has brought the Functional Safety Standard [11], which regulates the safe operation of safety-related electrical/electronic systems and broadly defines the hardware and software features of modern ECUs.

The complex monitoring system requires numerous sensors whose measured values can be retrieved from the vehicle's communication network, as multiple ECUs typically require these. In most cases, the in-vehicle communication is unknown, so the data are only interpretable after complex reverse engineering. However, retrieving data from specific ECUs using a diagnostic protocol is possible.

2.1. Protocols

Automotive protocols can be divided into two main categories: communication and diagnostic protocols. The former defines the rules for transmitting information, such as the number of wires and voltage levels, while the latter defines the structure of the information packages, i.e., messages.

Despite the increased amount of in-vehicle information in EVs, the well-known Controller Area Network (CAN) 2.0 [12] protocol is generally used by car manufacturers for communication between ECUs due to its low cost and simplicity. However, some Advanced Driver-Assistance Systems (ADAS) functions already require Automotive Ethernet [13] or FlexRay [14] for higher bandwidth.

An essential property of CAN is that it provides serial asynchronous data transmission. Therefore, the data recorded from the vehicle are very close in time (~50 ms), but there is still a time offset. This is a not negligible issue because if the data are not transformed to a shared time axis, there may be operating points where the vehicle speed is still increasing compared to the previous time point, but braking has already started.

The onboard diagnostic (OBD) standard [15] was the first official diagnostic standard, introduced in California in 1988. The purpose of this was to ensure that the vehicle constantly monitored the operation of the emission control systems. Later, the more specific OBD-II [16] standard was released, mandatory in the US since 1996 [17], for European petrol vehicles since 2001, and for European diesel vehicles since 2004 [18].

The OBD-II standard further includes that in the event of a malfunction or incorrect values, the vehicle must store the fault with the environmental parameters, which must be readable later. So, it has to support not only onboard but also off-board diagnostics, i.e., external devices must also be able to access the vehicle's communication network. If an external device can be connected to the car, it seems evident that the values of the sensors monitored by the vehicle should be accessible to external devices. This enabled data logging.

The OBD-II diagnostic protocol is public so that the parameters can be recorded and tested in the same form on all vehicles. This means only general information can be checked, such as speed, engine speed, or throttle position. Due to the growth of hybrid cars, OBD-II has been extended to provide not only emission-relevant data, for example, the hybrid battery pack's remaining life or the high-voltage battery SoC value.

For vehicle fault detection, not only the emission-relevant components are important, so the KeyWord Protocol 2000 (KWP) [19] was created as an extension of the OBD-II diagnostic standard, initially using the K-line bus system, and only a few ECUs were available with it. In the OBD-II standard, there was a connector [20] where SAE J1850 [21], K-line [22], and CAN bus systems are accessible, so KWP uses this connector as well.

KWP is not a generalized diagnostic, so it provides much more detailed data, such as steering wheel angle or brake system pressure. As a result, each vehicle has different ECU access addresses (identifiers) and values by which specific parameters can be accessed (local identifiers). Therefore, data logging is much more difficult with this protocol, as the software has to be parameterized for each vehicle.

The disadvantage of the K-line is that the maximum data rate is 10.4 Kbit/s, while with CAN, 1 Mbit/s is available. Therefore, the KWP2000 on CAN [23] standard has extended the diagnostic possibilities, such as flash programming of ECUs.

Unified diagnostic services (UDSs) [24] were created based on the KWP–CAN model to harmonize diagnostic standards. One of the advantages of UDS is that it is independent of the bus system, i.e., it works not only on CAN but also on Ethernet or FlexRay. Furthermore, in contrast to previous standards, periodic data requests are possible, so if continuous data logging is required, there is no need to send continuous data request messages to the ECUs.

2.2. Test Vehicle

The test vehicle was a 2011 Mitsubishi MiEV, produced from 2009 to 2021. The MiEV is one of the original first-generation electric vehicles, despite only around 32,000 being produced. The main parameters of the vehicle are shown in Table 1. The 47 kW power output may seem low, but the electric drive's characteristics and low weight make it dynamic to drive [25].

Table 1. Parameters of the test vehicles.

Main Group	Parameter	Value
Body	Dimensions (L × W × H)	3395 × 1475 × 1610 mm
	Mass	1100 kg
	Propulsion type	Rear wheel drive
Motor	Type	PMSM
	Max. power	47 kW
	Max. torque	180 Nm
	Transmission	6.11
Battery	Type	Lithium-ion
	Nominal voltage	330 V
	Max. output power	60 kW
	Nominal/Useable capacity	16 kWh/14.5 kWh
Range	EPA	95 km
	NEDC	160 km
Resistance	Drag coefficient	0.33
	Drag area	2.14 m^2
	Rolling resistance	0.011

2.3. Data Logging and Devices

As can be seen in Figure 1, the components of the test vehicle's powertrain communicate with each other via the CAN bus so that the relevant parameters can be retrieved via the OBD II connector using the KWP diagnostic protocol [26].

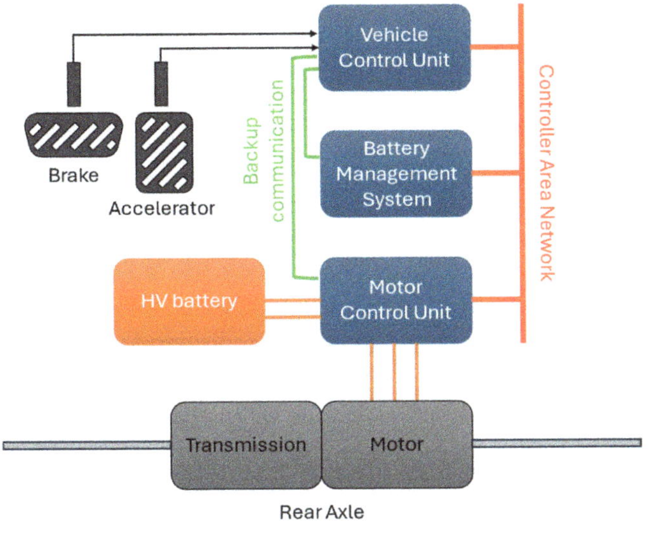

Figure 1. The communication network of the test vehicle.

A CSS CANedge 1 [27] standalone CAN bus data logger was used for data logging, for which the software was written by ourselves. Since gradient resistance has a major impact on consumption, the route must also be recorded. For this purpose, a u-blox NEO-M9V sensor module was used for the data logger, which has an accuracy of 2 m circular error probability and a sampling rate of 5 Hz.

The inputs of the vehicle model detailed in Section 3 are the parameters that the driver has direct control over; these are the accelerator, brake pedal, and steering wheel. Considering the additional driving resistance forces, the curvature resistance is negligible compared to the road resistance, so in this model, the curvature is not considered.

For data logging, it is important to note that for CAN 2.0, communication messages can be up to 8 bytes long, of which the diagnostics reserve 2–3 bytes as special values. In the case of KWP, typically, a local identifier can be used to retrieve data for an entire group of components, so the vehicle can send multiple response messages for a single data request; this is multi-frame messaging. However, in most cases, the sensor values are reduced to as few bytes as possible. For example, vehicle speed can be defined as an integer in 1 byte, from 0 to 255. In this example, the resolution of the speed is 1, since the minimum value change is 1. The SoC value can be from 0% up to 100%, so from 0 to 200 can be stored on 1 byte, and the minimum value difference, i.e., resolution, can be 0.5%.

The accelerator and brake pedals can be used to obtain the desired speed, but the test car's exact pedal characteristics were unknown, so the vehicle speed was used as a reference signal. The speed can be extracted from the test vehicle with a resolution of 1 km/h, so—taking advantage of the fixed-ratio non-disconnectable drive—the speed calculated from the engine speed was used instead of the vehicle speed, which has a much higher resolution converted back to km/h. The data retrieved from the vehicle are given in Table 2.

Table 2. List of logged parameters with the smallest value change.

Parameter	Unit	Resolution
Speed	km/h	1 km/h
Engine speed	RPM	1 RPM
Brake system pressure	Bar	0.0154 Bar
SoC	%	0.5%
HV battery voltage	V	2 V
HV battery current	A	1 A
Torque	Nm	0.1 Nm
Height above sea level	m	1 m

The status of the state of charge was also monitored to investigate energy consumption. Similar to the speed, the resolution of the charge is not sufficient because the minimum value change for the test car is 0.5%. Therefore, the current and voltage of the high-voltage battery were also retrieved from the CAN bus.

3. Model Structure

The measurement system development aims to verify and validate a longitudinal vehicle and powertrain model that can present the energetics of a battery electric vehicle (BEV). A new approach can be introduced in BEVs because there are three in-vehicle main onboard energy storage systems: the battery, the kinetic energy of the vehicle, and the gravitational potential energy of the vehicle. The novelty of these vehicles is that the energies can be converted back and forth in any way.

3.1. Vehicle Model

The motion Equation (1) of the vehicle represents the sum of the forces acting on the vehicle [28], considering the rolling weights with λ. The traction, the brake, and the resistances can change the vehicle's speed.

$$\lambda m_{veh} a_{veh} = F_{trac} - F_{brake} - F_{res}. \tag{1}$$

Due to the regenerative charging in electric vehicles, the traction force can be negative. Thus, the brake force, which is a loss, has a minor role in vehicles with combustion engines.

The resistance forces can be handled as follows: (2) They are the sum of the road and the drag resistances.

$$F_{res} = F_{road} + F_{drag}. \tag{2}$$

Furthermore, road resistance has two parts: rolling and grade resistance, where α is the road grade and f is the rolling resistance factor:

$$F_{road} = F_{roll} + F_{grade} = m_{veh} g (sin\alpha + f cos\alpha) \tag{3}$$

The drag resistance can be handled in the following well-known way:

$$F_{drag} = \frac{1}{2} \rho_{air} c_w A_{veh} v_{veh}^2. \tag{4}$$

The vehicle also has a conventional hydraulic brake system. Ignoring the geometrical parameters of the wheel brake systems, the brake force can be modeled as a function of the brake pressure multiplied by a constant coefficient:

$$F_{brake} = k_{brake} p_{brake} m g. \tag{5}$$

Due to the dual braking opportunity, the brake control should have a strategy:

- The regenerative braking does not operate below 10 km/h;
- The hydraulic brake system does not generate force on the stopped vehicle;
- In battery electric vehicles, during emergency braking, the regenerative braking also does not operate; finally, this function was unnecessary because emergency braking was not tested.

3.2. Powertrain Model

In a BEV, the powertrain system has two parts: an electric machine and a gearbox. Due to regenerative braking, the energy flow can go back and forth in them. The losses always decrease energy flow. Thus, there are two forms of the equations.

The connection between the electric machine's torque and the traction force on the wheels can be estimated in the following way, with positive torque:

$$F_{trac} = M_{mot}\eta_{mech}\frac{i_{pt}}{r_{wheel}}. \tag{6}$$

During regenerative braking, mechanical efficiency should be applied with its reciprocal:

$$F_{trac} = M_{mot}\frac{1}{\eta_{mech}}\frac{i_{pt}}{r_{wheel}}. \tag{7}$$

The mechanical efficiency represents the mechanical losses in the gearbox. In this paper, it is modeled as a constant. Without measurements, it can be estimated with the following equation [28]:

$$\eta_{mech} = 0.98^a 0.97^b 0.99^c. \tag{8}$$

A throttle input controls the actual torque of the electric machine:

$$M_{mot} = u_{throttle} Mmot_{max} \tag{9}$$

This throttle with a negative sign can also operate as an input for regenerative braking. The power consumption of the electric motor can be estimated by considering its efficiency map. The equations are also different with positive and negative torque. With positive torque:

$$P_{mot} = M_{mot}\omega_{mot}\frac{1}{\eta_{mot}(M_{mot}, \omega_{mot})}. \tag{10}$$

And with negative torque, i.e., with regenerative braking:

$$P_{mot} = M_{mot}\omega_{mot}\eta_{mot}(M_{mot}, \omega_{mot}). \tag{11}$$

The model does not consider the actual temperature of the electric machine. This would change its torque characteristics. Based on the measurements, it was not necessary to implement them into the model.

3.3. Inverter and Battery Model

The equation logic of the energy flow through the inverter is similar to the previous ones: the losses always decrease. Up-to-date inverters have high efficiencies. In this model, it is estimated as a constant.

With positive engine torque, it can be estimated as follows:

$$P_{inv} = P_{mot}\frac{1}{\eta_{inv}}. \tag{12}$$

Also, with negative engine torque, it is changed in the following way:

$$P_{inv} = P_{mot}\eta_{inv}. \tag{13}$$

The current consumption of the inverter can be calculated by the battery voltage:

$$I_{inv} = \frac{P_{inv}}{U_{bat}}. \tag{14}$$

The battery's main parameter is its state of charge. It can be estimated by the following equation [8]:

$$SoC = \frac{C_0 - \int_0^t P_{inv}(t)dt}{C_0}. \tag{15}$$

The model does not consider the battery's SoC-dependent charging behavior. The measurements did not start with a fully charged battery.

Other major in-vehicle energy consumers near the powertrain system include cabin heating/cooling systems, power steering systems, and other onboard electric consumers. The energy consumption of these systems can be at a wide interval. The heating/cooling system was turned off during the measurements, and only the necessary current consumers were used. Thus, the power demand of these systems was estimated to be constant.

3.4. Model Inputs and Driver Model

The longitudinal vehicle-end powertrain model has the following inputs:

- Vehicle speed demand, Matlab
- Hydraulic brake system actual pressure,
- Road grade.

The model can be calibrated and validated by the following signals:

- Electric motor's actual torque and actual speed,
- Inverter actual current consumption,
- Battery's actual voltage and state of charge.

Matlab Simulink's (version 2023b) longitudinal driver model was applied from the Powertrain Blockset in the control loop. It was set as a PI driver.

For the logically right operation, some complement functions have to be implemented into the model. At the same time, there should not be throttling or mechanical braking. Additionally, below 10 km/h, there should not be any regenerative braking. The block scheme of the model can be seen in Figure 2.

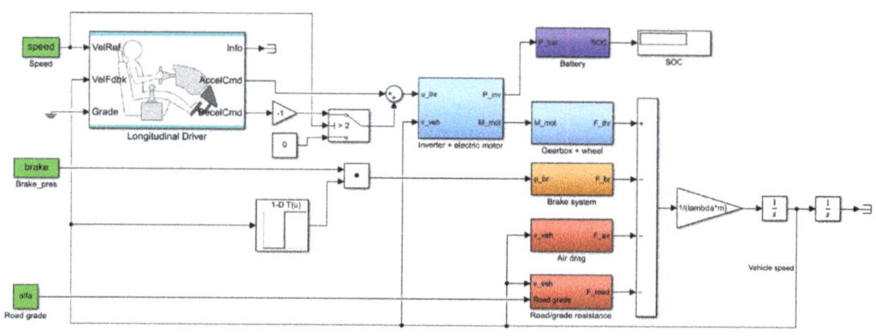

Figure 2. The block scheme of the longitudinal vehicle and powertrain model.

4. Measurements

For the model development, three routes were selected in Budapest. All of them are city roads, i.e., the maximum legal speed is 50 km/h. However, thanks to Budapest's topography, smaller and higher road grades can be selected. The vehicle's high load and regenerative braking can be analyzed by these.

The first applied route is called the "Erzsébet" cycle because it is between the BME University and the Erzsébet bridge along the river Duna. The route does not contain high elevation changes. The road grade changes are nearly negligible. The length is 4.7 km, and the maximum road grade is below 4%. Due to the traffic control lights, the cycle has several stops. The map of the route can be seen in Figure 3.

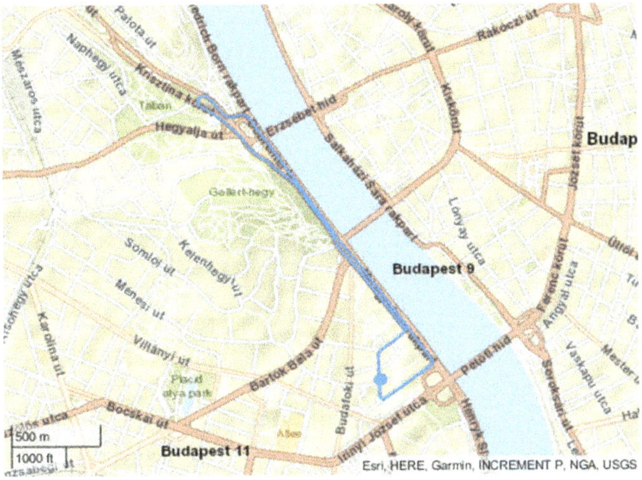

Figure 3. The route of the "Erzsébet" cycle.

The second route is called the "Citadella" cycle; the route connects the university and the Citadella castle on the top of Szent Gellért hill. The elevation change is 91 m, and the maximum road grade is 12%. The length is 5 km. The vehicle runs to the top of the cycle and then partially returns. Therefore, the cycle contains a higher load, as does the recuperation part. The map of the route can be seen in Figure 4.

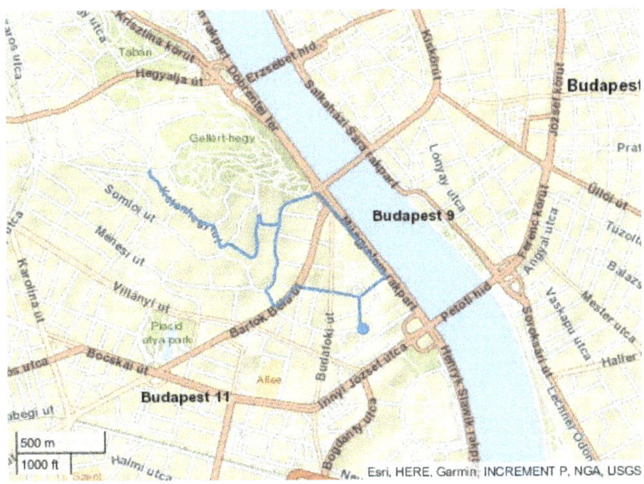

Figure 4. The route of the "Citadella" cycle.

The third and fourth routes are called "Normafa Up" and "Normafa Down". Normafa is one of Budapest's highest points. In "Normafa Up", the vehicle is driven to the hill. The way back on the same route is "Normafa Down". The altitude change is 326 m, and the maximum road grade is 11%. The length is nearly 6 km. The map of the route can be seen in Figure 5.

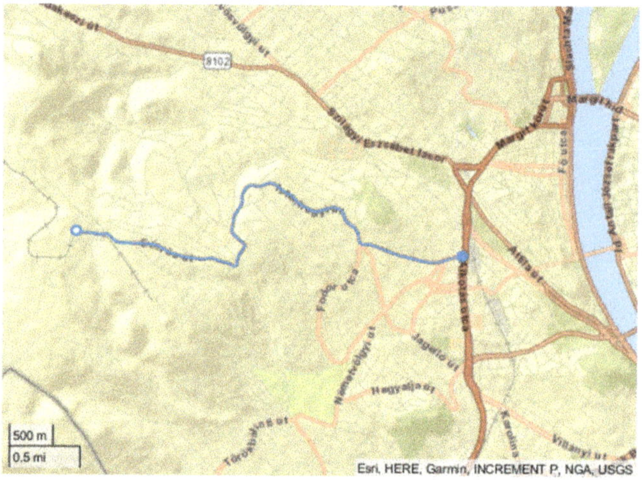

Figure 5. The route of the "Normafa" cycle.

The road grade differences between the three routes are depicted in Figure 6. Moreover, Table 3 summarizes the main properties of the selected routes.

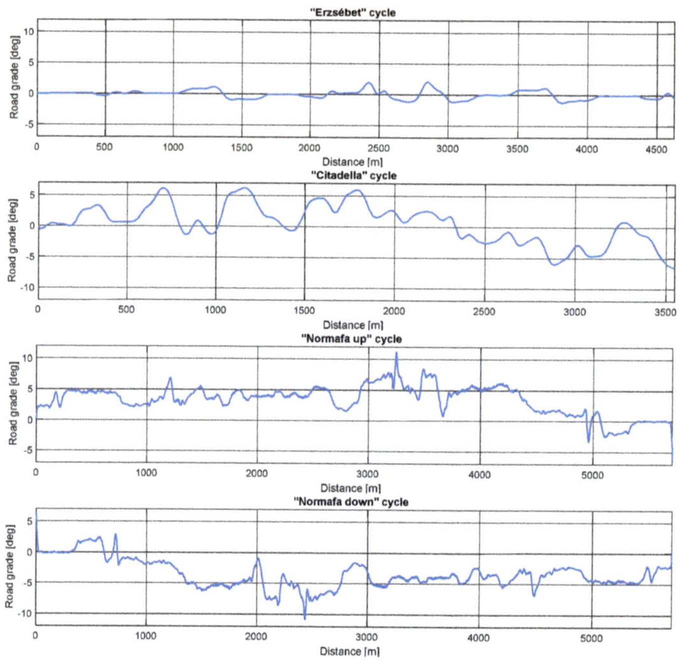

Figure 6. Road grade changes on the selected routes.

Table 3. Main properties of the selected routes.

	"Erzsébet" Cycle	"Citadella" Cycle	"Normafa" Cycle
Type	essentially flat road	uphill then downhill	same route twice: up and down
Distance	4.7 km	5 km	5.8 km
Time	~12 min	~17 min	up ~9 min, down ~12 min
Road grade	maximum 4%	maximum 12%	maximum 11%

Several data were recorded during the measurements. For the longitudinal EV model validation, the following signals are necessary in function of time:

- High-voltage battery state-of-charge (SoC),
- Inverter input power (the voltage and the current can also be measured),
- Motor speed,
- Motor torque, which is an estimated value by the ECU,
- Vehicle speed,
- Hydraulic brake pressure,
- Route's longitude, latitude, and altitude coordinates for estimating the elevation.

Figure 7 shows the recorded data during the measurements. In further research, the vehicle's thermal and lateral behavior can also be analyzed.

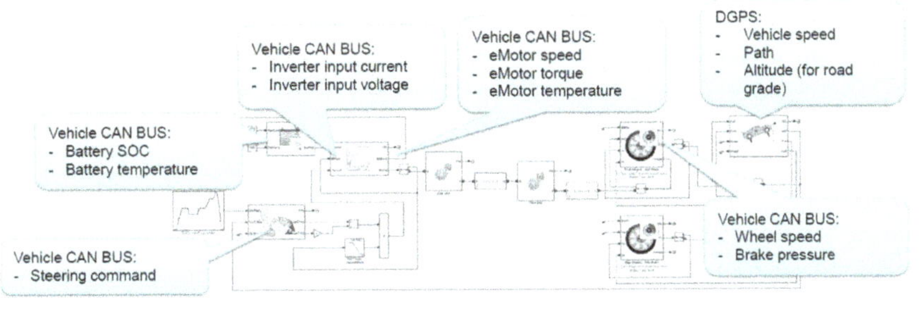

Figure 7. The recorded data during the measurements.

The pressure measurement of the hydraulic brake is a key moment in the model validation; it makes it easier. Electric vehicles at slow speeds (typically below 10 km/h) do not use recuperative braking due to weak motor efficiency. In this speed range, the conventional brake system does all the braking. During the deceleration, a pre-defined strategy gradually switches between the two braking possibilities. This strategy can be analyzed by the measurements, but it is not the aim of this research. By recording the brake pressure, the braking force on the vehicle can be estimated, and the driver controller should only complement it (if necessary) by the drivetrain.

In electric vehicles, while it stays and there are no pedal commands, there is some traction torque on the drivetrain. It can be ceased by putting the gear selector on neutral or pushing the brake pedal harder This behavior should have been applied during the measurements for more accessible model validation.

5. Model Verification and Validation Strategy

For the first approach, most parameters can be set by the literature. The most uncertain parameters are the rolling resistance factor, the frontal area, the rotational mass factor, and the efficiency maps. With enough measurement results, these parameters can be more accurate: rolling resistance can be determined at low speeds, and air drag parameters can be set at higher vehicle speeds. Electric vehicles typically have a one-speed gearbox; its mechanical efficiency can be estimated by the number of connected gear pairs [28].

The efficiency map of the electric motor is unknown. However, the efficiency map of synchronous electric motors is not as varied as it is for internal combustion engines [29].

For this aim, the efficiency map of the first-generation Nissan Leaf's motor characteristics [30] is used with smaller modifications: changing the speed, torque, and power range. Furthermore, smaller tuning was necessary at higher speeds and in the recuperation zone.

The wind magnitude and direction, as well as the road surface changes, are neglected in the model due to measuring issues. The measurements were made in September in good and calm weather. The most significant uncertainty is the road grade estimation. The altitudes are known from the DGPS recordings, but their accuracy is only 1 m. The data should be filtered (smoothed), and linear interpolation was used between the equidistant distances. In the presented model, the distance between two road grade points is 10 m to avoid high oscillation.

The driver model is a PI controller based on Matlab's built-in driver controller [31]. Validation criteria can be defined for speed tracking. The root mean square (RMS) error was selected for this purpose. The aim is to keep them below 10% [32].

$$RMS = \sqrt{\frac{1}{T}\int_0^T \left(\frac{x_{meas} - x_{sim}}{x_{meas}}\right)^2 dt}. \quad (16)$$

Tuning the P and I parameters of the driver model typically results in two things. With strict requirements for speed tracking, the controller will give us nervous throttle and brake commands, i.e., the torque signal cannot be validated due to the high peak values. The torque maximums can be decreased by reducing the strict speed track requirements with lower P and I parameters in the controller. In this case, the speed track will have a small delay, as will the torque input. Thus, the torque tracking cannot be validated by the RMS error. The case of the inverter input power is the same. After several iterations, the p value was selected at 10, and the I value was selected at 3. In further research, more complex driver controllers can be tested; however, it is not the aim of this paper—for parameter identification, the PI controller is the right choice [33].

Considering that the modeling aim is an energetic analysis of the electric vehicle, another validation criterion can be the relative tolerance of the cumulative input energy of the inverter, i.e., the consumed energy during the cycle:

$$x_{rel\_err} = \frac{\int_0^T x_{sim} dt - \int_0^T x_{meas} dt}{\int_0^T x_{meas} dt}. \quad (17)$$

The validation criteria can also be to keep this value below 10%.

The measured signals are used to check the energy transformation in three places in the chain:

- Vehicle speed,
- Electric motor output: speed and torque,
- Inverter input: power, voltage, and current,
- Battery SoC.

The model calibration and validation processes can be simplified by dividing the model along these chain links. For example, the electric motor's efficiency can be checked by its inputs and outputs without the driver and drivetrain models.

Considering these, the parameter calibration was conducted using the following steps:

- Battery nominal capacity calibration is conducted using the inverter input power. Additionally, the non-drivetrain consumer's consumption can also be set at a constant value.
- Electric motor efficiency map calibration is conducted using the input of the inverter and the motor output power (speed and torque). The inverter efficiency is estimated by a constant.
- Vehicle parameter calibration with the measured torque input. The realizable vehicle speed response should fit the measured vehicle speed. Without feedback and driver

control, perfect fitting cannot be achieved due to disturbances and uncertainties. However, for parameter calibration, it is the most helpful step because the controller's behavior can be neglected.

After these steps, the feedback model can be tested and finalized.

6. Model Calibration and Results

The achieved model parameters after the validation process are presented in Table 4. Most of them are in the typical range for a passenger vehicle. The rolling resistance factor is slightly lower than its usual value, probably caused by the very narrow wheels. The parameter nomenclature presents the changing parameters during the simulation in Appendix A.

Table 4. The values of the calibrated model parameters.

Name	Sign	Value	Unit
Air density (20 °C)	ρ_{air}	1.188	kg/m^3
Battery nominal capacity	C_0	10	kWh
Brake force coefficient	k_{brake}	0.012	1/bar
Drag coefficient	c_w	0.25	-
Electric machine's maximum power	P_{max}	45	kW
Electric machine's maximum speed	n_{max}	10,000	1/min
Electric machine's maximum torque	M_{max}	180	Nm
Gravitational acceleration	g	9.81	m/s^2
Inverter efficiency	η_{inv}	0.99	-
Number of loaded axle joints	c	2	-
Number of loaded bevel gears	b	0	-
Number of loaded spur gears	a	2	-
Road grade	α	0	°
Rolling resistance factor	f	0.005	-
Rotational mass factor	λ	1.2	-
Transmission gear ratio	i_{pt}	6.11	-
Transmission mechanical efficiency	η_{mech}	0.97	-
Vehicle frontal area	A	2	m^2
Vehicle mass	m_{veh}	1300	kg
Wheel radius	r_{wheel}	0.2784	m

The rotational mass factor is another important parameter that is not easily achievable. Its value is slightly higher than that of conventional (thermal engine) vehicles. The cause can be the different properties of electric drivetrains.

Figure 8 depicts the different cycles' speed profiles and speed tracking. As can be seen, the "Erzsébet" cycle is typical city driving, with long stays due to the traffic lights. It reaches only once the legal maximum speed. Thus, the dominant property of this cycle are the slow accelerations and decelerations. The motor load is low due to the small road gradients.

The "Citadella" cycle has a smaller rate of staying. It runs through a small-speed residential area, where the legal maximum speed is sometimes 30 km/h. The narrow lanes and the high road grade resulted in a smaller speed cycle with higher motor loads.

Finally, the "Normafa" cycle's speed results can be seen in Figure 8. For most of the cycle, the legal maximum speed is reachable, and traffic lights do not stop the vehicle too often. "Normafa Up" requires higher motor torques; "Normafa Down" provides more recuperation. It is interesting that "Normafa Down" has lower speeds.

The speed tracking can also be seen in the figure. The biggest differences are during braking. Furthermore, there are some overshoots during every acceleration.

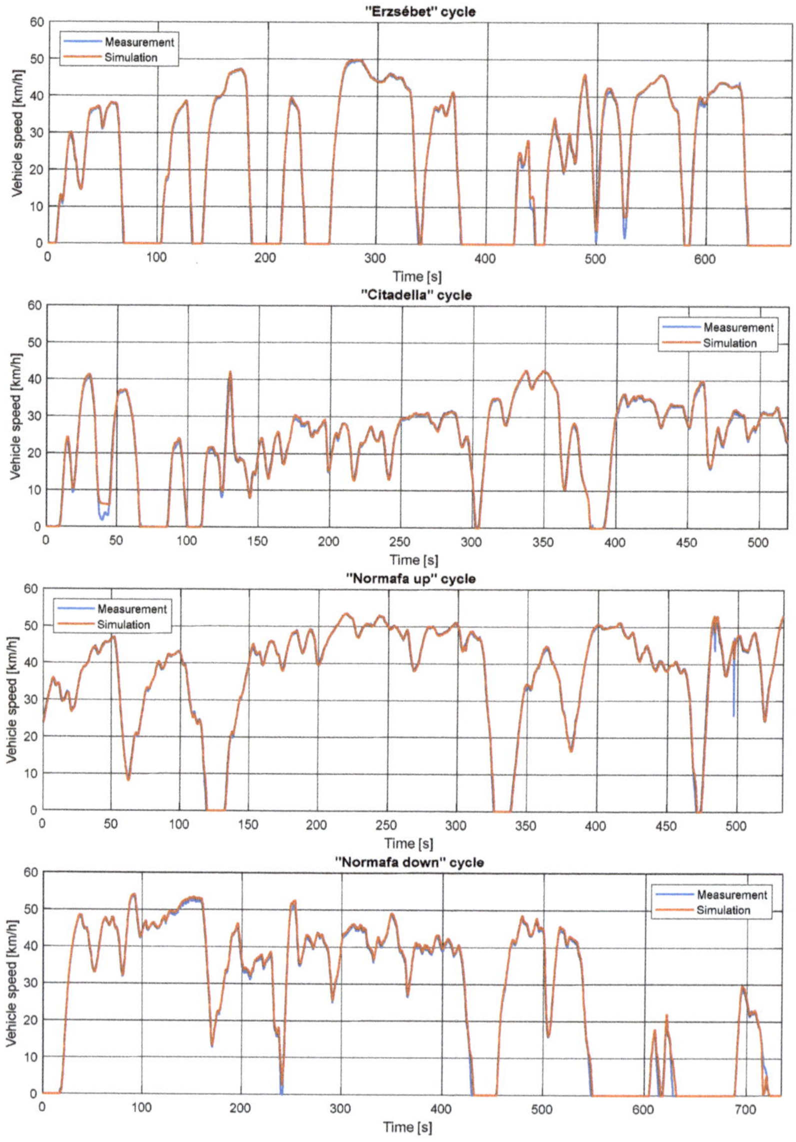

Figure 8. Measured and simulated speed profiles and speed tracking of the different cycles.

In the following figures, the verification will be demonstrated through examples. The other omitted diagrams can be found in the Appendices B and C.

Figure 9 shows the measured and simulated torque during the "Erzsébet" cycle. The signal tracking is verifiable; only the peak values have some inaccuracy. It is caused by the PI driver controller's reaction to the slight speed delay. Thus, the controller's behavior is more nervous than that of the real driver. In the other measurements, the magnitude is different due to the altitude changes: more significant for positive road grade, more regenerative torque for negative road grade. These results can be seen in Appendices B and C.

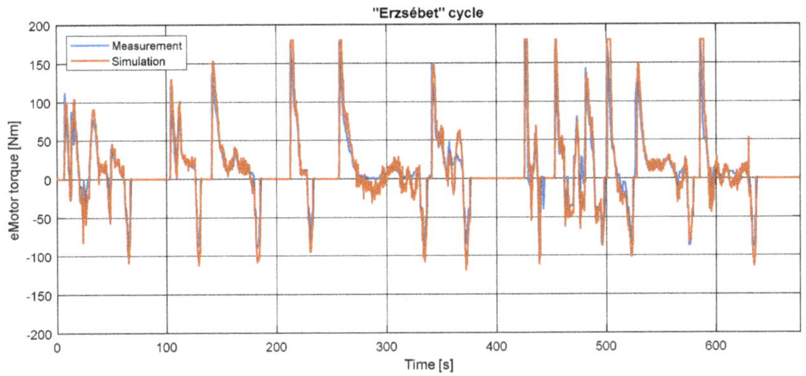

Figure 9. Measured and simulated torque of the electric motor during the "Erzsébet" cycle.

The vehicle has a permanent magnet synchronous motor, which usually has more than 95% top efficiency. In everyday use, it rarely operates at its maximum efficiency; Figure 10 shows the simulation results for the "Normafa Up" cycle. While the motor provides significant torque, its efficiency is never less than 60%. The typical magnitude is between 80 and 90%. During recuperation, it is slightly lower.

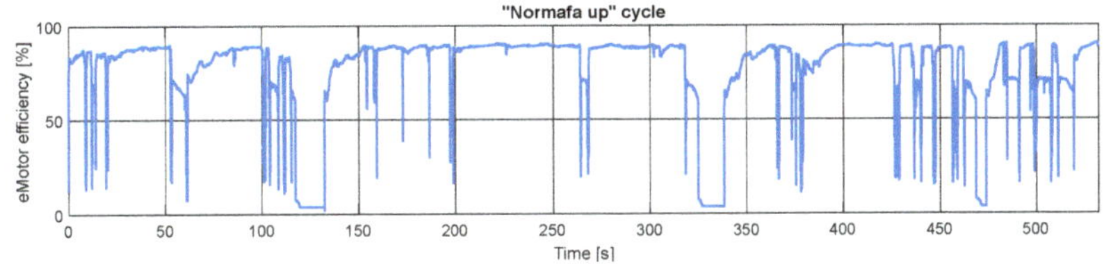

Figure 10. Measured and simulated efficiency of the electric motor during the "Erzsébet" cycle.

In the same measurement, the battery output power is depicted in Figure 11. The behavior is like the torque's: it is verifiable but has inaccurate peaks and a nervous oscillation. At the last phase of the cycle, while the driving style is more aggressive, this inaccuracy increases. As earlier, the other results can be seen in Appendix C; they have similar properties.

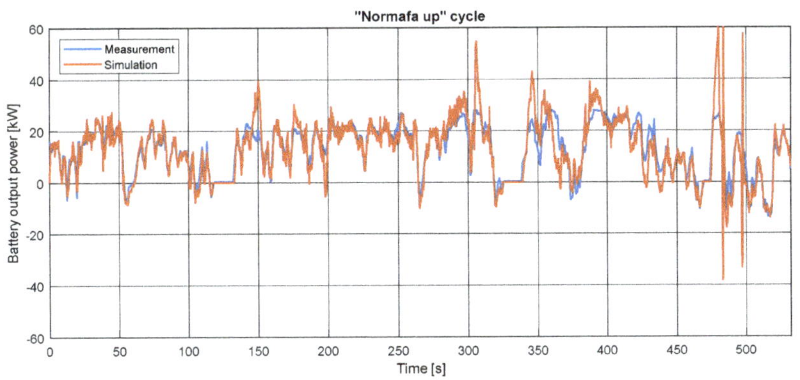

Figure 11. Measured and simulated battery output power during the "Normafa" cycle.

The validation criteria can be checked using the simulated speed and inverter input power results. After the iterative model development, Table 5. contains the values they achieved. All of them are below 10%, i.e., the model is acceptable for powertrain energetics analysis.

Table 5. RMS errors and cumulative relative errors for validation checking.

	"Erzsébet" Cycle		"Citadella" Cycle		"Normafa Up" Cycle		"Normafa Down" Cycle	
Speed validation RMS error	6.337	%	5.076	%	3.052	%	6.792	%
Inverter input power cumulative relative error	7.449	%	2.902	%	1.956	%	1.567	%

The energetic analysis is presented in Table 6. The first three cycles of the four are energy consumers, and the fourth one recuperates more energy than it consumes. Thus, the input energy in the first three cycles is from the battery (inverter input energy). In contrast, in the fourth cycle, road resistance works as an energy source due to the decreasing altitude.

Table 6. Energetic analysis of the pre-defined cycles.

	"Erzsébet" Cycle		"Citadella" Cycle		"Normafa Up" Cycle		"Normafa Down" Cycle	
	kWh	%	kWh	%	kWh	%	kWh	%
Inverter input power	0.594	100.00	0.612	100.00	1.888	100.00	−0.5248	46.28
eMotor losses	0.292	49.11	0.234	38.30	0.339	17.96	0.404	35.63
Gearbox losses	0.086	14.47	0.051	8.34	0.125	6.60	0.0399	3.52
Braking work	0.091	15.24	0.079	12.97	0.011	0.56	0.1021	9.00
Air resistance work	0.045	7.59	0.020	3.23	0.066	3.49	0.0634	5.59
Road resistance work	0.081	13.63	0.218	35.63	1.310	69.39	−1.134	100.00
Traction work	0.570		0.543		1.573		0.1783	
Recuperated energy	0.353		0.217		0.149		1.147	

In the "Erzsébet" cycle, the vehicle runs with low loads. This case does not use the electric motor's good efficiency zones. Thus, nearly half of the invested energy becomes motor losses. In this usage, the gearbox's losses also occur at a significantly high rate. Due to the low speeds, the air drag does the least work. During the cycle, the recuperated energy is 62% of the traction work, nearly six times higher than the braking work.

In the "Citadella" cycle, the motor load is higher: in the first 2/3, the vehicle runs up to the hill, and then it partially runs down. In these higher loads, the motor's efficiency is improved, and the losses decreased to 38.3%. In contrast, the road resistance work increases to 35.3%.

In the "Normafa Up" cycle, the traction work is much higher than the recuperated energy due to the high altitude increase. In this high-load motor operation, the electric motor losses decrease to 18%. Nearly 70% of the invested energy is used to overcome road resistance.

In the "Normafa Down" cycle, the behavior is the opposite. Due to the decrease in altitude, the vehicle's potential energy transforms into losses and recuperated energy. More than 46% of the "invested" energy was recoverable in this cycle. The recuperative braking motor efficiency is lower; thus, motor losses are more than 35%. Additionally, braking loss also increases to 9%. Recuperated energy is more than six times higher than traction work.

Finally, the high-voltage battery's measured and simulated SoC changes in the different cycles are shown in Figure 12. The results were not used for validation because the battery SoC estimation of the vehicle should not necessarily be linear, as its characteristics are unknown. For instance, it can depend on the actual driving style. Apart from that, the diagrams demonstrate that the model can be calibrated to follow this signal accurately.

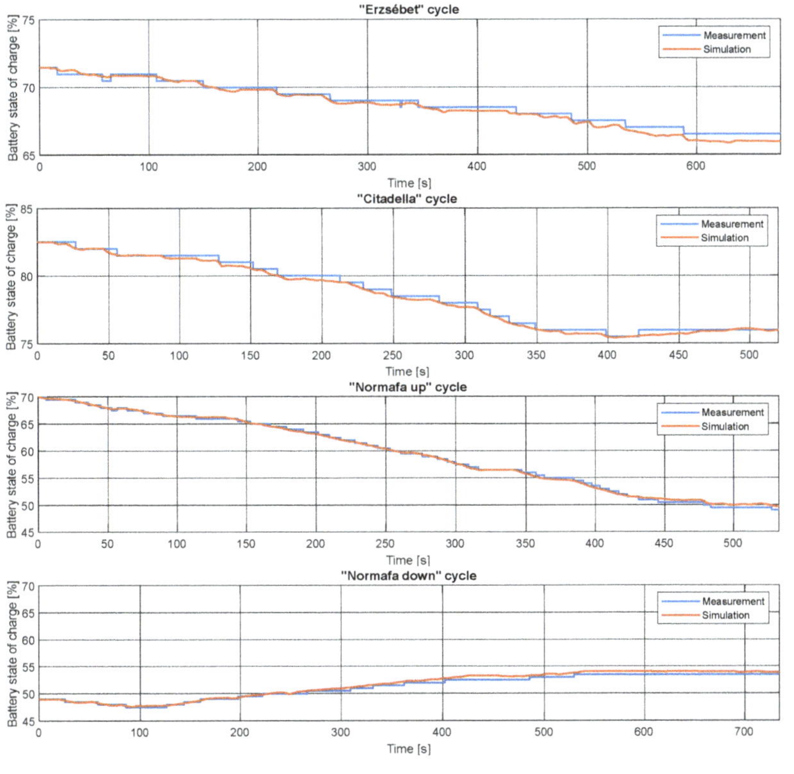

Figure 12. High-voltage batteries measured and simulated SoC changes during the cycles.

7. Conclusions

This paper aims to present a measurement system for BEV modeling and analysis. These vehicles' energy management, recuperation strategy, and range prediction are important research topics. The paper provides a solution to quickly process the real-world driving cycle's data and analyze it with simulation tools. The primary outcomes are the experiences with the measurement system and the model calibration. This paper provides the necessary signals for the analysis and gives suggestions for the sample time, too.

Due to the recuperation opportunity, knowing the road grade is particularly important. The only costly part of the measurement system is a GPS system that can record the actual altitude of the vehicle. It can be substituted with, for example, Google-based data, but its accuracy probably will not be enough. The simulation experiences confirmed that road grade accuracy is crucial for the model's accuracy.

This paper focuses on city driving with higher altitude changes. The simulation model is a longitudinal vehicle-drivetrain model investigating the energy transformation from the high-voltage battery to the vehicle's motion. Some typically hard-to-reach data also become estimable by the measurements, such as the additional rotational masses or the rolling resistance factor. The degradation of the battery can also be detected by the energy consumption and the HV battery SoC changes—in our case, the implemented nominal battery capacity is also significantly lower than the official data.

The presented measurement system can be the basis for several research studies on BEVs. The driving styles can also be analyzed by recording to develop the pedal characteristics and recuperation strategies for better vehicle efficiency. Additionally, by monitoring the motor and battery temperature signals, the thermal behavior can also be analyzed, and more complex battery models can be implemented. Furthermore, we also recorded the steering angle to investigate the lateral dynamics of the vehicle.

Author Contributions: Conceptualization, A.G. and M.Z.; methodology, D.T. and Á.N.; software, Á.N. and D.T.; validation, Á.N.; investigation, D.T. and M.S.J.; data curation, D.T.; writing—original draft preparation, M.S.J.; writing—review and editing, Á.N. and D.T.; visualization, Á.N. and D.T.; supervision, A.G. and M.Z. All authors have read and agreed to the published version of the manuscript.

Funding: This study was prepared with the professional support of the doctoral student scholarship program of the co-operative doctoral program of the Ministry for Innovation and Technology from the source of the national research, development, and innovation fund. The research was supported by the European Union within the framework of the National Laboratory for Autonomous Systems (RRF-2.3.1-21-2022-00002).

Institutional Review Board Statement: Not applicable.

Informed Consent Statement: Not applicable.

Data Availability Statement: The data presented in this study are available on request from the corresponding author.

Conflicts of Interest: The authors declare no conflicts of interest.

Appendix A

Table A1. Nomenclature.

Name	Unit
Battery state of charge	$SoC\%$
Battery voltage	U_{bat}
Braking force	F_{brake}
Drag resistance	F_{drag}
Electric machine angular velocity	ω_{mot}
Electric machine's efficiency map	η_{mot}
Electric machine's power	P_{mot}
Electric machine's torque	M_{mot}
Hydraulic brake pressure	p_{brake}
Inverter current consumption	I_{inv}
Inverter power	P_{inv}
Powertrain control input	$u_{throttle}$
Resistance forces	F_{res}
Road grade resistance	F_{grade}
Rolling resistance	F_{roll}
Traction force	F_{trac}
Vehicle acceleration	a_{veh}
Vehicle speed	v_{veh}

Appendix B

Measured and simulated motor torque results that were omitted from the discussion.

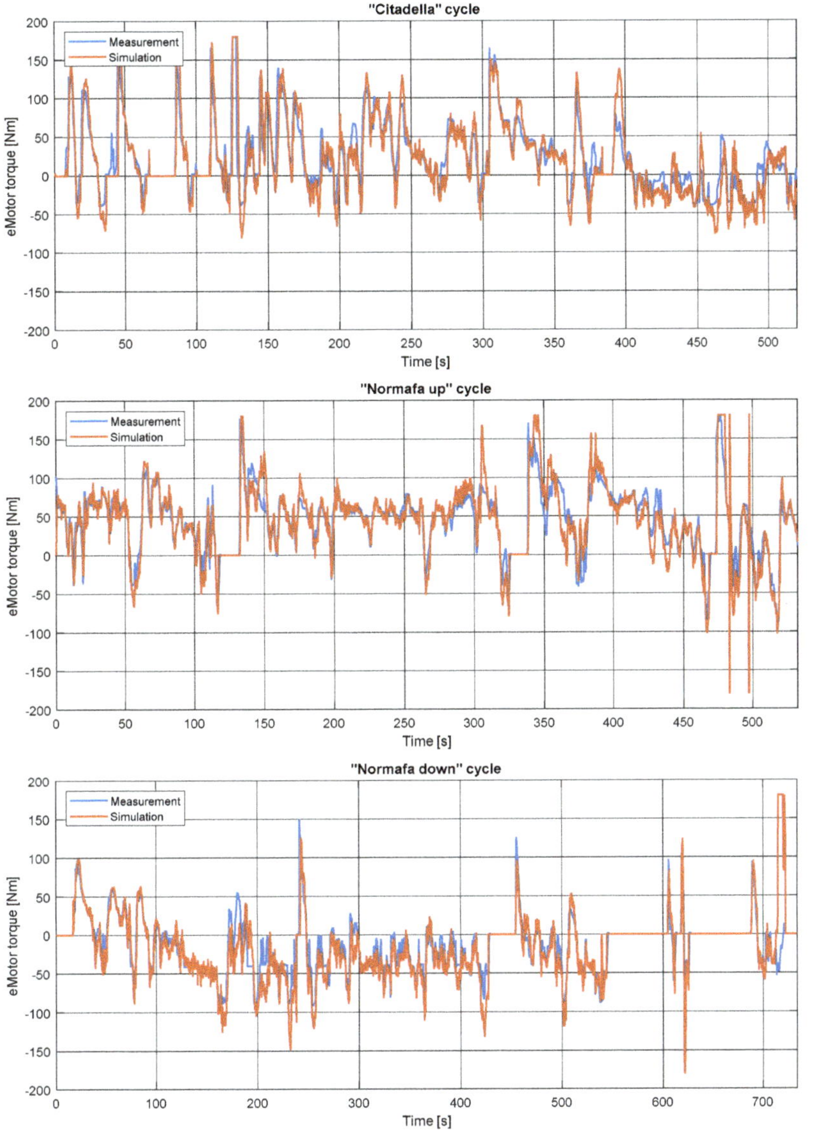

Figure A1. Measured and simulated the torque of the electric motor during the cycles.

Appendix C

Measured and simulated battery output power results that were omitted from the discussion.

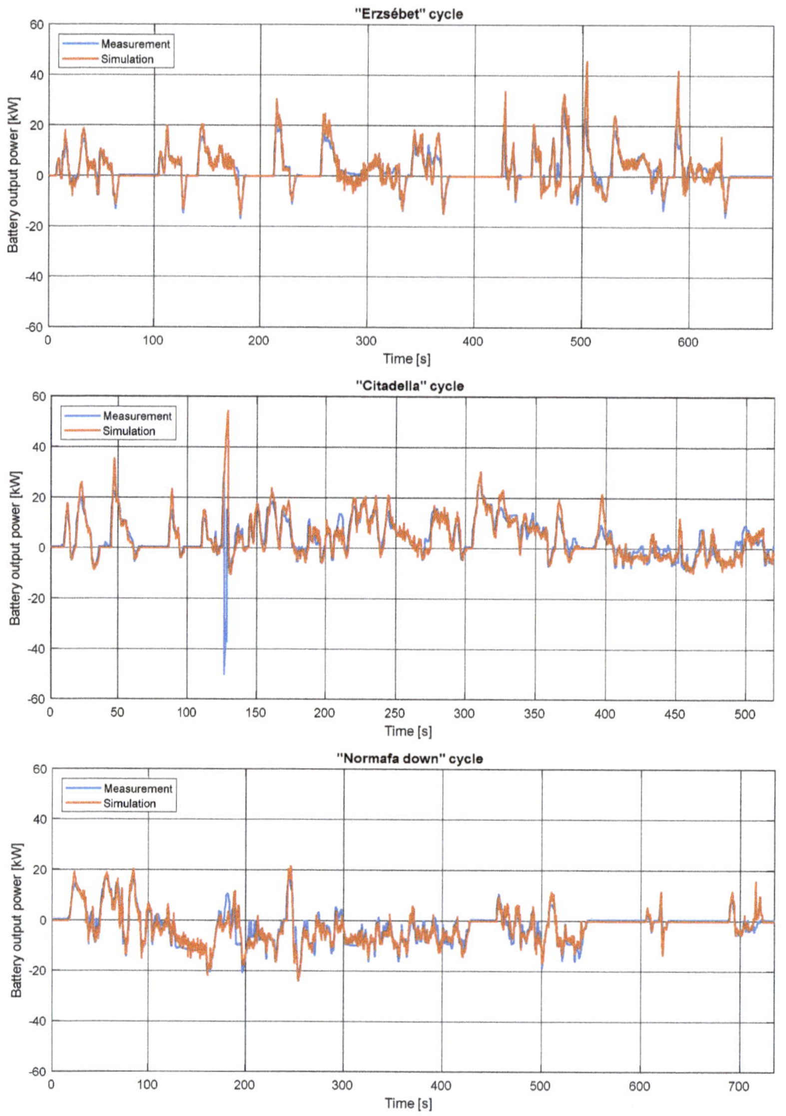

Figure A2. Measured and simulated battery output power during the cycles.

References

1. Zoldy, M.; Szalmane Csete, M.; Kolozsi, P.P.; Bordas, P.; Torok, A. Cognitive Sustainability. *Cogn. Sustain.* **2022**, *1*, 1. [CrossRef]
2. Bhushan, N.; Mekhilef, S.; Tey, K.S.; Shaaban, M.; Seyedmahmoudian, M.; Stojcevski, A. Overview of Model- and Non-Model-Based Online Battery Management Systems for Electric Vehicle Applications: A Comprehensive Review of Experimental and Simulation Studies. *Sustainability* **2022**, *14*, 15912. [CrossRef]
3. Giordano, G.; Klass, V.; Behm, M.; Lindbergh, G.; Sjöberg, J. Model-Based Lithium-Ion Battery Resistance Estimation from Electric Vehicle Operating Data. *IEEE Trans. Veh. Technol.* **2018**, *67*, 3720–3728. [CrossRef]
4. Manriquez-Padilla, C.G.; Cueva-Perez, I.; Dominguez-Gonzalez, A.; Elvira-Ortiz, D.A.; Perez-Cruz, A.; Saucedo-Dorantes, J.J. State of Charge Estimation Model Based on Genetic Algorithms and Multivariate Linear Regression with Applications in Electric Vehicles. *Sensors* **2023**, *23*, 2924. [CrossRef] [PubMed]
5. Feng, S.; Wang, A.; Cai, J.; Zuo, H.; Zhang, Y. Health State Estimation of On-Board Lithium-Ion Batteries Based on GMM-BID Model. *Sensors* **2022**, *22*, 9637. [CrossRef] [PubMed]

6. Basu, A.K.; Tatiya, S.; Bhattacharya, S. Overview of Electric Vehicles (EVs) and EV Sensors. In *Sensors for Automotive and Aerospace Applications*; Bhattacharya, S., Agarwal, A., Prakash, O., Singh, S., Eds.; Springer: Singapore, 2019.
7. Nikam, S.P.; Fernandes, B.G. Low cost electric drivetrain for conversion of ICE vehicle into an electric vehicle. In Proceedings of the 2014 IEEE International Conference on Power Electronics, Drives and Energy Systems (PEDES), Mumbai, India, 16–19 December 2014.
8. Nyerges, Á.; Zöldy, M. Longitudinal vehicle model development for range estimation in electric vehicles. *Tech. Rev. EMT* **2019**, *74*, 13–22.
9. Kubaisi, R.; Gauterin, F.; Giessler, M. A Method to analyze driver influence on the energy consumption and power needs of electric vehicles. In Proceedings of the 2014 IEEE International Electric Vehicle Conference (IEVC), Florence, Italy, 17–19 December 2014; ISBN 9781479960750.
10. Petri, R.; Springer, M.; Zelle, D.; McDonald, I.; Fuchs, A.; Krauß, C. Evaluation of lightweight TPMs or automotive software updates over the air. In Proceedings of the 4th ESCAR USA, Detroit, MI, USA, 5–6 June 2016.
11. *ISO 26262-1*; Road Vehicles—Functional Safety. International Organization for Standardization: Geneva, Switzerland, 2018. Available online: https://blog.ansi.org/2019/02/iso-26262-2018-road-vehicle-functional-safety (accessed on 30 June 2024).
12. Szydlowski, C. *CAN Specification 2.0: Protocol and Implementations*; SAE Technical Paper 921603; SAE: Warrendale, PA, USA, 1992. Available online: https://www.sae.org/publications/technical-papers/content/921603 (accessed on 30 June 2024).
13. *ISO/IEC/IEEE 8802-3*; Information Technology—Telecommunications and Information Exchange between Systems—Requirements for Local and Metropolitan Area Networks—Part 3: Standard for Ethernet. ISO: Geneva, Switzerland, 2021.
14. *ISO 17458-1*; Road Vehicles—FlexRay Communications System—Part 1: General Information and Use Case Definition. International Organization for Standardization: Geneva, Switzerland, 2013.
15. *SAE J1979*; E/E Diagnostic Test Modes. Society of Automotive Engineers International: Warrendale, PA, USA, 1991.
16. *ISO 15031-5*; Road Vehicles—Communication between Vehicle and External Equipment for Emissions-Related Diagnostics—Part 5: Emissions-Related Diagnostic Services. International Organization for Standardization: Geneva, Switzerland, 2006.
17. *40 CFR 86.305D*; Emission Regulations for New Gasoline-Fueled and Diesel-Fueled Heavy-Duty Engines, Gaseous Exhaust Test Procedures, Code of Federal Regulations. Legal Information Institute: Ithaca, NY, USA, 1996.
18. *Directive 98/69/EC*; Measures to be Taken against Air Pollution by Emissions from Motor Vehicles and Amending Council Directive 70/220/EEC. European Parliament, Council of the European Union: Brussels, Belgium, 1998.
19. *ISO 14230-4*; Road Vehicles—Diagnostic Systems—Keyword Protocol 2000. International Organization for Standardization: Geneva, Switzerland, 2000.
20. *SAE J1962*; Diagnostic Connector. Society of Automotive Engineers International: Warrendale, PA, USA, 1992.
21. *SAE J1850*; Class B Data Communication Network Interface. Society of Automotive Engineers International: Warrendale, PA, USA, 1988.
22. *ISO 9141-2*; Road Vehicles—Diagnostic Systems—Part 2: CARB Requirements for Interchange of Digital Information. International Organization for Standardization: Geneva, Switzerland, 1994.
23. *ISO 15765-4*; Road Vehicles—Diagnostics on Controller Area Networks (CAN). International Organization for Standardization: Geneva, Switzerland, 2004.
24. *ISO 14229*; Road Vehicles—Unified Diagnostic Services (UDS). International Organization for Standardization: Geneva, Switzerland, 2006.
25. Eckstein, L.; Göbbels, R.; Wohlecker, R. Benchmarking of the Electric Vehicle Mitsubishi i-MiEV. *ATZ Worldw. eMagazine* **2011**, *113*, 48–53. [CrossRef]
26. Kamachi, M.; Miyamoto, H.; Sano, Y. Development of power management system for electric vehicle "i-MiEV". In Proceedings of the 2010 International Power Electronics Conference—ECCE ASIA, Sapporo, Japan, 21–24 June 2010; pp. 2949–2955.
27. CSS CANedge1. Available online: https://www.csselectronics.com/products/can-logger-sd-canedge1-gnss-imu (accessed on 28 May 2024).
28. Kádár, L.; Varga, F.; Kőfalusi, P. Construction Theory of Road Vehicles. BME MOGI 2014. Available online: http://www.mogi.bme.hu/TAMOP/kozuti_jarmurendszerek_szerkezettana/book.html (accessed on 30 June 2024).
29. Pyrhönen, J.; Jokinen, T.; Hrabovcová, V. *Design of Rotating Electrical Engines*, 2nd ed.; Wiley: Hoboken, NJ, USA, 2014; ISBN 9781118581575.
30. Sato, Y.; Ishikawa, S.; Okubo, T.; Abe, M.; Tamai, K. Development of High Response Motor and Inverter System for the Nissan LEAF Electric Vehicle. *SAE Tech. Pap. Ser.* **2011**, *7*, 13.
31. Matlab Library, Powertrain Blockset/Vehicle Scenario Builder: Longitudinal Driver. 2017. Available online: https://ch.mathworks.com/help/vdynblks/ref/longitudinaldriver.html (accessed on 30 June 2024).
32. Hangos, K.; Bokor, J.; Szederkényi, G. *Analysis and Control of Nonlinear Process Systems*; Springer: New York, NY, USA, 2004.
33. Mohan, G.; Assadian, F.; Longo, S. Comparative analysis of forward-facing models vs backwardfacing models in powertrain component sizing. In Proceedings of the IET Hybrid and Electric Vehicles Conference 2013 (HEVC 2013), London, UK, 6–7 November 2013; pp. 1–6. [CrossRef]

Disclaimer/Publisher's Note: The statements, opinions and data contained in all publications are solely those of the individual author(s) and contributor(s) and not of MDPI and/or the editor(s). MDPI and/or the editor(s) disclaim responsibility for any injury to people or property resulting from any ideas, methods, instructions or products referred to in the content.

sensors

MDPI

Article

Hybrid MambaVision and Transformer-Based Architecture for 3D Lane Detection

Raul-Mihai Cap and Călin-Adrian Popa *

Department of Computers and Information Technology, Politehnica University of Timișoara, 300223 Timișoara, Romania; raul.cap@student.upt.ro
* Correspondence: calin.popa@cs.upt.ro

Abstract: Lane detection is an essential task in the field of computer vision and autonomous driving. This involves identifying and locating road markings on the road surface. This capability not only helps drivers keep the vehicle in the correct lane, but also provides critical data for advanced driver assistance systems and autonomous vehicles. Traditional lane detection models work mainly on the 2D image plane and achieve remarkable results. However, these models often assume a flat-world scenario, which does not correspond to real-world conditions, where roads have elevation variations and road markings may be curved. Our approach solves this challenge by focusing on 3D lane detection without relying on the inverse perspective mapping technique. Instead, we introduce a new framework using the MambaVision-S-1K backbone, which combines Mamba-based processing with Transformer capabilities to capture both local detail and global contexts from monocular images. This hybrid approach allows accurate modeling of lane geometry in three dimensions, even in the presence of elevation variations. By replacing the traditional convolutional neural network backbone with MambaVision, our proposed model significantly improves the capability of 3D lane detection systems. Our method achieved state-of-the-art performance on the ONCE-3DLanes dataset, thus demonstrating its superiority in accurately capturing lane curvature and elevation variations. These results highlight the potential of integrating advanced backbones based on Vision Transformers in the field of autonomous driving for more robust and reliable lane detection. The code will be available online.

Keywords: 3D lane detection; Mamba vision; transformer networks; deformable attention; autonomous driving

1. Introduction

Detecting and pinpointing lanes on roads is an important task for computer vision and self-driving vehicles. Lane detection systems can help drivers stay in their lane and provide crucial information for advanced driver assistance systems and autonomous vehicles. Traditional lane detection methods use image processing techniques to recognize lane markings in camera images. Recently, deep learning has led to the creation of more precise and durable lane detection models based on convolutional neural networks (CNNs) and Transformer models. However, lane detection remains difficult due to varying road conditions, lighting conditions, camera perspectives, and the need for real-time performance.

There are two main approaches to lane detection: 2D and 3D models. Two-dimensional methods use CNNs such as ResNet [1], EfficientNet [2], and MobileNet [3,4] to detect lane markings in monocular images. These models are fast and effective in ideal conditions, but may struggle with poor weather, lighting, or degraded lane markings. In contrast,

3D models estimate the position and orientation of lane markings in 3D space, offering improved accuracy and robustness. Although most 3D models use inverse perspective mapping (IPM) [5] to create a bird's-eye view (BEV), not all rely on this method. However, 3D models are generally more computationally intensive and depend on precise transformations.

Previous state-of-the-art methods hypothesized wrongly the construction of the BEV image, assuming that the lines have no height. This made the obtained image not accurate because they started from the premise that the world is flat. The authors of those works tried to determine the coordinates in the 3D space of the traffic lanes depending on the curvature of the lane lines. It is understandable that it was very difficult, nearly impossible, to generate a more accurate top-down view image because, at that time, there were no datasets in which the exact coordinates of the traffic lanes could be mapped in 3D space. At that time, there existed only synthetic datasets such as Apollo [6], which were obtained from various game engines.

Developing a 3D lane detection dataset poses a significant challenge, primarily due to the requirement of precise and comprehensive ground truth annotations. Typically, this process involves manually labeling lane markings in a 3D space, which can be time-consuming and requires specialized equipment, such as LiDAR sensors.

Another challenge relates to generating a diverse and representative set of scenes that accurately capture a wide range of real-world scenarios, including various road types, weather, and lighting conditions. This undertaking requires careful planning, execution, and access to appropriate data collection sites.

Fortunately, we now have at our disposal three large datasets that contain a multitude of driving scenarios from real situations, whose annotations are very well detailed. The datasets to which we refer are OpenLane [7], OpenLaneV2 [8], and ONCE-3DLanes [9]. One of the things that differentiates these datasets is the evaluation metrics. A well-constructed dataset plays a crucial role for lane detection tasks because, when annotations are inconsistent, training becomes noisy and slow, and the resulting model struggles to generalize. Imperfect ground truths such as missed or misaligned lane markings can degrade the performance of deep neural networks, which is supported by recent results from the PSSCL framework [10]. Furthermore, careful selection of cleaner data during training significantly improves the robustness to such label noise.

Certainly, there are several methods that can work alongside those mentioned in order to obtain better results. These methods involve the inclusion of several sensors, such as RADAR or LiDAR, in contemplation of better understanding the surrounding environment by generating point clouds or depth maps. The main disadvantage of those methods is that, as we add more sensors, the computational cost increases drastically and it becomes more and more difficult and expensive to be used in real life.

Transformers are increasingly being used in computer vision applications such as image classification, object detection, and semantic segmentation [11,12], as they are capable of capturing long-range dependencies and contextual information in the input data. Unlike CNNs, which process input data sequentially and extract features using local receptive fields, Transformers utilize a self-attention mechanism to compute global interactions between all input elements, thereby enabling them to capture both local and global contextual information. However, their computational requirements grow exponentially with sequence length, creating challenges for training and deployment.

For computer vision tasks, Transformer-based models such as Vision Transformers (ViT) [13], DETR [14], and SETR [15] have demonstrated state-of-the-art performance on challenging datasets such as ImageNet [16], COCO [17], and Cityscapes [18]. Transformers have several advantages over CNNs, including the ability to model long-range dependencies and context more effectively, less dependence on predefined architectures, and the ability to handle variable-sized inputs. However, with the recent introduction of Mamba Vision models, even better results have been achieved, surpassing the performance of ViTs in many computer vision tasks by combining the strengths of state-space models and Transformers for enhanced accuracy and efficiency.

The main objective of our research is to integrate the Transformer [19] and Mamba Vision [20] architectures into current state-of-the-art models and replace the classic convolutional neural networks in order to achieve better results.

The main contributions of the paper are as follows:

- We propose a simple hybrid backbone that mixes MambaVision-S-1K CNN blocks with a Transformer, so the model learns both fine details and global context.
- Our approach works directly on front-view images and skips any IPM or bird's-eye view step.
- We fine-tune and optimize the whole model so it can run well on systems which do not require the latest and best configuration.
- We achieve state-of-the-art results on ONCE-3DLanes and also very good results on all subsets of the Apollo 3D Lane Detection benchmark.

The code will be available at https://github.com/raul-cap/mamba-3d-lane-detection (accessed on 29 July 2025).

2. Related Work

A study on 3D-LaneNet [21] has applied two new concepts: an intra-network inverse perspective mapping that provides a dual representation information flow for both a regular image captured by the camera and its top-down view, and an anchor per column representation, which makes the lane detection task more like an object detection problem, replacing the common methods such as outlier rejection and clustering. This is the first work to make use of the top-view representation of the standard image to determine the lane in 3D space.

Being the first work to approach this topic, 3D-LaneNet inspired most of the works that followed, as well as the work entitled "Gen-LaneNet: A Generalized and Scalable Approach for 3D Lane Detection" [6]. With this paper, the authors noticed and tried to remove the drawbacks of 3D-LaneNet. One of the drawbacks is that the bird's-eye projection does not align with the image feature resulting from IPM in the presence of a non-zero slope. Ref. [21] uses an unsuitable coordinate frame in its anchor representation, causing ground truth lanes to be misaligned with visual features. On the other hand, the authors proposed the so-called "virtual top view", which aligns with the IPM image. Ref. [22] is also an extension of the first state-of-the-art methods [6,21]. The authors have introduced a new attention mechanism called Dual Attention. Its purpose is to help the model work better and improve accuracy in more complicated driving situations and scenarios. Besides this attention module, another change compared to the previous models is the use of the linear-interpolation loss function, which is used to locate the 3D lane markings with greater precision.

Although other previously presented works all had 3D-LaneNet as their starting point and each work represented an improvement of the basic one, in CLGo [23], the authors applied a different strategy. Among the only similarities to the other works is that its architecture has two stages. However, the first stage is no longer used for image segmentation, but instead for finding the best camera pose and polynomial parameters using geometric constraints. In this case, finding the best camera pose is equivalent to finding the camera height and the pitch angle of the camera. This is performed by using a Transformer encoder–decoder.

The backbone of this work replaces the classic backbones that used convolutional neural networks with one based on Transformers, which are mainly used to enhance image features. The convolution is only used to extract the convolutional features that will be inputs for the Transformer encoder (TRE). By decoding the characteristics using the Transformer decoder (TRD), the camera pose is obtained. The authors added a lane branch to interpret 3D lanes to aid in camera pose learning. First, the 3D lanes are obtained, and then they are projected into the 2D plane by conducting a homographic transformation. Thus, the model learns the best camera pose by comparing the results to the ground truths in both 3D and 2D planes.

In [7], the authors proposed a method entitled PersFormer, which is an end-to-end monocular 3D lane detector model that can simultaneously detect 2D and 3D lanes. Like the latest state-of-the-art models, the method is based on the Transformer architecture. However, the Transformer is not used as the backbone for feature extraction, which is the first part of this network. The second part of the model is the Perspective Transformer, which is relatively complex, and also the inspiration for the name of the model. The general idea behind the Perspective Transformer is to generate an accurate BEV of the input image by taking into consideration the front view features generated by the ResNet [1] backbone, the coordinates of the transformation matrix from IPM, and the camera intrinsic and extrinsic parameters, such as camera height, and the pitch angle of the camera. The authors used a Transformer to attend local context and aggregate global features to produce a solid representation in BEV rather than just projecting the one-to-one feature correspondence from front view to BEV. In addition to the proposed model, the authors highlighted one of the existing problems in the 3D lane detection task, namely, the lack of data for the learning process. They addressed this problem by introducing a new large-scale dataset called OpenLane.

With the same purpose as [7], previously presented, ONCE-3DLanes [9] addresses the problem of insufficient data for 3D lane detection and introduces a new real-world autonomous driving dataset to encourage development on this subject. In addition to this, the authors also proposed a model for 3D lane detection called SALAD that does not require human-made anchors and also does not require transformation to BEV or top-view.

In [24], a lane-aware query generator is introduced, which adapts to lane-specific characteristics and improves detection accuracy by jointly leveraging lane-level and point-level embeddings. Furthermore, LATR uses a Dynamic 3D Ground Positional Embedding that iteratively refines a 3D ground plane to better align with the actual road geometry. Thus, it addresses the limitations of traditional assumptions related to a fixed 3D space. Using these algorithms, the method achieved state-of-the-art results on the ONCE-3DLanes dataset and very impressive results on other datasets without using BEV image generation.

The state-space model (SSM) introduced by Mamba [25] addresses ViT problems by providing linear-time complexity. Its results match or even exceed the performance of Transformers in natural language processing tasks. Mamba's key innovation lies in the efficient handling of long sequences, making it a scalable solution compared to traditional Transformers.

Vision Mamba [26] further improves upon this by incorporating bidirectional SSMs to mitigate the weaknesses of Transformers in capturing global context and spatial relationships. However, the added complexity of processing sequences in bidirectional SSMs can lead to challenges such as increased latency, overfitting, and uncertainty in accuracy improvement. Despite these challenges, ViTs and CNNs often outperform Mamba-based models in visual tasks because of their efficiency and reliability.

Recently, hybrid models that combine Mamba and Transformer architectures have gained considerable attention. Ref. [20] introduced MambaVision, a novel hybrid model designed for computer vision tasks that combines the strengths of both Mamba and Transformers. It employs a hierarchical structure with multi-resolution CNN-based residual blocks to swiftly extract features across different resolutions.

Related to our design choices, we also consider advances in detection architectures beyond lane detection. Ref. [27] shows how combining multiple feature streams can remain robust even when one modality, for example, thermal or RGB, is noisy or unreliable. This idea is similar to our goal of making geometric cues robust against imperfect or uncertain signals. In another line of work, ref. [14] replaces dense attention with sparse, reference-point-based attention, which improves greatly the efficiency and accuracy in detection. This mechanism directly inspires the use of deformable attention in the decoder in our work.

3. Method Overview

Motivated by the success of the latest state-of-the-art LATR [24] model, we propose a 3D lane detection model with Dual Transformer. In this section, we will present the overall architecture and implementation details.

The objective of the proposed solution is to obtain the coordinates of the traffic lanes in a three-dimensional space, starting from an image that is captured by a camera mounted on the front of the vehicle. Because we do not generate a BEV image for the lane detection, we no longer require precise calibration of the well-defined intrinsic and extrinsic parameters of the camera, such as the height of the camera, the focal length, or the pitch angle, as we previously did in [6,21]. Concretely, we still use a simple pinhole projection to map 3D points to the image, but we let the ground plane be adjusted during training with two residuals corresponding to pitch $(\Delta\theta_x)$ and height (Δh), which are supervised in the lane pixels by L_{plane}. On ONCE-3DLanes, the camera extrinsics are unavailable. So, we follow previous works [7,24] and use approximate camera settings, and the learned plane compensates for the resulting mismatch. Thus, we can use this method to approximate the camera extrinsics on every dataset.

The architecture of our proposed model is shown in Figure 1. The process begins by extracting feature maps from the input image using the MambaVision backbone, which was chosen for its ability to capture both fine-grained details and broader contextual features effectively. After this important step, the incoming process is inspired and similar to LATR. Lane-aware queries are generated, where the number of queries corresponds to the number of lanes, and the number of points represents each lane. These queries are then designed to interact with the extracted feature maps using a deformable attention mechanism, which allows the model to dynamically focus on the most relevant parts of the image.

To further enhance the model's understanding of the scene, we employ a Dynamic 3D Ground Positional Embedding Generator. This component integrates three-dimensional spatial information into the two-dimensional features which ensure that the model accurately reflects lane geometry without relying on static 3D structures. Finally, a prediction head processes the enhanced queries and leads to the generation of the final lane predictions. Each component of this architecture will be explained in more detail in the following sections.

3.1. Backbone

MambaVision serves as the feature extraction backbone for the object detection and segmentation models. In these frameworks, the backbone network is responsible for extracting rich feature maps from the input image, which are then used by subsequent layers to detect objects and generate segmentation masks. The hierarchical architecture of MambaVision, which includes both CNN-based residual blocks and Transformer blocks, makes it highly effective for this role.

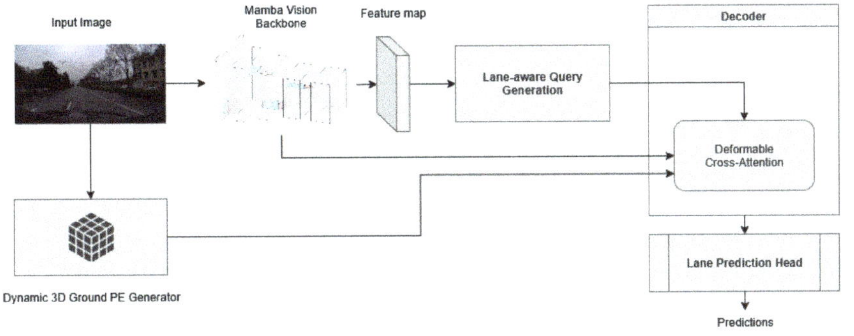

Figure 1. Proposed method architecture. The model begins by extracting a feature map X using the MambaVision backbone, followed by generating lane-aware queries Q that refine the feature map X' through a deformable attention mechanism. Next, the Dynamic 3D Ground Positional Embedding introduces 3D context, updating the feature map to X'', after which the dynamic plane is iteratively updated to align with real-world geometry using a transformation matrix, and, finally, the prediction head processes the refined feature map to produce accurate 3D lane predictions.

CNN layers in MambaVision, specifically in Stages 1 and 2, quickly process high-resolution features. Given an input image $\mathbf{X} \in \mathbb{R}^{H \times W \times 3}$, where H and W are the height and width of the input image, which is an RGB image with three channels, the number of channels is denoted with C. These layers extract feature maps $\mathbf{F} \in \mathbb{R}^{\frac{H}{2} \times \frac{W}{2} \times C}$ through a series of convolutions and pooling operations:

$$\mathbf{F} = \mathrm{CNN}(\mathbf{X}),$$

where $\mathrm{CNN}(\cdot)$ denotes the convolutional operations, including residual connections, which are formulated as

$$\mathbf{F}_{\mathrm{res}} = \mathrm{Conv}(\sigma(\mathrm{BN}(\mathbf{F}))) + \mathbf{F}.$$

Here, σ represents the activation function, typically ReLU, and BN denotes batch normalization. These residual connections help maintain spatial details crucial for accurately locating and identifying objects by capturing both low-level and high-level features.

In Stages 3 and 4, MambaVision employs a combination of MambaVision Mixer blocks and Transformer blocks. The MambaVision Mixer blocks utilize Structured SSMs to capture both short- and long-range dependencies. The SSM in MambaVision can be expressed using the following state-space equations:

$$\mathbf{h}'(t) = \mathbf{A}\mathbf{h}(t) + \mathbf{B}x(t), \quad y(t) = \mathbf{C}\mathbf{h}(t),$$

where $\mathbf{h}(t) \in \mathbb{R}^M$ is the hidden state, $x(t) \in \mathbb{R}$ is the input, and $y(t) \in \mathbb{R}$ is the output. The parameters $\mathbf{A} \in \mathbb{R}^{M \times M}$, $\mathbf{B} \in \mathbb{R}^{1 \times M}$, and $\mathbf{C} \in \mathbb{R}^{1 \times M}$ are learnable. These parameters are discretized for computational efficiency. This allows the MambaVision Mixer to process sequences effectively and capture complex dependencies.

The output of the MambaVision Mixer block can be represented as

$$\mathbf{F}_{\mathrm{mixer}} = \mathrm{Mixer}(\mathbf{F}_{\mathrm{low}}),$$

where $\mathbf{F}_{\text{low}} \in \mathbb{R}^{\frac{H}{4} \times \frac{W}{4} \times C'}$ are the lower-resolution feature maps. The Mixer block combines convolutional operations with SSMs, capturing both local and global context.

The Transformer blocks in Stages 3 and 4 further enhance the model's ability to maintain global context. Using the multi-head self-attention mechanism, the Transformer blocks process the feature maps $\mathbf{F}_{\text{mixer}}$ as

$$\text{Attention}(\mathbf{Q}, \mathbf{K}, \mathbf{V}) = \text{softmax}\left(\frac{\mathbf{Q}\mathbf{K}^T}{\sqrt{d_k}}\right)\mathbf{V},$$

where $\mathbf{Q}$, $\mathbf{K}$, and $\mathbf{V}$ are the query, key, and value matrices derived from $\mathbf{F}_{\text{mixer}}$, and d_k is the dimensionality of the key vectors. This self-attention mechanism allows the model to capture dependencies across the entire image to ensure a comprehensive understanding of object relationships and spatial configurations.

The combined output from the MambaVision Mixer and Transformer blocks provides enriched feature maps that are well-suited for generating accurate segmentation masks and object detections:

$$\mathbf{F}_{\text{out}} = \text{Transformer}(\mathbf{F}_{\text{mixer}}),$$

where Transformer$(\cdot)$ denotes the transformation applied by the Transformer block. This hierarchical combination of CNNs, SSMs, and Transformers within MambaVision makes it a powerful backbone for object detection and segmentation tasks, which balances both computational efficiency and accuracy.

The LATR model begins by processing the input image I using the MambaVision-S-1K backbone to extract a feature map X. The feature map captures the necessary visual details from the front view of the driving scene and provides a rich set of features for further processing:

$$X \in \mathbb{R}^{C \times H \times W} = \text{MambaVision}(I).$$

Here, C is the number of channels, H is the height, and W is the width of the feature map.

3.2. Lane-Aware Query Generation

After extracting the feature map, the model generates lane-aware queries Q using the lane-aware query generator. These queries are tailored to each detected lane and incorporate both lane- and point-level embeddings. The lane-aware queries Q are represented as

$$Q = Q_{\text{lane}} \oplus Q_{\text{point}},$$

where Q_{lane} captures the overall lane structure and is computed using instance activation maps (IAMs) derived from the feature map X, and Q_{point} captures detailed information at specific points along the lane, leveraging the learnable weights corresponding to the predefined longitudinal coordinates.

3.3. Interaction via Deformable Attention

The lane-aware queries Q then interact with the feature map X through a deformable attention mechanism. This mechanism allows the model to selectively focus on relevant regions in the image by dynamically adjusting its attention based on the queries:

$$X' = \text{DeformAttn}(Q, X),$$

where X' is the refined feature map after the deformable attention operation, which uses the lane-aware queries to capture relevant features for lane detection.

3.4. Dynamic 3D Ground Positional Embedding (PE) Generation

To integrate three-dimensional context, the model employs a Dynamic 3D Ground Positional Embedding Generator. This component enhances the feature map X' with 3D spatial information by projecting a hypothetical 3D ground plane into the 2D feature space:

$$d \cdot [u, v, 1]^T = T \cdot [x, y, z, 1]^T,$$

where (u, v) are the 2D coordinates on the feature map X' (in pixels), $d > 0$ is the homogeneous scale factor of the projection, $[x, y, z, 1]^\top$ is a 3D point on the ground plane, and T is the pinhole camera projection matrix.

A 3D ground plane P is initialized and iteratively refined using a transformation matrix D to align with the real-world ground. This embedding step enhances the feature map X'', which allows the model to accurately represent the lane geometry:

$$X'' = \text{PEGenerator}(X').$$

During training, the ground plane is adjusted step by step by adjusting the pitch and height. We train it together with the lane losses using a simple plane alignment loss (Section 3.7). This learned, adjustable plane works better than a fixed plane or a fixed frustum.

3.5. Iterative Plane Update

We describe the transformation matrix D using the pitch angle $(\Delta\theta_x)$ and the vertical shift (Δh). With these, we construct the matrix:

$$D = \begin{bmatrix} 1 & 0 & 0 & 0 \\ 0 & \cos\Delta\theta_x & -\sin\Delta\theta_x & 0 \\ 0 & \sin\Delta\theta_x & \cos\Delta\theta_x & \Delta\, h \\ 0 & 0 & 0 & 1 \end{bmatrix}.$$

D is used to update the 3D ground plane, adjusting its position to better fit the ground truth road surface. In the decoder layer t, the plane grid from the previous step, P_{t-1}, is updated by multiplying it by this matrix:

$$P = D \cdot P_{t-1}.$$

The update is based on the residuals calculated between the predicted 3D ground plane and the actual 3D lane annotations.

3.6. Prediction Head for Final Lane Predictions

Finally, the enhanced feature map X'' is fed into a prediction head that uses an MLP to output the final 3D lane predictions. This head estimates the 3D coordinates of the lane points and their visibility:

$$[\Delta x, \Delta z, v] = \text{MLP}(Q).$$

This final step produces the predicted 3D positions and classifications of the lanes, providing a comprehensive 3D lane model. This workflow avoids the use of traditional 3D surrogates like IPM and leverages Transformer-based attention mechanisms for efficient and accurate 3D lane detection.

3.7. Loss Function

The loss function is computed the same as in [24] and combines several components to ensure accurate 3D lane detection, focusing on regression, visibility, and classification.

The primary loss for 3D lane prediction $\mathcal{L}_{\text{lane}}$ consists of

$$\mathcal{L}_{\text{lane}} = w_x\mathcal{L}_x + w_z\mathcal{L}_z + w_v\mathcal{L}_v + w_c\mathcal{L}_c,$$

where $\mathcal{L}_x$ and $\mathcal{L}_z$ are regression losses for the x- and z-coordinates of the lane points, $\mathcal{L}_v$ is the visibility loss, using binary cross-entropy, and $\mathcal{L}_c$ is the classification loss, handled by focal loss to manage class imbalance.

The overall loss function is a weighted combination of 3D lane prediction, segmentation, and 3D ground plane alignment losses:

$$\mathcal{L} = w_s\mathcal{L}_{\text{seg}} + w_p\mathcal{L}_{\text{plane}} + w_l\mathcal{L}_{\text{lane}},$$

where $\mathcal{L}_{\text{seg}}$ trains a simple lane versus background mask obtained by drawing the ground-truth lane lines on the image and $\mathcal{L}_{\text{plane}}$ makes the learned ground plane, composed of pitch and height, to match the 3D lane annotations at labeled pixels.

We supervise the ground plane with a simple alignment loss computed only at lane pixels:

$$\mathcal{L}_{\text{plane}} = \sum_{(u,v)\in\mathcal{L}} \left\| M_p(:,u,v) - M_\ell(:,u,v) \right\|_2,$$

where $M_p \in \mathbb{R}^{3\times H\times W}$ is the 3D canvas induced by the current ground plane and M_ℓ stores the metric 3D of the annotated lane points. Training is end-to-end, so this loss is optimized together with the lane losses, and gradients pass through the plane update, so the plane aligns to the road surface over time.

The values we have used for the base weights are the following: $w_s = 5.0$, $w_x = 2.0$, $w_z = 10.0$, w_c=10.0, and all others =1.0. We find that they are stable across both datasets.

This structure ensures that the model optimizes not only for accurate lane positioning but also for visibility and correct ground plane alignment.

4. Experiments

In this section, we present the training details alongside the results of the proposed model on the ONCE-3DLanes (https://once-3dlanes.github.io/3dlanes/, accessed on 29 July 2025) and Apollo 3D Lane Detection (https://developer.apollo.auto/synthetic.html, accessed on 29 July 2025) datasets.

4.1. Training Details

In our experiments, we set the batch size to 2 and used three workers during training. The number of categories for classification is 2, one representing the lane and the other the background, and we defined a positive threshold of 0.3 for classification. The labels corresponding to the lane and background created labels by drawing the ground-truth lane lines on the image and making them a bit thicker, 3 pixels at 1/4 of the resolution of the feature map. The pixels on these thick lines are labeled 'lane' and all other pixels are 'background'. This auxiliary head is used only during training and it is dropped at inference.

Our model follows an encoder–decoder architecture based on [24]. We resize the input images to a resolution of 720 × 960 and they are then fed to the backbone. The encoder is initialized with the MambaVision-S-1K pretrained model, and we employ a Feature Pyramid Network (FPN) as the neck, with input channels [192, 384, 768] and output dimension of 192. Those input channels correspond to the feature map channels extracted from the last three stages of the MambaVision backbone, which represent different levels of spatial detail. The FPN outputs four feature maps and each one is passed through additional convolutions with the largest scale aggregated as input for the decoder.

For the decoder, we apply deformable attention with four heads, eight sample points, and 192-dimensional embeddings. A six-layer Transformer decoder is used for object detection and segmentation, with 12 queries and 20 anchor points per query distributed across the y axis.

All models are trained using the AdamW optimizer with a learning rate of 2×10^{-4} and a weight decay of 0.01. We set the learning rate multiplier for `sampling_offsets` to 0.1.

The proposed method was trained and evaluated on a system that has the following specifications: Intel 4790K CPU (Intel, Santa Clara, CA, USA), one NVIDIA RTX 3060 Ti GPU (NVIDIA, Santa Clara, CA, USA), 16GB DDR3 RAM, and 1TB SSD storage. We strongly believe that slightly better results can be achieved by fine-tuning the hyper-parameters, but due to our system limitation, these were the best results we could achieve.

4.2. Evaluation Metrics

The evaluation metrics used to evaluate the performance of the proposed model on the ONCE-3DLanes dataset are the F1 score, precision, recall, and CD error. Precision represents the ratio between all correctly identified positive instances and all instances that were predicted as positive. Recall refers to the ratio of true positives out of all the actual positive instances in the dataset. The F1 score represents the harmonic mean between precision and recall. Those three metrics can be mathematically expressed as

$$\text{Precision} = \frac{TP}{TP + FP}, \quad \text{Recall} = \frac{TP}{TP + FN}, \quad \text{F1} = 2\,\frac{\text{Precision} \times \text{Recall}}{\text{Precision} + \text{Recall}},$$

where TP is the number of true positives, FP the number of false positives, and FN the number of false negatives.

The CD error represents the Chamfer distance between the predicted lane line and the ground truth.

The Apollo dataset uses the F1 score and, besides it, errors that are computed using the Euclidean distance and measured in meters. It compares the matched lanes for the near range (0–40 m) and the far range (40–100 m).

4.3. Results on ONCE-3DLanes

The ONCE-3DLanes [9] dataset is a collection of annotated data for autonomous driving, which includes lane layout information in 3D space. To create this dataset, a pipeline was developed that can automatically generate precise 3D lane location data from 2D annotations. This is achieved by utilizing the explicit relationship between point clouds and image pixels, having over 211,000 road scenes, and resulting in a high-quality dataset. The model was trained for 20 epochs, with the best results being achieved in the 11th epoch.

The results obtained by our model on the ONCE-3DLanes dataset, as shown in Table 1, indicate a significant improvement over current state-of-the-art methods across all key metrics. Specifically, our model achieved an F1 score of 82.39%, outperforming the previous highest score of 80.59% achieved by LATR [24]. This indicates a superior balance between precision and recall in detecting lane markings.

4.4. Results on Apollo 3D Lane Detection

The Apollo 3D Lane Detection dataset, introduced in [6], is an extension of the original Apollo Synthetic Dataset. The dataset was built using the Unity game engine to encourage research and development in the field of autonomous driving. The main reason behind it was the lack of labeled real-world data. The Apollo 3D Lane Detection dataset contains a total of 6000 samples from the virtual highway map, 1500 samples from the urban map, and 3000 samples from the residential area, along with the corresponding depth map, semantic segmentation map, and 3D lane line information.

Table 1. Evaluation results on ONCE-3DLanes. Best values are highlighted in bold.

Model	F1 (%)	Precis. (%)	Recall (%)	CD Err. (m)
3D-LaneNet [21]	44.73	61.46	35.16	0.127
Gen-LaneNet [6]	45.59	63.95	35.42	0.121
SALAD [9]	64.07	75.90	55.42	0.098
PersFormer [7]	74.33	80.30	69.18	0.074
Anchor3DLane [28]	74.87	80.85	69.71	0.060
WS-3D-Lane [29]	77.02	84.51	70.75	0.058
LATR [24]	80.59	**86.12**	75.73	**0.052**
Ours	**82.39**	85.09	**79.85**	0.055

The dataset is split into three sets to evaluate the algorithms. The "balanced scenes" set follows a standard five-fold split for unbiased data training and testing. The "rarely observed scenes" set uses a subset of testing data from a complex urban map to evaluate generalization capabilities. The "scenes with visual variations" set evaluates methods under illumination changes, such as 3D examples from before dawn.

The performance of our model on the Apollo 3D Lane Detection dataset, particularly in the balanced scenes category, Table 2, highlights its strength compared to current state-of-the-art methods. Our model achieved an F1 score of **97.0%**, which is second after the best of **97.4%** achieved by [30]. This performance in F1 score indicates that our model strikes an excellent balance between precision and recall, outperforming models like [24,28], which also performed well but fell slightly short in this aspect.

Table 2. Evaluation results on Apollo 3D Lane Detection balanced scenes. Best values are highlighted in bold.

Model	F1 (%)	X Error Near (m)	X Error Far (m)	Z Error Near (m)	Z Error Far (m)
3D-LaneNet [21]	86.4	0.068	0.477	0.015	**0.202**
Gen-LaneNet [6]	88.1	0.061	0.469	0.012	0.214
CLGo [23]	91.9	0.061	0.361	0.029	0.250
PersFormer [7]	92.9	0.054	0.356	0.010	0.234
Anchor3DLane [28]	95.6	0.052	0.306	0.015	0.233
CurveFormer [31]	95.8	0.078	0.326	0.018	0.219
LATR [24]	96.8	0.022	0.253	**0.007**	**0.202**
BEV-LaneDet [32]	96.9	**0.016**	**0.242**	0.020	0.216
LaneCPP [30]	**97.4**	0.030	0.277	0.011	0.216
Ours	97.0	0.024	0.255	0.009	0.204

For the rarely observed and visual variant scenes, as shown in Tables 3 and 4, our model delivers solid performance. In the rarely observed scenes, while [32] achieves the highest F1 score, our model remains competitive with an F1 score of 95.1%, positioning it closely to other top-performing methods. Similarly, in the visual variant scenes, our model achieves an F1 score of 94.5%, indicating reliable performance across diverse environmental and visual conditions.

Table 3. Evaluation results on Apollo 3D Lane Detection rarely observed scenes. Best values are highlighted in bold.

Model	F1 (%)	X Error Near (m)	X Error Far (m)	Z Error Near (m)	Z Error Far (m)
3D-LaneNet [21]	72.0	0.166	0.855	0.039	**0.521**
Gen-LaneNet [6]	78.0	0.139	0.903	0.030	0.539
CLGo [23]	86.1	0.147	0.735	0.071	0.609
PersFormer [7]	87.5	0.107	0.782	0.024	0.602
Anchor3DLane [28]	94.4	0.094	0.693	0.027	0.579
CurveFormer [31]	95.6	0.182	0.737	0.039	0.561
LATR [24]	96.1	0.050	0.600	**0.015**	0.532
LaneCPP [30]	96.2	0.073	0.651	0.023	0.543
BEV-LaneDet [32]	**99.1**	**0.031**	**0.594**	0.040	0.556
Ours	95.1	0.076	0.626	0.022	0.536

Table 4. Evaluation results on Apollo 3D Lane Detection visual variant scenes. Best values are highlighted in bold.

Model	F1 (%)	X Error Near (m)	X Error Far (m)	Z Error Near (m)	Z Error Far (m)
3D-LaneNet [21]	72.5	0.115	0.601	0.032	0.230
Gen-LaneNet [6]	85.3	0.074	0.538	**0.015**	0.232
CLGo [23]	87.3	0.084	0.464	0.045	0.312
PersFormer [7]	89.6	0.074	0.430	**0.015**	0.266
LaneCPP [30]	90.4	0.054	0.327	0.020	**0.222**
CurveFormer [31]	90.8	0.125	0.410	0.028	0.254
Anchor3DLane [28]	91.4	0.068	0.367	0.020	0.232
LATR [24]	95.1	0.045	**0.315**	0.016	0.228
BEV-LaneDet [32]	**96.9**	**0.027**	0.320	0.031	0.256
Ours	94.5	0.067	0.337	0.018	0.231

Our backbone, MambaVision-S-1K, is composed from the following hybrid blocks: Conv, Mamba, Transformer. The Mamba and the self-attention blocks from the Transformer have the ability to capture and model long-term relationships, but weaker inductive biases than CNNs. Thus, they rely more on the amount and diversity of data to learn these relationships correctly. In [20], it was observed that, when trained only on ImageNet-1K, which has around 1.2 M images, MambaVision-S performs comparably to ResNet, but really gains advantage on larger sets such as ImageNet-22K. As the Apollo subsets have much fewer images than ONCE-3DLanes, it was expected that our configuration would not perform as well on this dataset as on ONCE-3DLanes. Although there is room for improvement in these more challenging scenarios, our model still shows good overall performance.

4.5. Runtime and Complexity Comparison

At our evaluation resolution, 720 × 960, our MambaVision-S-based model counts 62.6 M parameters and 175.5 GFLOPs (FLOPs $\times 10^9$), in comparison to LATR, which uses the ResNet-50 backbone and has 46.8 M parameters and 127.9 GFLOPs. Those results were not presented in [24], but we used their official GitHub repository to download a pretrained model and calculate the complexity under the same input. Although this may seem like an increase in complexity, our model remains substantially more efficient than PersFormer, which is the baseline model for LATR in this line of work, despite being moderately heavier than [24]. The authors of [7] did not publish the complexity of the model, but thanks to GroupLane's work [33], we were able to obtain them. PersFormer working with EfficientNet-B7 [2] is much heavier, having 572.4 GFLOPs, which is ∼3.26 times more than our computational cost.

We ran the inference on our system, which uses NVIDIA RTX 3060 Ti GPU (NVIDIA, Santa Clara, CA, USA), with batch size 1 and excluded the first five warmup iterations. We measure 14.44 FPS for our model versus 15.84 FPS for LATR under the same configuration. In terms of latency, it is ∼69.3 ms per frame for ours and ∼63.1 ms for LATR, which represents a difference of ∼6.1 ms (∼8.8% in FPS). In practice, this is a modest difference on our system, and both models operate in essentially almost the same real-time regime at this resolution.

5. Conclusions

In conclusion, we have proposed an innovative 3D lane detection model that uses the MambaVision backbone integrated with Transformer-based attention mechanisms. This approach significantly improves lane detection performance in 3D space and addresses the challenges posed by non-flat surfaces and variable road conditions. Using deformable attention and lane-aware queries, the model dynamically focuses on relevant regions of the image, resulting in enhanced accuracy for lane detection tasks. Our method achieved state-of-the-art results on the ONCE-3DLanes dataset and competed closely with other methods on the Apollo 3D Lane Detection dataset, demonstrating its superiority in terms of precision, recall, and F1 score. These results highlight the potential of Transformer-enhanced backbones for robust 3D lane detection in autonomous driving applications.

Author Contributions: Conceptualization, R.-M.C. and C.-A.P.; methodology, R.-M.C. and C.-A.P.; software, R.-M.C.; validation, R.-M.C.; formal analysis, R.-M.C.; investigation, R.-M.C.; resources, C.-A.P.; data curation, R.-M.C.; writing—original draft preparation, R.-M.C.; writing—review and editing, C.-A.P.; visualization, R.-M.C.; supervision, C.-A.P.; project administration, C.-A.P.; funding acquisition, C.-A.P. All authors have read and agreed to the published version of the manuscript.

Funding: This research received no external funding. The APC was funded by Politehnica University of Timișoara.

Institutional Review Board Statement: Not applicable.

Informed Consent Statement: Not applicable.

Data Availability Statement: Data are contained within the article.

Conflicts of Interest: The authors declare no conflicts of interest.

References

1. He, K.; Zhang, X.; Ren, S.; Sun, J. Deep Residual Learning for Image Recognition. In Proceedings of the 2016 IEEE Conference on Computer Vision and Pattern Recognition (CVPR), Las Vegas, NV, USA, 27–30 June 2016. [CrossRef]
2. Tan, M.; Le, Q. EfficientNet: Rethinking Model Scaling for Convolutional Neural Networks. In Proceedings of the 36th International Conference on Machine Learning Research, Long Beach, CA, USA, 9–15 June 2019; Chaudhuri, K., Salakhutdinov, R., Eds.; PMLR: Cambridge, MA, USA, 2019; Volume 97, pp. 6105–6114.

3. Howard, A.G.; Zhu, M.; Chen, B.; Kalenichenko, D.; Wang, W.; Weyand, T.; Andreetto, M.; Adam, H. MobileNets: Efficient Convolutional Neural Networks for Mobile Vision Applications. *arXiv* **2017**, arXiv:1704.04861. [CrossRef]
4. Sandler, M.; Howard, A.; Zhu, M.; Zhmoginov, A.; Chen, L.C. MobileNetV2: Inverted Residuals and Linear Bottlenecks. In Proceedings of the 2018 IEEE/CVF Conference on Computer Vision and Pattern Recognition, Salt Lake City, UT, USA, 18–22 June 2018. [CrossRef]
5. Wang, J.; Mei, T.; Kong, B.; Wei, H. An approach of lane detection based on Inverse Perspective Mapping. In Proceedings of the 17th International IEEE Conference on Intelligent Transportation Systems (ITSC), Qingdao, China, 8–11 October 2014. [CrossRef]
6. Guo, Y.; Chen, G.; Zhao, P.; Zhang, W.; Miao, J.; Wang, J.; Choe, T.E. Gen-LaneNet: A Generalized and Scalable Approach for 3D Lane Detection. In Proceedings of the Computer Vision—ECCV 2020, Glasgow, UK, 23–28 August 2020; Springer International Publishing: Berlin/Heidelberg, Germany, 2020; pp. 666–681. [CrossRef]
7. Chen, L.; Sima, C.; Li, Y.; Zheng, Z.; Xu, J.; Geng, X.; Li, H.; He, C.; Shi, J.; Qiao, Y.; et al. PersFormer: 3D Lane Detection via Perspective Transformer and the OpenLane Benchmark. In Proceedings of the Computer Vision—ECCV 2022, Tel Aviv, Israel, 23–27 October 2020; Springer Nature Switzerland: Cham, Switzerland, 2022; pp. 550–567. [CrossRef]
8. Wang, H.; Li, T.; Li, Y.; Chen, L.; Sima, C.; Liu, Z.; Wang, B.; Jia, P.; Wang, Y.; Jiang, S.; et al. Openlane-v2: A topology reasoning benchmark for unified 3D HD mapping. *Adv. Neural Inf. Process. Syst.* **2024**, *36*, 18873–18884.
9. Yan, F.; Nie, M.; Cai, X.; Han, J.; Xu, H.; Yang, Z.; Ye, C.; Fu, Y.; Mi, M.B.; Zhang, L. ONCE-3DLanes: Building Monocular 3D Lane Detection. In Proceedings of the 2022 IEEE/CVF Conference on Computer Vision and Pattern Recognition (CVPR), New Orleans, LA, USA, 18–24 June 2022. [CrossRef]
10. Zhang, Q.; Zhu, Y.; Cordeiro, F.R.; Chen, Q. PSSCL: A progressive sample selection framework with contrastive loss designed for noisy labels. *Pattern Recognit.* **2025**, *161*, 111284. [CrossRef]
11. Carion, N.; Massa, F.; Synnaeve, G.; Usunier, N.; Kirillov, A.; Zagoruyko, S. End-to-End Object Detection with Transformers. In Proceedings of the Computer Vision—ECCV 2020, Glasgow, UK, 23–28 August 2020; Springer International Publishing: Berlin/Heidelberg, Germany, 2020; pp. 213–229. [CrossRef]
12. Dosovitskiy, A.; Beyer, L.; Kolesnikov, A.; Weissenborn, D.; Zhai, X.; Unterthiner, T.; Dehghani, M.; Minderer, M.; Heigold, G.; Gelly, S.; et al. An Image is Worth 16 × 16 Words: Transformers for Image Recognition at Scale. In Proceedings of the International Conference on Learning Representations, Online, 3–7 May 2021.
13. Han, K.; Wang, Y.; Chen, H.; Chen, X.; Guo, J.; Liu, Z.; Tang, Y.; Xiao, A.; Xu, C.; Xu, Y.; et al. A Survey on Vision Transformer. *IEEE Trans. Pattern Anal. Mach. Intell.* **2023**, *45*, 87–110. [CrossRef] [PubMed]
14. Zhu, X.; Su, W.; Lu, L.; Li, B.; Wang, X.; Dai, J. Deformable DETR: Deformable Transformers for End-to-End Object Detection. In Proceedings of the International Conference on Learning Representations, Online, 3–7 May 2021.
15. Zheng, S.; Lu, J.; Zhao, H.; Zhu, X.; Luo, Z.; Wang, Y.; Fu, Y.; Feng, J.; Xiang, T.; Torr, P.H.; et al. Rethinking Semantic Segmentation from a Sequence-to-Sequence Perspective with Transformers. In Proceedings of the 2021 IEEE/CVF Conference on Computer Vision and Pattern Recognition (CVPR), Nashville, TN, USA, 19–25 June 2021. [CrossRef]
16. Russakovsky, O.; Deng, J.; Su, H.; Krause, J.; Satheesh, S.; Ma, S.; Huang, Z.; Karpathy, A.; Khosla, A.; Bernstein, M.; et al. ImageNet Large Scale Visual Recognition Challenge. *Int. J. Comput. Vis.* **2015**, *115*, 211–252. [CrossRef]
17. Lin, T.Y.; Maire, M.; Belongie, S.; Hays, J.; Perona, P.; Ramanan, D.; Dollár, P.; Zitnick, C.L. Microsoft COCO: Common Objects in Context. In Proceedings of the Computer Vision—ECCV, Zurich, Switzerland, 6–12 September 2014; Springer International Publishing: Berlin/Heidelberg, Germany, 2014; pp. 740–755. [CrossRef]
18. Cordts, M.; Omran, M.; Ramos, S.; Rehfeld, T.; Enzweiler, M.; Benenson, R.; Franke, U.; Roth, S.; Schiele, B. The Cityscapes Dataset for Semantic Urban Scene Understanding. In Proceedings of the 2016 IEEE Conference on Computer Vision and Pattern Recognition (CVPR), Las Vegas, NV, USA, 26 June–1 July 2016. [CrossRef]
19. Vaswani, A.; Shazeer, N.; Parmar, N.; Uszkoreit, J.; Jones, L.; Gomez, A.N.; Kaiser, L.u.; Polosukhin, I. Attention is All you Need. In *Advances in Neural Information Processing Systems*; Guyon, I., Luxburg, U.V., Bengio, S., Wallach, H., Fergus, R., Vishwanathan, S., Garnett, R., Eds.; Curran Associates, Inc.: New York, NY, USA, 2017; Volume 30.
20. Hatamizadeh, A.; Kautz, J. Mambavision: A hybrid mamba-transformer vision backbone. In Proceedings of the Computer Vision and Pattern Recognition Conference, Nashville, TN, USA, 11–15 June 2025; pp. 25261–25270.
21. Garnett, N.; Cohen, R.; Pe'er, T.; Lahav, R.; Levi, D. 3D-LaneNet: End-to-End 3D Multiple Lane Detection. In Proceedings of the 2019 IEEE/CVF International Conference on Computer Vision (ICCV), Seoul, Republic of Korea, 27 October–2 November 2019. [CrossRef]
22. Jin, Y.; Ren, X.; Chen, F.; Zhang, W. Robust Monocular 3D Lane Detection with Dual Attention. In Proceedings of the 2021 IEEE International Conference on Image Processing (ICIP), Anchorage, AK, USA, 19–22 September 2021. [CrossRef]
23. Liu, R.; Chen, D.; Liu, T.; Xiong, Z.; Yuan, Z. Learning to Predict 3D Lane Shape and Camera Pose from a Single Image via Geometry Constraints. In Proceedings of the AAAI Conference on Artificial Intelligence, online, 22 February–1 March 2022; Association for the Advancement of Artificial Intelligence (AAAI): Washington, DC, USA, 2022; Volume 36, pp. 1765–1772. [CrossRef]

24. Luo, Y.; Zheng, C.; Yan, X.; Kun, T.; Zheng, C.; Cui, S.; Li, Z. Latr: 3d lane detection from monocular images with transformer. In Proceedings of the IEEE/CVF International Conference on Computer Vision, Paris, France, 1–6 October 2023; pp. 7941–7952.
25. Gu, A.; Dao, T. Mamba: Linear-Time Sequence Modeling with Selective State Spaces. In Proceedings of the 1st Conference on Language Modeling, Philadelphia, PA, USA, 7–9 October 2024.
26. Zhu, L.; Liao, B.; Zhang, Q.; Wang, X.; Liu, W.; Wang, X. Vision Mamba: Efficient visual representation learning with bidirectional state space model. In Proceedings of the 41st International Conference on Machine Learning (ICML'24), Vienna, Austria, 21–27 July 2024.
27. Tang, H.; Li, Z.; Zhang, D.; He, S.; Tang, J. Divide-and-Conquer: Confluent Triple-Flow Network for RGB-T Salient Object Detection. *IEEE Trans. Pattern Anal. Mach. Intell.* **2025**, *47*, 1958–1974. [CrossRef] [PubMed]
28. Huang, S.; Shen, Z.; Huang, Z.; Ding, Z.h.; Dai, J.; Han, J.; Wang, N.; Liu, S. Anchor3dlane: Learning to regress 3D anchors for monocular 3D lane detection. In Proceedings of the IEEE/CVF Conference on Computer Vision and Pattern Recognition, Vancouver, BC, Canada, 18–22 June 2023; pp. 17451–17460.
29. Ai, J.; Ding, W.; Zhao, J.; Zhong, J. WS-3D-Lane: Weakly Supervised 3D Lane Detection With 2D Lane Labels, 2023. *arXiv* **2022**, arXiv:2209.11523. [CrossRef]
30. Pittner, M.; Janai, J.; Condurache, A.P. LaneCPP: Continuous 3D Lane Detection using Physical Priors. In Proceedings of the IEEE/CVF Conference on Computer Vision and Pattern Recognition, Seattle, WA, USA, 16–22 June 2024; pp. 10639–10648.
31. Bai, Y.; Chen, Z.; Fu, Z.; Peng, L.; Liang, P.; Cheng, E. CurveFormer: 3D Lane Detection by Curve Propagation with Curve Queries and Attention. In Proceedings of the 2023 IEEE International Conference on Robotics and Automation (ICRA), London, UK, 29 May–2 June 2023; pp. 7062–7068. [CrossRef]
32. Wang, R.; Qin, J.; Li, K.; Li, Y.; Cao, D.; Xu, J. Bev-lanedet: An efficient 3d lane detection based on virtual camera via key-points. In Proceedings of the IEEE/CVF Conference on Computer Vision and Pattern Recognition, Vancouver, BC, Canada, 17–24 June 2023; pp. 1002–1011.
33. Li, Z.; Han, C.; Ge, Z.; Yang, J.; Yu, E.; Wang, H.; Zhang, X.; Zhao, H. GroupLane: End-to-End 3D Lane Detection with Channel-wise Grouping. *IEEE Robot. Autom. Lett.* **2024**, *9*, 10487–10494. [CrossRef]

Disclaimer/Publisher's Note: The statements, opinions and data contained in all publications are solely those of the individual author(s) and contributor(s) and not of MDPI and/or the editor(s). MDPI and/or the editor(s) disclaim responsibility for any injury to people or property resulting from any ideas, methods, instructions or products referred to in the content.

sensors

MDPI

Article

Design and Validation of an Active Headrest System with Integrated Sensing in Rear-End Crash Scenarios

Alexandru Ionut Radu [1], Bogdan Adrian Tolea [2,*], Horia Beles [2], Florin Bogdan Scurt [2] and Adrian Nicolaie Tusinean [3]

1 Department of Automotive and Transport, Faculty of Mechanical Engineering, Transilvania University of Brasov, 500036 Brasov, Romania; alexandru.radu@unitbv.ro
2 Department of Mechanical Engineering and Automotive, Faculty of Managerial and Technological Engineering, University of Oradea, 410087 Oradea, Romania; horia.beles@uoradea.ro (H.B.); scurt.florinbogdan@didactic.uoradea.ro (F.B.S.)
3 Advanced Mechatronik Technologies SRL, 500025 Brasov, Romania; adriantusinean@gmail.com
* Correspondence: bogdan.tolea@uoradea.ro

Abstract: Rear-end collisions represent a major concern in automotive safety, particularly due to the risk of whiplash injuries among vehicle occupants. The accurate simulation of occupant kinematics during such impacts is critical for the development of advanced safety systems. This paper presents an enhanced multibody simulation model specifically designed for rear-end crash scenarios, incorporating integrated active headrest mechanisms and sensor-based activation logic. The model combines detailed representations of vehicle structures, suspension systems, restraint systems, and occupant biomechanics, allowing for the precise prediction of crash dynamics and occupant responses. The system was developed using Simscape Multibody, with CAD-derived components interconnected through physical joints and validated using controlled experimental crash tests. Special attention was given to modelling contact forces, suspension behaviour, and actuator response times for the active headrest system. The model achieved a root mean square error (RMSE) of 4.19 m/s^2 and a mean absolute percentage error (MAPE) of 0.71% when comparing head acceleration in frontal collision tests, confirming its high accuracy. Validation results demonstrate that the model accurately reproduces occupant kinematics and head acceleration profiles, confirming its reliability and effectiveness as a predictive tool. This research highlights the critical role of integrated sensor-actuator systems in improving occupant safety and provides a flexible platform for future studies on intelligent vehicle safety technologies.

Keywords: mathematical model; active headrest; integrated sensor; simulation; multibody; Simscape multibody; dynamic simulation; crash test

1. Introduction

Motor vehicle collisions, particularly frontal and rear-end impacts, continue to pose a significant risk to vehicle occupants and represent a major concern for automotive safety research.

According to global accident statistics, rear-end and frontal impacts remain among the most frequent types of vehicular collisions, often leading to whiplash and head–torso-related injuries. In the European context, rear-end collisions account for over 30% of road traffic incidents with injuries, while frontal collisions are associated with high fatality rates due to high-speed impacts [1]. These types of accidents are strongly correlated with occupant injury mechanisms, which this study aims to address by using biomechanical simulations. A detailed statistical review and classification of such accidents is available in [2].

Understanding the kinematics and dynamics of occupants during such collisions is critical for enhancing vehicle design and improving occupant protection systems.

Mathematical modelling has proven to be a powerful tool in this field, offering detailed insights into the behaviour of vehicle structures and human biomechanics during crash events [3–5].

Among these approaches, multibody system (MBS) models have gained prominence due to their ability to accurately simulate complex interactions between the vehicle, restraint systems, and human body segments [6–8]. Recent advancements have highlighted the effectiveness of MBS models validated by experimental crash tests, providing robust frameworks for analyzing occupant kinematics and injury mechanisms. For instance, Ionut et al. (2017) developed a frontal collision model validated with crash tests, while Radu et al. (2021) extended this methodology to side impacts [3,4].

Furthermore, Xu et al. (2023) introduced a refined three-dimensional multibody dynamics model for single-vehicle crashes against rigid barriers [5]. Complementary studies by Ryu et al. (2023) and An et al. (2023) expanded the application of MBS to multibody contact dynamics and vehicle–infrastructure interactions, respectively [6–8]. A growing body of research has also explored active safety systems aimed at mitigating occupant injuries. Kim et al. (2020) analyzed the behaviour of active and proactive headrests during low-speed rear-end collisions, demonstrating the benefits of dynamic head support [9]. Zhang et al. (2023) and Vychytil et al. (2020) developed and validated active headrest mechanisms to improve occupant biomechanics in rear-end impacts [10,11]. Other studies, such as He et al. (2022, 2024) and Yan and Chen (2023), proposed semi-active seat designs and adaptive headrest systems that respond to occupant posture and crash dynamics [12–14]. Furthermore, Lin et al. (2023) applied finite element modelling to study head–neck biomechanical responses during impacts, highlighting alternative high-fidelity modelling techniques [14]. Liu and Zhang (2024) enhanced the capabilities of multibody dynamic simulations specifically for rear-end collisions [15], while Yuan and Li (2023) focused on sensor-based active headrest deployment to improve real-time injury mitigation [16]. Additionally, complementary research has emphasized the role of road surface effects [17], machine learning-based crash severity estimation [18], and the influence of seat configurations on injury risk [19], underlining the multidisciplinary nature of modern vehicle safety studies [20].

Further advancements in occupant protection systems have focused on innovations in airbag and seatbelt technologies. Kitagawa et al. (2012) developed an enhanced airbag system specifically designed for far-side impact protection, addressing occupant movement in lateral collisions [21].

Similarly, Mizuno (2015) evaluated advanced seatbelt systems that demonstrated significant reductions in chest injury risks during frontal crashes, underlining the critical role of restraint system design [22–24]. Research on contact dynamics modelling has also gained importance in accident analysis.

Benea and Soica (2023) emphasized the significance of accurately simulating the contact phase in vehicle–pedestrian collisions, providing valuable insights for improving safety measures and reconstruction methodologies [25]. Additionally, Soica (2024) investigated the forces transmitted through pretensioner tubes during collisions, demonstrating how mechanical modelling can aid in accident reconstruction and injury analysis [26]. These studies collectively highlight the essential role of precise mathematical modelling, occupant–restraint interaction analysis, and advanced safety system development in improving overall vehicle safety performance.

Recent advances in Micro-Electro-Mechanical Systems (MEMS) technology have enabled the development of highly sensitive, compact, and cost-effective acceleration sensors suitable for automotive safety applications. MEMS accelerometers are widely used in vehicle crash-detection systems due to their ability to rapidly measure dynamic changes in acceleration with high precision [27,28]. In the context of active headrest systems, acceleration sensors play a critical role by detecting the onset of rear-end collisions. Upon registering a sudden longitudinal acceleration that exceeds a predetermined threshold—typically around 50 m/s^2—the MEMS sensor triggers the activation mechanism, propelling the headrest forward to mitigate the relative motion between the head and torso [17,29]. The compact design and robustness of MEMS sensors allow for seamless integration into vehicle structures, such as within the seat back or vehicle frame. Their high sampling rates and resilience to vibration and temperature variations ensure reliable operation, even under severe crash conditions [27,28]. The integration of sensor-based logic into the active headrest system thus provides a significant enhancement to occupant protection, reducing the risk of whiplash injuries by supporting the occupant's head earlier in the collision event [17,29].

Building upon this foundation, this work proposes an enhanced mathematical multibody model specifically designed for analyzing rear-end collisions involving integrated active headrest systems and embedded sensor-based logic. By incorporating realistic contact algorithms and detailed occupant kinematics, the model aims to predict crash outcomes and occupant responses with greater precision, addressing the limitations of current models, which often lack detailed contact mechanics and fail to accurately capture the complex kinematics of the head and neck during rear-end collisions.

The key contributions of this study include the following:

- The development of a comprehensive multibody simulation model integrating active headrest mechanisms.
- The validation of the model against experimental crash data.
- The demonstration of its potential for optimizing occupant protection systems and supporting the design of advanced automotive safety features.

This model offers a robust, flexible tool for simulating rear-end crash dynamics and evaluating the performance of sensor-driven active safety technologies, bridging the gap between traditional MBS techniques and emerging intelligent vehicle systems.

2. Background

An earlier version of the model was developed in analytical form, with the equations of motion derived and solved using the ODE45 numerical solver in MATLAB 2024b [30]. This initial model was designed to investigate frontal collision dynamics and occupant kinematics. However, due to its simplified structure, it exhibited limitations in versatility and the ability to represent more complex crash scenarios [3]. Despite these constraints, the results demonstrate that key dynamic parameters could be estimated with good accuracy using a relatively simple analytical approach.

Following the positive results obtained using the initial analytical model, the next development phase focused on creating a complete multibody system capable of producing highly accurate results when compared with experimental crash test data. To illustrate the workflow for building and optimizing the multibody system, a schematic diagram has been prepared, as shown in Figure 1. This diagram outlines the sequential process, starting from the analytical model, progressing through CAD modelling, and culminating in a validated Simscape multibody simulation environment, with iterative optimization based on input parameters and simulation outputs.

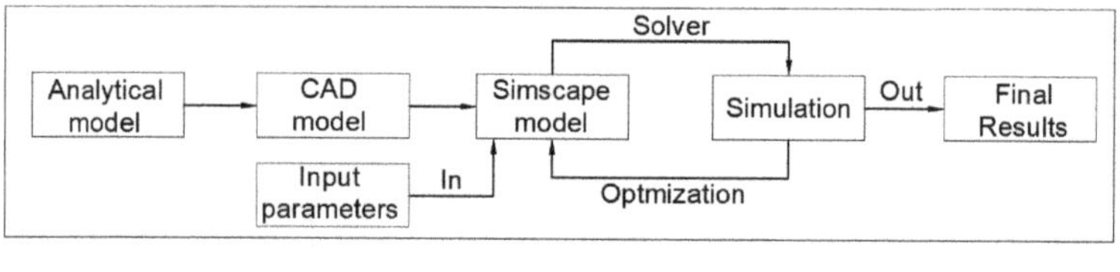

Figure 1. System diagram for model development.

The second development step involved creating a detailed CAD model of the components, using the real-world dimensions of both the vehicles and the occupant body segments. The resulting CAD model was then imported into Simscape Multibody, where additional information was incorporated, including mechanical joints, contact force models, body positioning, solver configurations, and dynamic input parameters. The multibody model was constructed using the physical network approach in Simscape, which employs physical blocks and realistic connections to ensure the accurate representation of the system's behaviour and practical usability. The numerical solver selected by the software was ODE3 (Bogacki–Shampine method), automatically determined based on the model's configuration, the specified initial conditions, and the defined simulation time step.

3. Methodology

3.1. Model Description

The multibody system model was developed using CAD-generated parts, with each component connected using appropriate mechanical joints to realistically replicate the physical interactions between the occupant and the vehicle structure. The initial step in the modelling process involved creating a two-dimensional schematic diagram to illustrate the connections between the bodies and the types of stiffness elements (springs and dampers) implemented in the system. Figure 2 presents this schematic representation, highlighting the interaction points between the various components and specifying whether the connection is modelled as a translational spring–damper (rigid connection) or a rotational spring–damper (torsional connection). This schematic formed the foundation for constructing the complete multibody model in the Simscape environment.

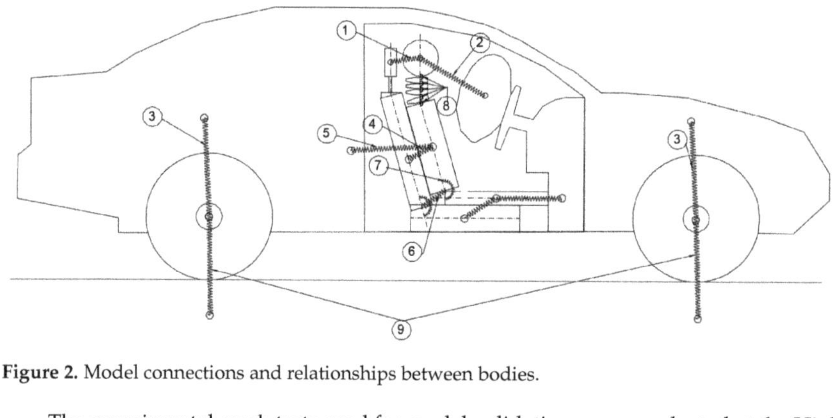

Figure 2. Model connections and relationships between bodies.

The experimental crash tests used for model validation were conducted at the High-Tech Product Development Research and Innovation Institute of Transilvania University of Brașov. To capture acceleration data, two types of acquisition systems were employed. A PIC DAQ system was mounted in the front floor section of the vehicle to measure accelerations at the vehicle's centre of mass. Additionally, for occupant motion analysis, a

PicDAQ MT5 system equipped with tri-axial accelerometers was installed in the dummy's torso and head. These sensors recorded accelerations along the X, Y, and Z axes with a sampling rate of up to 2 kHz, and data were logged to an SD card for further processing using PocketDAQ 1.0 software.

To support the kinematic analysis of the occupant, high-speed video recording equipment was deployed, including Casio Exilim EX-F1 and Fastec Hispec 5 cameras, capable of capturing up to 1400 frames per second. Still image documentation was performed using Nikon AW 100 and Samsung Galaxy S5 cameras. In addition, a DS5 GPS system was used to record vehicle speed, position, and impact velocity, utilising a GPS receiver connected via RS 232-USB interface to a notebook.

The testing area was secured using safety protocols such as isolation of the test zone, reliable towing and braking systems, and remote braking controls. The detailed setup ensured reproducibility and safe execution of crash scenarios that served to validate the developed multibody simulation model.

The complete model consists of several interconnected components designed to accurately replicate the dynamics of a rear-end collision scenario. The vehicle structure is modelled as a single rigid body, connected to the four wheels through suspension systems represented by prismatic joints with spring–damper elements (presented in Table 1).

Table 1. Figure notations.

Notation	Description
1	Rigidity between the head and upper seat (headrest)
2	Rigidity between the head and airbag
3	Rigidity between the wheels and the car (suspension)
4	Rigidity between the upper torso and the upper seat
5	Rigidity between the car and the upper body (seatbelt)
6	Rigidity between the lower body and the lower seat (lower seatbelt)
7	Torsion springs between the upper body and the lower body
8	Torsion spring in the neck
9	Rigidity between the wheels and the ground

The seat assembly is divided into two distinct components: the lower seat and the backrest. These two elements are connected by a revolute joint equipped with a torsional spring–damper, allowing for the realistic flexion of the seatback during impact.

The occupant is modelled as a simplified anthropomorphic structure without limbs, consisting of the lower body, torso, neck, and head. The neck itself is articulated using four sequential vertebrae, each connected by revolute joints to capture the flexibility of the cervical spine. All segments of the occupant are connected through revolute joints, enabling realistic relative rotations between body parts.

The occupant is positioned on the seat via direct body contacts, simulating gravitational preload and support. Additionally, the torso and lower body are connected to the vehicle body via linear springs, representing seatbelt forces that restrain the occupant during impact. Contact interactions between the lower body, torso, and head with the dashboard are also modelled. The airbag is represented as a single non-deformable body; although no visual deformation is simulated, dynamic interaction is modelled via contact forces. The deceleration effect of the airbag on the head is computed based on predefined stiffness and damping contact parameters. This detailed setup ensures that the occupant dynamics, restraint effects, and head–dashboard–airbag interactions are realistically captured within the simulation framework.

To expand the model and simulate rear-end collision dynamics more realistically, a second vehicle was incorporated into the system. A contact connection was defined between the front structure of the first vehicle and the rear structure of the second vehicle, enabling the evaluation of impact forces and deformation behaviour at the point of collision. This addition allows for the accurate modelling of the force transmission between the two vehicles during a rear-end impact and ensures that the dynamic responses of both vehicles are captured within the simulation framework. In this study, a simplified modelling approach is adopted where both vehicles are assumed to have identical geometry and dimensions. This assumption facilitates the analysis of occupant dynamics while focusing on the interaction forces during rear-end impacts (see Figure 3).

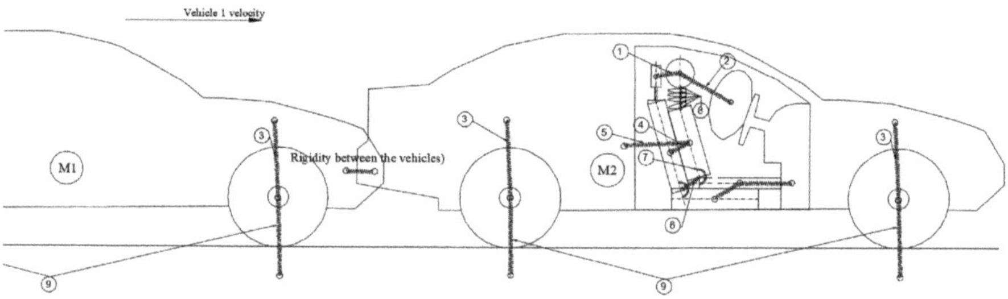

Figure 3. Collision model between the two vehicles.

Figure 4 presents the principal dimensions of the vehicle model along with the reference points used for positioning and alignment during simulation. The side view shows the longitudinal and vertical distances between key structural points, including the location of the prismatic joints used to simulate suspension movement. The suspension joints are positioned to accurately replicate the vehicle's real suspension geometry, and their relative heights and distances from the vehicle centre are specified.

Figure 4 highlights the vehicle's main reference point, which serves as a fixed origin for global positioning within the simulation environment. In the top view, the layout of the seat relative to the vehicle structure is illustrated, including the seat reference point positioned 100 mm rearward of the global reference point, ensuring the accurate placement of the occupant model within the cabin. The distances between the suspension mounting points across the vehicle width are also detailed, supporting the symmetrical construction of the chassis in the model.

This dimensional accuracy was essential for ensuring that the multibody system behaves realistically during impact scenarios, particularly in replicating suspension kinematics and occupant–seat interactions. The principal dimensions of the vehicle model and the associated reference points are presented in Figure 4. The side view details the longitudinal and vertical distances between key structural points, including the placement of the suspension prismatic joints and the inclination of the front suspension system. The top view illustrates the seat positioning relative to the main reference point of the vehicle, with the seat reference point positioned 100 mm rearward of the global reference. Symmetrical distances between suspension mounting points are also shown, ensuring consistent chassis construction. Accurate definition of these dimensions and reference points was essential to ensure the realistic simulation of suspension kinematics, occupant seating position, and impact dynamics.

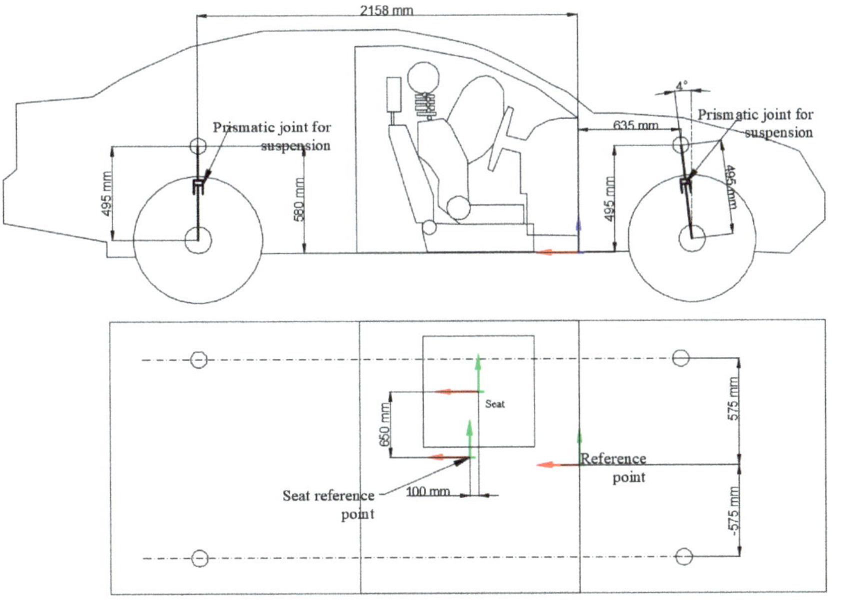

Figure 4. Model dimensions with respect to the reference points.

3.2. Contact Modelling

Each body in motion within the model is governed by its own differential equation of motion. The principle underpinning the model is based on the classical sprung mass system, described using the following general form [31]:

$$M\frac{d^2x}{dt^2} + C\frac{dx}{dt} + Kx = F \quad (1)$$

where

- x is the vector of displacements and rotational angles of the bodies;
- M is the mass matrix;
- C is the damping matrix;
- K is the stiffness matrix;
- F is the vector of external forces acting on the system.

This general equation is applied to all moving components within the model and the contact interactions between vehicles and between occupants and interior structures. It provides a unified framework for modelling both translational and rotational dynamics, as well as contact and spring–damper behaviours throughout the system.

3.3. Model Equations

Following the general formulation of the dynamic system, the specific equations of motion for the rear-end collision model are derived. The model incorporates two vehicles, suspension systems, and a biomechanical representation of the occupant within the front vehicle. Each body and joint are associated with translational and rotational degrees of freedom (3 DOF), as well as with corresponding mass, damping, and stiffness properties. The resulting system of equations, summarized in Table 2, describes the dynamic interaction between the vehicles and the occupant under rear-end impact conditions. In this formulation, each symbol denotes either a kinematic quantity, a physical parameter, or a degree of freedom relevant to the multibody system.

Table 2. Model parameters.

Notation	Description
$x1$	Position of Vehicle 1 (rear vehicle)
$x2$	Position of Vehicle 2 (front vehicle)
$xs1$	Suspension deflection of Vehicle 1
$xs2$	Suspension deflection of Vehicle 2
x_{UB}	Translational position of the upper body (torso)
θ_{UB}	Rotation angle of the upper body (torso)
x_{LB}	Translational position of the lower body
θ_{LB}	Rotation angle of the lower body
k_{LB}	Rotational stiffness of the lower body
d_{LB}	Rotational damping of the lower body
θ_H	Rotation angle of the head
θ_N	Rotational angle of the neck
$x_{\_H}$	Position of the head
$x_{\_hr}$	Position of the headrest (in Vehicle 2)
$k_{\_H}$	Head–neck stiffness
$d_{\_H}$	Head–neck damping
$k_{\_N}$	Neck-torso stiffness
$d_{\_N}$	Head–torso damping
$m_{\_h}$	Mass of the head
m_{UB}	Mass of the upper body (torso)
M_1	Mass of Vehicle 1
M_2	Mass of Vehicle 2
F_{act}	Active headrest actuator force
I_h	Moment of inertia of the head
I_N	Moment of inertia of the neck
I_{ub}	Moment of inertia of the upper body (torso)
k_{hr}	Headrest stiffness
d_{hr}	Headrest damping coefficient
k_{sb}	Seatbelt stiffness
d_{sb}	Seatbelt damping coefficient
k_{s1}	Suspension stiffness of Vehicle 1
c_{s1}	Suspension damping coefficient of Vehicle 1
k_{s2}	Suspension stiffness of Vehicle 2
c_{s2}	Suspension damping coefficient of Vehicle 2
k_c	Stiffness of the inter-vehicle contact
c_c	Damping of the inter-vehicle contact
k_{θ_neck}	Neck joint stiffness
d_{θ_neck}	Neck joint damping

Table 2. *Cont.*

Notation	Description
k_{θ_ub}	Seatback stiffness (torso rotation)
d_{θ_ub}	Seatback damping (torso rotation)
L	Effective distance from upper body to head centre
l_{hr}	Offset from vehicle 2 reference to headrest

While the model considers a simplified two-vehicle collision scenario, this approach can be extended to more complex events, such as multi-vehicle (mass) crashes, by expanding the dynamic boundary conditions and incorporating additional contact constraints.

This system models the interaction between two vehicles during a rear-end collision and includes a biomechanical model of the human occupant in the front vehicle [31–33].

Head:

- Translational motion:

$$m_h\ddot{x}_h = -k_{hr}(x_h - x_{hr}) - d_{hr}(\dot{x}_h - \dot{x}_{hr}) \quad (2)$$

- Rotational motion:

$$I_h\ddot{\theta}_h = -k_{\theta,neck}(\theta_h - \theta_N) - d_{\theta,neck}\left(\dot{\theta}_h - \dot{\theta}_N\right) \quad (3)$$

Neck:

- Rotational motion (neck stiffness and damping):

$$I_N\ddot{\theta}_N = k_H(\theta_h - \theta_N) + d_H\left(\dot{\theta}_h - \dot{\theta}_N\right) - k_N(\theta_h - \theta_{UB}) - d_N\left(\dot{\theta}_h - \dot{\theta}_{UB}\right) \quad (4)$$

Torso (upper body):

- Translational motion (forces from seatbelt and head–headrest coupling):

$$m_{UB}\ddot{x}_{UB} = k_{sb}(x_2 - x_{UB}) + d_{sb}(\dot{x}_2 - \dot{x}_{UB}) + k_{hr}(x_h - x_{hr}) + d_{hr}(\dot{x}_h - \dot{x}_{hr}) \quad (5)$$

- Rotational motion:

$$I_{UB}\ddot{\theta}_{UB} = k_N(\theta_N - \theta_{UB}) + d_N\left(\dot{\theta}_N - \dot{\theta}_{UB}\right) - k_{UB}\theta_{UB} - d_{UB}\dot{\theta}_{UB} \quad (6)$$

Torso (lower body):

- Rotational motion:

$$I_{LB}\ddot{\theta}_{LB} = -k_{LB}\theta_{LB} - d_{LB}\dot{\theta}_{LB} \quad (7)$$

Vehicle 1 (with occupant):

- Translational motion:

$$M_1\ddot{x}_1 = -k_{s1}x_{s1} - c_{s1}\dot{x}_{s1} - k_{sb}(x_1 - x_{UB}) - d_{sb}(\dot{x}_1 - \dot{x}_{UB}) - k_c(x_1 - x_2) - c_c(\dot{x}_1 - \dot{x}_2) \quad (8)$$

Vehicle 2:

- Translational motion:

$$M_2\ddot{x}_2 = -k_{s2}x_{s2} - c_{s2}\dot{x}_{s2} + k_c(x_1 - x_2) + c_c(\dot{x}_1 - \dot{x}_2) \quad (9)$$

3.4. Active Headrest Implementation

An active headrest system is integrated into the occupant model to lessen the whiplash effect during rear-end collisions. The headrest is actively moved by a linear actuator controlled by a microcontroller that receives data from an accelerometer, in response to detected deceleration. The translational equation of motion for the occupant's head is increased by the time-dependent force Fact(t) that the actuator applies to the head. As a result, the model can replicate both active and passive whiplash defences (minimizing the head–neck movement during collision).

The vehicle's rear-end collision-related abrupt deceleration is detected by an accelerometer installed near the bumper or chassis. This signal is processed by the microcontroller, which then turns on the linear actuator that is mounted behind the headrest. By applying a force Fact(t) forward to the headrest, the actuator lessens whiplash by decreasing the relative motion between the head and torso (see Figure 5).

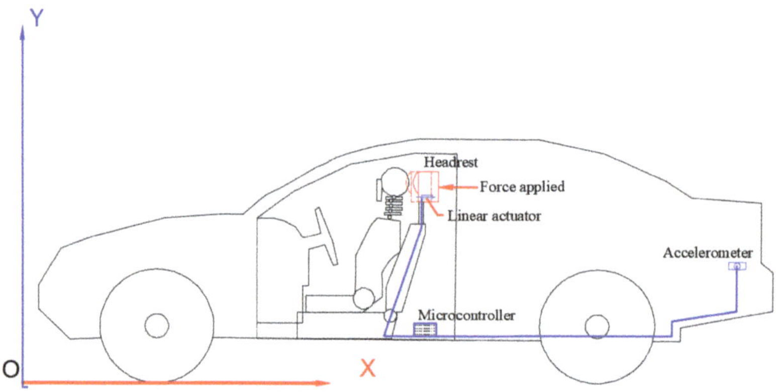

Figure 5. Active headrest schematic.

In response to a crash event, the headrest is activated by a linear actuator that is modelled as a prismatic joint with stiffness and damping. It is possible to derive a 1D translational equation of motion (EOM) for the headrest mass along the horizontal axis (X-direction). Using Newton's Second Law,

$$m_h(\ddot{x}_h + a_c(t)) = -k_{hr}x_h - d_{hr}\dot{x}_h \quad (10)$$

where m_h—is the headrest mass, k_{hr}—headrest linear stiffness, d_{hr}—headrest linear damper, and x_h—headrest displacement.

The force applied by the actuator is then

$$F_{act} = -k_{hr}x_h - d_{hr}\dot{x}_h \quad (11)$$

The headrest interaction force that already exists can be increased by the actuator force as follows:

$$m_h\ddot{x}_h = -k_{hr}(x_h - x_{hr}) - d_{hr}(\dot{x}_h - \dot{x}_{hr}) + F_{act}(t) \quad (12)$$

A threshold-based logic that tracks the longitudinal acceleration of the vehicle is used to control the active headrest. When the measured acceleration surpasses a certain threshold (e.g., 50 m/s^2), a linear actuator applies force to the headrest. This logic is implemented using a Simulink block diagram, as illustrated in Figure 6.

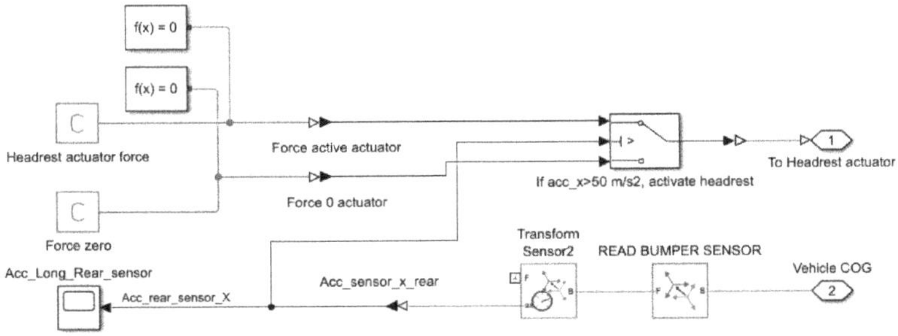

Figure 6. Implementation of active headrest in the Simscape model.

The active headrest system is modelled using a closed-loop Simulink control block. A longitudinal acceleration sensor is placed near the rear of the vehicle (referred to as Acc_Long_Rear_sensor) to detect the sudden deceleration caused by rear-end collisions. The sensor data is transformed into the vehicle's centre of gravity reference frame using a transform block (Transform Sensor2), and the longitudinal component ax,rear is extracted.

A conditional logic block compares the rear acceleration signal to a threshold value (in this case, 50 m/s^2). When this threshold is exceeded, a predefined actuator force Fact is applied to the headrest, pushing it forward toward the occupant's head. If the threshold is not exceeded, the actuator applies zero force.

The headrest force applied to the head can be expressed as follows:

$$F_{act}(t) = \begin{cases} F_{max}, \ if \ a_{x,\,rear}(t) > a_{thresh} \\ 0, \ if \ a_{x,\,rear}(t) < a_{thresh} \end{cases} \tag{13}$$

where $a_{x,rear}(t)$—longitudinal acceleration measured at the rear of the vehicle, $a_{thresh} = 50$ m/s^2—activation threshold and F_{max}—constant actuator force.

3.5. Simscape Model

The functional simulation model was developed using Simscape Multibody in MATLAB. CAD-derived parts were imported and interconnected using mechanical joints to accurately represent the dynamic behaviour of the system during a rear-end collision.The assembled Simscape model, illustrated in Figure 7, is structured into three principal components: Vehicle 1 (depicted in blue), Vehicle 2 (depicted in red), and the occupant positioned within Vehicle 1.

Each principal component is organized into a hierarchical subsystem mask containing multiple layers of individual elements, such as wheels, suspension components, body structures, and occupant segments. All components are connected through defined reference points and a combination of prismatic, revolute, and translational joints, ensuring accurate kinematic and dynamic interaction between the bodies.

This modular architecture allows for clear model organization, facilitates parameter updates, and provides the flexibility to adapt the system for future extensions or additional active safety systems.

Figure 7 illustrates the complete Simscape Multibody model developed for the simulation of rear-end collisions. The base body is connected to the real-world frame and serves as the global reference for all bodies in the system. Both vehicles are modelled with joints connecting their chassis to their respective wheels, allowing for independent suspension behaviour. The interaction between the vehicles and the ground is modelled by introducing spatial contact forces at each wheel, simulating tyre–ground contact and enabling the transmission of vertical and longitudinal forces during impact and rebound.

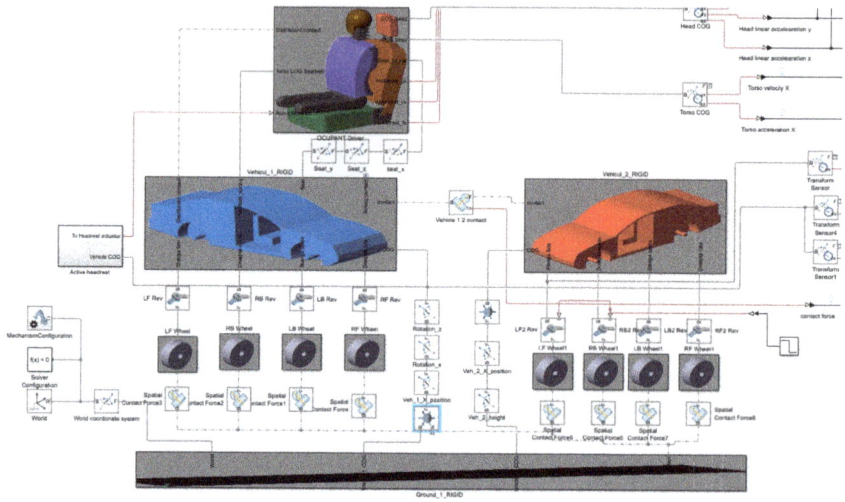

Figure 7. Simscape model for rear-end collisions.

An additional contact force is defined between the front of Vehicle 1 and the rear of Vehicle 2, representing the stiffness and damping characteristics of the front-end structures during collision.

To enable realistic vehicle motion,

- Vehicle 1 is connected to the ground through a 6-DOF joint, allowing for full three-dimensional movement with controlled constraints;
- Vehicle 2 is connected to the ground via a prismatic joint, permitting only longitudinal translation, with the initial velocity of Vehicle 2 adjustable to simulate different rear-end impact scenarios.

Within Vehicle 1, the occupant seat and chassis are connected by a prismatic joint combined with a spring–damper system to model the dynamic tension and restraint effect of the seatbelt during collision.

The neck segment of the occupant is connected to the upper body via a revolute joint and a spatial contact force, allowing for a realistic head–torso relative motion and enabling the simulation of neck flexion under impact conditions.

This model architecture ensures the accurate representation of vehicle and occupant dynamics, structural compliance, and contact interactions necessary for simulating rear-end crash scenarios involving active safety systems. The contact force between the components was calculated using Equation (2) [32,34,35].

$$f_n = s(d)\cdot\left(k\cdot d + b\cdot\dot{d}\right) \tag{14}$$

where

- f_n is the normal force applied in equal-and-opposite fashion to each contacting geometry.
- d is the penetration depth between two contacting geometries.
- d' is the first-time derivative of the penetration depth.
- k is the normal-force stiffness specified in the block.
- b is the normal-force damping specified in the block.
- $s(d)$ is the smoothing function.

A low-pass filter with a cut-off frequency of X Hz was used as the smoothing function to eliminate high-frequency noise from the sensor signal.

The position of the vehicles is shown in Figure 8. This representation shows the real-world frame of reference and the position of the two vehicles in relation to this frame of reference. This is crucial when analyzing the displacement parameters of individual components of the model. Note that the road reference point is positioned with respect to the real-world reference point and the vehicles are positioned with respect to the road reference point.

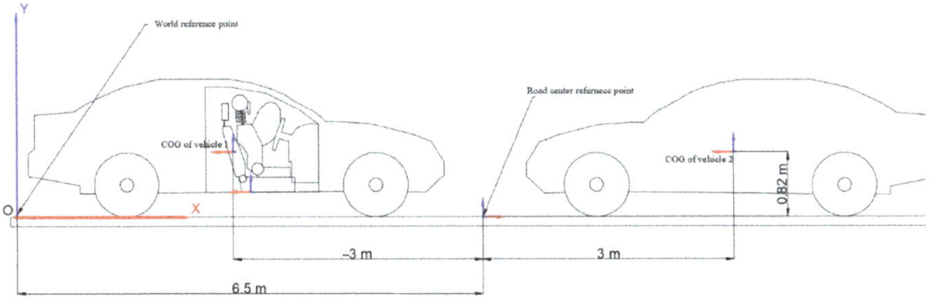

Figure 8. Vehicle position with respect to the real-world reference frame.

3.6. Model Parameters

In this section, the mathematical parameters for the model are presented and grouped into categories. These parameters define the functionality of the model and ensure that the model has a dynamic and kinematic response identical to real-world situations and multiple types of crash configurations. The first set of parameters are the mass parameters for each CAD component. These were used to calculate the inertia and contact forces during a crash (see Table 3).

1. Mass parameters

Table 3. Model mass parameters.

Body	Parameter	Mass Value	Unit
Vehicle 1	M_1	1236	Kg
Vehicle 2	M_2	1450	Kg
Wheel × 4	M_w	20	Kg
Dashboard	M_{dhb}	50	Kg
Airbag	M_{arb}	0.5	Kg
Lower seat	M_{ls}	46	Kg
Upper seat	M_{us}	35	Kg
Headrest	M_{hr}	3	Kg
Lower body	M_{LB}	25	Kg
Upper body	M_{UB}	40	Kg
Neck × 4	M_n	0.2	Kg
Head	M_h	5	Kg

The values chosen for the model are the average for each component. Vehicle 1 has a lower mass due to the interior bodies such as occupant, dashboard, and seat. The occupant weighs a total of 71 kg, which is the average for a man in the 50th percentile range. The mass values were average and identical to those used in the experimental tests to validate the model. The second set of parameters are the dynamic ones, such as velocities and gravitational acceleration (see Table 4).

2. Dynamic parameters

Table 4. Model dynamic parameters of the model.

Parameter	Parameter	Value	Unit
Vehicle 1 velocity	$\dot{x}_1$	0	Km/h
Vehicle 2 velocity	$\dot{x}_2$	33–35	Km/h
Gravitational acceleration	g	9.81	m/s^2
Front suspension caster angle	α_c	4	degrees

For validation purposes, Vehicle 1 had a speed of 0 km/h and Vehicle 2 had a speed between 33 and 35 km/h (33 km/h for the rear impact and 35 km/h for the frontal impact). Both vehicles have a positive caster angle of 4 degrees. Although Vehicle 1 is modelled as being stationary in this scenario, the resulting relative velocity simulates the real-world condition where the leading vehicle decelerates and the following driver fails to brake in time —a common cause of rear-end collisions in urban traffic. The third set of parameters are the spring values of the joints. There are 3 revolute joints on the occupant and a prismatic joint for the vehicle suspension. The occupant seatbelt is a linear spring. The values are shown in Table 5.

3. Spring and damper parameters

Table 5. Spring parameters.

Bodies	Type of Joint	Param.	Stiffness	Unit	Damping	Unit
Head–neck	Revolute	k	1.5	Nm/deg	0.02	Nm/(deg/s)
Upper body–Lower body	Revolute	k_U	0.5	Nm/deg	0.05	Nm/(deg/s)
Upper seat–Lower seat	Revolute	k_S	200	Nm/deg	10	Nm/(deg/s)
Vehicle suspension	Prismatic	k_{sus}	30,000	N/m	3000	N/(m/s)
Seatbelt	Linear spring	k_{st}	50,000	N/m	1000	N/(m/s)

The seatbelt connection is a linear spring that connects two reference points: one on the vehicle and one on the occupant. The values were determined in previous studies and according to standards [36–38].

The vehicle suspension stiffness and damping values correspond to previous research and standards related to vehicle dynamics [38–40].

Previous research shows that average neck stiffness is between 0.5 Nm/rad and 2 Nm/rad (0.03 Nm/deg), determined using biomechanical analyses. However, in the presented model, using such low values are not enough to keep the head in the upright position. By adjusting the stiffness and running the simulation several times, it was observed that the optimal value of neck stiffness is around 86 Nm/rad (approx. 1.5 Nm/degree). This value ensures the correct kinematic response compared to the crash test dummy used in the experiment [41–45].

The neck-damping values were also determined from biomechanical analyses, and are between 0.1 and 0.5 Nm/rad (0.017 Nm/deg). Here, too, running the simulation several times resulted in an optimal value of 0.02 for the model presented [46–49]. The fourth set of parameters are the connection types and degrees of freedom between them. These are presented in Table 6.

4. Connections between the model bodies

Table 6. Connection types of the model.

Bodies	Connection Type	Degrees of Freedom
World–Road surface	Fixed joint	0
Vehicles–Road surface	Six-degrees-of-freedom joint	6
Vehicle–Wheels	Prismatic joint	1
Wheels–Vehicle	Revolute joint	1
Dashboard–Vehicle 1	Fixed joint	0
Lower seat–Vehicle 1	Fixed joint	0
Lower seat–Upper seat	Revolute joint	1
Lower seat–Lower body	Six-degrees-of-freedom joint	6
Lower body–Upper body	Revolute joint	1
Headrest–Upper seat	Fixed joint	0
Upper body–Neck	Revolute joint	1
Head–Neck	Revolute joint	1
Model degrees of freedom		4

The last set of parameters are the contact stiffness and damping between the bodies that have contact with each other. These are presented in Table 7.

5. Contact parameters

Table 7. Model contact parameters.

Bodies in Contact	Stiffness	Unit	Damping	Unit
Wheels–Road surface	100,000	N/m	100,000	N/(m/s)
Vehicle 1–Vehicle 2	500,000	N/m	80,000	N/(m/s)
Head–Airbag	8000	N/m	10	N/(m/s)
Head–Headrest	8000	N/m	100	N/(m/s)
Head–Upper seat	10,000	N/m	1000	N/(m/s)
Head–Upper body	10,000	N/m	1000	N/(m/s)
Lower body–Lower seat	10,000	N/m	1000	N/(m/s)
Head–Dashboard	100,000	N/m	1000	N/(m/s)
Upper body–Dashboard	100,000	N/m	1000	N/(m/s)

The stiffness of the vehicle and the damping were determined experimentally and analytically. To better understand this, the absorbed kinetic energy equation can be used for the vehicle stiffness calculation [50]:

$$E_k = \frac{1}{2}mv^2 \tag{15}$$

where m is the vehicle mass (kg) and v is the collision velocity (m/s).

Stiffness is determined by the total deformation of the front surface of the vehicle; therefore, we can denote it as D_x. During collision, the kinetic energy is converted into elastic potential energy, which can be calculated using Equation (4).

$$E_p = \frac{1}{2}k(\Delta x)^2 \tag{16}$$

If the initial kinetic energy becomes potential energy, we use the following equation:

$$\frac{1}{2}mv^2 = \frac{1}{2}k(\Delta x)^2 \tag{17}$$

From Equation (5), we can determine the rigidity value k (N/m) in Equation (6):

$$k = \frac{mv^2}{(\Delta x)^2} \tag{18}$$

The value of the stiffness can differ from one case to another because of the frontal deformation value of each vehicle. The average value of k can be calculated by determining an average deformation value from experimental controlled crash tests or real vehicle crashes for a given velocity and one type of vehicle class. For the given model, the vehicle is a sedan type and the impact velocity is around 35 km/h (11.1 m/s). Documentation regarding the deformation values for frontal collision, known as the EES catalogue, can give us the possibility to determine the average deformation for a sedan-class vehicle at the equivalent energy speed (EES) of 15 km/h, which is about 35 km/h impact velocity, after the energy absorption is considered. The deformation values are presented in Table 8 [51,52].

Table 8. Deformation values for frontal collisions from the EES catalogue [51,52].

Impact Velocity [km/h]	EES [km/h]	Deformation [m]	Type of Collision
35	15	0.18	Frontal
35	15	0.19	Frontal
35	15	0.20	Frontal
35	15	0.21	Frontal
35	15	0.21	Frontal
35	15	0.22	Frontal
35	15	0.22	Frontal
35	15	0.23	Frontal
35	15	0.25	Frontal
35	15	0.26	Frontal
Average		**0.217**	

The damping value for the model was determined using the natural frequency and critical damping equations [50].

$$\omega_n = \sqrt{\frac{k}{m}} \tag{19}$$

where w_n is the natural frequency (rad/s), k is the stiffness, and m is the mass. To calculate the critical damping value c_c, the following equation can be used:

$$c_c = 2\sqrt{km} \tag{20}$$

Depending on the vehicle type and class, the contact between vehicles may be modified to better adapt to the frontal stiffness of each vehicle.

3.7. Model Dimensions

The vehicle is a standard sedan class. The measurements are average values for this class [50]. Depending on one's needs, these models can be modified and adapted for different crash scenarios. The occupant and seat have the dimensions of an average 50th percentile male body [53,54].

To simplify the representation of the crash phenomenon, the occupant was shown in a simplified manner without limbs. However, this simplification has no influence on the kinematics of the body in a frontal or rear-end crash.

4. Model Validation

To validate the virtual model and demonstrate that the output results of the model are correct and reflect the outcome of an accident, experimental tests were conducted in a controlled environment using average passenger vehicles and crash test dummies.

To quantitatively validate the virtual model against experimental crash test data, RMSE, MAPE, and MAE indicators were calculated using the next set of equations [55–58].

$$RMSE = \sqrt{\frac{1}{n}\sum_{i=1}^{n}\left(y_i^{sim} - y_i^{test}\right)^2} \quad (21)$$

$$MAPE = \frac{100\%}{n}\sum_{i=1}^{n}\left|\frac{y_i^{sim} - y_i^{test}}{y_i^{test}}\right| \quad (22)$$

$$MAE = \frac{1}{n}\sum_{i=1}^{n}\left|y_i^{sim} - y_i^{test}\right| \quad (23)$$

These statistical indicators were used to assess the level of agreement between the simulated outputs and experimental crash test data. According to commonly accepted validation criteria in occupant safety modelling, an average error below 5% is considered an indicator of good model fidelity. In this study, all three metrics—RMSE, MAPE, and MAE—were calculated using the resultant acceleration data of the head. The results confirm that the differences between the model and the experimental data remained within an acceptable threshold, thus validating the accuracy and reliability of the virtual multibody system.

4.1. Frontal Collision

The first validation concerned a frontal collision at 35 km/h with an airbag. Vehicle 1 was stationary, and the second vehicle accelerated, and the collision occurred with an overlap of 100%. The kinematics and impact phases are shown in Figure 9.

We can observe that, in the pre-impact phase, in a side-by-side comparison, the occupant and the vehicle perfectly match the position of the occupant of the experiment. In the impact phase, the kinematics of the occupant corresponds to the kinematics of the crash test dummy. The same applies to the post impact phase.

To validate the model numerically, a comparison of vehicle velocities and head acceleration values was carried out. The first result is the head acceleration, shown in Figure 10.

It can be observed that both acceleration values match, almost perfectly overlapping each other, granting a high level of similarity. In this type of collision, the acceleration is on the X-axis, the longitudinal axis of the vehicle, and two phases can be distinguished on the diagram: the head and neck flexion phase and the extension phase. The first phase lasts between 0.1 and 0.25 s, while the second phase lasts between 0.25 and 0.45 s. These values (80 m/s^2 and 65 ms^2, respectively) are low in terms of occupant injury due to two factors: the belt tensioner and the front airbag.

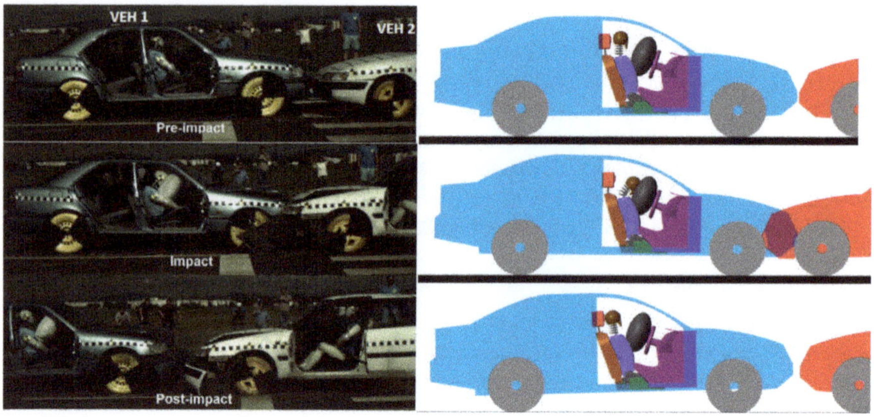

Figure 9. Kinematic comparison of the virtual model and the experimental test for a frontal collision.

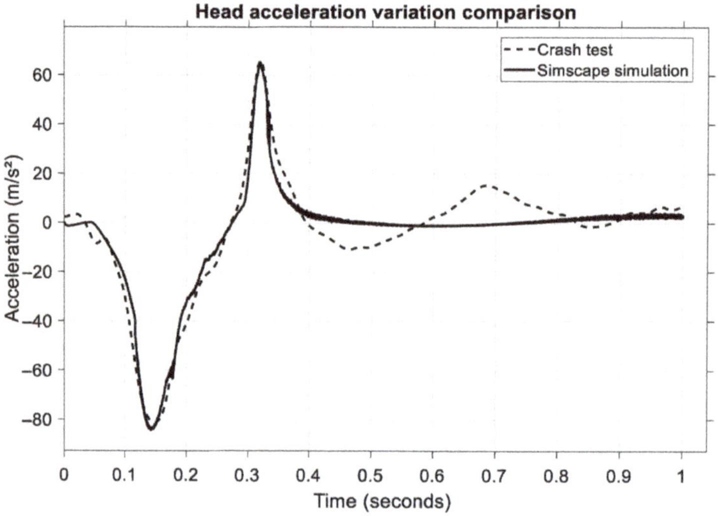

Figure 10. Comparison of head acceleration between the virtual model and the experimental test.

To quantitatively assess the agreement between the simulation results and the crash test data, three statistical indicators were calculated: Root Mean Square Error (RMSE), Mean Absolute Percentage Error (MAPE), and Mean Absolute Error (MAE). The results for head acceleration during the frontal collision scenario are summarized in Table 9.

Table 9. Statistic parameter for the head acceleration in frontal collision.

Parameter	Value	Unit
RMSE	4.19	m/s^2
MAPE	0.71	%
MAE	4.19	m/s^2

The low values obtained, particularly the MAPE of 0.71%, indicate a high level of accuracy between the simulated and experimental data. This confirms the validity of the multibody model in predicting occupant head dynamics under frontal collision conditions.

In the next figure, Figure 11, a comparison is made between the struck vehicle velocity.

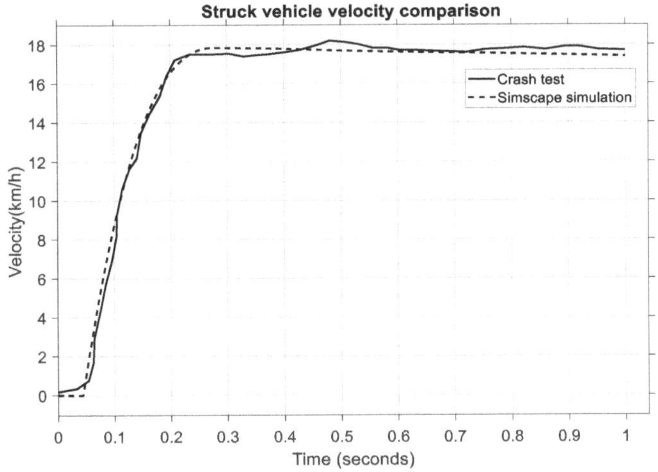

Figure 11. Comparison of vehicle velocities for the struck vehicle.

Both the crash test and the model have good similarity, which ensures the high accuracy of the mathematical model, although there is a slight difference of 0.3 s–0.9 s in the post-impact velocity. What is crucial here, however, is the impact phase between 0.05 and 0.3 s, in which an increase in velocity from 0 km/h to 18 km/h can be observed. To further strengthen the validity of the virtual model, a statistical comparison of the struck vehicle's velocity was also performed using RMSE, MAPE, and MAE indicators.

The results in Table 10 indicate a very high level of agreement between the simulated and experimental vehicle velocities, with all error metrics well below the 5% threshold, confirming the accuracy of the model's dynamic response during impact.

Table 10. Statistic parameters for the struck vehicle velocity comparison.

Parameter	Value	Unit
RMSE	0.29	m/s
MAPE	0.01	%
MAE	0.29	m/s

A similar velocity variation occurs for the second vehicle, the impact vehicle, whose speed is shown in Figure 12.

The velocity for the striking vehicle starts at 35 km/h, and the impact velocity for the crash test and the variation is observed to follow a similar timeframe as the struck vehicle, from 0.05 to 0.25 s. Even though these plots do not overlap perfectly, the variation during the critical phase, such as the impact phase, is almost identical, and this is the important aspect of the validation process for the mathematical model.

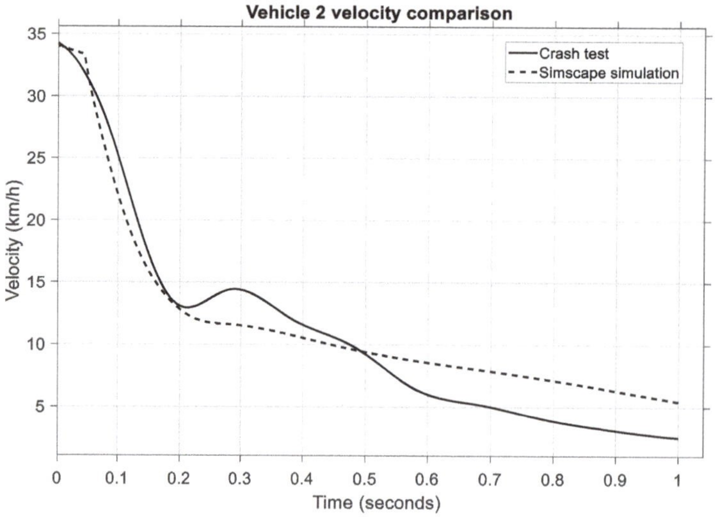

Figure 12. Comparison of vehicle velocities for the striking vehicle.

4.2. Rear-End Collision

The model can be easily configured for rear-end collisions by simply rotating the struck vehicle 180 degrees on the *Z*-axis. In such a scenario, the occupant's kinematics follow a similar pattern to frontal collisions, only this time, the headfirst contacts the headrest and then swings forward for the flexion phase. The validation of this crash was also achieved by comparing the kinematic and dynamic values of a real crash test, shown in Figure 13. The crash test was carried out at 33 km/h with the same vehicles and the same dummy as in the frontal impact with identical parameters. This comparison validates the mathematical model for rear-end collisions by comparing the kinematic and dynamic values.

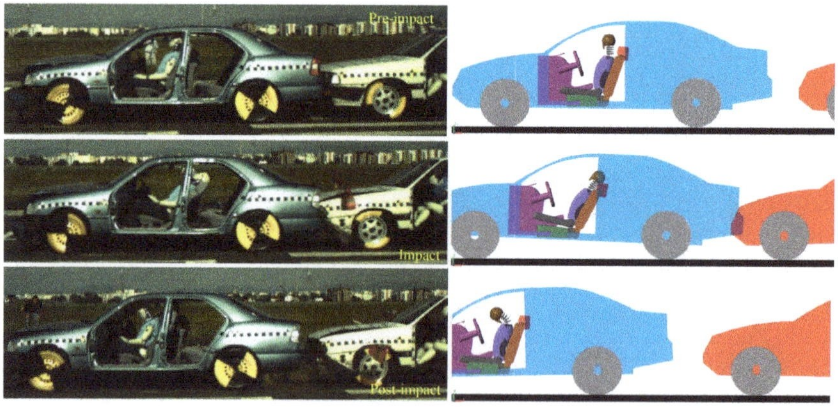

Figure 13. Kinematic comparison of the virtual model and the experimental test for the rear-end collisions.

Here, too, the kinematic comparison was carried out using the crash test and the model, which are compared side-by-side to better observe the occupant's kinematics. It can be clearly seen, in both the model and the crash test, that the occupant's head hits the headrest during the impact phase, and the head swings forward in the post-impact phase.

For numerical validation, in Figure 14, the head acceleration values are presented. For this test, the resultant acceleration is used.

In this scenario, the impact phase occurs between 0.1 and 0.2 s when the head hits the headrest, and a maximum head deceleration value of 200 m/s is reached. Both the crash test and the model have an almost identical pattern during the impact phase, thus ensuring a high level of accuracy for the model created. To support this observation quantitatively, statistical indicators were computed. Table 11 presents the RMSE, MAPE, and MAE values for the rear-end collision head acceleration comparison.

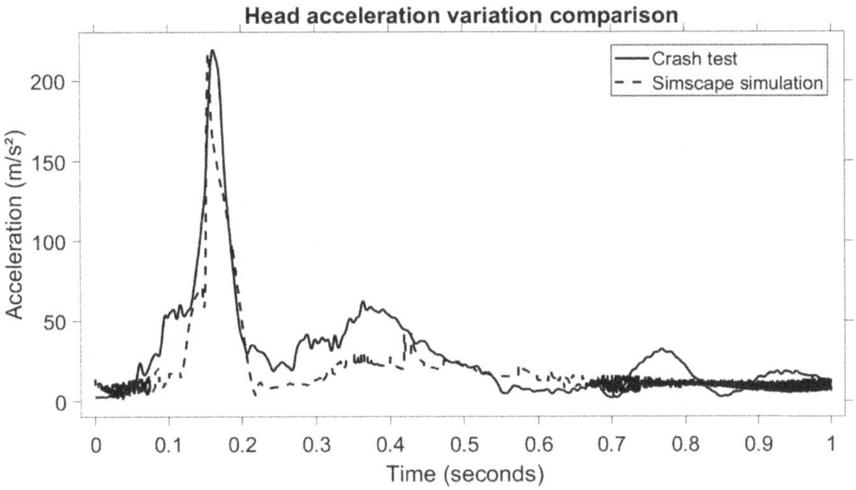

Figure 14. Comparison of head acceleration between the virtual model and the experimental test for the second scenario.

Table 11. Comparison of the statistic parameters of head acceleration for the rear-end collisions.

Parameter	Value	Unit
RMSE	0.14	m/s^2
MAPE	0.01	%
MAE	0.14	m/s^2

To further analyze this scenario, the speed of the struck vehicle is shown in Figure 15. We can see that the velocity of the struck vehicle increases from 0 to 18 km/h within 0.1–0.2 s, and the timeframe is almost identical to the head acceleration variation during the impact phase. To support this result quantitatively, Table 12 presents the statistical parameters (RMSE, MAPE, and MAE) used to evaluate the agreement between the experimental crash data and the simulation results. All indicators fall well below the 5% error threshold, confirming the reliability of the model in replicating the kinematic behaviour of the struck vehicle.

In Figure 16, we can observe the velocity comparison of the impact vehicle.

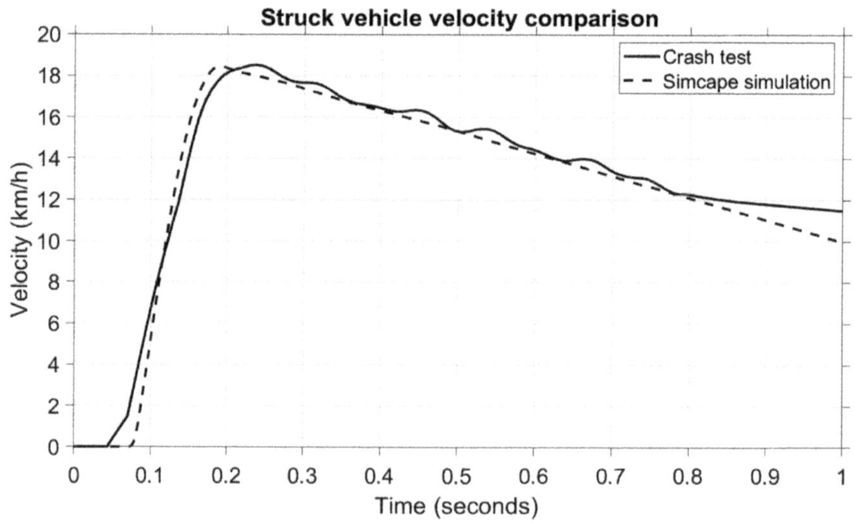

Figure 15. Comparison of vehicle velocities for the struck vehicle—second scenario.

Table 12. Comparison of the statistic parameters of vehicle velocity for the rear-end collisions.

Parameter	Value	Unit
RMSE	1.39	m/s
MAPE	0.12	%
MAE	1.39	m/s

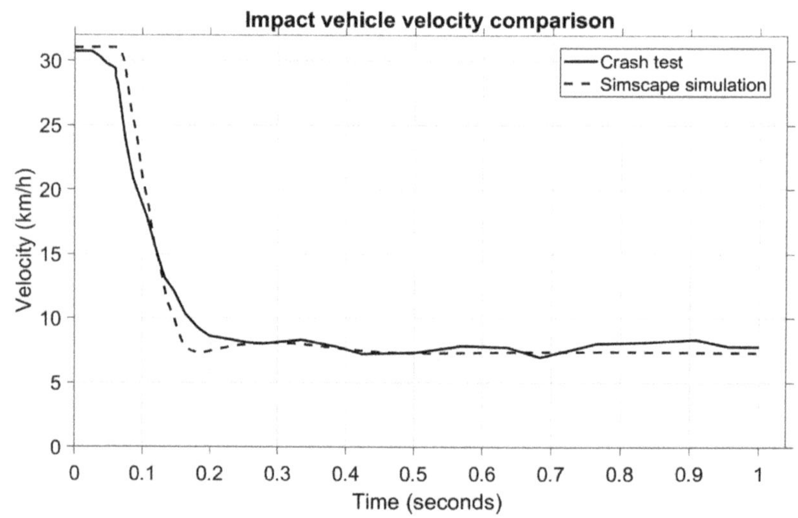

Figure 16. Comparison of vehicle velocities for the impact vehicle—second scenario.

In this case, a change in velocity is observed in the same period of 0.1–0.2 s when the value drops from 34 km/h to 5 km/h during the impact phase. Note that the model and the crash test only show an almost identical pattern in the mentioned time. In the post-impact phase, the crash test vehicle continued to roll at 15 km/h, while in the model, the velocity dropped to 5 km/h in the next 0.8 s. This is because, in the model, the brakes were applied, while in the crash test, the car rolled another second before the brakes were activated.

5. Active Headrest Simulation

The effectiveness of an active headrest in a rear-end collision was assessed using the validated dynamic model of the occupant–vehicle system. Three distinct collision speeds were modelled:

- 20 km/h (low severity);
- 31 km/h (validation speed, based on experimental data);
- 40 km/h (high severity).

A virtual sensor placed close to the rear bumper, which mimicked the actual vehicle sensor, was used to measure the rear acceleration. Figure 17 displays the longitudinal acceleration profiles that were captured at these speeds, with reference to the initial plot. One can observe the following:

The maximum rear acceleration increases with collision severity.

Every velocity causes the acceleration to approach or surpass values near 50 m/s^2, which was chosen as the active headrest system's activation threshold.

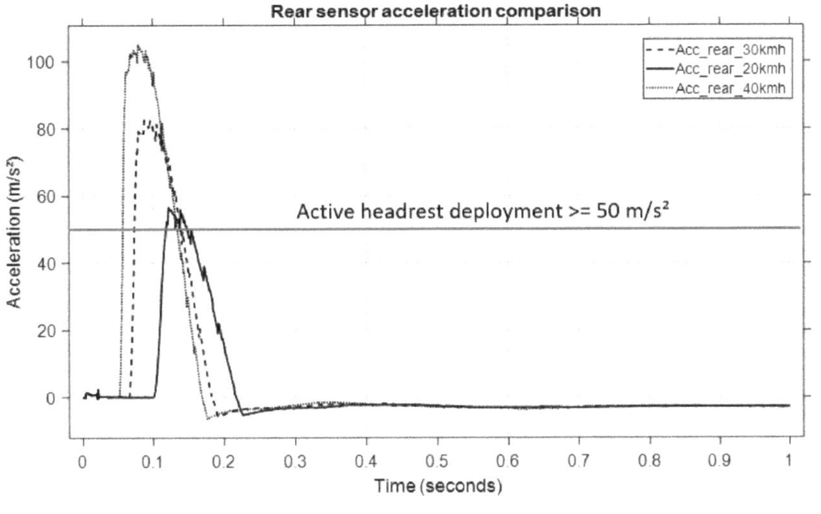

Figure 17. Struck vehicle acceleration comparison at different velocities.

The active headrest system was triggered when the measured rear acceleration exceeded 50 m/s^2, according to the logic described previously. Upon activation, a predefined force was applied to the headrest to push it forward and reduce the head–torso relative displacement (whiplash effect).

Without activation, the head moved backward freely relative to the torso (normal headrest scenario).

With activation, the headrest moved forward synchronously, limiting relative motion (active headrest scenario).

Figure 18 (second image) shows a comparison of occupant kinematics during the collision:

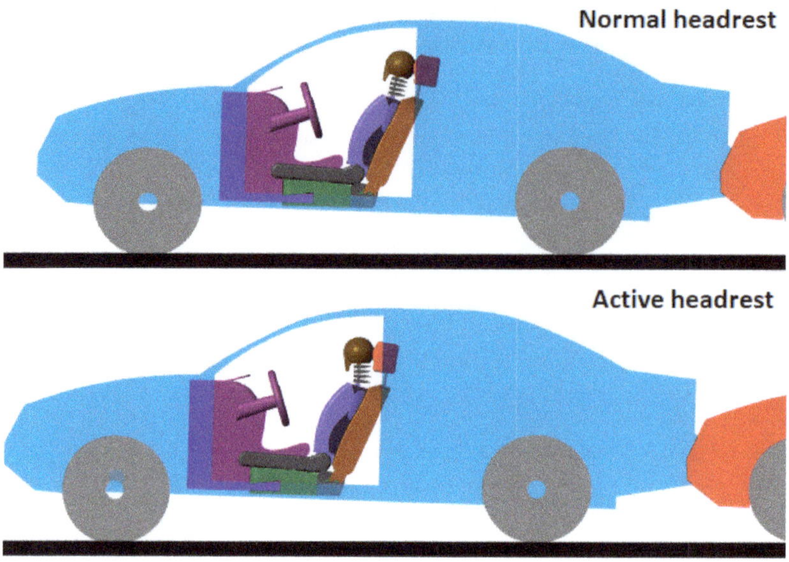

Figure 18. Kinematic comparison of the occupant with and without the active headrest.

Normal headrest (top figure):

- Large backward motion of the head relative to the torso.
- Neck hyperextension occurred.

Active headrest (bottom figure):

- Head motion relative to the torso was reduced.
- Neck bending was significantly limited, minimizing injury risk.

The active headrest effectively reduced the whiplash effect by proactively supporting the head during the critical early phase of the collision (approximately 0.1–0.2 s after impact).

To further assess the effectiveness of the active headrest, the head linear acceleration was analyzed for a low-speed rear-end collision at 20 km/h. Figure 19 presents the time histories of the head acceleration for both configurations: passive (normal) headrest and active headrest.

In the passive headrest case, the occupant's head experienced a sharp acceleration peak of approximately 175 m/s^2, occurring around 0.20 s after impact initiation.

By contrast, when the active headrest was deployed, the peak head acceleration was significantly reduced to approximately 110 m/s^2. Additionally, the active headrest led to a broader and smoother acceleration pulse, indicative of a more gradual deceleration of the head.

This reduction of approximately 37% in peak acceleration highlights the effectiveness of the active system in mitigating injurious head motions. Moreover, the earlier engagement of the headrest (approximately at 0.17 s) demonstrates the benefit of anticipative movement control, ensuring head support during the critical whiplash phase.

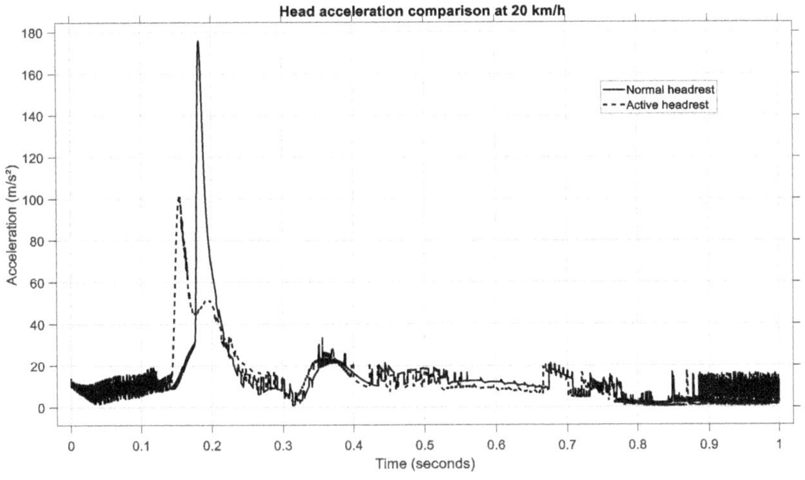

Figure 19. Head acceleration comparison at 20 km/h.

To validate the model under a medium-severity rear-end collision, simulations were performed at an impact speed of 31 km/h. The resulting head acceleration profiles for the passive and active headrest configurations are shown in Figure 20.

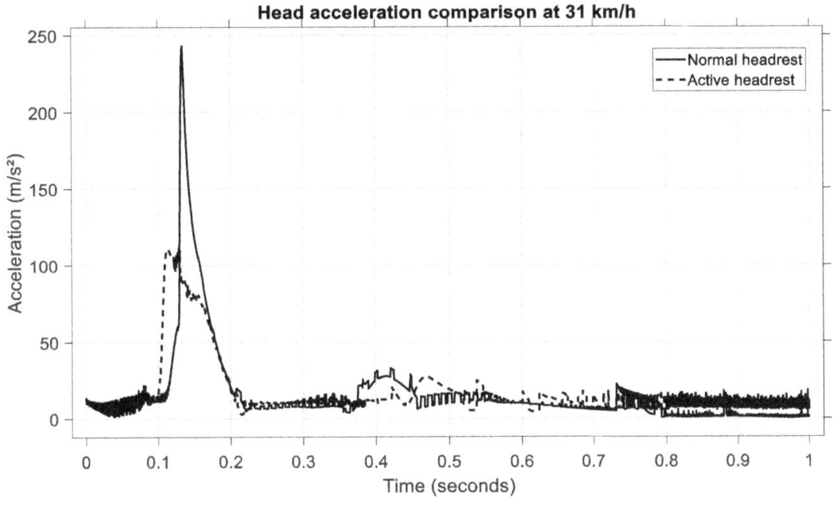

Figure 20. Head acceleration comparison at 31 km/h.

In the case of the passive headrest, the occupant's head experienced a pronounced acceleration peak, reaching approximately 240 m/s^2 and occurring at around 0.14 second's post-impact.

Conversely, when the active headrest was engaged, the maximum head acceleration was limited to approximately 105 m/s^2. The active configuration also resulted in a more gradual acceleration profile, characterized by a reduced rate of change (jerk) and a broader temporal distribution of the peak.

This corresponds to a reduction of approximately 56% in the peak head acceleration when using the active headrest. Such a significant reduction confirms the capacity of the active headrest to mitigate the head's backward motion, thereby diminishing the loads transmitted through the cervical spine.

To assess the performance of the active headrest under high-severity impact conditions, a rear-end collision at 40 km/h was simulated. The head acceleration time histories for both passive and active headrest configurations are illustrated in Figure 21.

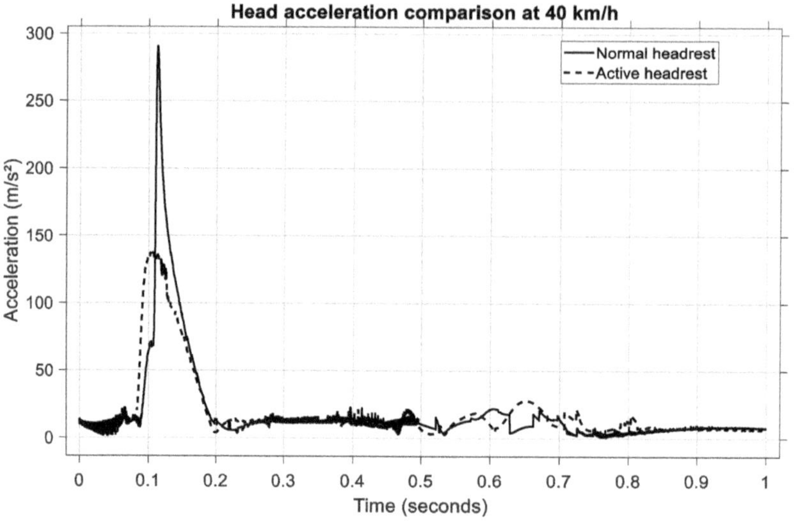

Figure 21. Head acceleration comparison at 40 km/h.

In the case of the normal (passive) headrest, the occupant's head exhibited a sharp acceleration peak, reaching approximately 290 m/s^2 and occurring at approximately 0.12 s following impact. When the active headrest was employed, the peak head acceleration was significantly attenuated, reaching a value of approximately 135 m/s^2. Additionally, the shape of the acceleration curve under active control is smoother, indicating reduced jerk levels and a more progressive head deceleration.

The active system therefore achieved a reduction of approximately 53% in the peak head acceleration compared to the passive configuration. This substantial improvement in head kinematic control, particularly under severe impact conditions, confirms the effectiveness of the active headrest in mitigating potential whiplash injuries by providing earlier and more controlled head support.

The numerical simulations conducted at 20 km/h, 31 km/h, and 40 km/h rear-end collision scenarios demonstrate the significant benefits of the active headrest system in mitigating head kinematics and reducing injury risk. Across all impact speeds, the active headrest effectively reduced the peak head acceleration by approximately 37%–56% depending on collision severity. Additionally, the active system enabled earlier and smoother deceleration of the head, limiting relative motion between the head and torso and thereby minimizing neck extension.

These results confirm that the implementation of an active headrest mechanism offers a substantial improvement in occupant protection compared to a conventional passive system, particularly during moderate- and high-severity rear-end impacts.

These findings support the use of active headrest technologies as a viable countermeasure for reducing whiplash-associated injuries in automotive collisions.

6. Discussion and Conclusions

This study presents the development of a comprehensive mathematical model using a multibody system (MBS) approach to analyze the dynamics of motor vehicles during frontal and rear-end collisions. By integrating rigid body dynamics, occupant biomechanics, vehicle suspension behaviour, and advanced modelling of restraint systems, the model provides a robust and flexible framework for crash analysis, including the design and validation of an active headrest system with integrated sensing.

A key innovation introduced in this work is the inclusion of an active headrest system, which is controlled via an integrated longitudinal acceleration sensor and a dedicated logic circuit. Upon detecting a rear-end impact exceeding a defined acceleration threshold, the system actively displaces the headrest to limit head–torso relative motion. Simulation results across various collision speeds (20, 31, and 40 km/h) confirmed the effectiveness of the active headrest in significantly reducing peak head accelerations and mitigating potential whiplash injuries.

The model was validated against experimental crash test data, demonstrating excellent agreement in both vehicle kinematics and occupant dynamics. The ability of the model to replicate complex interactions between vehicle structures, restraint systems, and occupant body segments confirms its reliability and applicability for real-world automotive safety research and vehicle design optimization.

The validation scenarios in this study focus on rear-end collisions at low to medium speeds ($\leq$40 km/h), which represent the majority of real-world rear-end impacts. At these speeds, occupant kinematics and acceleration profiles are predominantly influenced by seat restraint systems and headrest geometry, rather than extensive structural deformation. Also, the current validation includes collisions up to 40 km/h, focusing on urban scenarios, where whiplash injuries are most prevalent. Future work will involve extending the model to higher-speed crashes (e.g., 60 km/h) and coupling it with vehicle deformation models.

The presented model focuses on occupant dynamics without explicit structural deformation, a simplification which has been noted in the previous literature. For example, Przemysław P et al. [59] studied vehicle structural deformation and its effect on energy absorption and occupant deceleration using finite element models. While our multibody model cannot capture local deformation effects, it provides accurate timing and magnitude of head acceleration and body motion, which are critical for early whiplash prediction. The timing of peak head acceleration (~200 m/s^2 at ~0.18 s) aligns well with the values reported in validated rear-end impact simulations [60].

In addition to its predictive accuracy, the developed model offers substantial practical advantages. A MATLAB-based control script enables users to easily configure simulation parameters, run the Simulink model, and automatically export results to a spreadsheet format. This streamlined process enhances the model's accessibility and usability, allowing for rapid iteration and scenario testing without the need for extensive manual intervention.

The current model is based on a 50th percentile male occupant. Biomechanical variations associated with different occupant groups, such as females or children, were not considered. However, the parametric flexibility of the Simscape Multibody platform allows future inclusion of demographic variability through modifications of joint properties, mass distribution, and anthropometric dimensions.

Nevertheless, the model does present some limitations. It does not explicitly simulate large material deformations in components such as airbags, seats, or the vehicle's crumple zones. These phenomena, characterized by complex nonlinear behaviour, require finite element modelling (FEM) approaches, which were beyond the scope of the present study. However, it should be noted that the force and acceleration interactions between bodies are correctly computed based on equivalent stiffness and damping parameters, ensuring that dynamic responses remain physically meaningful, even without visual deformation.

Future research directions may involve extending the model to incorporate different impact configurations (e.g., oblique or lateral impacts), a wider variety of vehicle types, and more detailed interior component modelling. Furthermore, integrating additional active safety devices, such as pre-crash seatbelt systems or adaptive airbags, could enhance the model's capability to evaluate next-generation automotive safety technologies.

Overall, the development and validation of this multibody system model highlights the importance of multidisciplinary methodologies in advancing vehicle safety research. By combining MBS dynamics, integrated sensor-based control strategies, and experimental validation, this work establishes a solid foundation for future innovations in occupant protection and crash-mitigation systems.

Author Contributions: Conceptualization, A.I.R., B.A.T. and H.B.; Data curation, A.I.R., B.A.T. and F.B.S.; Formal analysis, A.I.R., B.A.T., H.B. and A.N.T.; Investigation, B.A.T., H.B. and F.B.S.; Methodology, A.I.R., B.A.T. and H.B.; Software, A.I.R.; Validation, F.B.S. and A.N.T.; Writing—original draft, A.I.R. and B.A.T.; Writing—review and editing, A.I.R., B.A.T., H.B., F.B.S. and A.N.T. All authors have read and agreed to the published version of the manuscript.

Funding: The research was funded by University Transilvania of Brașov. This research did not receive external funding. The experimental tests were conducted with the support of the University Transilvania of Brașov, which provided access to its research facilities. The APC was funded by the University of Oradea.

Institutional Review Board Statement: Not applicable.

Informed Consent Statement: Not applicable.

Data Availability Statement: Data are contained within the article.

Acknowledgments: The authors contributed equally to this work.

Conflicts of Interest: Author Adrian Nicolaie Tusinean was employed by the company Advanced Mechatronik Technologies SRL. The remaining authors declare that the research was conducted in the absence of any commercial or financial relationships that could be construed as a potential conflict of interest.

Abbreviations

The following abbreviations are used in this manuscript:

MBS	Multibody System
CAD	Computer Aided Design
MEMS	Micro-Electro-Mechanical Systems
ODE	Ordinary Differential Equations
EOM	Equation of Motion
RMSE	Root Mean Square Error
MAE	Mean Absolute Error
MAPE	Mean Absolute Percentage Error
MATLAB	Matrix Laboratory
DOF	Degrees of Freedom
EES	Equivalent Energy Speed
1D	One Dimension

References

1. European Commission. Annual Accident Report 2022. European Road Safety Data 2022. Available online: https://road-safety.transport.ec.europa.eu/document/download/287aa31e-48c2-4e04-a9cc-e2ca24d29cc2_en?filename=ERSO_annual_report_20220509.pdf (accessed on 4 April 2025).
2. Dąbrowski, A.; Stawiarski, A. Analysis of the causes of road accidents in Poland with particular attention to rear-end collisions. *Appl. Sci.* **2022**, *12*, 6371. [CrossRef]
3. Ionut, R.A.; Corneliu, C.; Bogdan, T. Mathematical model validated by a crash test for studying the occupant's kinematics and dynamics in a cars' frontal collision. *Int. J. Automot. Technol.* **2017**, *18*, 1017–1025. [CrossRef]
4. Radu, A.I.; Toganel, G.R.; Trusca, D.D. Mathematical model validated by a crash test to be used as kinematic and dynamic study for side impacts. *Int. J. Automot. Technol.* **2021**, *22*, 1267–1277. [CrossRef]
5. Xu, T.; Xu, P.; Zhao, H.; Yang, C.; Peng, Y. Vehicle running attitude prediction model based on Artificial Neural Network-Parallel Connected (ANN-PL) in the single-vehicle collision. *Adv. Eng. Softw.* **2023**, *175*, 103356. [CrossRef]
6. Ryu, S.K.; Kim, Y.R.; Park, J.H. Modelling and simulation of point contact multibody system dynamics. *ScienceDirect* **2023**, *67*, 321–334.
7. An, K.E.; Lee, S.W.; Ryu, S.K.; Seo, D. Detecting a pothole using deep convolutional neural network models for an adaptive shock observing in a vehicle driving. In Proceedings of the 2018 IEEE International Conference on Consumer Electronics (ICCE), Las Vegas, NV, USA, 12–14 January 2018; IEEE: Piscataway, NJ, USA, 2018; pp. 1–2.
8. Schiehlen, W. Research trends in multibody system dynamics. *Multibody Syst. Dyn.* **2007**, *18*, 3–13. [CrossRef]
9. Kim, J.; Lee, S.; Park, H. Behaviour Analysis of Active and Proactive Headrest during Low-Speed Rear-End Collisions. *Appl. Sci.* **2020**, *10*, 1451.
10. Zhang, Y.; Liu, J.; Zhao, Y. Study on Active Headrest Mechanism and Human Biomechanical Response in Rear-End Collisions. *Machines* **2023**, *11*, 1060.
11. Vychytil, J.; Hlucha, J.; Kovar, L.; Kostikova, M.; Bucsuhazy, K. Innovative Active Head Restraint System in a Car: Safety Assessment with Virtual Human Body Model. *SAE Int. J. Adv. Curr. Pract. Mobil.* **2020**, *2*, 2107–2124. [CrossRef]
12. He, Z.; Zhang, Y.; Feng, W. Semiactive Car-Seat System for Rear-End Collisions. *Machines* **2022**, *12*, 530.
13. Yan, J.; Chen, L. Adaptive Headrest Adjustment Considering Occupant Posture. *Machines* **2023**, *11*, 1076.
14. Lin, J.; Huang, W.; Su, H. Head-Neck Biomechanical Analysis Based on Finite Element Modelling in Rear-End Impacts. *Bioengineering* **2023**, *10*, 1059.
15. Liu, W.; Zhang, H. Advanced Multibody Dynamic Simulations for Rear-End Collisions. *Machines* **2024**, *12*, 626.
16. Yuan, C.; Li, Y. Sensor-Based Active Headrest Systems in Automotive Safety. *Machines* **2023**, *11*, 1060.
17. Ghafar-Zadeh, E.; Sawan, M. MEMS Sensors for Automotive Applications. *Sensors* **2015**, *15*, 14539–14565.
18. Beeby, S.; White, N. *MEMS Mechanical Sensors*; Artech House: Norwood, MA, USA, 2004.
19. Sharafeldin, M.; Farid, A.; Ksaibati, K. Injury severity analysis of rear-end crashes at signalized intersections. *Sustainability* **2022**, *14*, 13858. [CrossRef]
20. Lee, S.; Park, J.; Kim, Y. Rear-End Collision Severity Estimation Based on Machine Learning Approaches. *arXiv* **2023**, arXiv:2303.12159. Available online: https://arxiv.org/abs/2303.12159 (accessed on 8 April 2025).
21. Kitagawa, Y.; Yasuki, T.; Hasegawa, J. Development of a new airbag system for far-side impact protection. *SAE Tech.* Paper **2012**, *1*, 574.
22. Mizuno, K. Evaluation of an Advanced Seatbelt System in Reducing Chest Injuries in Frontal Crashes. *Traffic Inj. Prev.* **2015**, *16*, 94–101.
23. Tollner, D.; Zöldy, M. Road Type Classification of Driving Data Using Neural Networks. *Computers* **2025**, *14*, 70. [CrossRef]
24. Lengyel, H.; Maral, S.; Kerebekov, S.; Szalay, Z.; Török, Á. Modelling and simulating automated vehicular functions in critical situations—Application of a novel accident reconstruction concept. *Vehicles* **2023**, *5*, 266–285. [CrossRef]
25. Benea, B.; Soica, A. The Contact Phase in Vehicle—Pedestrian Accident Reconstruction. *Appl. Sci.* **2023**, *13*, 9404. [CrossRef]
26. Soica, A. Force in Cable of Pretensioner Tube—A Possibility of Car Accident Reconstruction. *Appl. Sci.* **2024**, *14*, 3087. [CrossRef]
27. Frej, D. The effect of changing the angle of the passenger car seat backrest on the head trajectories of the 50th percentile male dummy. *Sensors* **2024**, *24*, 3868. [CrossRef] [PubMed]
28. Gabauer, D.J.; Gabler, H.C. Comparison of Delta-V and Occupant Impact Velocity Crash Severity Metrics Using Event Data Recorders. *Ann. Adv. Automot. Med.* **2008**, *52*, 77–86.
29. Craig, G.L.; Arnoux, P.J.; Ijspeert, C.; Thollon, L.; Brunet, C. A Numerical Study of Abdomen Impact Responses and Injuries. In Proceedings of the IRCOBI Conference, Maastricht, The Netherlands, 19–21 September 2007; pp. 203–214.
30. Radu, A.I.; Cofaru, C.; Tolea, B.; Popescu, M. Development of a New Recline Mechanism in Order to Reduce the "Whiplash" Effect Using a Virtual Model. *Proc. Inst. Mech. Eng. Part D J. Automob. Eng.* **2018**, *232*, 1701–1712. [CrossRef]
31. Goodwine, B. *Engineering Differential Equations: Theory and Applications*; Springer Science & Business Media: Berlin, Germany, 2010.
32. Flores, P.; Lankarani, H.M. *Contact Force Models for Multibody Dynamics*; Springer: Berlin, Germany, 2016; Volume 226, pp. 28–51.

33. Schiehlen, W. (Ed.) *Advanced Multibody System Dynamics: Simulation and Software Tools*; Springer Science & Business Media: Berlin, Germany, 2013; Volume 20.
34. Johnson, K.L. *Contact Mechanics*; Cambridge University Press: Cambridge, UK, 1987.
35. Hariharesan, S.; Barhorst, A.A. Modelling, Simulation and Experimental Verification of Contact/Impact Dynamics in Flexible Multibody Systems. *J. Sound Vib.* **1999**, *221*, 709–732. [CrossRef]
36. Nahum, A.M.; Melvin, J.W. (Eds.) *Accidental Injury: Biomechanics and Prevention*; Springer: Berlin/Heidelberg, Germany, 2002.
37. Klima, M.E.; Toomey, D.E.; Weber, M.J. Seat belt buckle performance in high energy wheel-to-ground impacts. *SAE Tech. Paper* **2005**, *1*, 1709. [CrossRef]
38. *Federal Motor Vehicle Safety Standards (FMVSS) No. 209—Seat Belt Assemblies*; National Highway Traffic Safety Administration: Washington, DC, USA, 2004.
39. GmbH, R.B. *Automotive Engineering Handbook*; Bosch, J., Ed.; SAE International: Warrendale, PA, USA, 2018.
40. Wong, J.Y. *Theory of Ground Vehicles*; Wiley: Hoboken, NJ, USA, 2022.
41. Pacejka, H.B. *Tyre and Vehicle Dynamics*; SAE International: Warrendale, PA, USA, 2002.
42. Milliken, W.F.; Milliken, D.L. *Fundamentals of Vehicle Dynamics*; SAE International: Warrendale, PA, USA, 1992.
43. McGill, S.M.; Norman, R.W. Partitioning of the L4-L5 Dynamic Moment into Disc, Ligamentous, and Muscular Components During Lifting. *Spine* **1986**, *11*, 666–678. [CrossRef]
44. McGregor, A.H.; Hukins, D.W. Lower limb involvement in spinal function and low back pain. *J. Back Musculoskelet. Rehabil.* **2009**, *22*, 219–222. [CrossRef]
45. White, A.A.; Panjabi, M.M. *Clinical Biomechanics of the Spine, 2nd ed.*; Lippincott Williams & Wilkins: Philadelphia, PA, USA, 1990.
46. Nightlinger, C.; Alem, N.M. Human Head-Neck Dynamics: Modelling and Validation. In Proceedings of the IMAC-XIX, Kissimmee, FL, USA, 5–8 February 2001.
47. Myers, B.S.; Winkelstein, B.A. Epidemiology, Classification, Mechanism, and Tolerance of Human Cervical Spine Injuries. *Crit. Rev. Biomed. Eng.* **1995**, *23*, 307–409. [CrossRef] [PubMed]
48. Mansfield, N.J.; Griffin, M.J. Non-Linearities in Apparent Mass and Transmissibility During Exposure to Whole-Body Vertical Vibration. *J. Biomech.* **2000**, *33*, 933–941. [CrossRef]
49. Bogduk, N.; Yoganandan, N. Biomechanics of the cervical spine Part 3: Minor injuries. *Clin. Biomech.* **2001**, *16*, 267–275. [CrossRef] [PubMed]
50. Meriam, J.L.; Kraige, L.G.; Bolton, J.N. *Engineering Mechanics: Dynamics*; John Wiley & Sons: Hoboken, NJ, USA, 2020.
51. EES Catalogue, DSD Austria, 2023. Available online: www.dsd.at (accessed on 8 April 2025).
52. PC-Crash User Manual, DSD Austria, 2023. Available online: https://dsd.at/images/download/PCCManual100E.pdf (accessed on 8 April 2025).
53. Silvestri, C.; Ray, M.H. Development of a Finite Element Model of the Knee-Thigh-Hip of a 50th Percentile Male Including Ligaments and Muscles. *Int. J. Crashworthiness* **2009**, *14*, 215–229. [CrossRef]
54. Sobie, E.A. An introduction to MATLAB. *Sci. Signal.* **2011**, *4*, tr7. [CrossRef] [PubMed]
55. Chai, T.; Draxler, R.R. Root mean square error (RMSE) or mean absolute error (MAE)?—Arguments against avoiding RMSE in the literature. *Geosci. Model Dev.* **2014**, *7*, 1247–1250. [CrossRef]
56. Willmott, C.J.; Matsuura, K. Advantages of the mean absolute error (MAE) over the root mean square error (RMSE) in assessing average model performance. *Clim. Res.* **2005**, *30*, 79–82. [CrossRef]
57. De Myttenaere, A.; Golden, B.; Le Grand, B.; Rossi, F. Mean absolute percentage error for regression models. *Neurocomputing* **2016**, *192*, 38–48. [CrossRef]
58. Hyndman, R.J.; Koehler, A.B. Another look at measures of forecast accuracy. *Int. J. Forecast.* **2006**, *22*, 679–688. [CrossRef]
59. Kubiak, P.P.; Kazmierczak, K.M.; Caroline Mierzejewska, P.; Szosland, A.A. Investigation of high-precision non-linear crash reconstruction method basing on a 'Mini' vehicle class database. *Int. J. Crashworthiness* **2018**, *23*, 635–644. [CrossRef]
60. Zhang, Y.; Wang, Z. Multibody Simulation Study on Rear-End Collisions and Human Body Dynamics. *Bioengineering* **2024**, *11*, 129. [CrossRef]

Disclaimer/Publisher's Note: The statements, opinions and data contained in all publications are solely those of the individual author(s) and contributor(s) and not of MDPI and/or the editor(s). MDPI and/or the editor(s) disclaim responsibility for any injury to people or property resulting from any ideas, methods, instructions or products referred to in the content.

sensors

MDPI

Article

Driver Drowsiness Multi-Method Detection for Vehicles with Autonomous Driving Functions

Horia Beles, Tiberiu Vesselenyi, Alexandru Rus, Tudor Mitran *, Florin Bogdan Scurt † and Bogdan Adrian Tolea †

Department of Mechanical Engineering and Automotive, University of Oradea, Universitatii St. 1, 410087 Oradea, Romania; horia.beles@uoradea.ro (H.B.); vtiberiu@uoradea.ro (T.V.); alrus@uoradea.ro (A.R.); scurt.florinbogdan@didactic.uoradea.ro (F.B.S.); bogdan.tolea@uoradea.ro (B.A.T.)

* Correspondence: tudor_mitran@uoradea.ro; Tel.: +40-744-152-212

† These authors contributed equally to this work.

Abstract: The article outlines various approaches to developing a fuzzy decision algorithm designed for monitoring and issuing warnings about driver drowsiness. This algorithm is based on analyzing EOG (electrooculography) signals and eye state images with the aim of preventing accidents. The drowsiness warning system comprises key components that learn about, analyze and make decisions regarding the driver's alertness status. The outcomes of this analysis can then trigger warnings if the driver is identified as being in a drowsy state. Driver drowsiness is characterized by a gradual decline in attention to the road and traffic, diminishing driving skills and an increase in reaction time, all contributing to a higher risk of accidents. In cases where the driver does not respond to the warnings, the ADAS (advanced driver assistance systems) system should intervene, assuming control of the vehicle's commands.

Keywords: autonomous driving; ADAS; sensor integration; sensors fusion; deep learning; video signal processing

1. Introduction

Driving in a drowsy state is a major problem worldwide. According to a CDC study (Centers for Disease Control and Prevention) from the United States [1], an estimated 1 in 25 drivers older than 18 report falling asleep while driving in the last 30 days. The NHTSA (National Highway Traffic Safety Administration) estimates that in 2017, drowsy driving was responsible for 91,000 crashes. Due to the high number of crashes, there were a resulting 50,000 injuries and nearly 800 deaths [2,3]. According to the EU Mobility and Transport Road Safety Commission, the causes of 10 to 25% of all road accidents in Europe are due to the drowsy state of the driver [4]. It has been observed that the frequent situation in which fatigue affects the driver's attention is when they are driving on the highway or on roads between cities, due to the monotony of driving activities.

It is well known that technology has advanced at a rapid pace, making it possible to use it in various driving activities. The role of the most functions implemented in vehicles is to reduce these high numbers of accidents and fatalities by two main methods: by warning the driver of possible dangers and/or by direct intervention in the control of the vehicle (automated/autonomous driving).

Regarding the automation of driving functions, according SAE International (Society of Automotive Engineers) standards—the J3016 standard [5]—there are 6 levels, from Level 0 (no automation) to Level 5 (fully autonomous/automated vehicle). In order to take into account the differences in automation and to make a clearer distinction between the SAE levels shown in Figure 1, all 6 levels will be briefly explained in the following paragraphs.

<table>
<tr><th>SAE Level =></th><th>SAE Level 0</th><th>SAE Level 1</th><th>SAE Level 2</th><th>SAE Level 3</th><th>SAE Level 4</th><th>SAE Level 5</th></tr>
<tr><td rowspan="3">What does the human driver have to do?</td><td colspan="3">The human driver is driving whenever these driver support features are engaged.</td><td colspan="3">The human driver is not driving when these automated driving features are engaged.</td></tr>
<tr><td colspan="3" rowspan="2">The human driver must constantly supervise these support features; He/she must steer, brake or accelerate as when the situation requires in order to maintain safety</td><td>When the feature request,</td><td colspan="2" rowspan="2">These automated driving features will not require you to take over driving</td></tr>
<tr><td>HE/SHE MUST DRIVE</td></tr>
<tr><td></td><td colspan="3">These are driver support features</td><td colspan="3">These are automated driving features</td></tr>
<tr><td>What do these features do?</td><td>The equipment can only provide warnings and momentary assistance</td><td>The features provide steering
OR
brake/acceleration support to driver</td><td>These features provide steering
AND
brake/acceleration support to driver</td><td colspan="2">The vehicle can run under limited conditions and will not operate unless all required conditions are met</td><td>The vehicle can run with no restriction and under all conditions.</td></tr>
<tr><td>Example Features</td><td>- automatic emergency braking (AEB)
- blind spot warning (BSM)
- lane departure warning (LDW)</td><td>- lane centering (LCA)
OR
- adaptive cruise control (ACC)</td><td>- lane centering
AND
- adaptive cruise control at the same time</td><td>- traffic jam driver function</td><td>- local driverless taxi function

- pedals/steering wheel may not be installed</td><td>- same as level 4, but the vehicle can run everywhere in all conditions</td></tr>
</table>

Figure 1. SAE J3016 standard for levels of driving automation [5].

SAE Level 0: the systems installed on these vehicles are mostly for warning, not for intervention, so the level has no driving automation functions.

SAE Level 1: the system could take over the control of the acceleration OR brake OR the steering wheel (in order to maintain the direction of travel). The systems present in these vehicles could be the following: adaptive cruise control or lane centering. The driver must be capable of intervening if the traffic situation so requires, because the systems are only for driver assistance.

SAE Level 2: the system is able to control steering, acceleration AND braking, when the functions are active (lane centering and adaptive cruise control working together). At this level (partial automation or assistance) the driver should still monitor the environment and be prepared to intervene and take control of the vehicle when the situation calls for it.

SAE Level 3: the vehicle can monitor the surrounding area and control the steering wheel, brakes and acceleration in some cases of travel. At this level it is conditional automation, and the driver still must be able to take over the control if the situation requires that.

SAE Level 4: also known as 'high automation', this level is represented by an autonomous driving system that takes over the driving and monitoring activities during the movement (functions like a local driverless taxi). At this level, the pedals and/or steering wheel may or may not be present.

SAE Level 5 (full automation): this level consists of a full automation system which permits the vehicle to manage all the situations for safe driving in any conditions of traffic.

Based on the SAE International classification, it can be seen that up to level 3 of autonomy, the human driver cannot be eliminated, even if sometimes they are only responsible for monitoring the autonomous driving system, or intervening when it is required. For this reason, driver drowsiness state in a Level 3 or lower autonomous vehicle can be extremely dangerous and need to be monitored.

Detecting driver fatigue can be carried out in the early stages of drowsiness and the driver can be warned. Over time, several methods have been proposed to determine driver fatigue: analysis of the driver's movements and face expressions, or biological signals measures that relate to monitoring of the driver behavior and vehicle movement [6]. The method of analyzing the driver's facial movements and expressions includes techniques for measuring mouth movements, eye closing, eye blinking, and head position [7–10].

The methods based on biological measures that relate to driver bio-signals includes physiological signals such as electrooculography (EOG), electromyogram (EMG), electroencephalogram (EEG), and electrocardiogram (ECG) [11–19].

The methods based on behavioral monitoring and vehicle motion include driving behavior, such as time in lane crossing, speed, steering wheel angle, lane position, etc. [7,18]. Some of these methods and other have been studied in detail, and moreover the advantages and disadvantages of each have been addressed in other articles. Combining several methods in a hybrid system would lead to an efficient drowsiness detection system.

In our previous studies, a system for fatigue detection by acquiring and processing encephalogram (EEG) signals was developed, and it is presented in [9,20]. In early research [8], the possibilities of acquiring electromyogram (EMG) signals by three sensors are addressed, and artificial neural networks were used to classify driver fatigue states [9,20]. In [9,20], we presented the driver fatigue state detection system's performance in the analysis of the driver's eye position (open, closed, or half-open) based on images acquired while driving.

This paper presents the experiments performed by the authors in order to study the applicability of the integration of different biological parameters of a driver in a multi-sensor system for drowsiness detection. The paper presents a system which could use two signals obtained from biological sensors: EEG and EOG, and two kinds of images: driver face images and eye images. Images of the driver's face were used to distinguish between face images with closed and open eyes. EEG and EOG signals were acquired in laboratory settings, while driver face images were acquired during driving. In addition to the previous research, this article also presents some suggestions for the implementation of this driver drowsiness monitoring system in vehicles with ADAS.

2. Materials and Methods

Devices used for research: NI usb 6003 acquisition board, EMG MIKROE-2623 sensors, Matlab R2022b software version 9.13 and Blackbox L300-1 video camera.

2.1. Drowsiness Detection System Based on EOG Signals and Face Image Analysis

The main methods which are used in this case are those based on electrooculography (EOG) signals and image classification of the eyelid position (closed or open). For the eye movements, we used an EOG tracking method by measuring eye muscle signals. The state of the eye (closed or open) is monitored through eye image processing. The development of such a system was discussed by some of the authors in [8]. The system presented in [8] is shown in Figure 2. The methods used in the system presented in previous work has its advantages but also some disadvantages. For example, EEG (especially) and EOG sensors presents a major discomfort due to the fact they have to be fixed with a conductive gel, and in some cases, devices must transmit the signal by wire (for a better data transfer). Important research in advanced materials and MEMS (microelectromechanical system) technology is solving some of the problems using some kind of dry electrodes for EEG, presented in [21], or other types of electrodes used, such as, for example, in fitness bracelets. The main developments and findings in this field are encouraged by efforts and the necessity of creating brain–computer interfaces [22–24] for various applications, including systems that could allow doctors to monitor patients or help people with disabilities.

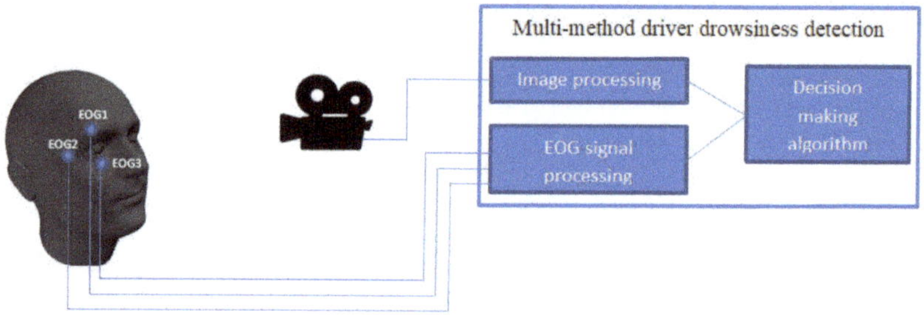

Figure 2. Schematics of a multi-method driver drowsiness detection system based on EOG signals and face image analysis [9].

2.2. System for Drowsiness Detection Based on Face Image Analysis and the Eye Aspect Ratio Algorithm (EAR)

In the case of this drowsiness detection system the driver's face image is used as input. The images are collected during the driving period and streamed to be analyzed [20,25]. After the camera captures the human driver's images, the images have to be processed by a face/eye recognition algorithm. After this, they are processed by two different variants (of algorithms): the first is an artificial neural network (ANN) which is trained to recognize the position of the driver's eyelids (open or closed), and the second variant is to use an algorithm for the eye aspect ratio (EAR), which can also provide data about the state of the eyelids (open/closed), but based on a different process. The final step in the image processing process, based on using a trained ANN to recognize a certain driver behavior, is the decision regarding whether the driver is tired and whether or not a warning is needed. For the decision, we used a well-known logistic regression, which is widely used in various fields (including medicine, finance, and machine learning) for tasks like prediction, and it has only two possible outcomes (usually 0 and 1, or 'yes' and 'no'). At this point, the frequency and distribution of blinks can be analyzed to determine if the driver is sleepy or in an alert state. The system's scheme is shown in Figure 3.

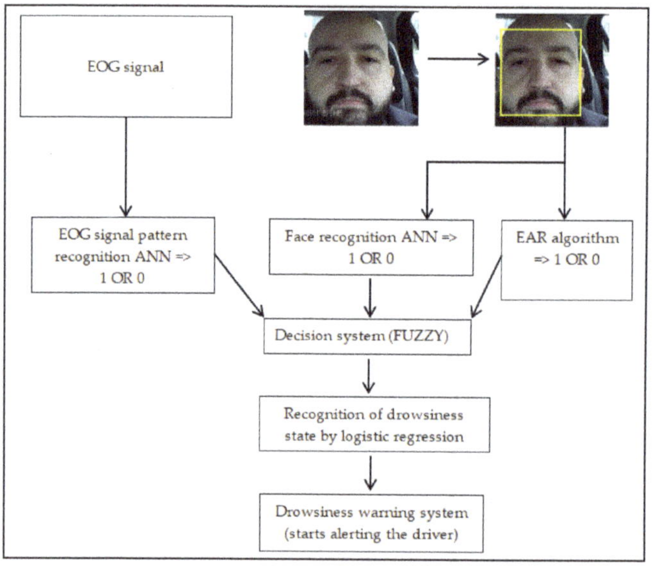

Figure 3. Schematics of a drowsiness detection system based on EEG signals; face image analysis and the EAR algorithm.

In the following paragraphs, the processing methods and the results obtained after testing and training are represented. MATLAB R2022b tools were used for the image (face/eyes) recognition through an ANN algorithm and also in order to develop the fuzzy decision algorithm [20,26,27]. To train the neural networks, more than 800 min were recorded while the vehicle was driving on public roads (municipal roads, expressways, and motorways), at different stages of driver fatigue and in different weather conditions, but ultimately, relevant sequences were extracted from this large amount of collected data. Thus, several 60-s video sequences (of 1975 frames each) recorded while driving were used to train and test the face image analysis system [20]. Information on driver monitoring in traffic is presented in Table 1.

Table 1. Information on the conduct of data acquisitions.

Day and Route No.	Total Sleep before Test	Route [km]	Driving (Recording) Time [min]	Weather	Time Interval	Comments
1	8 h	Oradea -> Turda -> Oradea (381 km)	332 min	cloudy	7:00 A.M.–2:00 P.M.	No/weak signs of fatigue
2	5 h 30 min	Oradea -> Arad -> Oradea (229 km)	200 min	cloudy/sunny	8:00 A.M.–11:45 A.M.	More frequent signs of fatigue
3	6 h	Oradea -> Carei -> Oradea (204 km)	175 min	rainy	11:00 A.M.–2:10 P.M.	Frequent signs of fatigue
4	7 h 30 min	Oradea -> Carei -> Oradea (204 km)	180 min	cloudy and rainy	11:00 A.M.–2:15 P.M.	Frequent signs of fatigue and lack of attention

Column one of the table shows the number of the day and the route on which the record was made. Column two shows how long the driver slept before driving. The third column shows the route followed by the driver; the route can be viewed using the GoogleMaps application, as a link is provided for each route. The fourth column shows how many minutes were recorded for each route. The fifth column describes the weather conditions under which the tests were carried out. The penultimate column describes the time intervals in which the recordings were made. The driver tried to follow his usual daily routine while including the 4 trips shown in the table. All trips were made without breaks for the driver. In addition to video monitoring, the driver was accompanied and supervised by a passenger, who tried to note as many signs of fatigue and inattention as possible during the journey. Thus, the last column in the table shows a summary of the passenger's notes. The first signs of fatigue started to appear only after about an hour of driving. Another quite important aspect noticed was that on the stretches of road where driving was monotonous (motorway—trip 1; road in alignment—trip 2, 3, 4; or road known by the driver), fatigue states appeared frequently. In addition to the monotonous driving (without much driving activity) which induced a feeling of fatigue/boredom, the driver's attention to the road decreased. This was especially evident on days 3 and 4, and more so on day 4, as the route was well known and the driver did not pay as much attention to driving. According to [28], due to the rainy weather during the last two days of data collection, the driver's state of fatigue may also have been accentuated by the use of windscreen wipers.

2.3. EEG and EOG Signal Analysis

In [20], an application of EEG system for acquisition and data processing is presented. The aims of this research [20] are to analysis the signal peaks (of Alpha rhythm) in order to determine if the driver is in an alert state or a drowsy state, see Figure 4. Drowsiness is often associated with an increase in alpha brain waves. These are relatively low-frequency waves (8–12 Hz) that are dominant in the posterior regions of the brain when a person is awake but relaxed. As a person transitions from wakefulness to drowsiness, theta waves

(4–7 Hz) might become more prominent. These are often associated with the early stages of sleep. Drowsiness can be characterized by an increase in slower frequencies, such as theta and delta waves, as the person moves closer to sleep.

Upon awakening, theta waves may still be present as the person transitions from sleep to wakefulness. Alpha waves might increase as the person becomes more alert. Beta waves (13–30 Hz) are associated with alert wakefulness. As the person becomes more awake and alert, beta activity in the EEG may increase. Delta waves (0.5–4 Hz), associated with deep sleep, should decrease as the person wakes up. It is important to note that these patterns may differ from person to person. The transition from sleep to wakefulness involves dynamic changes in EEG patterns.

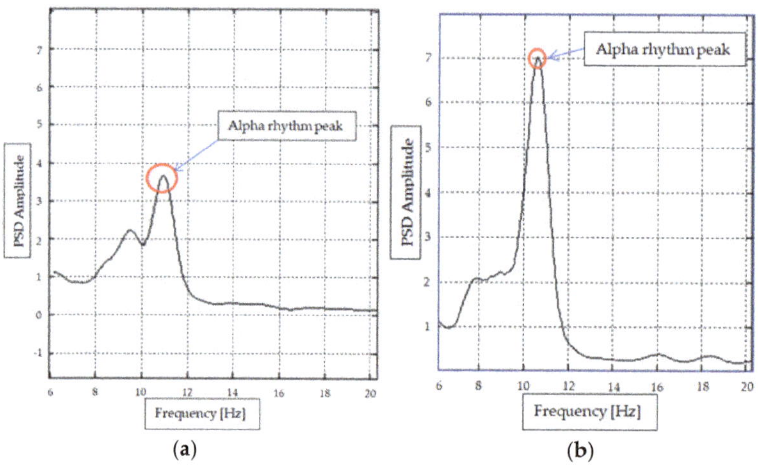

Figure 4. EEG signal diagrams, (PSD-power spectral density amplitude vs. frequency): (**a**)—driver in an alert state; (**b**)—driver in a drowsy state.

Measuring EEG signals (Figure 4.) in drivers can be difficult, so if we have to choose between EEG and EOG signals, EOG signals are easier to obtain. The possibility of using EOG signals, collected by three sensors mounted in the eye area (Figure 2), was analyzed in other research [7,11]. In Figure 5, the two types of signal (S1, S2) are shown, which were identified after the preprocessing step. Combining these two types of signals let us distinguish between different movements of the eye and blinking. Recording these different types of eye movements for normal behavior and drowsy behavior can give us information about the state of the driver. The signals provided by the EOG1, EOG2, and EOG3 sensors (sensor positions are shown in Figure 2) regarding the driver's eye blinking are as follows: S2 -> EOG1, S1 -> EOG2, S1 -> EOG3. The configuration of the 3 signals coming from the EOG sensors have unique configurations for each type of movement, so the configurations S2, S1, and S1 will not be encountered during any eye movement other than blinking. Typically, EOG1 and EOG2 are used to capture horizontal eye movements. The placement of electrodes allows for the measurement of electrical potential generated by the movement of the eyes from side to side. The positioning of the EOG3 electrode allows for the measurement of the electrical potential associated with up and down eye movement.

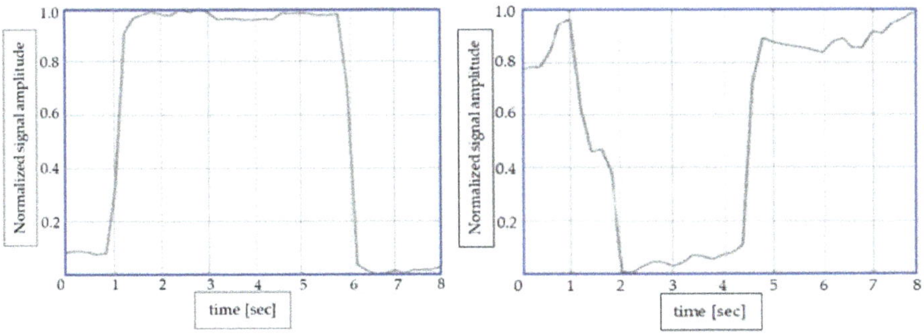

Figure 5. Types of signals recorded by EOG sensors (sensors' positions are shown in Figure 2 EOG1, EOG2, EOG3) denoting S1 (**left**) and S2 (**right**).

2.4. Face Image Analysis for Open or Closed Eye State Detection

For image classification of the driver state, a method using the neural network toolbox and the deep learning toolbox from MATLAB was studied [26,27]. The following premise was used: it assumes that drowsiness corresponds to the images where the driver's eyes are closed and the alert state of the driver corresponds to the images where the eyes are open [29].

Two hundred images were used to analyze the drowsiness of the driver during the driving process. Half of these images are of the driver with open/half-open eyes, and the other half are pictures of the driver with their eyes closed.

In Figure 6, some images used for processing are shown. In order to minimize the huge amount of input data for the artificial neural network, the images were cropped and down-sampled; moreover, the processing time was diminished, and the overload of memory was also avoided.

Figure 6. Images of the driver: (**a**)—with eyes open; (**c**)—with eyes closed. Images of the driver at low quality/resolution: (**b**)—with eyes open; (**d**)—with eyes closed [8].

In Figure 6b,d, images were down-sampled and sharpness was affected. This ensured a more robust recognition and classification. Furthermore, if the neural networks work well with low-quality images, it will surely work much better with high-quality images (Figure 6a,c).

2.5. Open or Closed Eye State Detection Using the EAR Algorithm

The ratio between eye width and length uses a point recognition algorithm concerning face characteristics (point coordinates in pixels) and learned 68-point iBUG 300-W dataset features. Using the eye detection algorithm, the driver's eyes were tracked as shown in Figure 7. In [29] the previous algorithm used in this case is described. The description is based on [30–32].

(**a**) (**b**)

Figure 7. (**a**)—Face detection and tracking; (**b**)—Face and eye detection in streamed video.

In Figure 8, the 6 characteristic points defined on the detected area of the eye are shown. Based on Relation (1), the EAR algorithm was computed and the results were split: high values mean the eyes were open, and low values mean the eyes were closed.

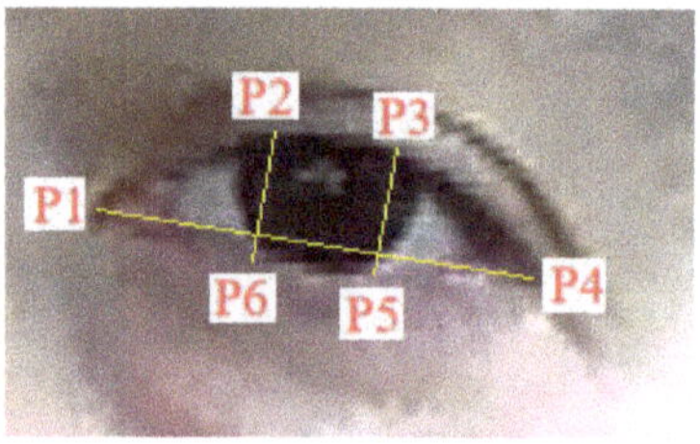

Figure 8. Eye aspect ratio (EAR) metric.

The Euclidian distance between the specified points are defined in Relation (1) with the double parting brackets [20].

$$EAR = \frac{||p_2 - p_6|| + ||p_3 - p_5||}{2||p_1 - p_4||} \quad (1)$$

where 'p_i' (i = 1...6) is one of the points presented in Figure 8. Every point 'p_i' has the coordinates given in pixels (x_i, y_i). The coordinates of the points are obtained from the image in Figure 7 by binarizing the image and then applying an edge detection algorithm (Sobel, Canny, Prewitt). The distance between the two points 'p_i' and 'p_k' is computed by the relation:

$$||p_i - p_k|| = \sqrt{(x_i - x_k)^2 + (y_i - y_k)^2} \quad (2)$$

Relation (1) will be written as:

$$EAR == \frac{\sqrt{(x_2 - x_6)^2 + (y_2 - y_6)^2} + \sqrt{(x_3 - x_5)^2 + (y_3 - y_5)^2}}{2\sqrt{(x_1 - x_4)^2 + (y_1 - y_4)^2}} \quad (3)$$

2.6. Face Detection and Tracking Algorithms

As an alternative to using the EAR algorithm trained on the iBUG 300-W dataset, another facial characteristic detection algorithm can be used, which has the potential to be more reliable. The algorithm is based on the Viola–Jones algorithm [33], which was developed to detect people's faces, noses, eyes, mouth, or upper body. The algorithm is implemented in MATLAB as a cascade object detector which is trained on positive and negative images and relays on a series of successive classifiers. The detected bounding box is shown in Figure 9a. Once the area is reduced by detecting the bounding box, characteristic points are selected using a minimum eigenvalue algorithm developed in [34] to find feature points in a greylevel image. The algorithm is implemented in MATLAB using the 'detectMinEigenFeatures' function, and result is shown in Figure 9b.

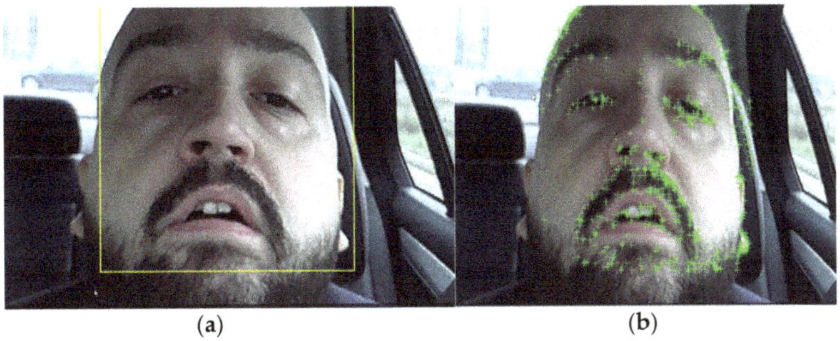

Figure 9. (**a**) Face detection using bounding box (yellow); (**b**) detected characteristic points (green +).

Having detected the feature points, we can apply a point tracker algorithm, which in our case was the Kanade–Lucas–Tomasi (KLT) algorithm [35]. The point tracker algorithm is implemented in the MATLAB R2022b v9.13 language as a software object that follows a group of points using the feature-tracking method. This can be used for camera motion stabilization and estimation, and object following. The algorithm is satisfactory for tracking faces that are not out of plane or changing characteristics related to the image plane. The point tracker can be used in the majority of cases as a subroutine in a larger context where successive detections are needed.

Due to points that can be lost because of lighting changes or out of plane movement, to track a face over a longer time, successive acquisitions of feature points must be made. This can be observed in Figure 10, where in Figure 10a, in the area of the eyes, there are a large set of feature points which were lost after the driver blinked (Figure 10b). Even so, the algorithm is much more reliable than a variant which uses the eye detection variant (using the classification model 'EyePairBig' or 'EyePairSmall') when the detection is much worse. The face tracking algorithm in this case is sufficiently robust to work properly even in relatively extreme cases (Figure 11) [36–38].

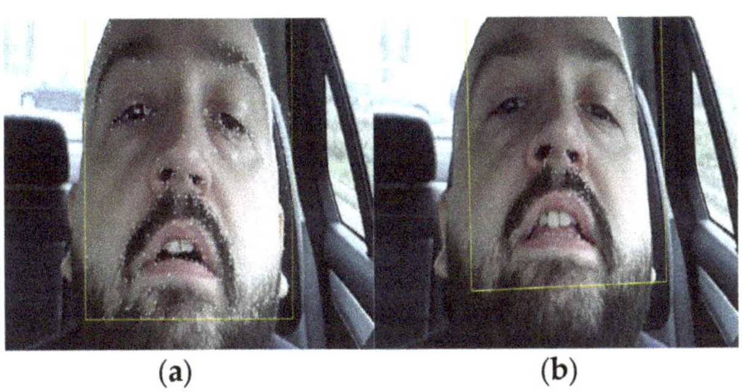

Figure 10. (**a**) Feature points before blinking; (**b**) loss of feature points after blinking.

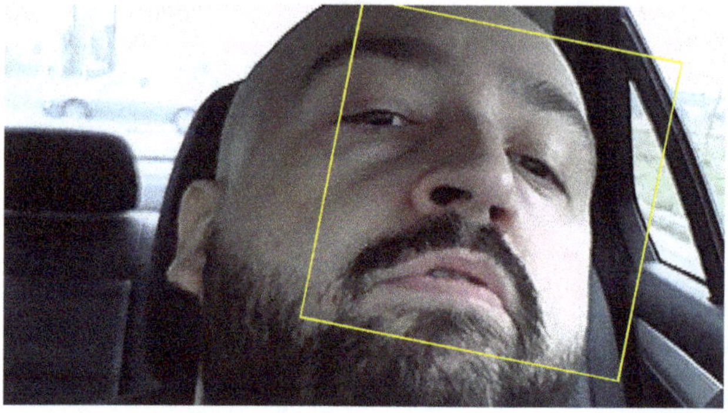

Figure 11. Tilted position of driver's face followed by the bounding box and still detected by the algorithm.

Once the bounding box is firmly established, the following step is to find the timeline of the open or closed eyes. This step is similar to that described in the following Section 3.1. for other detection methods. Along with 1 hidden layer or an autoencoder ANN for eye state detection, we can also use a deep learning ANN, like, for example, a CNN (convolutional neural network) which nowadays are largely applied in image recognition. A flowchart for the described algorithms is shown in Figure 12.

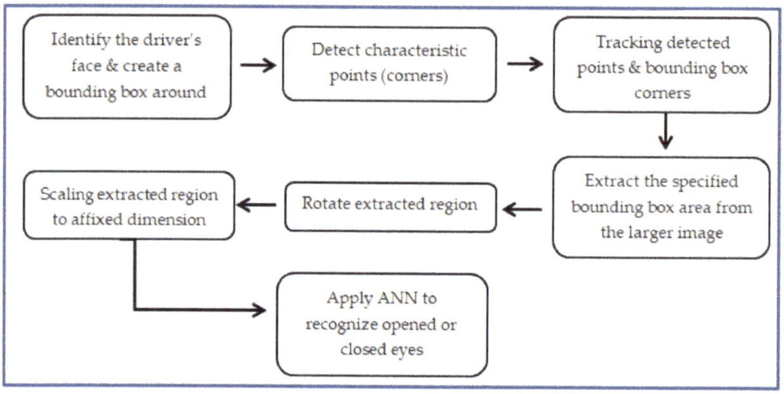

Figure 12. Block diagram of face detection and tracking algorithm.

2.7. The Use of Drowsiness Detection System in Autonomous Driving

Taking into account the classification based on SAE standards, it can be seen that only for the first 4 levels of autonomy is the driver's intervention required, while for the last two levels of autonomy (level 4 and 5), the driver can be totally exempted from specific driving activities. Autonomous vehicles that can be encountered in traffic most often are those of maximum level 3; this is due to several reasons: technological limitations, legislative limitations, limitations in terms of infrastructure, etc. Thus, the implementation of the driver drowsiness monitoring system is beneficial to these categories of vehicles.

Since, up to level 3 of autonomy, the driver must be fit to drive or take over the controls when the system requires it, monitoring of the driver's drowsiness must be carried out continuously, regardless of how the vehicle is used (autonomous mode or not). From this perspective, there are two different situations in which the detection system must work:

1. The vehicle is used conventionally, and the driver is in full control.

In this situation, if the system perceives that the driver is in a drowsy state, as a first measure, it will issue an acoustic warning and also a physical one through small vibrations of the devices that measure the biological signals, but also through vibrations in steering wheel. Next, there are two possibilities: the driver responds to the stimuli and tries to return to a state of heightened attention, or there is a situation where the driver does not respond to these stimuli (warnings). If the driver does not change his state following the warning signals, then through the ADAS system and based on the information provided by the fatigue monitoring system, the dynamic performance of the vehicle will be modified by reducing the speed and possibly by acting on the steering system if the level of autonomy allows this.

2. The vehicle (especially level 2 and 3) is operated with the autonomous driving function activated (the vehicle may have control over direction of travel, acceleration, and braking).

In the case of a level 3 autonomous car, even if the automatic driving system is in operation, the driver must be prepared to take control of the vehicle if the system requires this for various reasons (fails to correctly identify some obstacles due to weather conditions and thus fail to manage the situation). The driver must therefore be able to take over the controls. In such a scenario, the fatigue detection system can play an extremely important role as it detects the driver's state and can warn them so that they do not fall into a state of deep drowsiness and can be ready to take over at any time. If driver fatigue is detected, the system will work in the same way as before.

3. Results

Regarding the results of EEG spectrum analysis, the decision of drowsy or alert state can be made either by detecting the peak amplitude in the 10–12 Hz domain (see Figure 4) or by applying a method as described in [19]. Both methods have been tested with fairly good results.

Regarding the results for the experiments using the EOG signals, the complete study is presented in [10]. Here, the results were very good with a good repeatability and consistency.

3.1. Results for the Face Image Analysis

The obtained results were processed using ANN and the fuzzy algorithm for the system to give a clear answer to the question of whether the state of drowsiness is reached or not. These algorithms are presented in the following sections.

3.1.1. Artificial Neural Network with One Hidden Layer

A total of 70 images with eyes open or half-open as well as another 70 images with eyes closed were used to teach, validate, and also test the neural network. The total number of collected images was 200. The rest of the images (60 images) were kept for testing the working of the neural network after the training process was finished with the previous 140 images. The NN (neural network) was trained based on the structure presented in Figure 13. This structure was constructed based on 2601 neurons in the input layer, 10 neurons in the hidden layer, and 2 neurons in the output layer. Each neuron has its scope, so the neurons of the input layer match the number of elements of the input vectors (images) and the neurons in the output layer (two neurons) correspond to the possibilities of image classification. The inputs of the neural network are the images, which are treated as matrices and are read row by row. The input image for this case is Figure 6a, which is 416 $\times$ 416 in size, and Figure 6b, which has 2601 pixels (51 $\times$ 51 pixels), so the inputs for the network will be equal to 2601. There are two classes: drowsy/sleepy state or alert state [20].

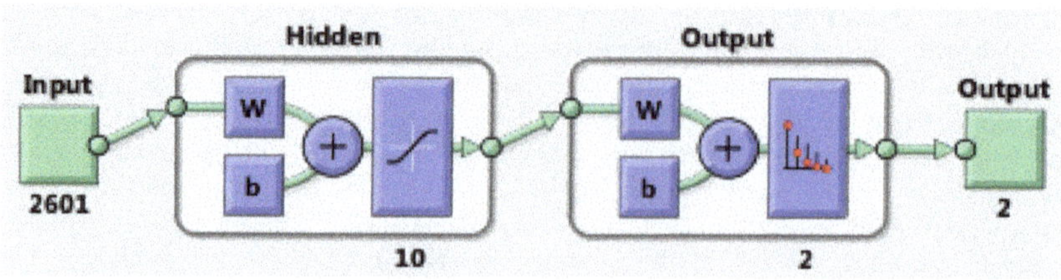

Figure 13. The structure of the network with 1 hidden layer [8].

In [9,39] the performance diagram and the error histogram are presented; both are obtained as a result of the neural network training process. From an analysis of the values from that performance training diagram and from the histogram boundaries, it can be said that there are some good results.

The Confusion matrix from Figure 14 shows the results of learning, validation, and testing. In such a matrix, the rows and the columns are used in the following way: a predicted class is represented by the each column and an actual class is presented by each row. The meaning of the colors in the matrix is as follows: green means a correctly classified sample and red means an incorrectly classified sample. In the case of the confusion matrix, the 3 × 3 configuration is due to the following fact: column 3 and row 3 of the matrix represent the sum of the two types of classified cases (wrongly classified and correctly classified). Analyzing the image, there can be seen that all have been correctly classified.

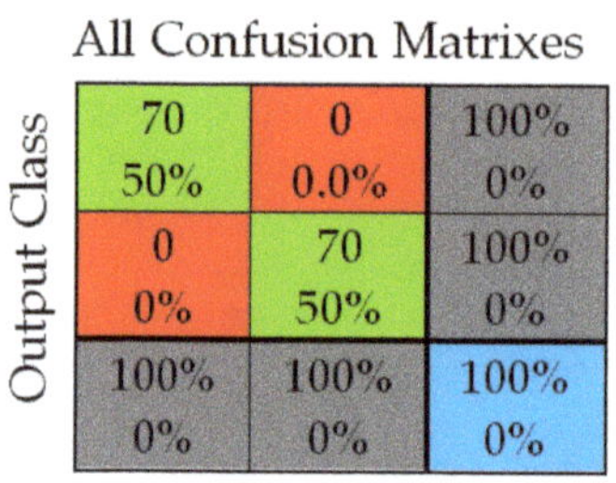

Figure 14. The training, validation, and testing matrix (confusion matrix) for one hidden layer network.

3.1.2. Deep Learning Autoencoder Neural Networks

For the autoencoder, the input and target vectors were similar, as in the case of the 1 hidden layer ANN. In the case of the autoencoder network, each layer was trained separately and then brought together in one network but with more layers, and the next step included the final training for the final entire network. The structure of this network is shown in Figure 15.

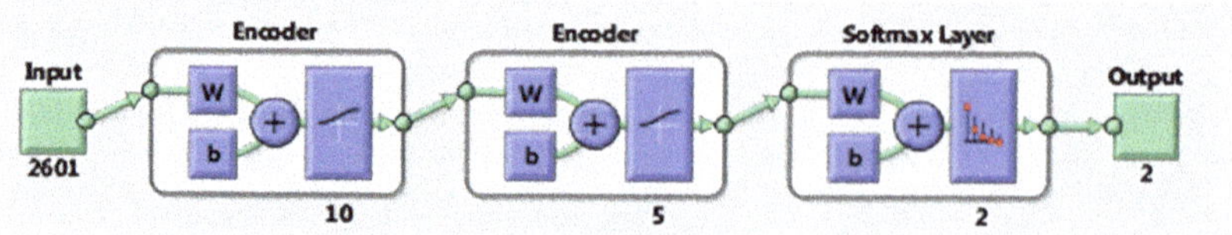

Figure 15. Network structure with 2 hidden layers [9].

The training results of the autoencoder reveal promising outcomes. The training performance emphasizes the achieved minimum value, which signifies improved performance. It is evident that there were no occurrences of false positives or false negatives. This indicates precise classification across all tested images (30 with closed eyes and 30 with open/half-open eyes). In summary, these findings denote a satisfactory performance.

3.2. Results for the EAR (Eye Aspect Ratio) Algorithm

For the EAR algorithm, the same video was used (as for the ANN analysis), and that way, it was able to obtain a value of EAR for every frame, as presented in Figure 16. Also in Figure 16, an area of interest is marked, which will be analyzed in Figure 17.

In Figure 17, the area of the graph which corresponds to the different states of the eyes (closed or open) in the video is marked. Using a simple threshold method, the intervals can be more easily detected.

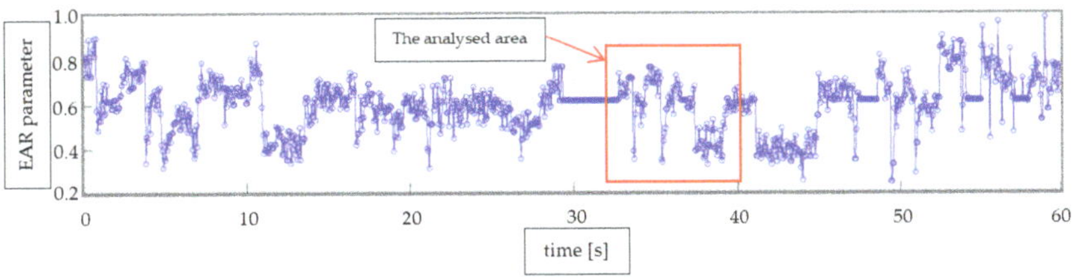

Figure 16. EAR algorithm results from video streaming.

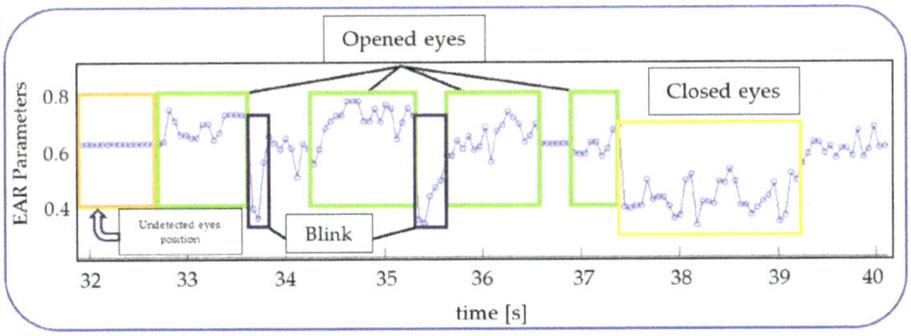

Figure 17. Detection of open or closed eye state by the EAR algorithm.

3.3. Results for the Face Image Analysis

In the case of image processing algorithms, there is a predisposition to errors due to the complexity of the input information (images). The detection accuracy could be seriously affected by many factors like: driver movements, environment influence, interior of the car, etc. In [8] was shown that the EEG signal detection, is influenced by the noise and sensor's accuracy. In addition, it is difficult to use for driver drowsiness detection because it is uncomfortable. One of the methods to overcome the inconvenience and to improve accuracy is to use more systems in parallel, like the system presented above, and employ a decision system. One well-known multi-criteria decision method is the fuzzy algorithm, which can be used to create such a complex system. In this case, the input of the system is the result of the EEG/EOG processing together with the results of EAR image processing and face recognition. The input membership functions are shown in Figure 18.

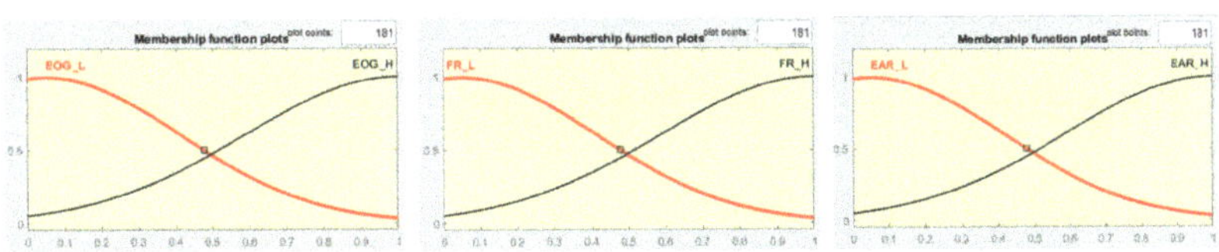

(a) (b) (c)

Figure 18. Input functions: (**a**)—EOG; (**b**)—FR (face recognition); (**c**)—EAR (eye aspect ratio); L = low value; H = high value.

Figure 19 presents the shape of the output membership function, using a Mamdani type output method. The system rule is defined thus: it is decided that the driver is in a drowsy state if at least two of the inputs are images with a closed eye state. The rules are presented more clearly in Figure 20 and Table 2. Figure 21 shows one of the decision surface models for processed inputs.

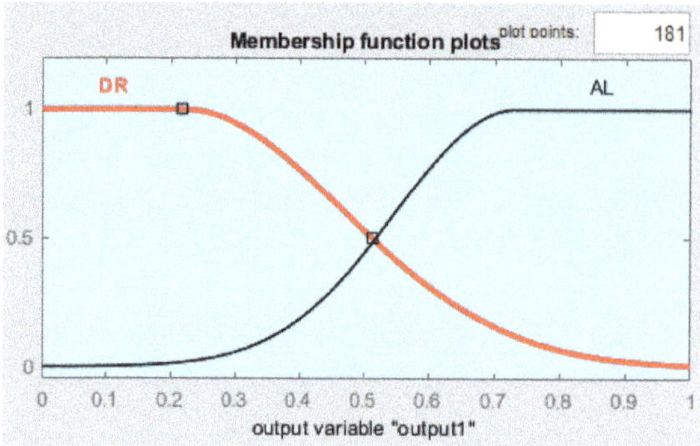

Figure 19. Output membership functions of the fuzzy system: DR—drowsy state; AL—alert state.

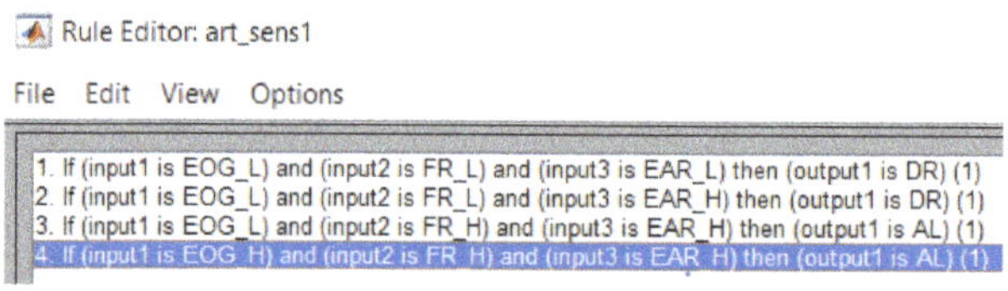

Figure 20. Rules that are applied in the system.

Table 2. The decision rule for fuzzy algorithm.

EEG/EOG		Face Tracking-Recognition		EAR	Result
L	and	L	and	L	drowsy
L	and	L	and	H	drowsy
L	and	H	and	H	alert
H	and	H	and	H	alert

H = high(1); L = low(0); they represent the values of the measurement systems introduced in the decision algorithm.

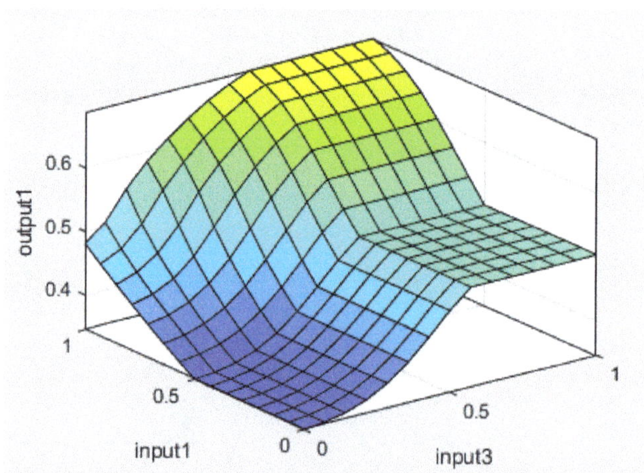

Figure 21. Decision surface for processing inputs [19].

Since the decision of the detection system is based on a fuzzy algorithm, if we look at Figure 3, it is obvious that the three input elements provide information to the decision algorithm simultaneously. The first input element can be considered, the result of analyzing the patterns (signals) either from EEG in the case of Figure 3 or from EOG sensors, this depends on the method chosen. The other two inputs to be analyzed in the decision algorithm are the result of the facial recognition algorithm and the EAR algorithm or even the facial recognition and tracking algorithm in Section 2.6.

Table 2 shows a model of the rules applied by the fuzzy algorithm used to analyze the signals from at least 3 detection systems, in order to provide a clearer result on the drowsiness of the driver.

Considering recently published articles by other researchers in this field, we can see a high interest in this topic. For example, in [40,41], most of the methods that have been tested for determining driver fatigue are mentioned and presented. Another example of a system similar to the one presented in this article can be viewed in [42], where Igor et al. provide a more detailed analysis on the method of determining fatigue focusing on EEG signal processing.

4. Discussion

In this paper, a drowsiness detection system for drivers and its possibilities was presented. The system combines three types of signal (sensors): EEG, EOG, and image processing. The methods were described in more detail in previous works. Regarding the disadvantages of such systems, even though the results are good in laboratory settings, difficulties appear at the vehicle driver stage due to the reluctance of drivers to wear some electrodes (for EEG and/or EOG) on their bodies while driving. That is why the systems would be hard to implement in all cars in this configuration. Moreover, the biological parameters are different for every person, which means the system must be calibrated for every driver separately. There are some possibilities to increase the performance of every single system or make it more convenient for the driver to use it. Regarding the systems for monitoring the biological parameters (EEG, EOG), other types of electrodes (dry electrodes) could be implemented, which could be integrated into sunglasses, (fitness) bracelets, or other type of usual accessories, with wireless communication. In the case of image processing, both networks used and presented in previous section have a good results, but should be tested in more and different conditions to ensure a good accuracy and functionality in almost all possible condition.

This driver drowsiness monitoring system is similar to the perception, classification, and decision system used to identify the driving environment, based also on sensor fusion. Thus, for image processing, an algorithm similar to that for analyzing images taken by the front camera of the vehicle can be used to detect the position of the eyelids. The notable difference between the sign recognition algorithm and the drowsiness recognition algorithm is the database used to train and validate it.

Based on the results provided by the driver drowsiness detection system, control of the vehicle if the driver does not respond to warnings is achieved by means of the systems found in the standard vehicle construction specific to each level. Important changes to the autonomy system are in the decision and control algorithms.

5. Conclusions

As mentioned in the previous section, the system works well under laboratory conditions, and it is now necessary to develop a more convenient design solution for the driver. This system solution must have the property of self-calibrating according to the driver using the vehicle. The main purpose of such a system is to avoid accidents caused by drowsiness, when the ADAS is on or off. In order to be able to develop and implement such tools in the construction of the vehicle, legislation should be drawn up to encourage the use of such an equipment, even if it would only be used for professional drivers who usually drive for long periods of time. This would be an intermediate solution between the classic vehicle and the high-level autonomous vehicle (level 4 or 5 according with SAE standard). A brief analysis of the papers [43,44] shows that the systems needed to create an autonomous bus or an autonomous vehicle are extremely complex and numerous, involving technologies that still need to be validated before they can be said to be safe, but above all before they can take full control of the vehicle without the driver having to intervene in any way (this type of autonomy corresponds to the highest level of SAE classification, level 5).

To reduce the numbers of accidents due to the drowsy state of the driver and also to encourage this technology, the European Commission for Transport (European Parliament) elaborated a legislative framework for this problem. It gave the following regulation for the car manufacturers: 'starting with 6 July 2022, all new model of vehicles should be equipped with a system for driver detection system, and starting with 7 July 2024, all new cars should have such a system' [4,45].

In conclusion, until autonomous vehicles reach a stage of maturity, the driver monitoring and warning option, combined with other prevention/warning systems, would be a beneficial solution in terms of reducing the number of accidents with or without victims. The same is desired with the development of autonomous driving technology. Moreover, it is hoped that (autonomous) vehicles can be used by all people (including people with disabilities) to avoid traffic jams, etc. However, for all these stages, it is necessary to adapt the legislative framework in addition to developing the technology.

Author Contributions: Conceptualization, T.V., F.B.S. and H.B.; data curation, H.B., T.V., A.R., T.M. and F.B.S.; formal analysis, T.V., A.R. and B.A.T.; investigation, T.V., B.A.T. and F.B.S.; methodology, T.V. and H.B.; visualization, H.B., A.R., T.M., T.V., B.A.T. and F.B.S.; writing—original draft preparation, H.B., T.M., B.A.T. and F.B.S.; writing—review and editing, H.B., T.M., B.A.T. and F.B.S. All authors have read and agreed to the published version of the manuscript.

Funding: This research was funded by the University of Oradea from the scientific research budget for 2024.

Informed Consent Statement: Informed consent was obtained from the subject involved in the study.

Data Availability Statement: The data presented in this study are available on request from the corresponding author.

Conflicts of Interest: The authors declare no conflicts of interest.

References

1. Centers for Disease Control and Prevention. Available online: https://www.cdc.gov/sleep/features/drowsy-driving.html (accessed on 3 November 2023).
2. National Highway Traffic Safety Administration. Drowsy Driving. Available online: https://www.nhtsa.gov/risky-driving/drowsy-driving (accessed on 3 November 2023).
3. Tefft, B.C. *Prevalence of Motor Vehicle Crashes Involving Drowsy Drivers, United States, 2009–2013 (Technical Report)*; AAA Foundation for Traffic Safety: Washington, DC, USA, 2014. Available online: https://aaafoundation.org/prevalence-motor-vehicle-crashes-involving-drowsy-drivers-united-states-2009-2013/ (accessed on 3 November 2023).
4. European Union Mobility & Transport—Road Safety. Available online: https://road-safety.transport.ec.europa.eu/index_en (accessed on 12 December 2023).
5. Shuttleworth, J. SAE Standards News: J3016 Automated-Driving Graphic Update. SAE International, 7 January 2019. Available online: https://www.sae.org/news/2019/01/sae-updates-j3016-automated-driving-graphic (accessed on 3 November 2023).
6. Marouf, M.; Saranovac, L.; Vukomanovic, G. Algorithm for EMG noise level approximation in ECG signals. *Biomed. Signal Process. Control* **2017**, *34*, 158–165. [CrossRef]
7. Nagy, R.B.; Vesselenyi, T.; Popentiu-Vladicescu, F. An analysis of electro-oculogram signals processing using an artificial neural network. In Proceedings of the 13th International Conference eLearning and Software for Education, Bucharest, Romania, 27–28 April 2017; Volume III, pp. 560–567.
8. Vesselenyi, T.; Rus, A.; Mitran, T.; Tataru, B.; Moldovan, O. Vehicle driver drowsiness monitoring and warning system. In Proceedings of the 13th Edition of the International Congress of Automotive and Transport Engineering, CONAT 2016 International Congress of Automotive and Transport Engineering, Brasov, Romania, 26–29 October 2016; Springer International Publishing: Berlin/Heidelberg, Germany, 2017; pp. 873–880.

9. Vesselenyi, T.; Moca, S.; Rus, A.; Mitran, T.; Tataru, B. Driver drowsiness detection using ANN image processing. 11th Edition of The International Congress of Automotive and Transport Engineering, Pitesti, Romania. *Proc. IOP Conf. Ser. Mater. Sci. Eng.* **2017**, *252*, 012097. [CrossRef]
10. Vesselenyi, T.; Dzitac, I.; Dzitac, S.; Hora, C.; Porumb, C. Preliminary Issues on Brain-Machine Contextual Communication Structure Development. In Proceedings of the IEEE Conference, 3rd International Workshop on Soft Computing Applications, Szeged, Hungary, 29 July–1 August 2009; pp. 35–40.
11. Nagy, R.B.; Vesselényi, T.; Popentiu-Vladicescu, F. Results Regarding an Eog-Based Assistive Wireless Robot Control System with Visual Feedback. In Proceedings of the International Scientific Conference eLearning and Software for Education, Bu-charest, Romania, 19–20 April 2018; Volume III, pp. 453–460.
12. Cardone, D.; Perpetuini, D.; Filippini, C.; Mancini, L.; Nocco, S.; Tritto, M.; Rinella, S.; Giacobbe, A.; Fallica, G.; Ricci, F.; et al. Classification of Drivers' Mental Workload Levels: Comparison of Machine Learning Methods Based on ECG and Infrared Thermal Signals. *Sensors* **2022**, *22*, 7300. [CrossRef]
13. Ebrahimian, S.; Nahvi, A.; Tashakori, M.; Salmanzadeh, H.; Mohseni, O.; Leppänen, T. Multi-Level Classification of Driver Drowsiness by Simultaneous Analysis of ECG and Respiration Signals Using Deep Neural Networks. *Int. J. Environ. Res. Public Health* **2022**, *19*, 10736. [CrossRef]
14. Arefnezhad, S.; Eichberger, A.; Frühwirth, M.; Kaufmann, C.; Moser, M.; Koglbauer, I.V. Driver Monitoring of Automated Vehicles by Classification of Driver Drowsiness Using a Deep Convolutional Neural Network Trained by Scalograms of ECG Signals. *Energies* **2022**, *15*, 480. [CrossRef]
15. Huang, J.; Luo, X.; Peng, X. A Novel Classification Method for a Driver's Cognitive Stress Level by Transferring Interbeat Intervals of the ECG Signal to Pictures. *Sensors* **2020**, *20*, 1340. [CrossRef]
16. Cardone, D.; Perpetuini, D.; Filippini, C.; Spadolini, E.; Mancini, L.; Chiarelli, A.M.; Merla, A. Driver Stress State Evaluation by Means of Thermal Imaging: A Supervised Machine Learning Approach Based on ECG Signal. *Appl. Sci.* **2020**, *10*, 5673. [CrossRef]
17. Awais, M.; Badruddin, N.; Drieberg, M. A Hybrid Approach to Detect Driver Drowsiness Utilizing Physiological Signals to Improve System Performance and Wearability. *Sensors* **2017**, *17*, 1991. [CrossRef] [PubMed]
18. Purnamasari, P.D.; Hazmi, A.Z. Heart beat based drowsiness detection system for driver. In Proceedings of the 2018 Interna-tional Seminar on Application for Technology of Information and Communication, Yogyakarta, Indonesia, 6–7 March 2018; pp. 585–590.
19. Schmidt, J.; Laarousi, R.; Stolzmann, W.; Karrer-Gauß, K. Eye blink detection for different driver states in conditionally auto-mated driving and manual driving using EOG and a driver camera. *Behav. Res. Methods* **2017**, *50*, 1088–1101. [CrossRef] [PubMed]
20. Vesselenyi, T.; Rus, A.; Mitran, T.; Moca, S.; Lehel, C. Fuzzy Decision Algorithm for Driver Drowsiness Detection. In Proceedings of the 30th SIAR International Congress of Automotive and Transport Engineering: Science and Management of Auto-Motive and Transportation Engineering, SMAT 2019, Craiova, Romania, 23 October 2019; Publisher Springer Link: Berlin/Heidelberg, Germany, 2020; pp. 458–467.
21. Sullivan, T.J.; Deiss, S.R.; Jung, T.P.; Cauwenberghs, G. A brain-machine interface using dry-contact, low-noise EEG sensors. In Proceedings of the 2008 IEEE International Symposium on Circuits and Systems, Seattle, WA, USA, 18–21 May 2008; pp. 1986–1989.
22. Daly, J.J.; Wolpaw, J.R. Brain–computer interfaces in neurological rehabilitation. *Lancet Neurol.* **2008**, *7*, 1032–1043. [CrossRef]
23. Ting, J.A.; D'Souza, A.; Yamamoto, K.; Yoshioka, T.; Hoffman, D.; Kakeif, S.; Sergio, L.; Kalaska, J.; Kawato, M.; Strick, P.; et al. Variational Bayesian least squares: An application to brain—Machine interface data. *Neural Netw.* **2008**, *21*, 1112–1131. [CrossRef]
24. Cvetkovic, D.; Übeyli, E.D.; Cosic, I. Wavelet transform feature extraction from human PPG, ECG, and EEG signal responses to ELF PEMF exposures: A pilot study. *Digit. Signal Process.* **2008**, *18*, 861–874. [CrossRef]
25. Ting, W.; Guo-Zheng, Y.; Bang-Hua, Y.; Hong, S. EEG feature extraction based on wavelet packet decomposition for brain computer interface. *Measurement* **2008**, *41*, 618–625. [CrossRef]
26. MATLAB—Mathworks, Users Manual, Neural Network Toolbox, Deep Learning, Autoencoders. 2016. Available online: https://www.mathworks.com/help/ (accessed on 12 February 2024).
27. Hinton, G.E.; Osindero, S.; Teh, Y.W. A fast learning algorithm for deep belief nets. *Proc. Neural Comput.* **2006**, *18*, 1527–1554. [CrossRef] [PubMed]
28. Camden, M.C.; Hickman, J.S.; Soccolich, S.A.; Hanowski, R.J. Identifying Equipment Factors Associated with Snowplow Operator Fatigue. *Safety* **2019**, *5*, 62. [CrossRef]
29. Bengio, Y.; Lamblin, P.; Popovici, D.; Larochelle, H. Greedy layer-wise training of deep networks. *Proc. Adv. Neural Inf. Process. Syst.* **2006**, *19*. [CrossRef]
30. Rosebrock, A. Drowsiness Detection with OpenCV. 8 May 2017. Available online: https://www.pyimagesearch.com/2017/05/08/drowsiness-detection-opencv (accessed on 10 December 2023).
31. Kazemi, V.; Sullivan, J. One millisecond face alignment with an ensemble of regression trees. In Proceedings of the 2014 IEEE Conference on Computer Vision and Pattern Recognition, Columbus, OH, USA, 23–28 June 2014; pp. 1867–1874.
32. Soukupova, T.; Cech, J. Eye blink detection using facial landmarks. In Proceedings of the 21st Computer Vision Winter Work-shop, Rimske Toplice, Slovenia, 3–5 February 2016.
33. Viola, P.; Jones, M. Rapid object detection using a boosted cascade of simple features. In Proceedings of the 2001 IEEE Computer Society Conference on Computer Vision and Pattern Recognition (CVPR), Kauai, HI, USA, 8–14 December 2001; Volume 1, pp. 1–9.

34. Jianbo, S.; Tomasi, C. Good features to track. In Proceedings of the Computer Society Conference on Computer Vision and Pattern Recognition, Seattle, WA, USA, 21–23 June 1994; pp. 593–600.
35. Tomasi, C.; Kanade, T. Detection and tracking of point. *Int. J. Comput. Vis.* **1991**, *9*, 137–154. [CrossRef]
36. Kalal, Z.; Mikolajczyk, K.; Matas, J. Forward-backward error: Automatic detection of tracking failures. In Proceedings of the 20th International Conference on Pattern Recognition, Istambul, Turkey, 23–26 August 2010; pp. 2756–2759.
37. Mansour, H.B.; Slimane, A.B.; Fournier, R.; Nait-Ali, A. 3D geometric human face modeling. In Proceedings of the International Conference on Human-Machine Interaction, Chennai, India, 27–30 August 2013; pp. 183–187.
38. Mohamed, G.M.; Patel, S.S.; Naicker, N. Data Augmentation for Deep Learning Algorithms that Perform Driver Drowsiness Detection. *Int. J. Adv. Comput. Sci. Appl.* **2023**, *14*, 233–248. [CrossRef]
39. Vesselenyi, T.; Rus, A.; Mitran, T.A.; Tătaru, M.B.; Moca, S. Monitoring and alert system: Neuro-fuzzy decision algorithm based on biological signals acquired from vehicle drivers. *Rom. J. Automot. Eng.* **2019**, *25*, 77–82.
40. Albadawi, Y.; Takruri, M.; Awad, M. A review of recent developments in driver drowsiness detection systems. *Sensors* **2022**, *22*, 2069. [CrossRef]
41. Doudou, M.; Bouabdallah, A.; Berge-Cherfaoui, V. Driver drowsiness measurement technologies: Current research, market solutions, and challenges. *Int. J. Intell. Transp. Syst. Res.* **2020**, *18*, 297–319. [CrossRef]
42. Stancin, I.; Cifrek, M.; Jovic, A. A review of EEG signal features and their application in driver drowsiness detection systems. *Sensors* **2021**, *21*, 3786. [CrossRef] [PubMed]
43. Denton, T. *Automated Driving and Driver Assistance Systems*, 1st ed.; Routledge: Abingdon-on-Thames, UK, 2019.

44. Maurer, M.; Gerdes, J.C.; Lenz, B.; Winner, H. *Autonomous Driving: Technical, Legal and Social Aspects*; Springer Open: Berlin/Heidelberg, Germany, 2016.
45. Regulation (EU) 2019/2144 of the European Parliament and of the Council. Explanatory Memorandum. Available online: https://eur-lex.europa.eu/legal-content/EN/TXT/HTML/?uri=PI_COM:C(2021)2639&from=EN (accessed on 14 December 2023).

Disclaimer/Publisher's Note: The statements, opinions and data contained in all publications are solely those of the individual author(s) and contributor(s) and not of MDPI and/or the editor(s). MDPI and/or the editor(s) disclaim responsibility for any injury to people or property resulting from any ideas, methods, instructions or products referred to in the content.

sensors

MDPI

Article

Point Cloud Painting for 3D Object Detection with Camera and Automotive 3+1D RADAR Fusion

Santiago Montiel-Marín *, Ángel Llamazares, Miguel Antunes, Pedro A. Revenga and Luis M. Bergasa

Department of Electronics, Universidad de Alcalá, 28805 Alcalá de Henares, Spain;
angel.llamazares@uah.es (Á.L.); miguel.antunes@uah.es (M.A.); pedro.revenga@uah.es (P.A.R.);
luism.bergasa@uah.es (L.M.B.)
* Correspondence: santiago.montiel@uah.es

Abstract: RADARs and cameras have been present in automotives since the advent of ADAS, as they possess complementary strengths and weaknesses but have been underlooked in the context of learning-based methods. In this work, we propose a method to perform object detection in autonomous driving based on a geometrical and sequential sensor fusion of 3+1D RADAR and semantics extracted from camera data through point cloud painting from the perspective view. To achieve this objective, we adapt PointPainting from the LiDAR and camera domains to the sensors mentioned above. We first apply YOLOv8-seg to obtain instance segmentation masks and project their results to the point cloud. As a refinement stage, we design a set of heuristic rules to minimize the propagation of errors from the segmentation to the detection stage. Our pipeline concludes by applying PointPillars as an object detection network to the painted RADAR point cloud. We validate our approach in the novel View of Delft dataset, which includes 3+1D RADAR data sequences in urban environments. Experimental results show that this fusion is also suitable for RADAR and cameras as we obtain a significant improvement over the RADAR-only baseline, increasing mAP from 41.18 to 52.67 (+27.9%).

Keywords: autonomous driving; sensor fusion; RADAR; camera; point cloud painting; object detection

1. Introduction

The progress of autonomous driving (AD) technologies involves the development of perception techniques that enhance the scene-understanding capabilities of the vehicles. The perception stage of an autonomous driving stack (ADS) is in charge of interpreting the information from the sensors to understand the environment in which the vehicle is moving. This is accomplished in three steps: obtaining the information from onboard sensors, processing it, and interpreting the environment around the autonomous vehicle (AV). The information gathered by the sensors is presented in the form of point clouds, images or spectral signals.

The processing stage transforms these raw data into more refined information, such as the location, dimensions, and speed of the rest of the road users (pedestrians, cyclists, and other vehicles) or a semantic definition of the world in which every piece of information is assigned to an inanimated category (drivable road, vegetation or road obstacles). The literature reflects that learning-based algorithms have been predominant in this field since the advent of Deep Learning (DL). Once this elaboration is finished, it is possible to interpret the scene by adding a temporal and spatial coherence between the processed information in a timestamp and the processed information of past timestamps. This temporal–spatial association of information allows the system to predict which possible scenarios are likely to happen within the successive timestamps. The final output of the perception processing pipeline is used as the input of the upstream or downstream stages of the ADS.

Since the advent of Advanced Driver Assistance Systems (ADASs), conventional 2+1D RADAR and camera sensors have played a major role in the automotive industry as integral parts of these systems, such as Adaptive Cruise Control (ACC), Lane Departure Warning (LDW) or Assistance Emergency Braking (AEB). The strengths and weaknesses of both RADAR and camera sensors are complementary; thus, the mass implementation of ADAS systems has been fulfilled with this combination of sensors. Whilst RADAR provides sparse range, angle and velocity measurements, cameras provide pixel-level density representations with richer semantic features.

Nevertheless, in the transition from ADAS to AD technologies, academia and industry have shifted their main interests to perception systems based on LiDARs and cameras. LiDAR has replaced conventional 2+1D RADAR due to its better spatial and angular resolution and height measurement capabilities. In this situation, a new generation of high-resolution RADARs (3+1D) has emerged. Providing richer point cloud representations than their predecessors, and with a higher density of range, angle, height and velocity measurements, these new RADARs show potential for being reintroduced into the perception sensor suite in the context of AD applications. A visual comparison of the density of the point clouds obtained by a LiDAR and a 3+1D RADAR can be seen in Figure 1. Furthermore, the 3+1D RADAR provides 1D Doppler velocity measurements for each reflection, which are projected over the radial component of the detections. Doppler velocity be used to enhance downstream tasks such as object detection, linear velocity estimation and motion tracking. In contrast, LiDAR-based systems require the inference of any velocity component by analyzing the evolution over time of the objects of interest. The inclusion of radial velocity components enables the point cloud to be categorized into two distinct subsets: static and dynamic objects. The set of static objects can be leveraged to obtain 3D grid maps through the accumulation of multiple RADAR frames over time.

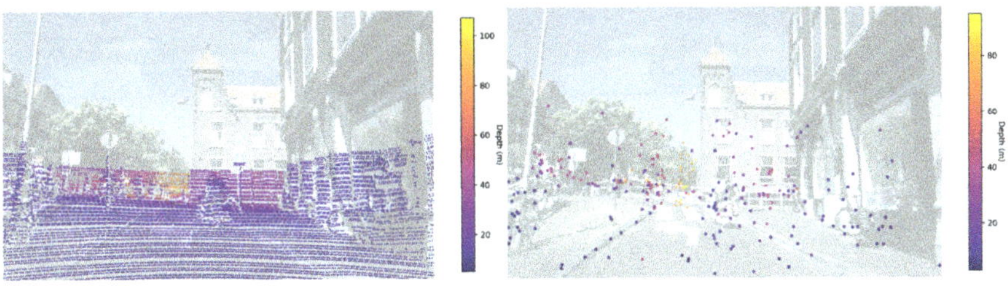

(a) (b)

Figure 1. Comparison of a LiDAR point cloud (**a**) and 3+1D RADAR point cloud (**b**) projected to the image plane for an automotive scene of View of Delft dataset.

In this work, we aim to explore a geometrical data fusion approach to detect and classify objects as 3D bounding boxes in the environment surrounding the vehicle. Overall, the main contribution of this study is the implementation and evaluation of a multi-stage DL system for 3D object detection in automotive applications. The system utilizes data from 3+1D RADAR point clouds and monocular camera images, aligning with contemporary trends in sensor fusion in the perspective view (PV) with semantics. The primary goal is to achieve object detection and classification around the vehicle, combining the strengths of automotive RADAR and monocular cameras in a geometrical–sequential sensor fusion scheme. We summarize the rest of the contributions of our work as follows:

- We reformulate **PointPainting** [1] and adapt it to the domain of RADAR point clouds and camera images using **YOLOv8** [2] as the instance segmentation network and **PointPillars** [3] as the detection network, respectively.
- We design a set of rules based on heuristics to deal with the main disadvantage of the previous method, the *smearing* effect, to mitigate the propagation of errors from the segmentation stage to the detection stage.

- We evaluate our system in the novel **View of Delft** (VOD) [4] dataset, which includes 3+1D RADAR point clouds and a monocular camera image to perform sensor fusion for object detection, and we obtain significant improvements over the RADAR-only object detection method.

2. Related Works

Sensor fusion aims to fuse the data of different modalities or sensors, taking advantage of the synergies that can result from the combination of them to fulfill a specific task. In the context of AD, sensor fusion improves the robustness and reduces the uncertainty of perception systems. In this Section, we define how sensor fusion can be performed by answering four main questions: **what to fuse**, **where to fuse**, **when to fuse** and **how to fuse** [5]. Then, a set of selected methods based on camera and automotive RADAR is revised.

What to Fuse? In determining the fusion targets, RADAR presents a diverse array of modalities. These modalities span from the raw Analog-to-Digital Converter (ADC) data to point cloud representations, encompassing spectrograms and RADAR cubes. Each modality bears distinct advantages and limitations. Conversely, in the domain of cameras, representation diversity is diminished. Common modalities include RGB or gray-scale images for visible light cameras alongside infrared or thermal representations.

Where to Fuse? Spatial coherency must exist when fusing data from various sources. Data must be referenced to the same coordinate frame. To tackle this, there are two main options in RADAR–camera fusion: the perspective projection of the image plane or the bird's eye view from space. Fusing in the image plane involves the projection of the RADAR tensors or point clouds via calibration matrices, forming sparse RADAR images. Fusing in the BEV plane involves the lifting of camera-based features inside a neural network since geometric projections under-perform in this task.

How to Fuse? Fusing two sensor modalities involves keeping spatial and temporal coherence or alignment. Temporal misalignment can occur when the latencies are not controlled or compensated. Spatial misalignment occurs when calibration methods are not accurate. Fulfilling these conditions, fusion is accomplished by means of fusion operators. These include transformation matrices for data-level fusions as well as additions, multiplication, concatenation and attention mechanisms for feature-level fusions.

When to Fuse? Depending on the stage of the perception pipeline, we can classify fusion schemes in these levels: object-level fusion, data-level fusion, feature-level fusion and any combination of them (hybrid-level fusion). Object-level fusion involves two separate independent methods that return their respective detections and a later fusion mechanism. In this case, Bayesian theory, Kalman filters and matching algorithms are used to associate detections spatially. The data-level fusion involves the creation of a new representation of the environment before any detection network or method, i.e., projecting the RADAR features to the image plane and concatenating those channels with the original RGB image, resulting in a different tensor that will feed a Neural Network (NN). The feature-level fusion involves the combination of RADAR feature maps and camera feature maps inside an NN applying any fusion operator (keeping spatial coherence). Then, hybrid approaches can make combinations of any of the previous ones applying different fusion operators.

Methods. Having reviewed the main components of sensor fusion, it is possible to review the recent literature in terms of RADAR–camera fusion for perception.

nuScenes [6] has become the dataset of reference for AV applications and has been the first big-scale benchmark to introduce conventional (2D+1) RADAR data. One of the first methods ranked in nuScenes for the camera and RADAR Detection Track was **CenterFusion** [7]. Nabati and Qi propose a feature-level fusion scheme with two main branches. The imaging branch uses **CenterNet** [8] as the backbone. The RADAR branch opts for a pillar rendering mechanism, which is similar to KPConv. The head is split into two parts: primary and secondary. The primary head obtains preliminary 3D bounding boxes. Pillar features and preliminary boxes are associated through a frustum-based mechanism, and a RADAR feature map is generated. Image and RADAR feature maps are concatenated and are the input to the secondary head, which refines the preliminary proposals. They performed a performance comparison in which the camera-only KPConv scored an NDS = 0.328 and the fusion of both sensors scored an NDS = 0.453, which is an improvement of 0.125 due to the fact including RADAR data information.

Nobis et al. [9] proposed a method to fuse RADAR and camera information in the image plane, extending the image with RADAR channels that include range, velocity and RCS information. The method is built based on **RetinaNet** [10] with two branches. The camera image follows a VGG [11] backbone, and the RADAR branch applies max pooling operations to preserve the maximum information possible when the dimensionality is reduced as it is a sparse image. Then, feature maps from RADAR and image branches are concatenated at multiple scales. An FPN and a detection head end the architecture. The method introduced a novel augmentation method called *BlackIn*. *BlackIn* shut down the neurons at the input of the camera image, so the network must rely on the RADAR channels to obtain the information.

A remarkable network that opted for a data-level approach is **RadSegNet** [12]. The method uses a 2D detector for oriented bounding boxes. In this case, a BEV grid map is generated from the RADAR point cloud. The grid is augmented with semantic information coming from the camera. A 2D semantic segmentation network is applied to the image. Then, a semantics-to-RADAR module projects the information from the image plane to the BEV grid map. The final grid map is composed of multiple channels: the grid map, point-based feature maps and semantic maps, leading to a 22-dimensional tensor. The architecture for semantic segmentation is a DeepLabV3+ with a ResNet-101 backbone trained in CityScapes [13] due to the absence of semantic labels in RADAR–camera-based datasets. The architecture for object detection is based on RADAR, treating detection as a per-pixel task, which was trained on the Astyx dataset [14].

Revisiting the existing literature and the methods that have been previously explained, it is possible to conclude that the future direction of sensor fusion is shifting from object or data-level fusion schemes to relying on BEV fusion mechanisms for perception [15]. Combining both modalities in the bird's eye view space leads to the design of novel mechanisms to either lift camera features to 3D space solely [16] or assisted by RADAR [17] via attention mechanisms.

Table 1 summarizes the methods revisited in this section. All methods are based on conventional 2+1D RADAR, and neither of them fuses 3+1D RADAR and image semantics in the image plane as our proposal.

Table 1. Summary of revisited state-of-the-art methods.

Method	Fusion Type	Description	Dataset	Year
Nobis et al. [9]	Data	Fusion in the image plane and extension in height of RADAR data.	nuScenes [6]	2019
CenterFusion [7]	Feature	Dual-branch backbone and feature fusion for detection.	nuScenes [6]	2021
RadSegNet [12]	Data	Fusion-then-detection in BEV. RADAR point cloud and image semantics projected.	Astyx [14]	2022

3. Method

The problem dealt with is 3D object detection, which is typically formulated in the literature as the regression and classification of 3D-oriented bounding boxes that enclose objects of interest for the autonomous navigation task. To address this task, we propose a pipeline based on [1] to perform 3D object detection based on painted automotive RADAR point clouds via segmented camera images. The method is composed of four stages:

instance segmentation, point cloud painting, rule-based cluster refinement and 3D object detection. A diagram representing the whole pipeline is shown in Figure 2.

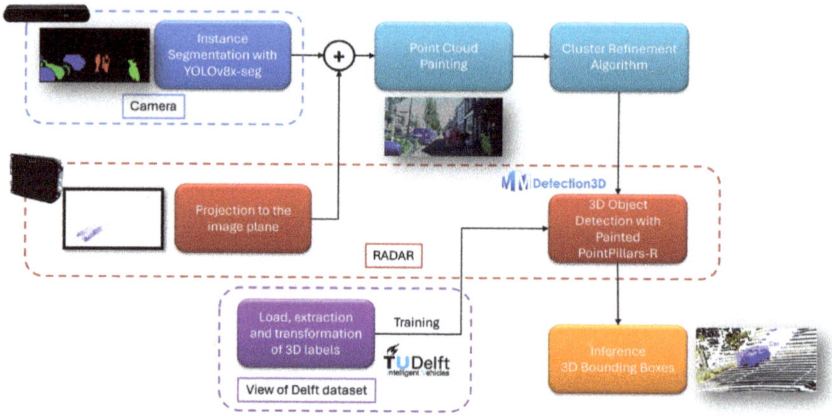

Figure 2. Overview of the complete pipeline.

3.1. *Two-Dimensional (2D) Instance Segmentation*

The first step of the framework consists of the application of a 2D instance segmentation network to the camera image to obtain the semantic category of the objects of interest that are present in the scene. We employ YOLOv8x-seg as an instance segmentation neural network to mitigate the performance drop that occurs due to the absence of specific semantic labels and a domain shift in the dataset studied, as this network has been extensively pre-trained on the COCO dataset [18].

YOLOv8 formulates the instance segmentation task as a "segmentation-by-detection" problem. It first detects an object of interest and then adjusts a binary mask that indicates which pixels in the image belong to that corresponding instance. The structure of the network is as follows: a CSP-Darknet53 backbone [19], a PAFPN-based neck, a decoupled head for detection and Proto heads for upscaling the images to their original size and extracting the segmentation mask per instance. Therefore, the output of this network is a set of instances where each instance is characterized by a bounding box, a semantic category, an objectness score that ranks from 0 to 1, which are output by the decoupled detection head and a binary segmentation mask output by the Proto segmentation heads. The main innovation in this stage is the use of instance segmentation networks, which focus on VRUs, instead of semantic segmentation networks as were in PointPainting [1] in which also background elements are segmented.

We run inference for each image, and after that, we apply the following processing. We filter the masks that belong to the classes: vehicle, person and bicycle. For each filtered instance, we multiply its segmentation mask by its objectness score. We sum all the masks, encoding each one of the semantic categories in an image channel. Finally, we saturate the channels to 1 to cover the edge case in which various masks within the same semantic categories overlap in the same area of the image. We call this representation the instance map, and an example can be seen in Figure 3. In this figure, an urban scenario is represented. In the left part of the image, multiple bicycles (green) and cars (blue) are segmented correctly. It can be noticed that the bluer tone is decreasing to distance, meaning that the confidence in the segmentation network is decreasing following the same pattern. In the center of the road, multiple pedestrians are correctly segmented (red). As previously stated, it can be seen that cyclists are segmented as a person (red) and bicycle (green) independently.

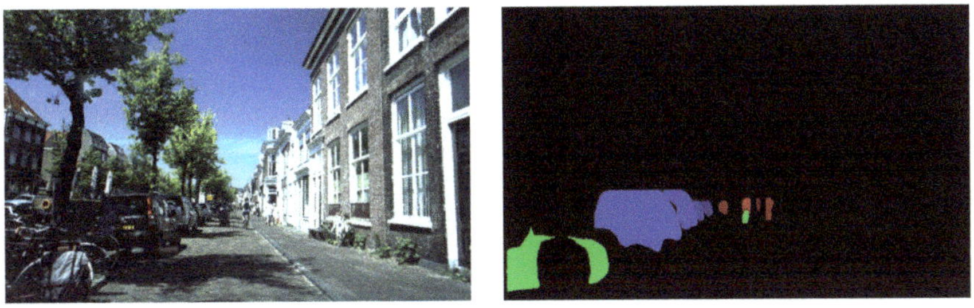

(a) (b)

Figure 3. Monocular camera image (**a**) and instance map highlighting pedestrians, bicycles and cars (**b**) for frame 03456 in the View of Delft dataset.

3.2. Point Cloud Painting

To perform sensor fusion between camera and RADAR data coherently, both sources of data must be spatially aligned. Painting a point cloud implies fusing features that are located in pixels of an image plane into their corresponding points in 3D space. Therefore, we follow [1] to perform data fusion. Nevertheless, we apply painting with both RGB images and instance maps.

We apply a homogeneous transformation from the RADAR coordinate frame to the image coordinate frame and project all RADAR points into the image plane via Equation (1), where P_{XYZ1} is an $\mathcal{R}^{4\times N}$ matrix of N homogeneous points in the RADAR coordinate frame, $T_{radar\,to\,cam}$ is a homogeneous transform from RADAR to the camera ($\mathcal{R}^{4\times 4}$), $R_{0,rect}$ is a rectification matrix ($\mathcal{R}^{4\times 4}$), P_2 is a pseudo-intrinsic parameters matrix for stereo systems ($\mathcal{R}^{3\times 4}$) and P_{UVW} is the output result, a matrix.

$$P_{UVW} = P_2 \cdot R_{0,rect} \cdot T_{radar\,to\,cam} \cdot P_{XYZ1} \quad (1)$$

Once the points are located in the image plane, we can establish a point–pixel correspondence for all the elements of the point cloud. After this process, we obtain a painted point cloud whose dimensionality has been increased, $\mathcal{R}^{N\times C+3+3}$, where C is the original number of features in the RADAR point cloud, 3 additional channels encode the RGB color of the image, and the last three channels encode the semantic category assigned to that point at the instance segmentation stage. The inclusion of color RGB channels in the painting mechanism is the main innovation of this stage.

3.3. Cluster Refinement

The main weakness of the painting method is the *smearing* effect. It is an effect that occurs when the segmentation of an object of interest is imperfect and produces a trace of colored or segmented points in the angle bin in which this object is projected. Objects of interest affected by this effect will show a different spatial distribution than objects that are correctly segmented. This potentially leads to the propagation of errors from the segmentation stage of the pipeline to the detection stage.

We identify four potential causes that could lead to the appearance of *smearing* phenomena: the quality imperfection of the segmentation masks, the calibration and temporal misalignment between the point cloud and the image, the absence of motion compensation and the propagation scheme of RADAR reflections, since multiple reflections can appear in the same angle bin.

As the instance segmentation networks allow us to propagate the instance concept, it is possible to discern which RADAR points belong to which segmented instance. We design a set of heuristic rules to identify objects affected by smearing and mitigate the propagation of erroneous paintings into the detection stage. We approach this problem by analyzing the radial dispersion of the detected objects, $\Delta r = max(r) - min(r)$. We obtain the radial distance to all points in an instance of interest and determine their difference. If the dispersion is greater than a certain threshold, the object is considered to be affected by *smearing*. To set these thresholds, we double the anchors set by the object detection network that will be used in the detection stage for the semantic categories we are coping with. The reason behind doubling the anchors and taking them as thresholds is to detect the most affected cases in which *smearing* will contaminate the data fusion process. The doubled anchors chosen are 7.8 m for the *Car* category, 3.52 m for the *Cyclist* category and 1.6 m for the *Pedestrian* category.

Once an object is considered to be affected by smearing, we determine if the object is dynamic or static by analyzing the presence of points with a component of compensated radial velocity in absolute value, $|V_{r\,comp}| \geq 0.3$ m/s. If an object is considered to be dynamic, we apply DBSCAN [20] clustering over the radial velocity features and we extract the cluster with more detections and mean velocity different from zero. If an object is considered static, we apply DBSCAN clustering over the spatial dimensions, and we retain the closest cluster in range, i.e., the first one encountered in the line of sight. An example of the application of this cluster refinement can be seen in Figure 4. In this qualitative scenario, the smearing effect is present as a large set of reflections that range from 10 to 30 m, which is produced due to the effect of RADAR scattering and noisy propagation inside a shop. As a pedestrian is standing in front of the shop door and multiple reflections are lying behind in the same radial direction, various points are incorrectly assigned to this instance when the painting procedure is performed. Applying the cluster refinement algorithm, we can identify the blob of points that belong to the object and filter noise that is not propagated into the following stage of the system.

3.4. Three-Dimensional (3D) Object Detection

The final step of the pipeline consists of the application of a 3D object detection network over the painted point cloud, by using a version of PointPillars [3] which is adapted for RADAR data.

The architecture of PointPillars can be seen in Figure 5 and is structured as follows: a data preprocessing module, which is responsible for voxelizing point clouds with a random sampling strategy to augment the training data; a Pillar Feature Net extracts features per pillar using the arithmetic mean of points and their offset from the center; a Middle Encoder converts learned features to a sparse pseudo-image, facilitating subsequent processing; then, the architecture comprises a backbone, neck, and head, following the typical scheme of a 2D object detection network. The backbone is a convolutional network, SECOND [21], producing features with decreasing resolutions. The neck performs upsampling and concatenates features from different resolutions. The head incorporates prior information about bounding box sizes and generates anchors for pedestrians, cyclists, and cars. Object detection is treated as a matching problem using Intersection over Union (IoU), with additional regression tasks for height and elevation, forming a complete 3D object detection network.

The main innovation in this stage is the adaptation of PointPillars to fit RADAR data. For this purpose, we perform the following modifications to the previously explained scheme:

- We adapt the point cloud range to the dataset we are coping with, which is explained in Section 4.1. We modify as a consequence the anchor generation range.
- We increase the voxel size. Since the RADAR sensors are placed in the front bumper of the vehicle, we explore the height range of $[-2, 3]$ m.
- We reduce the maximum number of points per voxel. Since RADAR point clouds are sparser than LiDAR ones, we decrease this threshold from 32 to 10 points.
- We delete data augmentation that deforms or rotates point clouds, since the RADAR information would become physically incoherent.
- We modify all the components that load point clouds and adapt them to the new dimensionality from LiDAR ($\mathcal{R}^{N\times4}$) to RADAR ($\mathcal{R}^{N\times7}$) or Painted RADAR ($\mathcal{R}^{N\times13}$).

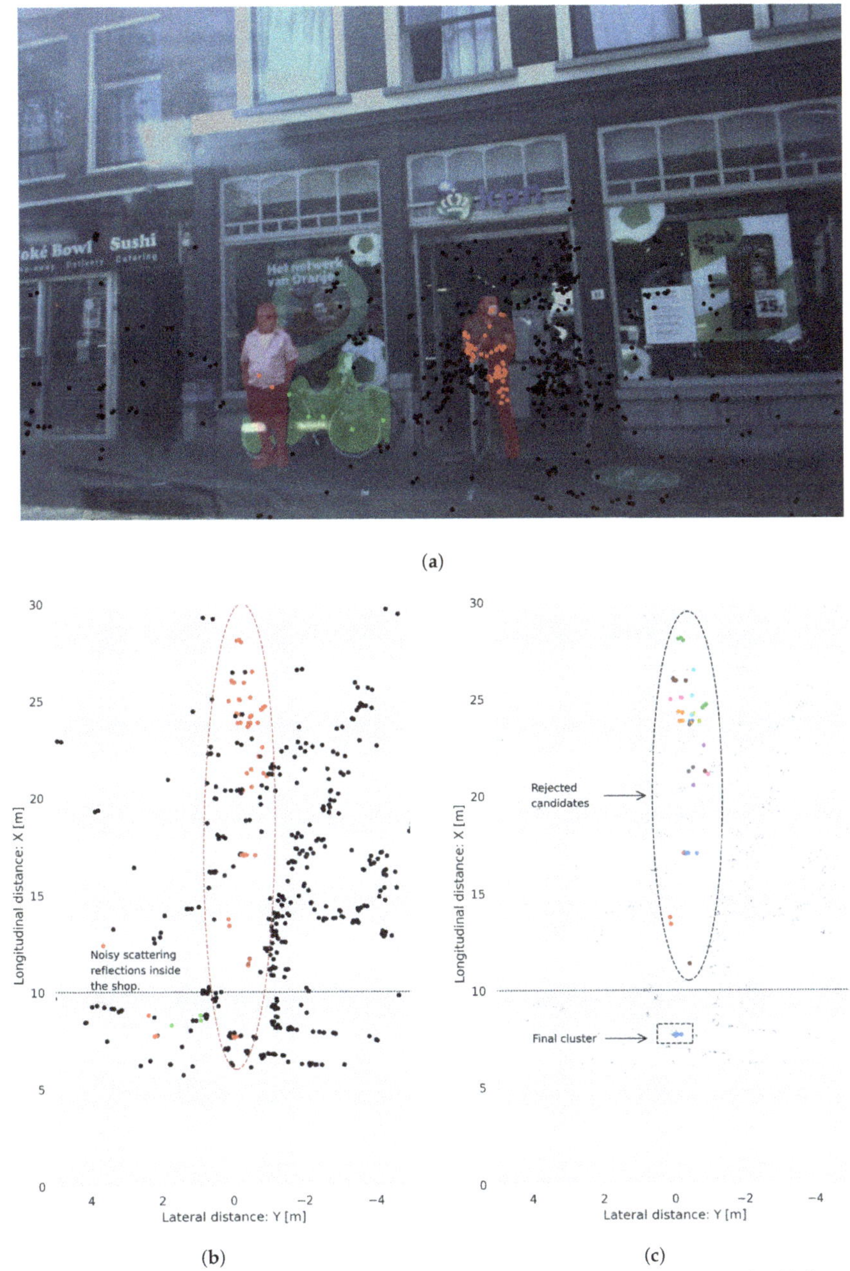

Figure 4. Example of cluster refinement algorithm applied in frame 00125 of the View of Delft dataset. A large set of reflections produced by scattering is generated inside the shop and projected into the pedestrian standing at the door in image (**a**), producing a large smearing that can be seen in (**b**). The cluster refinement algorithm detects the radial dispersion and corrects the probabilities assigned to the points in the point cloud (**c**).

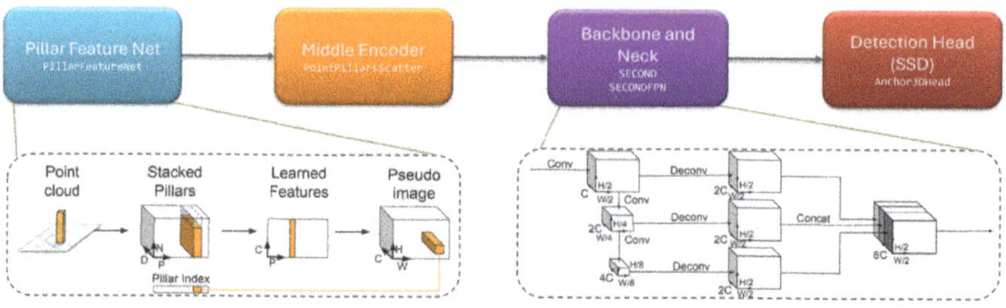

Figure 5. Overview of the PointPillars architecture. Adapted from [3].

With these modifications taken into consideration, PointPillars can perform inference and detect 3D objects in modified and painted RADAR point clouds.

4. Experiments and Results

4.1. Dataset

The experiments are carried out on the **View of Delft** (VoD) dataset. It was released in Q2 2022 and includes a collection of urban sequences recorded in Delft, The Netherlands. These sequences lead to 8682 individual frames with an official training–validation–testing split of 5139, 1296, and 2247, respectively. Each frame contains readings from multiple sensor modalities, of which LiDAR, RADAR, camera and their respective calibration and transform matrices are of particular interest for the validation of this work. Moreover, annotations for 2D and 3D object detection tasks are provided.

Delving into the sensor readings, LiDAR point clouds are captured at 10 Hz by a Velodyne HDL-64 S3. Every LiDAR point cloud is a vector $L_{PC} \in \mathcal{R}^{N\times 4}$, where N denotes the number of points in a point cloud and every point is a tuple $P_L = (x,\ y,\ z,\ i)$ that represents its spatial location $(x,\ y,\ z)$ and its reflection intensity (i). RADAR is captured by a ZF FRGen21 3+1D sensor at 13 Hz. Analogously, a RADAR point cloud is a vector $R_{PC} \in \mathcal{R}^{N\times 7}$ with N points and seven features per point. Then, $P_R = \left(x,\ y,\ z,\ vr,\ vr_{comp},\ \sigma,\ t\right)$, where $(x,\ y,\ z)$ is the spatial location of the reflection, $(vr,\ vr_{comp})$ is the radial velocity without and with ego-motion compensation, respectively, (σ) is the RADAR cross-section and (t) is a time component that allows the accumulation of multiple timestamps in a single point cloud. Lastly, a stereo vision system provides a single color image in the form $I \in \mathcal{Z}^{3\times H\times W}$, where the height and width are $H = 1216$ and $W = 1936$, respectively.

The annotations are provided in a KITTI-like style and are provided for each LiDAR scan in the dataset and cover an area of 50×50 m on both lateral and longitudinal axes. A set of 15 semantic classes is contained within the dataset with a focus on Vulnerable Road Users (VRUs), such as *pedestrians* and *cyclists*. For evaluation purposes, we map the original classes to *Pedestrian*, *Car* and *Cyclist* to follow the KITTI evaluation protocol.

4.2. Implementation Details

The components of the detection pipeline are trained and/or executed on an Intel i9-12700 CPU and an NVIDIA RTX4090 24 GB GPU; therefore, the training and inference times are reported concerning these hardware specifications. The implementation of the instance segmentation network is based on the `ultralytics` (Ultralytics on GitHub: https://github.com/ultralytics/ultralytics, accessed on date: 27 December 2023) package and the 3D object detection is implemented modifying PointPillars for LiDAR at `mmdetection3d` (mmdetection3d on GitHub: https://github.com/open-mmlab/mmdetection3d, accessed on date: 27 December 2023).

4.3. Performance on 3D Object Detection

We evaluate the performance of the pipeline on the 3D object detection task. For this procedure, we train **PointPillars** models following a KITTI standard training procedure: 80 epochs, where the learning rate is configured using around an initial value of $\eta = 1 \times 10^{-3}$

and is modified every epoch using a learning rate scheduler: **Cosine Annealing** with two stages. The first stage covers from the beginning of the training to epoch 32 and increases from 10^{-2} to 10^{-3} in a linear trend. The second stage covers from epoch 32 to the end of the training in epoch 80 and decreases η from 10^{-3} to 10^{-7} following Equation (2). The optimizer is AdamW, which is an adaptive momentum optimizer with weight decay.

$$\eta_t = \eta_{min} + \frac{1}{2}(\eta_{max} - \eta_{min})\left(1 + \cos\left(\frac{T_{cur}}{T_{max}}\pi\right)\right) \tag{2}$$

As for the loss function, we employ a weighted sum of three different components: focal loss [10], smoothed L1 [22] and cross-entropy, as Equation (3) states.

$$\mathcal{L} = \alpha_1 \cdot \mathcal{L}_{FL} + \alpha_2 \cdot \mathcal{L}_{L1_{smooth}} + \alpha_3 \cdot \mathcal{L}_{CE} \tag{3}$$

Focal loss is in charge of bounding box classification. It aims to address class imbalance during training, adding a modulating term to cross-entropy loss to focus on hard misclassified examples, following Equation (4). For our experiments, $\alpha_1 = 1.0$ and $\gamma = 2.0$.

$$\mathcal{L}_{FL} = -(1 - p_t)^{\gamma} \cdot \log p_t \tag{4}$$

Bounding box regression is dealt with a Smooth L1 term that uses a square penalty term term if an error falls below beta and an L1 otherwise. It prevents exploding gradients and follows (5). For our experiments, we set $\alpha_2 = 2.0$ and $\beta = 1.0/9.0$.

$$\mathcal{L}_{L1_{smooth}} = \begin{cases} 0.5 \cdot (x_n - y_n)^2/\beta & \text{if } |x_n - y_n| < \beta \\ |x_n - y_n| - 0.5 \cdot \beta & \text{otherwise} \end{cases} \tag{5}$$

Finally, bounding box orientation is estimated formulating the problem as a classification, where the goal is to categorize each box into predefined bins. For our experiments, we set $\alpha_3 = 0.2$.

We use Mean Average Precision (mAP) and Average Orientation Similarity (AOS) as metrics, which are defined as follows:

- Mean Average Precision (mAP) assesses the performance on axis-aligned 3D bounding box detection and weights the Average Precision (AP) results across all n categories. It is defined in Equation (6).

$$mAP = \frac{1}{n}\sum_{k=1}^{k=n} AP_k \tag{6}$$

  Average Precision for a single category is defined at Equation (7). It is a discrete evaluation over the recall curve in which detections are iteratively assigned to a ground truth object and assigned as true positive (TP) if their Intersection over Union (IoU) is more than a certain threshold. Multiple detections over a single ground truth object are considered false positives.

$$AP = \frac{1}{40}\sum_{k=0}^{k=40} Precision\left(\frac{k}{40}\right) \tag{7}$$

  Concerning the selection of IoU thresholds for every category, we follow the recent literature for RADAR 3+1D detection [4], which defines the minimum IoU thresholds necessary to consider an object as TP: IoU = 0.5 for *Car* category and IoU = 0.25 for *Pedestrian* and *Cyclist* categories.
- Average Orientation Similarity (AOS) assesses the performance of estimating the 3D orientation of the bounding box. Moreover, it is related to the recall at the object detection stage, from which multiple thresholds are set to perform a discrete evaluation over the recall curve, which is expressed as shown in Equation (8).

$$AOS = \frac{1}{40} \sum_{k=0}^{k=40} \max s\left(\frac{k}{40} > r\right) \tag{8}$$

It is derived from a normalized variant of cosine similarity, $s \epsilon (0,1)$, and computes the error from the difference between an estimated and ground-truth orientation for every 2D detection in the image plane, as Equation (9) states.

$$s(r) = \frac{1}{\mathcal{D}(r)} \cdot \sum_{i} \frac{1 + \cos \Delta_{\theta}^{(i)}}{2} \cdot \delta_i \tag{9}$$

where $\mathcal{D}(r)$ is the set of detections for a certain recall value r and $\Delta_{\theta}^{(i)}$ is the difference between angle θ estimation and ground truth for a matched detection i. Additionally, δ_i acts as a penalization mechanism: $\delta_i = 1$ if $\mathrm{IoU}_i \geq IoU_{min}$ or $\delta_i = 0$ otherwise.

The comparison will be performed between **PointPillars** for RADAR, which we will refer to as **PointPillars-R**, and the colorized version, which we will refer to as **Painted PointPillars-R**. The first one will serve as a baseline to compare with the latter, our approach. Moreover, we will experiment with the temporal aggregation of RADAR point clouds among frames, and we will train for each of the versions of the VoD dataset: single frame, 3 frames and 5 frames. As a remark, the metrics will be obtained over the validation set of VoD.

Table 2 shows the results of training the baseline **PointPillars-R** over the validation set of the VoD dataset. The table is split among three different criteria: metric, difficulty and categories. Among metrics, from left to right: mAP in the BEV plane, mAP in the 3D space, and AOS. Among difficulties: easy, medium and hard, depending on the size of the bounding box in the image plane and its occlusion level (moreover, in KITTI, the difficulty includes the level of truncation as an additional criterion, but this attribute is not labeled in the View of Delft dataset). Among categories, the three evaluated ones are *Car*, *Pedestrian* and *Cyclist*.

Table 2. Quantitative results of PointPillars-R with IoU thresholds (0.5, 0.25, 0.25) over the validation set of View of Delft dataset. Best results in **bold**.

	BEV			3D			AOS		
	Easy	**Med**	**Hard**	**Easy**	**Med**	**Hard**	**Easy**	**Med**	**Hard**
	AP40 Car (IoU @ 0.5)								
1 frame	29.63	**40.10**	**33.21**	24.20	30.42	**27.40**	20.14	**28.28**	**23.60**
3 frames	**32.78**	36.62	30.41	25.33	**30.52**	24.95	**21.52**	25.94	21.47
5 frames	31.69	32.31	26.44	25.85	28.77	21.13	21.06	21.02	17.07
	AP40 Pedestrian (IoU @ 0.25)								
1 frame	37.46	33.36	**30.61**	30.45	27.37	24.18	19.22	16.88	15.25
3 frames	**38.20**	**33.62**	29.85	**31.76**	**28.26**	24.88	**21.27**	**18.45**	**16.55**
5 frames	36.09	33.13	29.41	30.44	27.81	**24.97**	20.04	17.85	15.91
	AP40 Cyclist (IoU @ 0.25)								
1 frame	67.53	63.79	57.81	63.25	58.29	53.42	48.28	44.61	39.74
3 frames	71.53	66.19	59.94	66.88	61.50	55.40	51.71	46.71	42.26
5 frames	**76.41**	**70.12**	**63.60**	**73.15**	**66.97**	**60.31**	**58.22**	**53.51**	**47.56**

The performance results of the network indicate that RADAR-based methods have difficulties extracting bounding boxes for the *Car* and *Pedestrian* categories. Poor spatial resolution leads to scarce reflections in the targets of interest and difficulties in estimating the shape of the objects. The network shows successful performance in the *Cyclist* category, ranging from 63.25 3D AP40 in easy difficulty and single frame to 73.15 in easy difficulty and five accumulated frames. This is an atypical situation in automotive datasets, where VRUs are detected with difficulty. This may be due to the appearance of radial velocity and RCS as additional channels in the RADAR point clouds. After all, these features can be an excellent source of information to distinguish the few reflections that fall within cyclists from the static or different-in-terms-of-reflectivity background. Moreover, comparing single-frame and multi-frame versions, the increment between 1 and 3 frames is more notorious than the one between 3 and 5 scans for 3D AP40 across all categories. However, a rising trend among frame versions is perceived. Then, the network is taking advantage of the temporal aggregation of scans.

After this, we evaluate the performance of our proposal **Painted PointPillars-R** following the same schema as the baseline, whose results can be seen in Table 3. We train the single-frame and multi-frame versions following the same procedure. At a glance, it is observed that there is a clear improvement in both ways: comparing version to version with the baseline and comparing between the number of scans used for temporal aggregation. For example, in the case of 3D AP40 for *Cars*, the metrics increase for the baseline: 32.45 vs. 30.42 in medium difficulty; and they increase with respect to temporal aggregation within the proposed architecture: from 32.45 in single-frame mode to 39.64 in five-frames aggregation mode.

Table 3. Quantitative results of Painted PointPillars-R with IoU thresholds (0.5, 0.25, 0.25) over the validation set of View of Delft dataset. Best results in **bold**.

	BEV			3D			AOS		
	Easy	Med	Hard	Easy	Med	Hard	Easy	Med	Hard
	AP40 Car (IoU @ 0.5)								
1 frame	45.27	41.03	33.85	32.95	32.45	26.00	24.65	25.79	20.80
3 frames	51.41	39.59	32.87	43.05	33.56	**32.87**	33.55	26.23	**27.15**
5 frames	**62.59**	**51.95**	**43.16**	**47.21**	**39.64**	32.00	**33.83**	**30.20**	24.73
	AP40 Pedestrian (IoU @ 0.25)								
1 frame	39.62	34.90	31.29	32.37	28.64	25.31	20.38	18.35	16.17
3 frames	44.51	40.85	36.94	40.06	35.66	31.98	28.71	25.60	23.11
5 frames	**53.80**	**50.08**	**44.65**	**47.23**	**43.33**	**38.85**	**31.90**	**29.52**	**26.26**
	AP40 Cyclist (IoU @ 0.25)								
1 frame	75.76	71.01	64.25	68.27	62.28	55.76	46.75	41.94	36.95
3 frames	79.70	75.69	69.31	77.36	73.24	66.67	65.23	60.42	54.43
5 frames	**81.97**	**77.71**	**70.85**	**79.44**	**75.03**	**67.91**	**64.21**	**59.36**	**53.34**

It is observed that for the proposed version, mAP40 for all classes increases in both spaces, BEV and 3D, with respect to the baseline results. This means that the semantic information not only improves the height estimation but also contributes to the estimation of the width and length of the objects of interest, which leads to an overall better shape estimation of the bounding box.

Lastly, as the main objective of this pipeline is to perform 3D object detection, we compare the results on 3D mAP (medium difficulty) for **PointPillars-R** (in blue dots) and **Painted PointPillars-R** (in orange dots) in Figure 6. In the X-axis, we plot the three different versions depending on the temporal aggregation horizon (number of frames). In the Y-axis, we plot the metrics extracted from taking the average of all categories in medium difficulty. We remark on the incremental improvement between the baseline and our proposal for all temporal-aggregated versions of the dataset: ΔmAP = (+3.23, +6.57, +14.73) for BEV space and ΔmAP = (+2.43, +7.39, +11.49) for 3D space. Moreover, it is shown that this delta increases with respect to the temporal aggregation of scans.

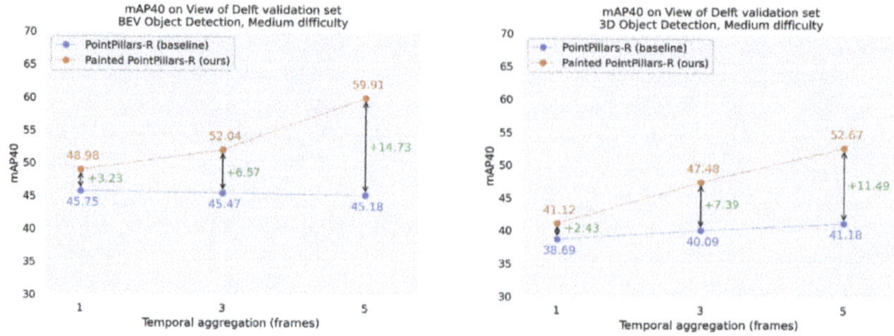

(**a**) (**b**)

Figure 6. Comparison on Mean Average Precision (40 points) between PointPillars-R and Painted PointPillars-R over the View of Delft dataset validation set for bird's eye view (**a**) and 3D object detection (**b**).

4.4. Runtime Performance

In this section, an analysis of the runtime performance will be conducted. We measure the execution time of the components of the pipeline following the experimental setup described in Section 4.2. We will test the inference time of all the components separately, performing cycles of 1000 executions and time measurements for each of them.

Firstly, we will test the inference time of the segmentation network, **YOLOv8-seg**. For this study, we will compare all versions: from *nano* (n) to *extra-large* (x). Three processes are taken into account: preprocessing (resize operation), inference and postprocessing (NMS). We use version **x** in the final pipeline as it offers the best performance, and as VoD does not have semantic labels, we minimize the impact of the domain shift. Table 4 reflects the results. Processing steps consume a fixed amount of time, while inference time increases from 1.7 ms in the lightest version to 6.8 ms in the heaviest one.

Table 4. Inference time for YOLOv8-seg series.

Model	Preprocessing	Inference	Postprocessing	Total
YOLOv8n-seg	0.8 ms	1.7 ms	0.5 ms	3.0 ms
YOLOv8s-seg	0.7 ms	2.1 ms	0.5 ms	3.3 ms
YOLOv8m-seg	0.7 ms	3.9 ms	0.5 ms	5.1 ms
YOLOv8l-seg	0.7 ms	5.5 ms	0.5 ms	6.7 ms
YOLOv8x-seg	0.8 ms	6.8 ms	0.5 ms	8.1 ms

Subsequently, the following steps in the pipeline are the **PointPainting** geometric fusion and the cluster refinement algorithm. The first one contains an *upsampling* operation, so we measured a mean time of 7.28 ms with a standard deviation of 0.25 ms. The latter one contains several runs of DBSCAN clusterings. We measured a mean time of 14.3 ms and a standard deviation of 0.36 ms. Then, we test the inference time of **PointPillars-R** and **Painted PointPillars-R** for the three versions of the VoD dataset. The results are reported in Table 5. It is observed that all models infer at an approximate time of 6 ms, and there are no variations among the baseline and our proposal or among dataset versions.

Finally, choosing **YOLOv8x-seg** and **Painted PointPillars-R** (the five frames temporal-aggregated dataset version) as components of our proposed framework, the total inference time is **35.94 ms** or 27.82 Hz. Therefore, the solution is suitable for real-time applications in the context of automotive perception systems, even when there is room for further optimization in future works.

Table 5. Inference time for PointPillars-R and Painted PointPillars-R models.

Model	Dataset Version	Mean Inference Time	Std. Deviation Inference Time
PointPillars-R	1 frame	6.15 ms	0.27 ms
	3 frames	6.15 ms	0.25 ms
	5 frames	6.14 ms	0.28 ms
Painted PointPillars-R	1 frame	6.27 ms	0.29 ms
	3 frames	6.29 ms	0.30 ms
	5 frames	6.34 ms	0.31 ms

4.5. Comparison with PointPainting for LiDAR

To set a fair comparison with LiDAR-based methods, we perform an experiment in which we train PointPillars for LiDAR point clouds. We follow the same training and evaluation procedures previously explained in Section 4.3, and transfer learning from other datasets is not performed. We name these methods **PointPillars-L** and **Painted PointPillars-L**, respectively. Table 6 shows the performance in medium difficulty for the validation set of the VoD dataset, in which LiDAR-based methods outperform (69.40 and 73.14 mAP) RADAR-based methods (41.18 and 52.67 mAP). A remarkable fact is that the camera contribution is far more significant from non-painted to painted methods in the RADAR domain (+11.19 mAP) than in LiDAR (+3.74 mAP).

Table 6. Performance comparison between PointPillars and Painted PointPillars for RADAR (R) and LiDAR (L).

	AP40 Car	AP40 Ped.	AP40 Cyc.	mAP40@3D
PointPillars-R (5f)	28.77	27.81	66.97	41.18
Painted PointPillars-R (5f)	39.64	43.33	75.03	52.67
PointPillars-L	70.80	56.79	80.59	69.40
Painted PointPillars-L	75.07	62.04	82.32	73.14

4.6. Ablation Study

In this section, we want to remark on the contribution of every single component of the system: from the baseline, **PointPillars-R**, to the proposed system, **Painted PointPillars-R** in an accumulated five-frames settings. For all the experiments, we follow the training and evaluation procedures previously explained in Section 4.3.

Table 7 shows the results of the ablation study. Starting from the baseline (38.69 mAP), the inclusion of the painting mechanism implies a significant performance boost (+2.17 mAP). Then, the cluster refinement algorithm allows noise filtering and contributes a minor improvement. Finally, the accumulation of frames contributes another significant performance boost. Comparing with the results in Section 4.3, it can be observed that the painting mechanism improves also its contribution concerning the number of frames.

Table 7. Ablation study for the pipeline of Painted PointPillars-R, reflecting the contribution of every element of the system individually.

Baseline	+ Painting	+ Cluster Refinement	+ 3 Frames	+ 5 Frames	mAP
✓					38.69
✓	✓				40.86
✓	✓	✓			41.12
✓	✓	✓	✓		47.48
✓	✓	✓		✓	52.67

4.7. Quantitative Results

This section presents qualitative results for the model **Painted PointPillars-R** in the version of five frames accumulated for the VoD dataset.

In Figure 7, a common use case within the dataset is presented. A narrow urban street with one single direction with vehicles parked at the left hand of the ego vehicle and pedestrians walking in front of the ego vehicle on the drivable road. In Figure 7b, a trace can be observed behind the pedestrians, showing that are moving frontwards. The algorithm can perceive correctly in this case. Few points with the semantics of *Pedestrians* are enough to propose bounding boxes at coordinates $\approx (10, 0)$ m and $\approx (12, 0)$ m. *Cars* at long range are perceived correctly.

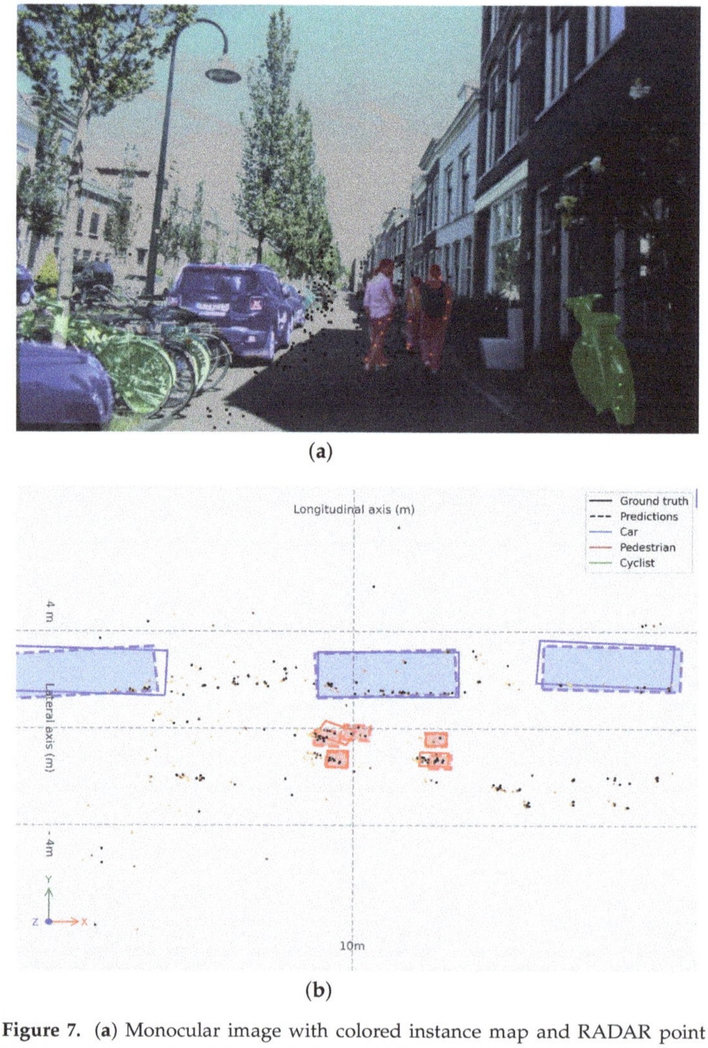

Figure 7. (**a**) Monocular image with colored instance map and RADAR point cloud projected. (**b**) Object detection in BEV space for frame 03970 with **Painted PointPillars-R**. Bounding boxes are: solid for ground truth and dashed and filled for predictions; red for *Pedestrians*, green for *Cyclists*, and blue for *Cars*. Older points are lighter orange and more recent points are darker orange.

Figure 8 justifies the good performance on the *Cyclist* category. One of the main characteristics of the VoD dataset is that this category is well represented and there are frames in which it is the only category present. This is an atypical use case in automotive datasets that takes place due to the location in which VoD is recorded is The Netherlands.

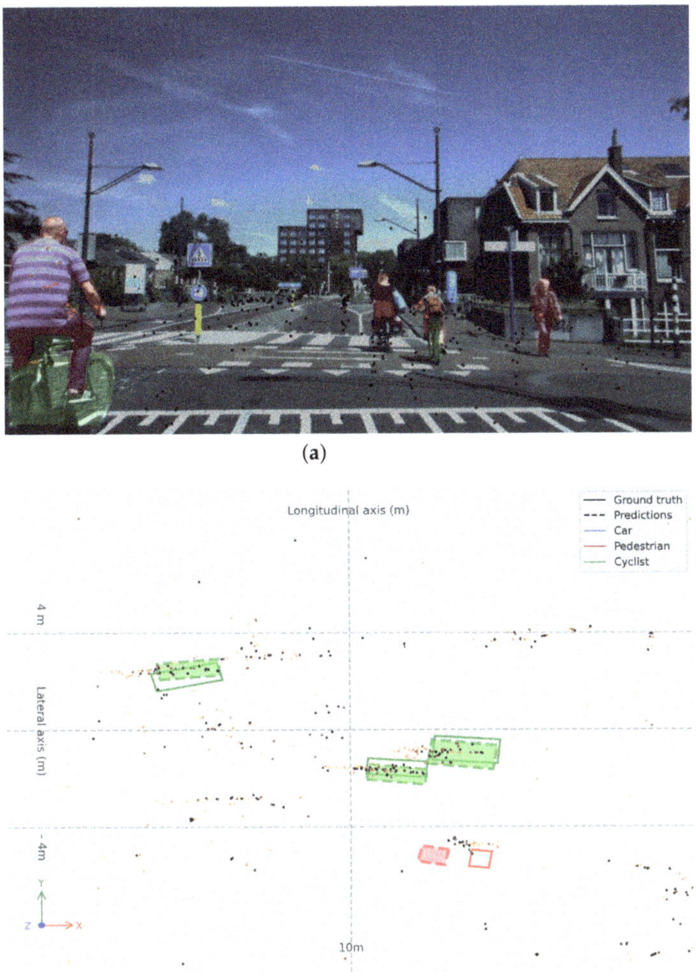

Figure 8. (**a**) Monocular image with colored instance map and RADAR point cloud projected. (**b**) Object detection in BEV space for frame 04362 with **Painted PointPillars-R**. Bounding boxes are solid for ground truth and dashed-and-filled for predictions; red for *Pedestrians*, green for *Cyclists*, and blue for *Cars*. Older points are lighter orange, and more recent points are darker orange.

Figure 9 shows a crowded environment with multiple *Pedestrians*. Most of them are detected correctly. It is remarkable to notice that most *Pedestrian* instances have few points inside their ground-truth bounding box, so the network may show difficulties. The *Cyclist* and *Car* are detected correctly. In the case of the *Cyclist*, a trace in the point cloud can be observed. Most of the *Cyclists* in this dataset are moving objects, so the accumulation of the frame plays a vital role in inferring the position of these VRUs.

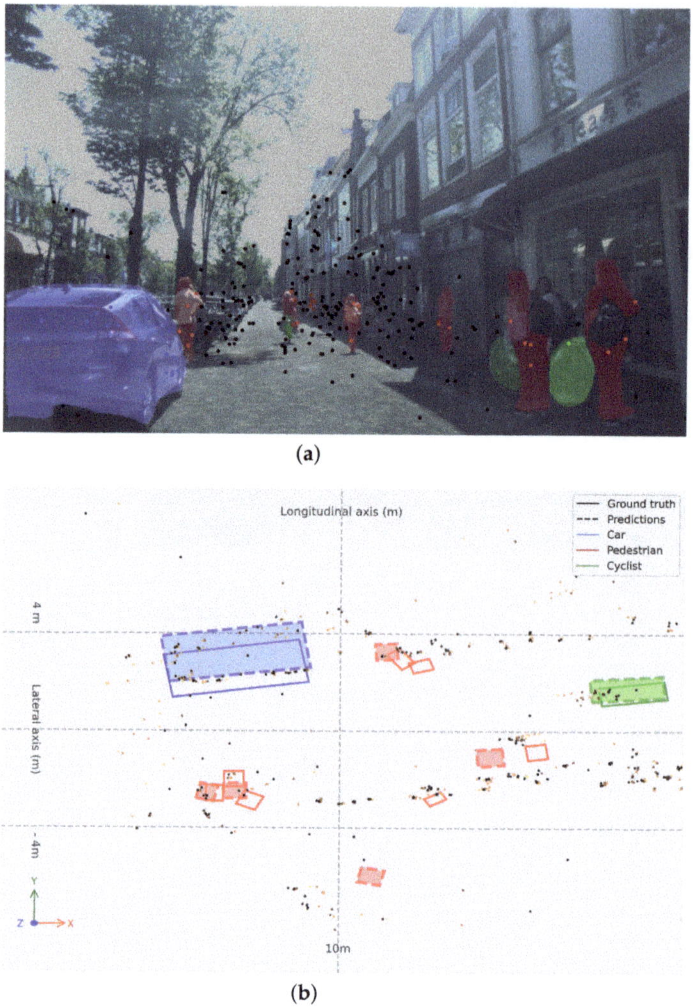

Figure 9. (**a**) Monocular image with colored instance map and RADAR point cloud projected. (**b**) Object detection in BEV space for frame 00180 with **Painted PointPillars-R**. Bounding boxes are solid for ground truth and dashed-and-filled for predictions; red for *Pedestrians*, green for *Cyclists*, and blue for *Cars*. Older points are lighter orange, and more recent points are darker orange.

5. Conclusions and Future Works

In this paper, we validate the performance of a sensor fusion algorithm for 3D object detection which originally was designed for LiDAR and a camera in the domain of 3+1D RADAR and camera data fusion in the View of Delft dataset.

The research introduces compelling insights, particularly highlighting the potential of RADAR–camera fusion for 3D object detection. This fusion significantly outperforms the original RADAR-only baseline across all classes, showcasing the efficacy of projection-based fusion methods. The study also emphasizes the positive impact of temporal information aggregation on neural networks, elevating their precision.

Notably, the analysis spotlights the unexpected performance boost in detecting cyclists, which is a crucial aspect for the safety of vulnerable road users. The study challenges conventional notions, indicating that an increased representation of cyclists enhances network performance, which can be potentially attributed to the unique information provided by radial velocity and RCS features.

Experiments were conducted to compare the performance of RADAR-based PointPainting and LiDAR-based PointPainting. Two main conclusions were drawn. Firstly, LiDAR outperforms RADAR due to the higher density of reflections in the point cloud. As a consequence, more research is needed to fully utilize RADAR data and shorten this gap. Unique features such as RCS and 1D Doppler velocity have not received much attention in the existing literature. Both the inclusion of these components and the investigation of RADAR-specific network architectures could help to achieve LiDAR-like performance for object detection networks. Secondly, it is observed that the contribution of the camera sensor to the RADAR–camera fusion schemes is comparatively higher than that of LiDAR-camera fusion methods, as the improvements from their respective baselines methods are +11.19 mAP and +3.74 mAP. This experiment supports the claim that RADAR and camera sensors have complementary strengths and weaknesses, making them suitable for an intelligent transportation sensor suite for perception.

However, challenges emerge, such as the negative impact of eliminating data augmentation on convergence during training. This underscores the importance of devising new augmentation techniques that consider the physical nuances related to radial velocity. Additionally, the study sheds light on the sensitivity of the validation set, particularly in urban scenes, necessitating improvements for generalization with DL-based techniques.

As a result of this work, we identify future lines of research in the domain of RADAR and camera fusion for perception techniques applied to the automotive domain.

- Comprehensive Object Detector Study: Incorporating additional 3D object detectors from the state of the art aims to glean insights into the compatibility of different detectors that were originally designed for LiDAR data with RADAR data.
- Temporal Aggregation Investigation: The research aims to delve deeper into multi-frame aggregated networks specifically designed for temporal aggregation. The demonstrated benefits of scan accumulation in the proposed architecture prompt further exploration of its potential and optimization.
- Feature-Level Fusion Architectures: Novel approaches within the domain are contemplated, focusing on fusing information at the feature level within NNs. Attention mechanisms, such as cross-attention modules, are envisioned to facilitate the interaction between features from different data modalities.
- Optimization with Acceleration Frameworks: The proposal's optimization journey involves leveraging DL acceleration frameworks and transitioning CPU processes to GPU implementations. This optimization is a necessary step toward integrating and testing algorithms within the real AD stack of the electric AV at the RobeSafe Research Group (University of Alcalá).

Author Contributions: Conceptualization, S.M.-M. and Á.L.; methodology, S.M.-M.; software, S.M.-M.; validation, S.M.-M. and M.A.; investigation, S.M.-M.; resources, S.M.-M. and M.A.; data curation, S.M.-M.; writing—original draft preparation, S.M.-M. and Á.L.; writing—review and editing, S.M.-M., L.M.B. and Á.L.; visualization, S.M.-M. and M.A.; supervision, Á.L., P.A.R. and L.M.B.; project administration, L.M.B., Á.L. and P.A.R.; funding acquisition, L.M.B., Á.L. and P.A.R. All authors have read and agreed to the published version of the manuscript.

Funding: This work has been supported from the Spanish PID2021-126623OB-I00 project, funded by the MICIN/AEI and FEDER, the TED2021-130131A-I00 and PDC2022-133470-I00 projects, funded by the MICIN/AEI and the European Union NextGenerationEU/PRTR, and ELLIS Unit Madrid funded by the Autonomous Community of Madrid.

Institutional Review Board Statement: Not applicable.

Informed Consent Statement: Not applicable.

Data Availability Statement: In this manuscript, View of Delft dataset has been used. It can be accessed via request to their original authors. The procedure and elegibility to use data can be found at their official website (https://tudelft-iv.github.io/view-of-delft-dataset/, accessed on 27 December 2023).

Conflicts of Interest: The authors declare no conflicts of interest.

Abbreviations

The following abbreviations are used in this manuscript:

ACC	Adaptive Cruise Control
AD	Autonomous Driving
ADAS	Advanced Driver Assistance Systems
AEB	Automatic Emergency Braking
AOS	Average Orientation Similarity
AV	Autonomous Vehicle
BEV	Bird's Eye View
DBSCAN	Density-Based Spatial Clustering for Applications with Noise
IoU	Intersection over Union
LDW	Lane Departure Warning
LiDAR	Lidar Detection and Ranging
mAP	Mean Average Precision
NMS	Non-Maximum Suppression
NN	Neural Network
RADAR	Radio Detection and Ranging
VOD	View of Delft dataset
VRU	Vulnerable Road User

References

1. Vora, S.; Lang, A.H.; Helou, B.; Beijbom, O. Pointpainting: Sequential fusion for 3D object detection. In Proceedings of the IEEE/CVF Conference on Computer Vision and Pattern Recognition, Seattle, WA, USA, 13–19 June 2020; pp. 4604–4612.
2. Jocher, G.; Chaurasia, A.; Qiu, J. YOLO by Ultralytics. 2023. Available online: https://github.com/ultralytics/ultralytics (accessed on 3 January 2024).
3. Lang, A.H.; Vora, S.; Caesar, H.; Zhou, L.; Yang, J.; Beijbom, O. Pointpillars: Fast encoders for object detection from point clouds. In Proceedings of the IEEE/CVF Conference on Computer Vision and Pattern Recognition, Long Beach, CA, USA, 15–20 June 2019; pp. 12697–12705.
4. Palffy, A.; Pool, E.; Baratam, S.; Kooij, J.F.; Gavrila, D.M. Multi-class road user detection with 3+1D radar in the View-of-Delft dataset. *IEEE Robot. Autom. Lett.* **2022**, *7*, 4961–4968. [CrossRef]
5. Yao, S.; Guan, R.; Huang, X.; Li, Z.; Sha, X.; Yue, Y.; Lim, E.G.; Seo, H.; Man, K.L.; Zhu, X.; et al. Radar-camera fusion for object detection and semantic segmentation in autonomous driving: A comprehensive review. *arXiv* **2023**, arXiv:2304.10410.
6. Caesar, H.; Bankiti, V.; Lang, A.H.; Vora, S.; Liong, V.E.; Xu, Q.; Krishnan, A.; Pan, Y.; Baldan, G.; Beijbom, O. nuscenes: A multimodal dataset for autonomous driving. In Proceedings of the IEEE/CVF Conference on Computer Vision and Pattern Recognition, Seattle, WA, USA, 13–19 June 2020; pp. 11621–11631.
7. Nabati, R.; Qi, H. Centerfusion: Center-based radar and camera fusion for 3d object detection. In Proceedings of the IEEE/CVF Winter Conference on Applications of Computer Vision, Waikoloa, HI, USA, 3–8 January 2021; pp. 1527–1536.
8. Duan, K.; Bai, S.; Xie, L.; Qi, H.; Huang, Q.; Tian, Q. Centernet: Keypoint triplets for object detection. In Proceedings of the IEEE/CVF International Conference on Computer Vision, Seoul, Republic of Korea, 27 October–2 November 2019; pp. 6569–6578.
9. Nobis, F.; Geisslinger, M.; Weber, M.; Betz, J.; Lienkamp, M. A deep learning-based radar and camera sensor fusion architecture for object detection. In Proceedings of the 2019 Sensor Data Fusion: Trends, Solutions, Applications (SDF), Bonn, Germany, 15–17 October 2019; pp. 1–7.
10. Lin, T.Y.; Goyal, P.; Girshick, R.; He, K.; Dollár, P. Focal loss for dense object detection. In Proceedings of the IEEE International Conference on Computer Vision, Venice, Italy, 22–29 October 2017; pp. 2980–2988.
11. Simonyan, K.; Zisserman, A. Very deep convolutional networks for large-scale image recognition. *arXiv* **2014**, arXiv:1409.1556.
12. Bansal, K.; Rungta, K.; Bharadia, D. Radsegnet: A reliable approach to radar camera fusion. *arXiv* **2022**, arXiv:2208.03849.

13. Cordts, M.; Omran, M.; Ramos, S.; Rehfeld, T.; Enzweiler, M.; Benenson, R.; Franke, U.; Roth, S.; Schiele, B. The cityscapes dataset for semantic urban scene understanding. In Proceedings of the IEEE Conference on Computer Vision and Pattern Recognition, Seattle, WA, USA, 17–21 June 2016; pp. 3213–3223.
14. Meyer, M.; Kuschk, G. Automotive radar dataset for deep learning based 3d object detection. In Proceedings of the 2019 16th European Radar Conference (EuRAD), Paris, France, 2–4 October 2019; pp. 129–132.
15. Li, H.; Sima, C.; Dai, J.; Wang, W.; Lu, L.; Wang, H.; Zeng, J.; Li, Z.; Yang, J.; Deng, H.; et al. Delving into the devils of bird's-eye-view perception: A review, evaluation and recipe. *IEEE Trans. Pattern Anal. Mach. Intell.* **2023**. [CrossRef] [PubMed]
16. Philion, J.; Fidler, S. Lift, splat, shoot: Encoding images from arbitrary camera rigs by implicitly unprojecting to 3d. In Proceedings of the Computer Vision–ECCV 2020: 16th European Conference, Glasgow, UK, 23–28 August 2020; pp. 194–210.
17. Kim, Y.; Shin, J.; Kim, S.; Lee, I.J.; Choi, J.W.; Kum, D. Crn: Camera radar net for accurate, robust, efficient 3d perception. In Proceedings of the IEEE/CVF International Conference on Computer Vision, Paris, France, 2–6 October 2023; pp. 17615–17626.
18. Lin, T.Y.; Maire, M.; Belongie, S.; Hays, J.; Perona, P.; Ramanan, D.; Dollár, P.; Zitnick, C.L. Microsoft coco: Common objects in context. In Proceedings of the Computer Vision–ECCV 2014: 13th European Conference, Zurich, Switzerland, 6–12 September 2014; pp. 740–755.
19. Bochkovskiy, A.; Wang, C.Y.; Liao, H.Y.M. Yolov4: Optimal speed and accuracy of object detection. *arXiv* **2020**, arXiv:2004.10934.
20. Ester, M.; Kriegel, H.P.; Sander, J.; Xu, X. A density-based algorithm for discovering clusters in large spatial databases with noise. *kdd* **1996**, *96*, 226–231.
21. Yan, Y.; Mao, Y.; Li, B. Second: Sparsely embedded convolutional detection. *Sensors* **2018**, *18*, 3337. [CrossRef] [PubMed]
22. Girshick, R. Fast r-cnn. In Proceedings of the IEEE International Conference on Computer Vision, Santiago, Chile, 7–13 December 2015; pp. 1440–1448.

Disclaimer/Publisher's Note: The statements, opinions and data contained in all publications are solely those of the individual author(s) and contributor(s) and not of MDPI and/or the editor(s). MDPI and/or the editor(s) disclaim responsibility for any injury to people or property resulting from any ideas, methods, instructions or products referred to in the content.

sensors

MDPI

Article

YOLOv7-TS: A Traffic Sign Detection Model Based on Sub-Pixel Convolution and Feature Fusion

Shan Zhao, Yang Yuan *, Xuan Wu, Yunlei Wang and Fukai Zhang

School of Software, Henan Polytechnic University, Jiaozuo 454000, China
* Correspondence: 361709010220@home.hpu.edu.cn

Abstract: In recent years, significant progress has been witnessed in the field of deep learning-based object detection. As a subtask in the field of object detection, traffic sign detection has great potential for development. However, the existing object detection methods for traffic sign detection in real-world scenes are plagued by issues such as the omission of small objects and low detection accuracies. To address these issues, a traffic sign detection model named YOLOv7-Traffic Sign (YOLOv7-TS) is proposed based on sub-pixel convolution and feature fusion. Firstly, the up-sampling capability of the sub-pixel convolution integrating channel dimension is harnessed and a Feature Map Extraction Module (FMEM) is devised to mitigate the channel information loss. Furthermore, a Multi-feature Interactive Fusion Network (MIFNet) is constructed to facilitate enhanced information interaction among all feature layers, improving the feature fusion effectiveness and strengthening the perception ability of small objects. Moreover, a Deep Feature Enhancement Module (DFEM) is established to accelerate the pooling process while enriching the highest-layer feature. YOLOv7-TS is evaluated on two traffic sign datasets, namely CCTSDB2021 and TT100K. Compared with YOLOv7, YOLOv7-TS, with a smaller number of parameters, achieves a significant enhancement of 3.63% and 2.68% in the mean Average Precision (mAP) for each respective dataset, proving the effectiveness of the proposed model.

Keywords: object detection; traffic sign detection; YOLOv7; sub-pixel convolution; feature fusion

1. Introduction

With the rapid development of Artificial Intelligence (AI) and Deep Learning (DL) technology, traffic sign detection, as a part of intelligent transportation, has become a popular research topic. In real-world scenes, traffic sign detection can assist drivers in accurately assessing information and reduce cases of neglect and misidentification caused by fatigue or inattention to avoid accidents. However, practical traffic sign detection can be affected by several factors, such as illumination, bad weather, and occlusion. Therefore, it is important to design a robust traffic sign detector with high detection accuracy.

For increased driver perceptibility, traffic signs usually have striking colors and special shapes. Traditional traffic sign detection focuses on feature extraction and classification. Color or shape features are extracted by different approaches for traffic sign recognition through a classifier. Benallal et al. [1] performed traffic sign detection through color segmentation by comparing the differences between the RGB components. However, the RGB color space is highly sensitive to changes in the lighting environment, making it difficult to achieve stable results. Ruta et al. [2] employed a normalization process to enhance the RGB color space, to better extract and detect the differences between the three colors (red, yellow, and blue). Nguwi et al. [3] converted traffic sign images from the RGB to the HSI color space and segmented the images with a fixed threshold due to the insensitivity of the latter to illumination changes. Fleyeh et al. [4] employed an AdaBoost binary classifier-based color segmentation approach. Yang et al. [5] proposed a model based on a color probability map in the Ohta space. Loy et al. [6] attempted to establish the possible center of mass location through the symmetry of the image and the edge information to detect traffic signs. However, this method does not apply to all shapes and has poor generalization. The Hough transform approach utilized by Zaklouta et al. [7] to extract specific shape features achieved better detection results. Abukhait et al. [8] utilized both traffic sign color and shape features and demonstrated a strong detection performance, even when the traffic sign shape was distorted or obscured by foreign objects. Although traditional methods for traffic sign detection demonstrated satisfactory performance at the time, they still face the following limitations: Firstly, traditional approaches often rely on manually designed features to represent traffic signs. This may overlook some complex textures, shapes, or background variations since manually designed features struggle to capture intricate linear relationships. Secondly, factors such as lighting conditions, different perspectives, and occlusion pose challenges to traditional traffic sign detection. Under these circumstances, traffic signs exhibit varying appearances, making it difficult for traditional methods to adapt to such changes. Thirdly, due to the diversity in the appearance of traffic signs and backgrounds, traditional methods struggle to generalize across different scenes and datasets, challenging their robustness in practical applications. Lastly, the computational complexity of traditional methods is relatively high, rendering them unsuitable for meeting real-time requirements.

Since the achievements of AlexNet [9] in the field of visual recognition in 2012, an increasing number of DL methods have been proposed and applied to object detection, replacing the traditional algorithms. These DL-based object detection methods can be broadly categorized into two groups, namely Two-Stage and One-Stage algorithms. The first ever Two-Stage object detection algorithm with industrial-grade accuracy, Regions with CNN features (R-CNN), was introduced by Girshick et al. [10]. A selective search approach was utilized by R-CNN to extract 2000 candidate regions, which were subsequently predicted individually. However, despite its breakthrough accuracy, R-CNN is plagued by significant speed limitations due to the generation of candidate regions and extensively repetitive computations. The efficiency of R-CNN has been optimized by Spatial Pyramid Pooling in Deep Convolutional Networks for Visual Recognition (SPP-Net) [11], which extracts features from the entire feature map and uses a spatial pyramid pooling layer to handle features of arbitrary sizes. The SPP-Net training process remains complex, although it has been improved compared with R-CNN. Fast R-CNN [12] unifies the classification and regression problems with a deep network implementation, eliminating the need for additional storage and dramatically improving the training speed. Faster R-CNN [13] became the first object detector with near real-time performance, integrating the four main steps, i.e., feature extraction, candidate region generation, regression of predicted edge position parameters, and object category determination, into the Region Proposal Network (RPN). Libra R-CNN [14] solves the imbalance problem during training with high accuracy improvement by introducing Intersection over Union (IoU) balanced sampling, balanced pyramid, and balanced Mean Absolute Error (L1 loss). Despite the significant enhancements in detection performance brought about by these Two-Stage algorithms, their speed still does not meet the demands of real-time object detection. This issue is tackled by the introduction of One-Stage object detection algorithms. You Only Look Once (YOLO) [15] transforms object detection into a regression problem and adopts CNNs to complete the prediction of bounding boxes and category determination, truly ushering in the era of real-time object detection. A detection accuracy on par with Faster R-CNN is achieved by Single Shot MultiBox Detector (SSD) [16], while maintaining the detection speed through the prediction of multi-scale features and the introduction of an anchor mechanism. In YOLOv2 [17], batch normalization is incorporated into the network to expedite model convergence, and the K-Means algorithm is utilized to determine the size and position of prior frames, resulting in better-matching detection frames for real objects. Darknet-53 is employed as the YOLOv3 [18] backbone, and the concept of Feature Pyramid Network (FPN) [19] is introduced for feature fusion at different scales, further enhancing detection performance. YOLOv4 [20] employs a new backbone, CSPDarknet53, by combining the

Cross Stage Partial Network (CSPNet) [21] with Darknet-53. This model additionally incorporates Spatial Pyramid Pooling (SPP) and Path Aggregation Network (PANet) [22] to expand the receptive field of features and improve feature fusion. In YOLOv5 [23], the feature fusion capabilities are enhanced by incorporating the CSP structure into the fusion network. Simultaneously, the Focus structure is added to the backbone network to perform slicing operations on feature maps, reducing algorithm computation and accelerating processing speed. Gui et al. [24] achieved a more lightweight model in YOLOv5 by introducing lightweight convolutional operations. Simultaneously, the incorporation of the bottleneck attention module was employed to extract effective information, thereby enhancing the accuracy of the model. Recently, breakthroughs have been achieved in the field of computer vision with models based on Transformer [25] architecture. The Vision Transformer (ViT) [26] applies the Transformer architecture to sequences of image patches, further advancing image classification tasks. ViT-FRCNN [27] integrates ViT with RPN to accomplish object detection tasks based on the transformer backbone. Xie et al. [28] proposed a novel integrated network, ViT-MVT, designed for joint optimization of image classification tasks and other downstream tasks, such as object detection and semantic segmentation. ViT-MVT demonstrates exceptional performance across multiple visual tasks, surpassing existing methods while requiring considerably less total storage. Wang et al. [29] proposed an MDL-NAS framework based on ViT, achieving high performance across various vision tasks and maintaining storage efficiency for model deployment through a coarse-to-fine searching space design and a joint-subnet search algorithm.

As a sub-type of object detection, traffic sign detection can benefit from the application of DL-based algorithms. However, this task has higher requirements for detection accuracy and speed. As a result, various studies have attempted to improve these algorithms in order to achieve improved traffic sign detection. Gavrilescu et al. [30] revealed that compared with Fast R-CNN, Faster R-CNN exhibited superior real-time traffic sign detection performance. Based on Faster R-CNN, Yang et al. [31] proposed a method utilizing an attention network to replace the RPN, improving the accuracy for both the TT100k [32] and BTSD [33] datasets. Additionally, Zhang et al. [34] introduced a cascaded R-CNN [35] approach to acquire multi-scale features from the pyramid and weighted them through dot products and softmax operations, thereby emphasizing traffic sign features and enhancing detection accuracy. The aforementioned algorithms are all formulated based on the Two-Stage object detection approach, and despite the significant improvement in accuracy, meeting real-time requirements is challenging due to the slower speed inherent in Two-Stage object detection. Mohd-Isa et al. [36] incorporated SPP into the YOLOv3 framework, enabling traffic sign recognition for real-world images of varying sizes through multi-scale pooling. This modification results in improved detection accuracy along with increased detection speed. Wang et al. [37] developed a lightweight YOLOv4 algorithm by replacing the backbone with a lightweight architecture, MobileNetv2 [38], and attention mechanisms were also employed to enhance the model's detection capabilities. In summary, DL-based traffic sign detection algorithms are characterized by increased robustness and higher detection accuracy compared with traditional methods. Simultaneously, One-Stage algorithms offer advantages in real-time processing, better aligning with the requirements of traffic sign detection.

Despite the better traffic sign detection results and faster speeds offered by the YOLO-series-based One-Stage algorithms, the problem of the high leakage rates of small objects persists. Detecting small objects has always been a challenging task in object detection. Firstly, small objects typically exhibit low-resolution with limited visual information, making the extraction of effective features more challenging. Secondly, detection methods based on neural networks typically extract deep-level features with rich semantic information for object detection. The dimensions of these features progressively diminish, causing the network to gradually lose spatial resolution and detailed information during the forward propagation process, thereby affecting the detection of small objects. In recent years, several methods have been proposed to achieve breakthroughs in the field of small object detection.

Liang et al. [39] introduced FS-SSD as a method for small object detection in UAV images. However, the algorithm introduces hyperparameters that are not conducive to optimization. Liu et al. [40] employed contextual information, encompassing object and scene features, as well as relationships between coexisting objects, to enhance the performance of small object detection. The approach relying on context fusion enhances the accuracy of object detection to a certain extent. However, identifying contextual information from a global scene that is beneficial for enhancing small object detection remains a challenging research problem. Li et al. [41] employed generative adversarial learning by mapping the features of small low-resolution objects into features equivalent to those of high-resolution objects to attain the same detection performance as that of larger-size objects. Nevertheless, training generative adversarial networks poses challenges, and achieving a well-balanced relationship between generators and discriminators is not easily realized. Zhu et al. [42] utilized Transformer Prediction Heads (TPH) to globally model the feature maps. While it enhances the detection capabilities for small objects, the complex structure of the Transformer also introduces a certain degree of redundancy to the model. In addition to the problem of small object detection, the number of feature channels can be directly reduced by the feature map extraction process using a 1×1 convolution, potentially leading to the loss of feature channel information. At the same time, the detection accuracy can be affected by insufficient feature fusion. Therefore, a traffic sign detection model, YOLOv7-Traffic Sign (YOLOv7-TS), based on sub-pixel convolution [43] and feature fusion is proposed in this study. Firstly, the up-sampling capability of sub-pixel convolution is utilized to integrate the channel dimension pixels and preserve channel information, thus extracting feature maps with richer features. Additionally, a small object detection layer is introduced to the feature fusion network, and the integration map is employed to enhance information interaction among all feature layers, which reduces the rate of small object omissions and improves the effectiveness of feature fusion. Furthermore, the highest-layer feature enhancement is accomplished by the addition of global average pooling and global max pooling, which provide the highest-layer feature with global information and significant information. Feature fusion is then employed to impart the benefits of the highest-layer feature to every other layer, ultimately enhancing the detection accuracy of the model. The major contributions of this study can be summarized as follows:

1. A Feature Map Extraction Module (FMEM) is devised to mitigate channel information loss during feature map extraction by employing sub-pixel convolution.
2. A Multi-feature Interactive Fusion Network (MIFNet) is developed to leverage the rich details of the small object detection layer and employ the integration map to enhance the interaction of information across all feature layers. The perceivability for small traffic sign objects is enhanced, and the effect of feature fusion is improved while reducing the model parameters.
3. A Deep Feature Enhanced Module (DFEM) is established, in which the highest feature is enriched by integrating global average pooling and global max pooling within the Spatial Pyramid Pooling Cross-Stage Partial Channel (SPPCSPC). Simultaneously, pooling is accelerated without altering the size of the original receptive field, thereby speeding up the inference of DFEM.
4. To validate the effectiveness of YOLOv7-TS, a substantial number of experiments are conducted on the CCTSDB2021 [44] and TT100K traffic sign datasets. These experiments demonstrate the commendable detection performance of YOLOv7-TS.

2. Related Work

2.1. YOLOv7

The latest object detector in the YOLO series is YOLOv7 [45], and the structure is depicted in Figure 1. It primarily consists of four components, namely, input, backbone, feature fusion network, and prediction. The complete YOLOv7 detection process is as follows: Initially, the images are preprocessed by YOLOv7, resizing them to a fixed size of $640 \times 640 \times 3$ pixels to meet the input requirements of the backbone. Subsequently, deeper features are continuously extracted within the backbone using CBS modules, Efficient Long-range Attention Network (ELAN) [46] modules, and Max Pooling Reduction (MP) modules. Among these components, CBS comprises the convolution, batch normalization, and Sigmoid Linear Units (SiLU) activation function. The ELAN is constructed by stacking different CBS modules, employing a denser residual structure to aid optimization. The four feature layers, each processed through the different CBS modules, are stacked and subsequently integrated through another CBS module. This integrated result in the output feature has twice the number of channels as the input feature. The MP module employs max pooling in the upper branch and convolution in the lower branch. In both branches, the number of channels is halved by a 1 × 1 convolution, and both branches are spliced to produce down-sampled output feature with the same number of channels as the input feature. The YOLOv7 backbone outputs three feature layers of different sizes: 20 × 20 × 1024, 40 × 40 × 1024, and 80 × 80 × 512. These feature layers are then subjected to channel number adjustment through a 1 × 1 convolution and subsequently input into the Path Aggregation FPN (PAFPN) for feature fusion. The optimization process is aided by the SPPCSPC module, which leverages residual edges while enabling the highest feature to acquire receptive field information at different scales. The number of channels of the three feature layers of varying sizes is adjusted using RepConv, which draws inspiration from the concept of Re-parameterization Visual Geometry Group (RepVGG) [47] and introduces three residual branches to assist in training. These channel-adjusted feature layers are then fed into the YoloHead for prediction. In the deployment phase, the intricate residual structure is reparameterized into a 3 × 3 convolution, reducing the network complexity while maintaining high prediction performance. YOLOv7 outperforms all known object detection models in both speed and accuracy, ranging from 5 to 160 Frames Per Second (FPS). Its exceptional performance makes it suitable to be used as a baseline model for traffic sign detection. However, issues such as the loss of channel information and detection leakage of small traffic sign objects still exist. Therefore, this study focuses on improving YOLOv7 to create an object detector specifically tailored for traffic sign detection.

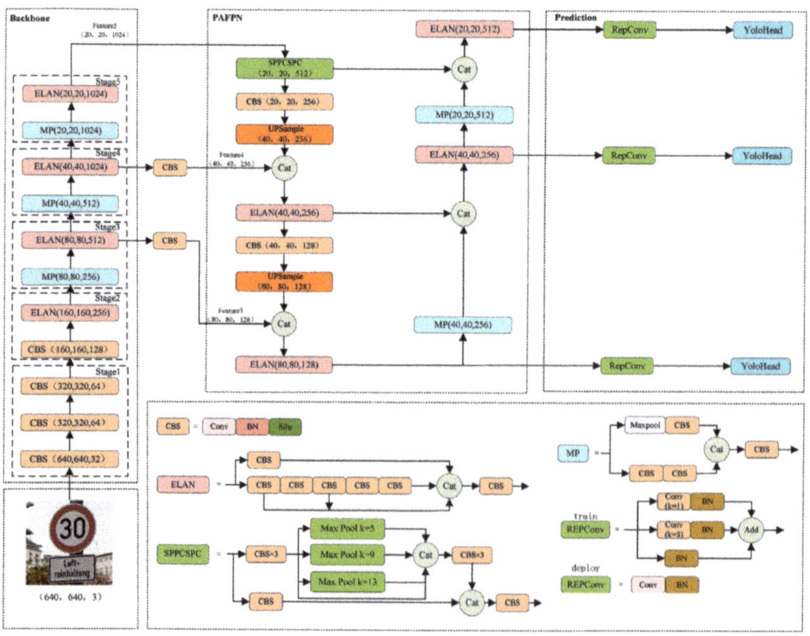

Figure 1. Overall architecture of YOLOv7.

2.2. Sub-Pixel Convolution

Sub-pixel convolution is proposed for image super-resolution tasks, in which low-resolution images are super-resolved by generating high-resolution images through sub-pixel convolution. Therefore, sub-pixel convolution is an up-sampling method, and its sampling process is shown in Figure 2. For the input image $W \times H \times C$ (where W, H, and C refer to the width, height, and channel dimensions, respectively), the number of image channels is firstly expanded to $C \times r^2$ with a 1×1 convolution, where r is the expansion factor. The expanded image becomes $W \times H \times C \times r^2$. Subsequently, the pixels in C are shuffled by the sub-pixel convolution to enlarge W and H by r times and obtain the up-sampled image, $r\,W \times r\,H \times C$. Mathematically, the pixel shuffle operator, PS, can be expressed as:

$$PS(T)_{x,y,c} = T\lfloor x/r\rfloor, \lfloor y/r\rfloor, C \cdot r \cdot \mathrm{mod}(y,r) + C \cdot \mathrm{mod}(x,r) + c \tag{1}$$

where T denotes the input feature and $PS(T)_{x,y,c}$ denotes the output feature pixels at the (x, y, c) coordinates. By initially adjusting the number of channels and subsequently applying sub-pixel convolution, it is ensured that the number of channels of the output image is consistent with that of the input image, while simultaneously increasing the resolution.

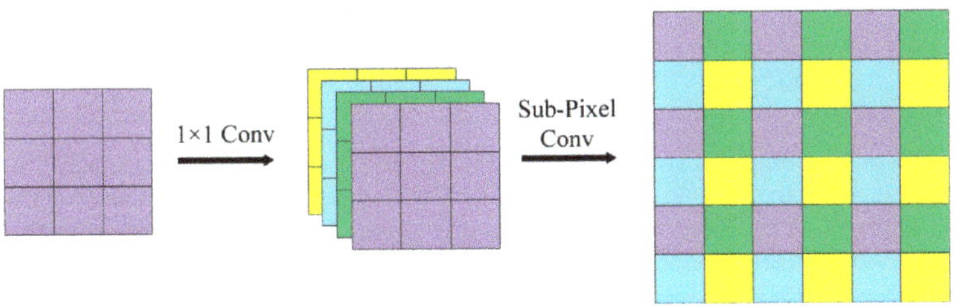

Figure 2. The process of sub-pixel convolution up-sampling.

2.3. Integration Map

A variety of semantic information is contained within the various feature layers of a neural network. Rich semantic information is encompassed by deep features, which is utilized for recognizing object types; rich detail information is abundant in shallow features, which is utilized to clearly localize objects. Due to their high-resolution, small objects can be more effectively perceived using shallow features. The fusion of both deep and shallow features is common in digital image processing tasks. Fan et al. [48] achieved good performance by employing a feature fusion module to receive and process different levels of features. An effective framework was established using FPN for merging feature maps at different scales to accomplish visual tasks through a top-down pathway, which has subsequently been widely adopted and subjected to extensive research [49,50]. An additional bottom-up pathway has been explored using PAFPN to further enhance the low-level information in the deep layers. A weighted bidirectional FPN has been proposed by Scalable and Efficient Object Detection (EfficientDet) [51] for executing simple and rapid feature fusion. Nevertheless, the fused features are required to possess balanced information from each level of resolution. The fused features are influenced by the aforementioned methods, placing greater emphasis on neighboring resolutions and reducing the significance of features with farther resolutions. As a result, the semantic information within the non-neighboring layers becomes diluted during the continuous feature fusion.

Integrating and refining multi-scale features through the introduction of a balanced feature pyramid is achieved by Libra R-CNN. The main structure utilized for this purpose is the integration map, which serves to enhance the interaction of information between all feature layers and mitigate the loss of semantic information in non-adjacent feature layers during the fusion process. The application process of the integration map is depicted in Figure 3. The low-resolution features are subjected to up-sampling through linear interpolation, while the high-resolution features are adjusted to the same resolution size via adaptive max pooling. Subsequently, the integration map is generated through simple averaging. At this point, rich information in multi-scale feature layers is incorporated into the integration map, with due consideration given to the information from non-adjacent layers. Following this, the restoration of the original resolution features is facilitated by the integration map through linear interpolation up-sampling and adaptive max pooling, enhancing the original features across all layers.

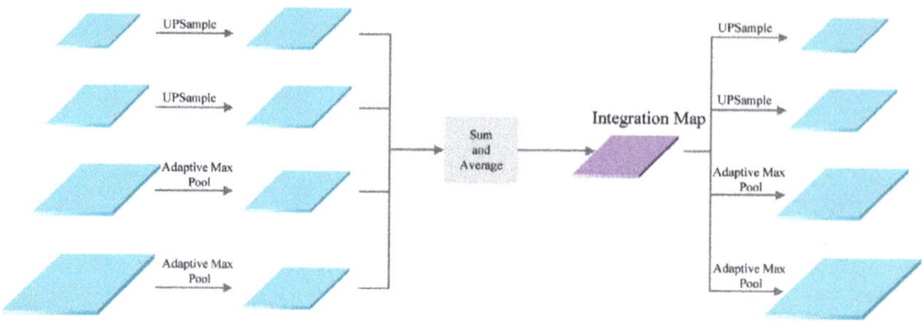

Figure 3. The application process of integration map.

3. YOLOv7-Traffic Sign (YOLOv7-TS)

3.1. Overall Architecture

While traffic sign detection algorithms based on the YOLO series have indeed achieved remarkable results in terms of both detection accuracy and speed, they still encounter three significant challenges. The first issue pertains to the loss of channel information during the feature map extraction process, while the second relates to the high detection leakage rate of small objects. The third issue concerns both the loss of information in the fusion process of non-adjacent feature layers and the inadequacy of the fusion. To address these issues, YOLOv7-TS is proposed as an extension of YOLOv7, and its structure is depicted in Figure 4. The YOLOv7-TS primarily comprises four components: input, backbone, feature fusion network, and prediction. The backbone is employed to extract features at various resolutions. The Feature Map Extraction Module (FMEM) is positioned between the backbone and the feature fusion network to facilitate feature map extraction, ensuring the retention of rich channel information from different feature layers. The feature maps extracted by FMEM contain both high-level semantic information and low-level detail information, which has a positive role in promoting target recognition and localization. Within the feature fusion network, the Multi-feature Interactive Fusion Network (MIFNet) is utilized to facilitate information interaction among all features, thereby enhancing the effectiveness of feature fusion. Simultaneously, the integration of detailed information from the high-resolution feature map aids in the precise localization of small objects, mitigating the issue of missed detections for small objects. Additionally, the highest-layer feature is enhanced using the Deep Feature Enhanced Module (DFEM), ensuring the provision of both global and significant information, thereby benefiting all layers through feature fusion. Ultimately, the fused features are subjected to classification and regression within the final prediction component.

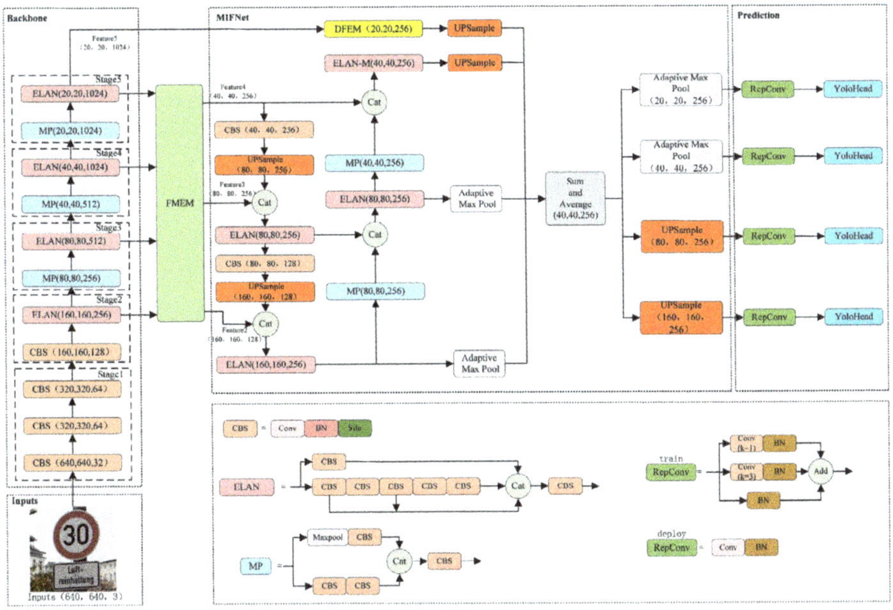

Figure 4. The architecture of our proposed YOLOv7-TS.

The complete process for YOLOv7-TS is as follows: The input images are initially adjusted to a fixed size of 640 × 640 × 3 and fed into the YOLOv7-TS backbone. Then, four feature layers, S_2, S_3, S_4, and S_5, are generated by the backbone, corresponding to the outputs of stages 2 to 5, with 256, 512, 1024, and 1024 channels, respectively. The output, S_2, of stage 2 is added as the small object detection layer. Subsequently, FMEM performs the sub-pixel convolution for the inputs $S_{2\text{–}5}$ to transform the channel dimension information of the different feature layers into spatial dimensions. The up-sampled images are then added element-by-element, with the initial feature layers of the same resolution, to obtain the feature maps $F_{2\text{–}4}$. The feature map F_5 is generated directly from S_5 without any other operations. Further, $F_{2\text{–}5}$ are processed by MIFNet for feature fusion, where $F_{2\text{–}4}$ are channeled into PAFPN for pyramid feature fusion, while F_5 is directly involved in the generation of the integration map after feature enhancement and no longer enters PAFPN. The purpose of this step is to alleviate the increase in the number of parameters brought about by the four-layer feature fusion, and simultaneously, avoid excessive feature fusion that could potentially affect the fusion effect. The DFEM is used to perform feature enhancement for F_5. Building upon the original SPPCSPC, global average pooling and global max pooling are incorporated in the residual path to provide global information and significant information to the highest-layer feature, respectively. Simultaneously, the original pooling structure is altered by DFEM, and pooling continues based on the result of the previous layer using a small pooling kernel. This achieves pooling acceleration with the same receptive field to enhance the overall module speed. The feature-enhanced F_5 and the pyramidal feature-fused $F_{2\text{–}4}$ are jointly involved in generating the integration map for the interaction among all feature layers and fusion effect enhancement. Finally, each feature layer is restored to its original resolution for subsequent predictions by max pooling and linear interpolation up-sampling. The details and design philosophy of this network will be elaborated in Sections 3.2–3.4.

3.2. Feature Map Extraction Module (FMEM)

As representatives of the One-Stage object detection algorithms, the YOLO series algorithms have been widely employed for real-time detection due to their exceptional detection speed. In pursuit of computational efficiency, the YOLO series algorithms employ a simple 1 × 1 convolution to extract features with a specific number of channels for feature fusion, as demonstrated in Figure 5. However, this practice results in the loss of channel information, even with the latest iteration, i.e., YOLOv7. Recently, numerous FPN-based networks have been predominantly applied for the development of efficient modules within the pyramid to enhance fusion results. However, these approaches do not inherently tackle the issue of channel information loss and fail to effectively harness the rich channel information present in the backbone output features.

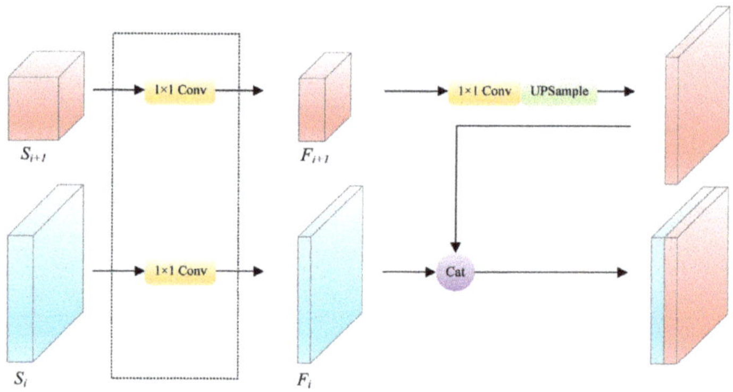

Figure 5. The process of using 1 × 1 convolution to extract specific channel number features in YOLO series algorithms.

The FMEM is designed to solve this issue by leveraging the abundant channel information within S_i through sub-pixel convolutional up-sampling, ultimately enhancing the detection capabilities. Sub-pixel convolution, as a form of up-sampling, contributes to the model in the following aspects: Firstly, it aids in recovering the lost detailed information in deep features while preserving rich semantic information within channels. Secondly, the feature maps extracted through sub-pixel convolution encompass both high-level semantic information and low-level detailed information, actively facilitating simultaneous object localization and recognition. Thirdly, by increasing the spatial dimension, the model fully learns the positional information of small objects, resulting in a more precise localization of the bounding boxes for small objects. This up-sampling method firstly requires the image to expand the number of channels by a 1 × 1 convolution, and then, increase the spatial dimension by shuffling the channel dimension pixels. The feature channel information is not lost, but transformed into spatial dimension information and retained. Typically, additional computational overhead is introduced by the process of expanding the number of channels. However, the sufficient number of channels of $S_{3\text{–}5}$ in this study allows the images to be up-sampled directly using a sub-pixel convolution, and the need to expand the number of channels using a 1 × 1 convolution is nullified. Therefore, the feature channel information is retained without requiring additional computation. The structure of FMEM is shown in Figure 6. Once $S_{2\text{–}5}$ are input into the FMEM, S_2 is solely involved in channel adjustment through a 1 × 1 convolution; S_5 exclusively employs sub-pixel convolution; whereas S_3 and S_4 simultaneously engage in sub-pixel and 1 × 1 convolutions. Subsequently, the features obtained through sub-pixel convolution and channel number adjustment are added element-wise, followed by a 1 × 1 convolution, resulting in feature maps $F_{2\text{–}4}$. The $F_{2\text{–}4}$ feature maps are generated by fusing the rich channel information in $S_{3\text{–}5}$, based on $S_{2\text{–}4}$. The FMEM process can be expressed using the following equations:

$$F_2 = C^{1\times1}\left(C^{1\times1}(S_2) + SPC(S_3)\right) \quad (2)$$

$$F_3 = C^{1\times1}\left(C^{1\times1}(S_3) + SPC(S_4)\right) \quad (3)$$

$$F_4 = C^{1\times1}\left(C^{1\times1}(S_4) + SPC(S_5)\right) \quad (4)$$

where S_i and F_i denote the input and output of FMEM, $C^{1\times1}$ represents the 1×1 convolution, and SPC is the sub-pixel convolution. Subsequently, the obtained feature maps are input into MIFNet for feature fusion.

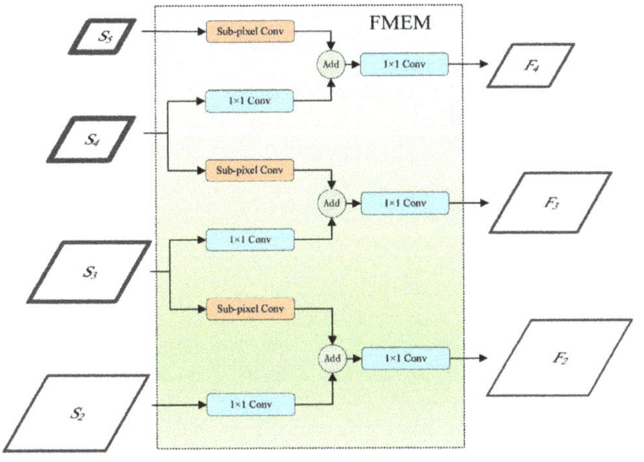

Figure 6. The structure of the proposed FMEM.

3.3. Multi-Feature Interactive Fusion Network (MIFNet)

Traffic sign detection mainly occurs in autonomous driving and assisted driving scenarios, and there are more small objects at a distance. This is primarily because there is a braking distance following emergency braking maneuvers, and early detection of traffic signs is essential to alert the driver and mitigate potential risks. YOLOv7 possesses the detection speed and model size requirements for traffic sign detection scenarios. However, the down-sampling multiple of YOLOv7 is relatively large, feature maps with the resolution of 20×20, 40×40, 80×80 in the lower three layers are usually selected, and it is difficult for the deeper feature maps to learn the features of small objects. Furthermore, YOLOv7 utilizes the PAFPN structure for feature fusion, merging semantic information from higher levels to naturally impart distinct contextual information to lower-level feature maps. Nonetheless, the feature fusion process exclusively involves the combination of neighboring feature layers, neglecting direct interactions of non-neighboring feature layers, and diluting their semantic information over multiple iterations.

In this study, we establish MIFNet to improve the detection performance of YOLOv7 for small objects and facilitate information interaction across all feature layers. Firstly, four feature layers are fed as inputs to MIFNet. Incorporating a high-resolution feature map to provide additional detailed information and employing feature fusion benefits the overall model, thus enhancing the perception of small objects. Inspired by Libra R-CNN, we further combine the integration map with PAFPN to augment multi-level features using the same depth-integrated balanced semantic feature map to enhance the information interaction between all feature layers. Lastly, to reduce the model parameters, the highest-level feature is directly utilized for integration map generation after feature enhancement, without utilizing the pyramid feature fusion of PAFPN. The rich contextual information in the highest-level feature is not lost but fully harnessed in generating the integration map, benefiting the other layers. Moreover, when the highest-level feature is fully utilized, excessive fusion can lead to unnecessary aliasing effects, which can have a negative impact on the fusion results. The structure of MIFNet is illustrated in Figure 7, where F_{3-5} are the original input feature layers and F_2 is the added small object detection layer. PAFPN is utilized to fuse the features $F_{2\text{–}4}$ to obtain the $P_{2\text{–}4}$ feature layers with increased perception of small objects. Additionally, F_5 alone uses the DFEM for feature enhancement to generate feature P_5, and the details of this module are described in Section 3.4. The integration map, I ($40 \times 40 \times 256$; consistent with F_4), is created from the four feature layers after pyramid feature fusion or enhancement by linear interpolation up-sampling and adaptive max pooling, respectively. The information interaction between features at different scales is facilitated by this process, thus ensuring the full utilization of features at each layer. The process of generating an integration map can be given by the following equation:

$$C_2 = MaxPool(P_2) \tag{5}$$

$$C_3 = MaxPool(P_3) \tag{6}$$

$$C_4 = UPSample(P_4) \tag{7}$$

$$C_5 = UPSample(P_5) \tag{8}$$

$$I = \frac{1}{4}\sum_{i=2}^{5} C_i \tag{9}$$

where P_i denotes the features generated from S_i after PAFPN feature fusion or DFEM feature enhancement, C_i represents the result of P_i after max pooling or up-sampling, *MaxPool* stands for the adaptive max pooling operation, and *UPSample* denotes linear interpolation up-sampling. The feature layers are then restored to the original resolution by adaptive max pooling and linear interpolation up-sampling for subsequent prediction.

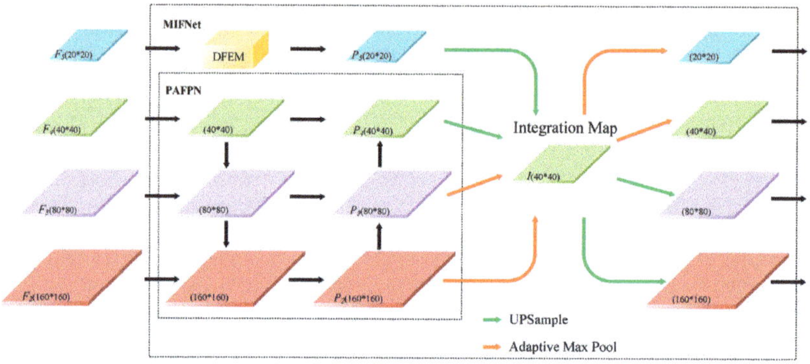

Figure 7. The structure of the proposed MIFNet.

3.4. *Deep Feature Enhanced Module (DFEM)*

Widely employed for enhancing the contextual information of features, SPP entails the extraction of various receptive field features to output fixed-size feature vectors. In YOLOv7, the SPPCSPC structure imparts distinct contextual information to the highest-level feature, enhancing every feature layer through the feature pyramid. The structure of SPPCSPC is illustrated in Figure 8. In the SPPCSPC structure, three independent pooling layers with varying kernels are employed to capture the contextual information from different receptive fields in the highest-level feature. The entire process can be represented by the following equation:

$$X = C^{1\times1}\left(C^{3\times3}\left(C^{1\times1}(F_5)\right)\right) \tag{10}$$

$$R_1 = MaxPool^{k=5}(X) \tag{11}$$

$$R_2 = MaxPool^{k=9}(X) \tag{12}$$

$$R_3 = MaxPool^{k=13}(X) \tag{13}$$

$$P5 = C^{1\times1}\left(Cat\left(C^{3\times3}\left(C^{1\times1}(Cat(R_i, X))\right), C^{1\times1}(F_5)\right)\right), i = 1,2,3 \tag{14}$$

where F_5 denotes the input feature layer, X represents the feature that will undergo pooling operations, $C^{1\times1}$ stands for the 1×1 convolution, $C^{3\times3}$ denotes the 3×3 convolution, k is the size of the max pooling kernel, R_i signifies the result after max pooling, P_5 symbolizes the total output of SPPCSPC, and *Cat* is the concatenation operation.

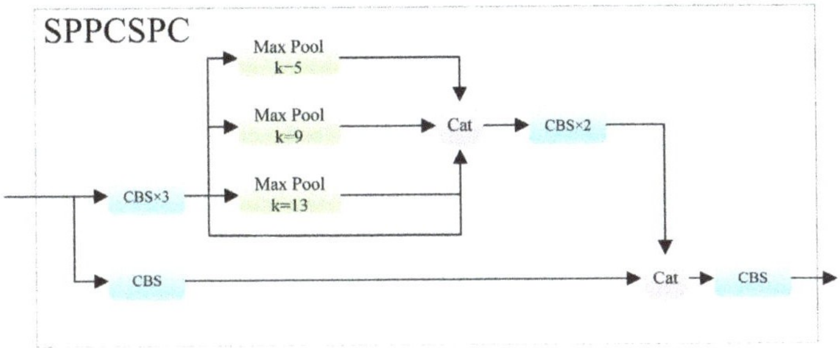

Figure 8. The structure of SPPCSPC.

To enhance the highest-level feature and improve the fusion effect of MIFNet, DFEM is introduced on the basis of SPPCSPC in this study, as illustrated in Figure 9. In the separate residual path, global average pooling and global max pooling are introduced initially. The outcomes of these pooling operations are subsequently broadcast to align with the resolution of the feature in the residual branch and added element-by-element to enhance the original feature. Simultaneously, inspired by SPPF in YOLOv5, the three independent pooling layers are improved in DFEM by connecting them in a way that small pooling kernel layers are utilized for output. This approach not only achieves pooling results equivalent to those obtained with a larger pooling kernel, preserving the receptive field, but also facilitates accelerated pooling, thereby enhancing the overall inference speed of the entire module. Ultimately, the result from the residual branch is concatenated with the accelerated pooling result to obtain the final output. This process can be represented using the following equations:

$$X = C^{1\times1}\left(C^{3\times3}\left(C^{1\times1}(F_5)\right)\right) \tag{15}$$

$$R_1 = MaxPool^{k=5}(X) \tag{16}$$

$$R_2 = MaxPool^{k=5}(R_1) \tag{17}$$

$$R_3 = MaxPool^{k=5}(R_2) \tag{18}$$

$$Y = C^{1\times1}(F_5) \tag{19}$$

$$P5 = C^{1\times1}\left(Cat\left(C^{3\times3}\left(C^{1\times1}(Cat(R_i, X))\right), \theta(GMP(Y)) + Y, \theta(GAP(Y)) + Y\right)\right), i = 1,2,3 \tag{20}$$

where F_5 denotes the input feature layer, X represents the feature to undergo the max pooling operation, $C^{1\times1}$ stands for the 1×1 convolution, $C^{3\times3}$ stands for the 3×3 convolution, k means the max pooling kernel size, R_i represents the results after max pooling, Y represents the feature that will undergo global pooling, *GMP* signifies the global max pooling, *GAP* signifies the global average pooling, θ symbolizes the broadcasting operation, P_5 manifests the total output of DFEM, and *Cat* is the concatenation operation. Through DFEM, the overall pooling process is accelerated while retaining the original receptive field.

Simultaneously, the original feature layer is endowed with global information and crucial information, enhancing the overall features.

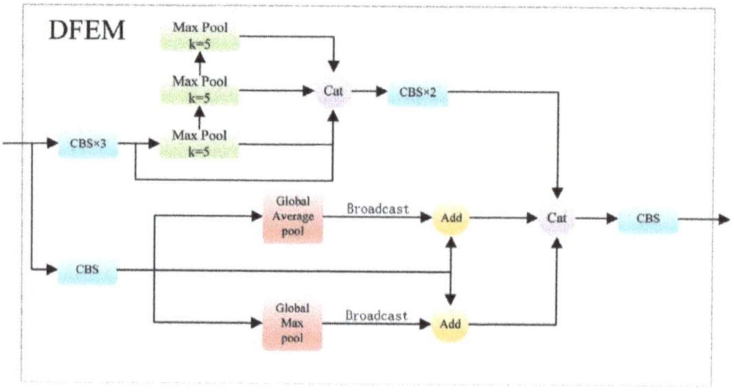

Figure 9. The structure of the proposed DFEM.

4. Experiments and Discussion

To demonstrate the effectiveness of the proposed model in this study, the experimental results are discussed in this section. To start, the datasets used in the experiments are introduced. Following this, a comprehensive presentation of the experiment setup and evaluation criteria is provided. Subsequently, we discuss the ablation experiments conducted to validate the effectiveness of the introduced modules. Finally, to verify the advantages of the proposed model, the model is compared with the baseline model and other common models.

4.1. Datasets

4.1.1. CCTSDB2021

The CCTSDB2021 dataset, created by Changsha University of Science and Technology, China, is one of the most recognized traffic sign datasets. The dataset is subdivided into three important categories, namely, "mandatory", "prohibitory", and "warning". The training and validation sets consist of 16,356 and 1500 images, respectively. To enrich the training dataset, images captured under a variety of weather conditions, such as foggy weather and snowy conditions, are also included in the CCTSDB2021 dataset. Examples of the three different categories of traffic signs and sample images from the CCTSDB2021 dataset are illustrated in Figures 10 and 11, respectively.

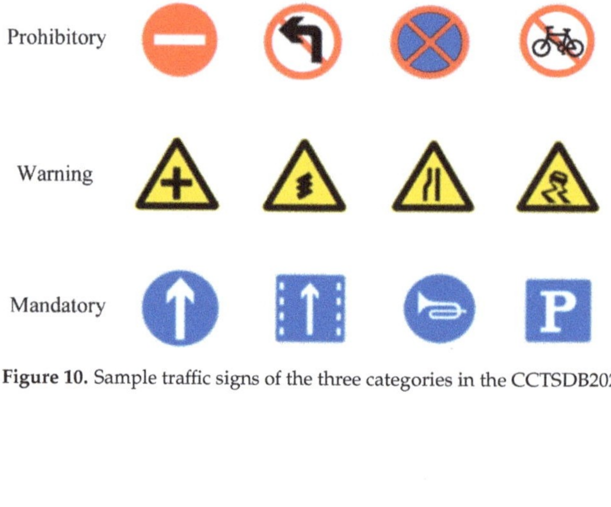

Figure 10. Sample traffic signs of the three categories in the CCTSDB2021 dataset.

Figure 11. Sample images of the CCTSDB2021 dataset.

4.1.2. TT100K

The TT100K dataset, curated and released by a joint laboratory between Tsinghua University and Tencent, comprises ~100,000 street-level images from various cities in China, and the traffic signs in these images are annotated with bounding boxes. A comprehensive range of traffic sign categories is covered by the dataset, totaling 221 different classes of traffic signs, comprising a total of 9176 images. The training and validation sets comprise 6105 and 3071 images, respectively. A significant data imbalance issue is encountered due to only 45 classes having over 100 instances and nearly half of the classes having only single-digit instances. Consequently, for our experiments, only these 45 traffic sign categories are selected, and all other classes in the TT100K dataset are excluded. The training and validation sets are subsequently reconfigured to a 9:1 ratio. The 45 traffic sign classes with over 100 instances in the TT100K dataset and sample images from the TT100K dataset are shown in Figures 12 and 13.

Figure 12. The 45 traffic sign classes with more than 100 instances in the TT100K dataset.

Figure 13. Sample images of the TT100K dataset.

4.2. Evaluation Metrics and Implementation Details

4.2.1. Evaluation Metrics

To quantitatively evaluate the performance of proposed model, precision (P), recall (R), F1 score (F1), mean average precision (mAP), the number of parameters (Params), and frames per second (FPS) are adopted as the evaluation metrics. The proportion of true positive samples among samples detected as positive by the model is denoted as P. R represents the proportion of true positive samples detected by the model among all positive samples. F1 combines both precision and recall to reflect the model's detection performance, with a higher value indicating better detection performance. The mAP is the average value of AP for all categories, where AP means the accuracy for a single class. A higher mAP value indicates better overall detection accuracy for the model. Params is employed to measure the model's complexity, and it is related to the amount of computer memory resources the model occupies. Smaller Params values indicate fewer model parameters and less memory usage. FPS is the number of images the model can detect per second. A higher FPS indicates faster model speed and the ability to process a larger number of images per second. These evaluation metrics can be calculated using the following formulas:

$$P = \frac{TP}{TP + FP} \tag{21}$$

$$R = \frac{TP}{TP + FN} \tag{22}$$

$$F1 = \frac{2PR}{P + R} \tag{23}$$

$$AP = \int_0^1 P(R)dR \tag{24}$$

$$mAP = \frac{1}{N}\sum_{i=1}^{N} AP_i \tag{25}$$

where TP represents the number of samples correctly detected as traffic signs, FP means the number of detected traffic signs with incorrect labels compared with the ground truth, FN denotes the number of samples where traffic signs are present in reality but not detected, and N is the total number of detected categories.

4.2.2. Implementation Details

To maintain objectivity, all the experiments in this study are performed on a setup with the following specifications: Linux operating system; PyTorch 1.7.1, CUDA version 11.7; and NVIDIA A100 40 G GPU. The reported FPS values in this study are obtained directly in the GPU environment. The IoU threshold for calculating mAP is set to 0.5. We employe the Adam training optimizer with an initial learning rate of 0.001. To save memory space, the batch size is set to 8. During training, the number of epochs is set to 100 and 300 for the CCTSDB2021 and TT100K datasets, respectively.

4.3. Ablation Experiments

To confirm the effectiveness of the modules proposed in YOLOv7-TS, extensive ablation experiments are conducted on both datasets and the experimental results are analyzed.

4.3.1. Ablation Study of FMEM

With YOLOv7 adopted as the baseline model, the FMEM input feature layer is modified to include three layers to match the three-layer output of the YOLOv7 backbone. FMEM is then incorporated into YOLOv7 to extract feature maps while reducing channel number loss. The specific experimental results of these two methods on the CCTSDB2021 and TT100K datasets are displayed in Table 1.

Table 1. Results of the ablation study for FMEM.

Method	P (%)	R (%)	F1 (%)	mAP (%)
Methods on CCTSDB2021 dataset				
YOLOv7	92.11	70.45	79.84	82.39
YOLOv7 + FMEM	93.12	72.69	81.65	83.70
Methods on TT100K dataset				
YOLOv7	91.46	83.01	87.03	89.77
YOLOv7 + FMEM	91.78	84.17	87.81	90.71

YOLOv7 obtains the values of 92.11%, 70.45%, 79.84%, and 82.39% for P, R, F1, and mAP, respectively, for the CCTSDB2021 dataset. However, after addressing the channel information loss, YOLOv7 + FMEM achieves P, R, F1, and mAP values of 93.12%, 72.69%, 81.65%, and 83.70%, respectively, which are 1.01%, 2.24%, 1.81% and 1.31% higher than the respective values for YOLOv7. Similarly, on the TT100K dataset, YOLOv7 + FMEM achieves P, R, F1, and mAP increments of 0.32%, 1.16%, 0.78%, and 0.94%, respectively, compared with YOLOv7. This comparative analysis of experimental data for the two methods demonstrates the superior detection performance and accuracy of YOLOv7 + FMEM. This effect is due to a reduction in the channel information loss by FMEM during feature map extraction, improving the overall detection performance.

To present the results of the ablation experiments more clearly, examples of the results with both datasets are provided for the two methods in Figures 14 and 15. From these figures, it is evident that YOLOv7 detects objects with relatively low confidence scores. However, after incorporating FMEM, the confidence scores for all detected objects increase due to its ability to preserve the channel information lost in basic YOLOv7. Among the objects missed by YOLOv7, some are detected after the addition of FMEM, but a few individual objects remain undetected, which indicates that while the preserved channel information may not completely resolve the issue of missing small objects, it still has a positive impact. Meanwhile, in the second set of comparison images in Figure 14, YOLOv7 misclassifies a "speed limit 30" traffic sign as "speed limit 50", which is corrected upon FMEM incorporation. In summary, these experimental results collectively provide evidence of the effectiveness of FMEM.

(a)

(b)

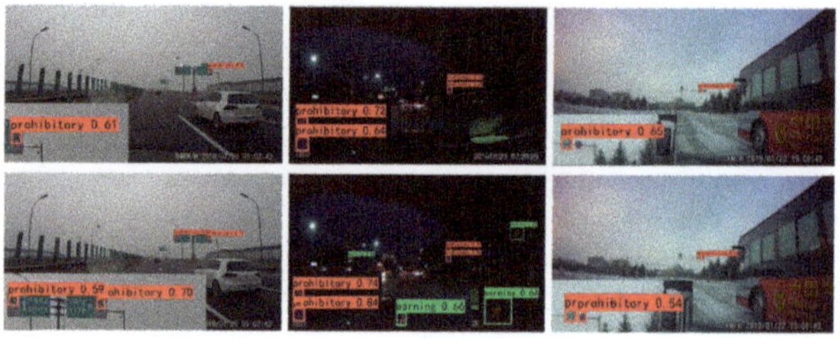

Figure 14. Ablation results for FMEM tested on the CCTSDB2021 dataset. (**a**) YOLOv7; (**b**) YOLOv7 + FMEM.

(a)

(b)

Figure 15. Ablation results for FMEM tested on the TT100K dataset. (**a**) YOLOv7; (**b**) YOLOv7 + FMEM.

4.3.2. Ablation Study of MIFNet

The impact of the MIFNet on traffic sign detection performance is also investigated. Ablation experiments are conducted on both datasets to assess the following models: (a) baseline YOLOv7, (b) YOLOv7 + integration map, (c) YOLOv7 + integration map + small object detection layer, and (d) YOLOv7 + MIFNet. The experimental results for these four methods are presented in Table 2.

Table 2. Results of the ablation study for MIFNet.

Method	P (%)	R (%)	F1 (%)	mAP (%)
Methods on CCTSDB2021 dataset				
YOLOv7	92.11	70.45	79.84	82.39
YOLOv7 + integration map	93.87	71.89	81.42	82.99
YOLOv7 + integration map + small object detection layer	90.34	78.21	83.84	83.91
YOLOv7 + MIFNet	92.23	78.36	84.73	84.02
Methods on TT100K dataset				
YOLOv7	91.46	83.01	87.03	89.77
YOLOv7 + integration map	92.89	83.78	88.10	90.12
YOLOv7 + integration map + small object detection layer	92.25	86.96	89.53	91.45
YOLOv7 + MIFNet	92.13	87.12	89.56	91.48

Using the CCTSDB2021 dataset, YOLOv7 + integration map achieves P, R, F1, and mAP values of 93.87%, 71.89%, 81.42%, and 82.99%, respectively, demonstrating an improvement of 1.76%, 1.44%, 1.58%, and 0.60% compared with YOLOv7. This improvement can be attributed to the enhanced feature fusion effect due to the substantial information interaction among all feature layers. Further, with the addition of the small object detection layer, the obtained values are 90.34%, 78.21%, 83.84%, and 83.91%, respectively. While the P value is slightly lower than that of YOLOv7, the R, F1, and mAP values exhibited improvements of 7.76%, 4.00%, and 1.52%, respectively. This indicates that the high-resolution feature provides more detailed information, facilitating the model recognition of small objects. Finally, YOLOv7 + MIFNet achieves P, R, F1, and mAP values of 92.23%, 78.36%, 84.73%, and 84.02%, respectively, with the CCTSDB2021 dataset, demonstrating an improvement of 0.12%, 7.91%, 4.89%, and 1.63%, respectively, compared with YOLOv7, and 1.89%, 0.15%, 0.89%, and 0.11%, respectively, compared with YOLOv7 + integration map + small object detection layer. Although the number of feature layers used for the pyramid feature fusion is reduced with MIFNet, the detection accuracy does not decrease, but rather shows a slight improvement. This indicates that when feature information is already fully utilized, additional rounds of feature fusion may not necessarily result in performance gains and may even exacerbate model parameter redundancy. Using the TT100K dataset, excellent performance is also demonstrated by YOLOv7 + MIFNet, and all indicators are improved, verifying the superiority of MIFNet for small object detection and feature fusion.

The comparative sample results of YOLOv7 and YOLOv7 + MIFNet using both datasets are displayed in Figures 16 and 17. From these figures, it is noticeable that after using MIFNet for feature fusion, the detected objects exhibit higher confidence scores. This can be attributed to the excellent feature fusion capability of MIFNet, which enhances the model detection performance. Additionally, YOLOv7 + MIFNet successfully detects and accurately categorizes the objects that are missed or misclassified by YOLOv7 due to the former's effective utilization of detailed information in the high-resolution feature layer.

(a)

(b)

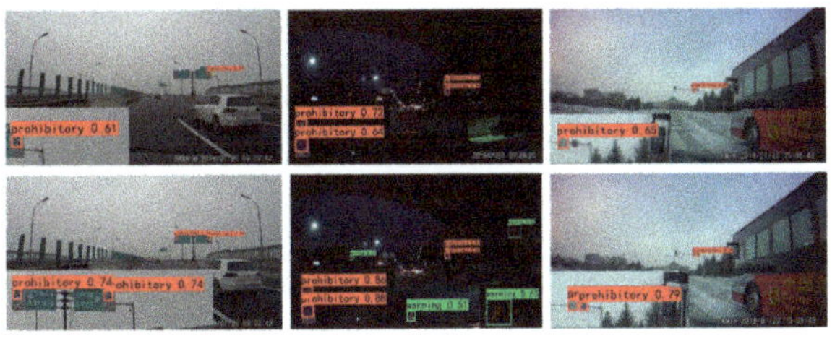

Figure 16. Ablation results for MIFNet tested on the CCTSDB2021 dataset. (**a**) YOLOv7; (**b**) YOLOv7 + MIFNet.

(a)

(b)

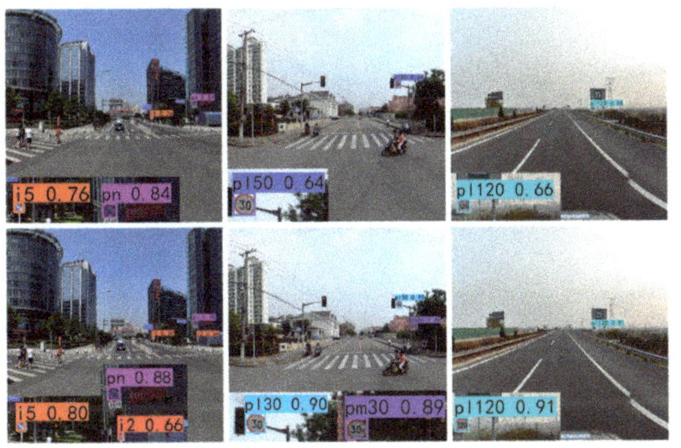

Figure 17. Ablation results for MIFNet tested on the TT100K dataset. (**a**) YOLOv7; (**b**) YOLOv7 + MIFNet.

To demonstrate that MIFNet can achieve favorable feature fusion effects, attain higher detection accuracy, and simultaneously reduce the model parameters, we present a parameter comparison in Table 3. With the utilization of the integration map, the model parameters decrease by 1.8 M compared with YOLOv7. Upon the addition of the small object detection layer and using the four-layer structure, the model parameters increase by 1.3 M. When applying MIFNet to YOLOv7, the model parameters reduce to 28.2 M, representing a decrease of 8.7 M compared with YOLOv7. Through the experimental data and visualization of detection results, it is conclusively demonstrated that MIFNet effectively reduces the model parameters while simultaneously enhancing accuracy.

Table 3. Comparison of the parameters among several models.

Method	Params (M)
YOLOv7	36.9
YOLOv7 + integration map	35.1
YOLOv7 + integration map + small object detection layer	36.4
YOLOv7 + MIFNet	28.2

4.3.3. Ablation Study of DFEM

To validate the effectiveness of DFEM, we conduct ablation experiments with both datasets for the YOLOv7 + DFEM model. The experimental results are presented in Table 4.

Table 4. Results of the ablation study for DFEM.

Method	P (%)	R (%)	F1 (%)	mAP (%)
Methods on CCTSDB2021 dataset				
YOLOv7	92.11	70.45	79.84	82.39
YOLOv7 + DFEM	94.24	71.23	81.14	83.71
Methods on TT100K dataset				
YOLOv7	91.46	83.01	87.03	89.77
YOLOv7 + DFEM	92.36	84.12	88.05	90.96

Applied to the CCTSDB2021 dataset, YOLOv7 + DFEM obtains P, R, F1, and mAP values of 94.24%, 71.23%, 81.14%, and 83.71%, respectively, demonstrating improvements of 2.13%, 0.78%, 1.30%, and 1.32%, respectively, compared with YOLOv7. Similarly, with the TT100K dataset, YOLOv7 + DFEM demonstrates improvements of 0.90%, 1.11%, 1.02%, and 1.19% in the P, R, F1, and mAP values, respectively, compared with YOLOv7. The highest-level feature is enhanced by providing it with global and crucial information. After combining it with the different receptive fields provided by the upper branch pooling, all feature layers benefit from feature fusion.

The comparative detection results of YOLOv7 and YOLOv7 + DFEM using the CCTSDB2021 and TT100K datasets are shown in Figures 18 and 19. Compared with YOLOv7, the enhancement of the highest-level feature by DFEM improves the model detection performance, and the confidence of all detected objects is improved. Some small objects that are initially missed due to the low confidence scores of YOLOv7 are now detected with higher confidence after feature enhancement with DFEM, indicating its positive impact on small object detection. Additionally, objects incorrectly detected by YOLOv7 are also correctly identified by YOLOv7 + DFEM. These results collectively demonstrate the effectiveness of DFEM in enhancing the highest-layer feature.

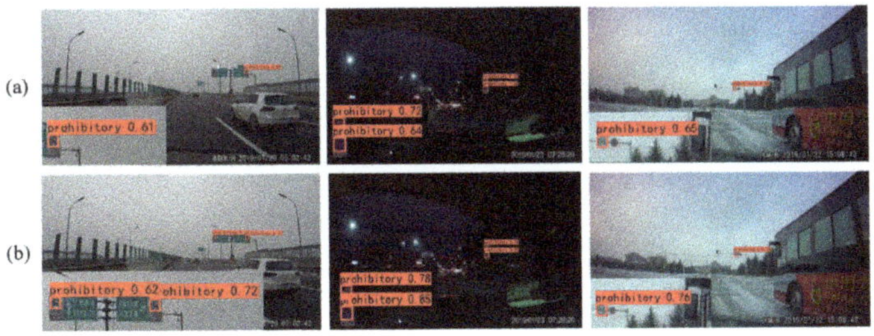

Figure 18. Ablation results for DFEM tested on the CCTSDB2021 dataset. (**a**) YOLOv7; (**b**) YOLOv7 + DFEM.

Figure 19. Ablation results for DFEM tested on the TT100K dataset. (**a**) YOLOv7; (**b**) YOLOv7 + DFEM.

To demonstrate that the introduction of pooling acceleration in DFEM can speed up the inference of the whole module, we present a parameter and inference time comparison in Table 5 among SPPCSPC, DFEM (without pooling acceleration), and DFEM. By comparing these, it is evident that DFEM (without pooling acceleration) increases the module parameters compared with SPPCSPC, simultaneously extending the inference time. This is due to the fact that DFEM (without pooling acceleration) adds global max pooling and global average pooling to the residual branch on the basis of SPPCSPC. After introducing pooling acceleration in DFEM, the module parameters remain unchanged compared with DFEM (without pooling acceleration). However, the inference time is reduced. Simultaneously, there is a significant improvement in inference speed compared with SPPCSPC.

Table 5. Comparison of the parameters and inference time among several modules.

Method	Params (M)	Inference Time (s)
SPPCSPC	7.6	1.2542
DFEM (without pooling acceleration)	8.1	1.2787
DFEM	8.1	1.0588

4.3.4. Ablation Study between Different Modules

The modules proposed in this study achieve varying degrees of performance improvement on the YOLOv7 baseline network. We further conduct ablation experiments to demonstrate that the concurrent application of the three proposed modules on YOLOv7 does not have a negative impact and can yield favorable results, as shown in Table 6.

Table 6. Results of the ablation study between the components in the proposed method.

Method	F1 (%)	mAP (%)
Methods on CCTSDB2021 dataset		
YOLOv7	79.84	82.39
YOLOv7 + FMEM + MIFNet	84.34	85.08
YOLOv7 + FMEM + DFEM	82.25	84.59
YOLOv7 + MIFNet + DFEM	83.96	84.83
YOLOv7+ FMEM + MIFNet + DFEM	84.90	86.02
Methods on TT100K dataset		
YOLOv7	87.03	89.77
YOLOv7 + FMEM + MIFNet	89.36	91.80
YOLOv7 + FMEM + DFEM	88.79	91.33
YOLOv7 + MIFNet + DFEM	89.50	91.95
YOLOv7+ FMEM + MIFNet + DFEM	89.92	92.45

It can be observed that for the CCTSDB2021 dataset, YOLOv7 + FMEM + MIFNet, which simultaneously focuses on rich channel information and feature fusion, achieves an F1 value of 84.34% and mAP of 85.08%. Values of 82.25% and 84.59%, respectively, are obtained by YOLOv7 + FMEM + DFEM, which benefits from the preservation of channel information and the enhancement of high-level features. YOLOv7 + MIFNet + DFEM, in which high-level features are enhanced and the enhanced features are propagated to other layers through feature fusion, achieved values of 83.96% and 84.83%. Significant improvements in the F1 score and mAP are exhibited by these three methods, indicating that model performance and detection accuracy can be enhanced by any combination of the proposed components in this study. When introducing all the proposed components into YOLOv7 simultaneously, the F1 and mAP values obtained are 84.90% and 86.02%, which are the highest values among the five methods. In addition, the experimental results using TT100K dataset also reach the same conclusion. This ablation experiment fully demonstrates the effectiveness of simultaneously introducing the components proposed in this study into the benchmark model.

4.4. Comparison Experiments

To validate the superiority of the YOLOv7-TS model, a comparison is conducted between YOLOv7-TS and the baseline model YOLOv7, as well as other common models.

4.4.1. Comparison Experiment between YOLOv7-TS and YOLOv7

Table 7 presents the comparative experimental results of YOLOv7 and YOLOv7-TS tested on both datasets to demonstrate the superiority of the latter, which is based on the YOLOv7 baseline model and incorporates FMEM, MIFNet, and DFEM.

Table 7. Results of the YOLOv7 and YOLOv7-TS.

Method	P (%)	R (%)	F1 (%)	mAP (%)	Params (M)	FPS
Methods on CCTSDB2021 dataset						
YOLOv7	92.11	70.45	79.84	82.39	36.9	42.3
YOLOv7-TS	92.69	78.31	84.90	86.02	34.7	37.0
Methods on TT100K dataset						
YOLOv7	91.46	83.01	87.03	89.77	36.9	43.1
YOLOv7-TS	92.36	87.60	89.92	92.45	34.7	37.3

A noteworthy trend is that YOLOv7-TS makes substantial strides in both model performance and detection accuracy, albeit at the cost of a modest reduction in speed. Specifically, using the CCTSDB2021 dataset, YOLOv7-TS demonstrates improvements of 0.58%, 7.86%, 5.06%, and 3.63% in P, R, F1, and mAP, respectively, compared with YOLOv7.

Similarly, with the TT100K dataset, it demonstrates enhancements of 0.90%, 4.59%, 2.89%, and 2.68% in P, R, F1, and mAP, respectively. These advancements can be attributed to the emphasis of YOLOv7-TS on leveraging rich channel information, bolstering feature quality, and refining feature fusion techniques. Moreover, YOLOv7-TS achieves a reduction of 2.2 million parameters relative to YOLOv7, underscoring the efficacy of employing a three-layer feature pyramid fusion approach in MIFNet for mitigating the redundancy of model parameters.

For the sake of a clearer comparison of the detection results and regions of interest between the two models, the experimental results and heatmaps are visualized in Figures 20–23. In Figures 20 and 21, it is revealed that when traffic signs are very small, issues such as missing objects and low confidence in the detected objects are exhibited by YOLOv7. Conversely, higher confidence levels in object detection are consistently yielded by YOLOv7-TS, thus attesting to the successful mitigation of channel information loss and the enhancement of feature fusion. Furthermore, the objects that are overlooked by YOLOv7 are accurately detected by YOLOv7-TS, signifying its superior capability in perceiving small objects. In Figures 22 and 23, the regions of interest and perceptual abilities of both models are depicted. The red boxes in the figures reflect that YOLOV7-TS has a stronger sensing ability than YOLOv7, enabling easier detection of nearby objects. With the incorporation of MIFNet, commendable perception capability for distant small objects is exhibited by YOLOv7-TS, whereas it remains a challenge for YOLOv7. In summary, the effectiveness of YOLOv7-TS in the realm of traffic sign detection is confirmed by the presented experimental results.

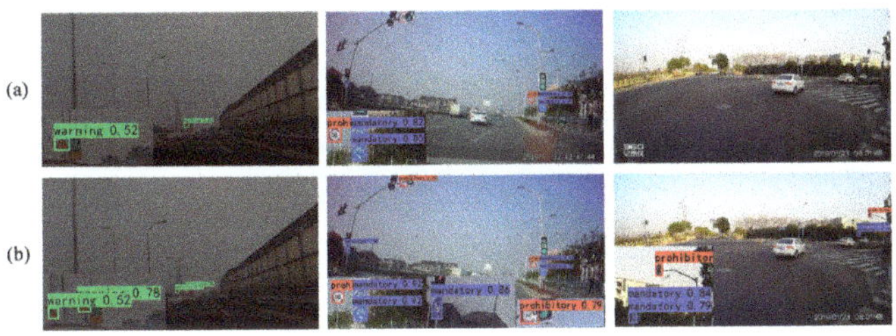

Figure 20. Comparison results of YOLOv7 and YOLOv7-TS tested on the CCTSDB2021 dataset. (**a**) YOLOv7; (**b**) YOLOv7-TS.

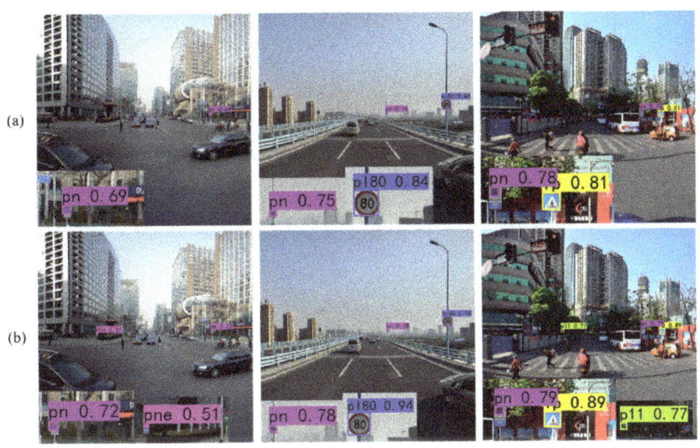

Figure 21. Comparison results of YOLOv7 and YOLOv7-TS tested on the TT100K dataset. (**a**) YOLOv7; (**b**) YOLOv7-TS.

(a)
(b)
(a)

Figure 22. Heatmaps of YOLOv7 and YOLOv7-TS tested on the CCTSDB2021 dataset. (**a**) Original images; (**b**) YOLOv7. (**c**) YOLOv7-TS.

(a)
(b)
(c)

Figure 23. Heatmaps of YOLOv7 and YOLOv7-TS tested on the TT100K dataset. (**a**) Original images; (**b**) YOLOv7; (**c**) YOLOv7-TS.

4.4.2. Comparison Experiment between YOLOv7-TS and Other Popular Models

Comparative traffic sign detection experiments are also conducted between YOLOv7-TS and other common Two-Stage and One-Stage object detection algorithms using the CCTSDB2021 dataset, as shown in Table 8.

Table 8. Results of the and YOLOv7-TS and different methods tested on the CCTSDB2021 dataset.

Method	F1 (%)	mAP (%)	FPS (%)
Faster R-CNN	66.52	56.58	4.9
Libra R-CNN	69.93	61.35	8.8
Dynamic R-CNN	69.83	60.01	9.0
SSD	42.00	49.20	22.3
YOLOv3	56.77	50.48	20.3
YOLOv4	62.15	51.69	16.6
RetinaNet	65.69	57.78	8.9
YOLOv5s	77.23	80.92	41.3
YOLOv5m	78.59	81.65	30.8
SC-YOLO	84.50	84.30	-
YOLOv7-TS	84.90	86.02	37.0

A significant advantage in terms of model performance and detection accuracy is exhibited by YOLOv7-TS compared with all other models. Specifically, Faster R-CNN, which utilizes a single feature layer without employing multi-scale feature fusion, achieves F1 and mAP values of 66.52% and 56.58%, respectively. Libra R-CNN, which is designed to pursue balance across three different levels during the training process, achieves F1 and mAP values of 69.93% and 61.35%, respectively. Dynamic R-CNN [52], which leverages dynamic label assignment strategies to adaptively adjust label assignment criteria, achieves F1 and mAP values of 69.83% and 60.01%, respectively. Although these Two-Stage algorithms show improved detection performance compared with R-CNN, they overlook feature enhancement, resulting in F1 and mAP values lower than those achieved by YOLOv7-TS. Furthermore, Two-Stage detection algorithms necessitate the generation of candidate regions prior to detection, leading to significantly slower detection speeds. In the case of One-Stage algorithms, the lowest performance is achieved by SSD, with F1 and mAP values of only 42.00% and 49.20%, respectively. This can be attributed to its direct use of multi-scale features for prediction by SSD while neglecting feature fusion and enhancement. The classic YOLOv3 algorithm achieves F1 and mAP values of 56.77% and 51.69%, respectively, and YOLOv4, achieves further improvements of 5.38% and 1.21%, respectively. Although FPN is employed for feature fusion in YOLOv3, it overlooks feature enhancement. Information with varying receptive fields is introduced through SPP in YOLOv4, but it does not match the feature enhancement achieved by DFEM. Class imbalance issues during the training of One-Stage algorithms are addressed by RetinaNet [53] through the introduction of focal loss, resulting in F1 and mAP values of 65.69% and 57.78%, respectively, surpassing those of YOLOv3 and YOLOv4. However, its detection speed struggles to meet real-time requirements. YOLOv5s achieves F1 and mAP values of 77.23% and 80.92%, respectively, while maintaining a high speed, while YOLOv5m sacrifices speed for higher detection accuracy. However, the issue of channel information loss during feature map extraction is not addressed by the YOLOv5 series, resulting in experimental results inferior to those of YOLOv7-TS. SC-YOLO [54], an improvement upon YOLOv5s, uses cross-stage attention networks and enhanced feature fusion, effectively improving small object detection capabilities and achieving F1 and mAP values of 84.50% and 84.30%, respectively. However, it still does not surpass the proposed YOLOv7-TS model. In summary, in testing on the CCTSDB2021 dataset, an F1 score of 84.90% and an mAP of 86.02% are achieved by YOLOv7-TS at 37.0 FPS, surpassing all other common algorithms. While it may not match the speed of YOLOv5s, YOLOv7-TS vastly outperforms it in terms of detection performance and accuracy.

5. Conclusions

In this study, we propose a traffic sign detection model named YOLOv7-TS, building upon YOLOv7 as its foundation. Firstly, the issue of feature channel information loss caused by the 1×1 convolution for feature map extraction is addressed through the introduction of

FMEM. The feature maps extracted through FMEM simultaneously encompass high-level semantic information and low-level detailed information, actively promoting both object recognition and localization. Secondly, MIFNet is designed to enhance the feature fusion by promoting information interaction among all feature layers. Simultaneously, the rich detail information in the high-resolution feature map is combined to locate small objects and alleviate the problem of missing detection of small objects. Thirdly, DFEM enriches the highest-layer feature and improves the inference speed of the whole module. The highest-level feature containing global information and crucial information improves the detection accuracy of the model through feature fusion. In summary, YOLOv7-TS leverages the rich channel semantic information and spatial detail information in multi-scale features. Through feature enhancement and a more effective feature fusion network, the entire model benefits from these aspects. The analysis in the experimental section shows that YOLOv7-TS achieves higher F1 scores and mAP values compared with YOLOv7 and other common models. This fully demonstrates the excellent performance and highest detection accuracy of the model in traffic sign detection. The comparison of the results and heatmaps between YOLOv7-TS and YOLOv7 indicates that YOLOv7-TS alleviates the shortcomings of YOLOv7 in small object detection, showing a stronger ability to perceive small objects. However, in terms of FPS, our model lags behind YOLOv7 and YOLOv5s. This is attributed to the more complex architecture of YOLOv7-TS, requiring larger computational resources, resulting in a relatively slower inference speed. Our future work will focus on optimizing the model architecture, exploring the use of lightweight backbone networks and lightweight convolutions, while refining module details to reduce computational overhead and improve model speed.

This study was conducted based on traffic sign datasets. There are still many potential challenges and factors that should be considered when applying the model to real-world scenarios for traffic sign detection: Firstly, real-world environments exhibit diverse conditions. Changes in lighting due to different times of day and weather conditions may lead to variations in the appearance of traffic signs. Adverse weather, such as rain, snow, or haze, can blur or partially obstruct signs, making them more challenging to detect. Seasonal changes may alter road and environmental conditions, impacting sign visibility. Different road situations, such as wet, slippery, uneven, or dirty roads, may affect the clarity of traffic signs. Secondly, specific regions in the dataset may have distinct geographical and cultural characteristics compared with the target deployment area, posing challenges in detection due to regional variations. Thirdly, dynamic context adaptation is essential. Real-world scenarios often encompass dynamic contexts, such as urban versus rural environments or highways versus local roads. The model should dynamically adjust based on specific circumstances, considering factors such as traffic density, speed limits, and road infrastructure. Addressing these potential challenges requires the model to have higher robustness and stronger generalization capabilities when detecting in real-world scenarios. Several strategies can be employed: Firstly, data augmentation is crucial. Utilizing diverse data augmentation techniques, including random rotation, flipping, scaling, and brightness adjustments, helps simulate the diversity present in real-world scenarios. This aids in enhancing the model's robustness to different environments and backgrounds. Secondly, training with multi-source data is essential. Integrating data from various regions and diverse scenarios ensures that the model encounters a broader range of contexts during the training phase. This contributes to improving the model's generalization performance in different geographical and cultural environments. Thirdly, transfer learning is a valuable approach. Employing transfer learning methods allows the model to transfer knowledge learned in one region or scenario to another. This accelerates the adaptation process of the model to new environments. Fourth, it is beneficial to include an environment awareness module and real-time feedback. An environment awareness module is introduced into the model to capture and deal with environmental changes and adjust the detection strategy. The real-time feedback mechanism continuously monitors the performance of the model in real-world scenarios, enabling online adjustment based on actual feedback. Beyond adapting to diverse environments and backgrounds, deploying the model in real-world scenarios necessitates considerations of integration with existing systems, ethical concerns, and adherence to legal regulations. Seamless integration ensures effective deployment and collaboration with other components of traffic infrastructure. As artificial intelligence applications become more prevalent in real-world settings, compliance with ethical standards and legal regulations becomes paramount. Ensuring the model's privacy compliance in public spaces is essential. In our future research, we will focus on addressing the potential challenges associated with deploying the model in real-world scenarios, continually enhancing the model's robustness and generalization to deliver outstanding performance in various environments and contexts.

Author Contributions: Conceptualization, S.Z. and Y.Y.; methodology, S.Z. and Y.Y.; software, Y.W. and Y.Y.; validation, X.W. and F.Z.; formal analysis, X.W. and F.Z.; investigation, Y.Y.; resources, S.Z. and F.Z.; data curation, X.W. and Y.W.; writing—original draft preparation, Y.Y.; writing—review and editing, S.Z. and X.W.; visualization, Y.Y.; supervision, Y.W.; project administration, S.Z.; funding acquisition, S.Z. All authors have read and agreed to the published version of the manuscript.

Funding: This research was funded by the National Natural Science Foundation of China (grant number 61602157) and Henan Science and Technology Planning Program (grant number 202102210167).

Institutional Review Board Statement: Not applicable.

Informed Consent Statement: Not applicable.

Data Availability Statement: The data presented in this study can be requested from the corresponding author, and these data are not currently available for public access.

Acknowledgments: The authors thank the editor and anonymous reviewers for providing helpful suggestions for improving the quality of this manuscript.

Conflicts of Interest: The authors declare no conflicts of interest.

References

1. Benallal, M.; Meunier, J. Real-time color segmentation of road signs. In Proceedings of the CCECE 2003-Canadian Conference on Electrical and Computer Engineering toward a Caring and Humane Technology (Cat. No. 03CH37436), Montreal, QC, Canada, 4–7 May 2003; pp. 1823–1826.
2. Ruta, A.; Li, Y.; Liu, X. Real-time traffic sign recognition from video by class-specific discriminative features. *Pattern Recognit.* **2010**, *43*, 416–430. [CrossRef]
3. Nguwi, Y.-Y.; Kouzani, A.Z. Automatic road sign recognition using neural networks. In Proceedings of the The 2006 IEEE International Joint Conference on Neural Network Proceedings, Vancouver, BC, Canada, 16–21 July 2006; pp. 3955–3962.
4. Fleyeh, H.; Biswas, R.; Davami, E. Traffic sign detection based on AdaBoost color segmentation and SVM classification. In Proceedings of the Eurocon 2013, Zagreb, Croatia, 1–4 July 2013; pp. 2005–2010.
5. Yang, Y.; Wu, F. Real-time traffic sign detection via color probability model and integral channel features. In Proceedings of the Pattern Recognition: 6th Chinese Conference, CCPR 2014, Changsha, China, 17–19 November 2014; Proceedings, Part II 6, 2014. pp. 545–554.
6. Loy, G.; Barnes, N. Fast shape-based road sign detection for a driver assistance system. In Proceedings of the 2004 IEEE/RSJ International Conference on Intelligent Robots and Systems (IROS) (IEEE Cat. No. 04CH37566), Sendai, Japan, 28 September–2 October 2004; pp. 70–75.
7. Zaklouta, F.; Stanciulescu, B. Real-time traffic sign recognition in three stages. *Robot. Auton. Syst.* **2014**, *62*, 16–24. [CrossRef]
8. Abukhait, J.; Abdel-Qader, I.; Oh, J.-S.; Abudayyeh, O. Road sign detection and shape recognition invariant to sign defects. In Proceedings of the 2012 IEEE International Conference on Electro/Information Technology, Indianapolis, IN, USA, 6–8 May 2012; pp. 1–6.
9. Krizhevsky, A.; Sutskever, I.; Hinton, G.E. Imagenet classification with deep convolutional neural networks. *Adv. Neural Inf. Process. Syst.* **2012**, *25*, 84–90. [CrossRef]
10. Girshick, R.; Donahue, J.; Darrell, T.; Malik, J. Rich feature hierarchies for accurate object detection and semantic segmentation. In Proceedings of the IEEE Conference on Computer Vision and Pattern Recognition, Columbus, OH, USA, 23–28 June 2014; pp. 580–587.
11. He, K.; Zhang, X.; Ren, S.; Sun, J. Spatial pyramid pooling in deep convolutional networks for visual recognition. In *IEEE Transactions on Pattern Analysis and Machine Intelligence*; IEEE: New York, NY, USA, 2015; Volume 37, pp. 1904–1916.
12. Girshick, R. Fast r-cnn. In Proceedings of the IEEE International Conference on Computer Vision, Santiago, Chile, 7–13 December 2015; pp. 1440–1448.

13. Ren, S.; He, K.; Girshick, R.; Sun, J. Faster r-cnn: Towards real-time object detection with region proposal networks. In Proceedings of the Advances in Neural Information Processing Systems 28 (NIPS 2015), Montreal, QC, Canada, 7–12 December 2015; Volume 28.
14. Pang, J.; Chen, K.; Shi, J.; Feng, H.; Ouyang, W.; Lin, D. Libra r-cnn: Towards balanced learning for object detection. In Proceedings of the IEEE/CVF Conference on Computer Vision and Pattern Recognition, Long Beach, CA, USA, 15–20 June 2019; pp. 821–830.
15. Redmon, J.; Divvala, S.; Girshick, R.; Farhadi, A. You only look once: Unified, real-time object detection. In Proceedings of the IEEE Conference on Computer Vision and Pattern Recognition, Las Vegas, NV, USA, 27–30 June 2016; pp. 779–788.
16. Liu, W.; Anguelov, D.; Erhan, D.; Szegedy, C.; Reed, S.; Fu, C.-Y.; Berg, A.C. Ssd: Single shot multibox detector. In Proceedings of the Computer Vision–ECCV 2016: 14th European Conference, Amsterdam, The Netherlands, 11–14 October 2016; Proceedings, Part I 14, 2016. pp. 21–37.
17. Redmon, J.; Farhadi, A. YOLO9000: Better, faster, stronger. In Proceedings of the IEEE Conference on Computer Vision and Pattern Recognition, Honolulu, HI, USA, 21–26 July 2017; pp. 7263–7271.
18. Redmon, J.; Farhadi, A. Yolov3: An incremental improvement. *arXiv preprint* **2018**, arXiv:1804.02767.
19. Lin, T.-Y.; Dollár, P.; Girshick, R.; He, K.; Hariharan, B.; Belongie, S. Feature pyramid networks for object detection. In Proceedings of the IEEE Conference on Computer Vision and Pattern Recognition, Honolulu, HI, USA, 21–26 July 2017; pp. 2117–2125.
20. Bochkovskiy, A.; Wang, C.-Y.; Liao, H.-Y.M. Yolov4: Optimal speed and accuracy of object detection. *arXiv preprint* **2020**, arXiv:2004.10934.
21. Wang, C.-Y.; Liao, H.-Y.M.; Wu, Y.-H.; Chen, P.-Y.; Hsieh, J.-W.; Yeh, I.-H. CSPNet: A new backbone that can enhance learning capability of CNN. In Proceedings of the IEEE/CVF Conference on Computer Vision and Pattern Recognition Workshops, Seattle, WA, USA, 14–19 June 2020; pp. 390–391.
22. Liu, S.; Qi, L.; Qin, H.; Shi, J.; Jia, J. Path aggregation network for instance segmentation. In Proceedings of the IEEE Conference on Computer Vision and Pattern Recognition, Salt Lake City, UT, USA, 18–23 June 2018; pp. 8759–8768.
23. Ultralytics/YOLOv5. Available online: https://github.com/ultralytics/yolov5 (accessed on 26 June 2020).
24. Gui, Z.; Chen, J.; Li, Y.; Chen, Z.; Wu, C.; Dong, C. A lightweight tea bud detection model based on Yolov5. *Comput. Electron. Agric.* **2023**, *205*, 107636. [CrossRef]
25. Vaswani, A.; Shazeer, N.; Parmar, N.; Uszkoreit, J.; Jones, L.; Gomez, A.N.; Kaiser, Ł.; Polosukhin, I. Attention Is All You Need. In Proceedings of the Advances in Neural Information Processing Systems 30 (NIPS 2017), Long Beach, CA, USA, 4–9 December 2017; Volume 30.
26. Dosovitskiy, A.; Beyer, L.; Kolesnikov, A.; Weissenborn, D.; Zhai, X.; Unterthiner, T.; Dehghani, M.; Minderer, M.; Heigold, G.; Gelly, S. An image is worth 16x16 words: Transformers for image recognition at scale. *arXiv preprint* **2020**, arXiv:2010.11929.
27. Beal, J.; Kim, E.; Tzeng, E.; Park, D.H.; Zhai, A.; Kislyuk, D. Toward transformer-based object detection. *arXiv preprint* **2020**, arXiv:2012.09958.
28. Xie, T.; Dai, K.; Jiang, Z.; Li, R.; Mao, S.; Wang, K.; Zhao, L. ViT-MVT: A Unified Vision Transformer Network for Multiple Vision Tasks. In *IEEE Transactions on Neural Networks and Learning Systems;* IEEE: Piscataway Township, NJ, USA, 2023.
29. Wang, S.; Xie, T.; Cheng, J.; Zhang, X.; Liu, H. MDL-NAS: A Joint Multi-Domain Learning Framework for Vision Transformer. In Proceedings of the IEEE/CVF Conference on Computer Vision and Pattern Recognition, Vancouver, BC, Canada, 17–24 June 2023; pp. 20094–20104.
30. Gavrilescu, R.; Zet, C.; Foșalău, C.; Skoczylas, M.; Cotovanu, D. Faster R-CNN: An approach to real-time object detection. In Proceedings of the 2018 International Conference and Exposition on Electrical And Power Engineering (EPE), Iasi, Romania, 18–19 October 2018; pp. 0165–0168.
31. Yang, T.; Long, X.; Sangaiah, A.K.; Zheng, Z.; Tong, C. Deep detection network for real-life traffic sign in vehicular networks. *Comput. Netw.* **2018**, *136*, 95–104. [CrossRef]
32. Zhu, Z.; Liang, D.; Zhang, S.; Huang, X.; Li, B.; Hu, S. Traffic-sign detection and classification in the wild. In Proceedings of the IEEE Conference on Computer Vision and Pattern Recognition, Las Vegas, NV, USA, 27–30 June 2016; pp. 2110–2118.
33. Mathias, M.; Timofte, R.; Benenson, R.; Van Gool, L. Traffic sign recognition—How far are we from the solution? In Proceedings of the 2013 International Joint Conference on Neural Networks (IJCNN), Dallas, TX, USA, 4–9 August 2013; pp. 1–8.
34. Zhang, J.; Xie, Z.; Sun, J.; Zou, X.; Wang, J. A cascaded R-CNN with multiscale attention and imbalanced samples for traffic sign detection. *IEEE Access* **2020**, *8*, 29742–29754. [CrossRef]
35. Cai, Z.; Vasconcelos, N. Cascade r-cnn: Delving into high quality object detection. In Proceedings of the IEEE Conference on Computer Vision and Pattern Recognition, Salt Lake City, UT, USA, 18–23 June 2018; pp. 6154–6162.
36. Mohd-Isa, W.-N.; Abdullah, M.-S.; Sarzil, M.; Abdullah, J.; Ali, A.; Hashim, N. Detection of Malaysian traffic signs via modified YOLOv3 algorithm. In Proceedings of the 2020 International Conference on Data Analytics for Business and Industry: Way Towards a Sustainable Economy (ICDABI), Sakheer, Bahrain, 26–27 October 2020; pp. 1–5.
37. Wang, H.; Yu, H. Traffic sign detection algorithm based on improved YOLOv4. In Proceedings of the 2020 IEEE 9th Joint International Information Technology and Artificial Intelligence Conference (ITAIC), Chongqing, China, 11–13 December 2020; pp. 1946–1950.
38. Sandler, M.; Howard, A.; Zhu, M.; Zhmoginov, A.; Chen, L.-C. Mobilenetv2: Inverted residuals and linear bottlenecks. In Proceedings of the IEEE Conference on Computer Vision and Pattern Recognition, Salt Lake City, UT, USA, 18–23 June 2018; pp. 4510–4520.

39. Liang, X.; Zhang, J.; Zhuo, L.; Li, Y.; Tian, Q. Small object detection in unmanned aerial vehicle images using feature fusion and scaling-based single shot detector with spatial context analysis. *IEEE Trans. Circuits Syst. Video Technol.* **2019**, *30*, 1758–1770. [CrossRef]
40. Liu, S.; Huang, D. Receptive field block net for accurate and fast object detection. In Proceedings of the European Conference on Computer Vision (ECCV), Munich, Germany, 8–14 September 2018; pp. 385–400.
41. Li, J.; Liang, X.; Wei, Y.; Xu, T.; Feng, J.; Yan, S. Perceptual generative adversarial networks for small object detection. In Proceedings of the IEEE Conference on Computer Vision and Pattern Recognition, Honolulu, HI, USA, 21–26 July 2017; pp. 1222–1230.
42. Zhu, X.; Lyu, S.; Wang, X.; Zhao, Q. TPH-YOLOv5: Improved YOLOv5 based on transformer prediction head for object detection on drone-captured scenarios. In Proceedings of the IEEE/CVF International Conference on Computer Vision, Montreal, BC, Canada, 11–17 October 2021; pp. 2778–2788.
43. Shi, W.; Caballero, J.; Huszár, F.; Totz, J.; Aitken, A.P.; Bishop, R.; Rueckert, D.; Wang, Z. Real-time single image and video super-resolution using an efficient sub-pixel convolutional neural network. In Proceedings of the IEEE Conference on Computer Vision and Pattern Recognition, Las Vegas, NV, USA, 27–30 June 2016; pp. 1874–1883.
44. Zhang, J.; Zou, X.; Kuang, L.-D.; Wang, J.; Sherratt, R.S.; Yu, X. CCTSDB 2021: A More Comprehensive Traffic Sign Detection Benchmark. *Hum. Cent. Comput. Inf. Sci.* **2022**, *12*, 23.
45. Wang, C.-Y.; Bochkovskiy, A.; Liao, H.-Y.M. YOLOv7: Trainable bag-of-freebies sets new state-of-the-art for real-time object detectors. In Proceedings of the IEEE/CVF Conference on Computer Vision and Pattern Recognition, Vancouver, BC, Canada, 18–22 June 2023; pp. 7464–7475.
46. Wang, C.-Y.; Liao, H.-Y.M.; Yeh, I.-H. Designing network design strategies through gradient path analysis. *arXiv preprint* **2022**, arXiv:2211.04800.
47. Ding, X.; Zhang, X.; Ma, N.; Han, J.; Ding, G.; Sun, J. Repvgg: Making vgg-style convnets great again. In Proceedings of the IEEE/CVF Conference on Computer Vision and Pattern Recognition, Nashville, TN, USA, 20–25 June 2021; pp. 13733–13742.
48. Fan, H.; Ling, H. Siamese cascaded region proposal networks for real-time visual tracking. In Proceedings of the IEEE/CVF Conference on Computer Vision and Pattern Recognition, Long Beach, CA, USA, 15–20 June 2019; pp. 7952–7961.
49. Guo, C.; Fan, B.; Zhang, Q.; Xiang, S.; Pan, C. Augfpn: Improving multi-scale feature learning for object detection. In Proceedings of the IEEE/CVF Conference on Computer Vision and Pattern Recognition, Seattle, WA, USA, 13–19 June 2020; pp. 12595–12604.
50. Cao, J.; Chen, Q.; Guo, J.; Shi, R. Attention-guided context feature pyramid network for object detection. *arXiv preprint* **2020**, arXiv:2005.11475.
51. Tan, M.; Pang, R.; Le, Q.V. Efficientdet: Scalable and efficient object detection. In Proceedings of the IEEE/CVF Conference on Computer Vision and Pattern Recognition, Seattle, WA, USA, 13–19 June 2020; pp. 10781–10790.
52. Zhang, H.; Chang, H.; Ma, B.; Wang, N.; Chen, X. Dynamic R-CNN: Towards high quality object detection via dynamic training. In Proceedings of the Computer Vision–ECCV 2020: 16th European Conference, Glasgow, UK, 23–28 August 2020; Proceedings, Part XV 16, 2020. pp. 260–275.
53. Lin, T.-Y.; Goyal, P.; Girshick, R.; He, K.; Dollár, P. Focal loss for dense object detection. In Proceedings of the IEEE International Conference on Computer Vision, Venice, Italy, 22–29 October 2017; pp. 2980–2988.
54. Shi, Y.; Li, X.; Chen, M. SC-YOLO: A Object Detection Model for Small Traffic Signs. *IEEE Access* **2023**, *11*, 11500–11510. [CrossRef]

Disclaimer/Publisher's Note: The statements, opinions and data contained in all publications are solely those of the individual author(s) and contributor(s) and not of MDPI and/or the editor(s). MDPI and/or the editor(s) disclaim responsibility for any injury to people or property resulting from any ideas, methods, instructions or products referred to in the content.

sensors

MDPI

Article

Identification of Driver Status Hazard Level and the System

Jiayuan Gong [1,2,3], Shiwei Zhou [2,3,*] and Wenbo Ren [2,3]

[1] Harbin Engineering University, Harbin 150001, China; 20160013@huat.edu.cn
[2] Institute of Automotive Engineers, Hubei University of Auomotive Technology, Shiyan 442002, China; 202111205@huat.edu.cn
[3] Shiyan Industry Technique Academy of Chinese Academy of Engineering, Shiyan 442002, China
* Correspondence: 20220054@huat.edu.cn; Tel.: +86-187-7284-6185

Abstract: According to the survey statistics, most traffic accidents are caused by the driver's behavior and status irregularities. Because there is no multi-level dangerous state grading system at home and abroad, this paper proposes a complex state grading system for real-time detection and dynamic tracking of the driver's state. The system uses OpenMV as the acquisition camera combined with the cradle head tracking system to collect the driver's current driving image in real-time dynamically, combines the YOLOX algorithm with the OpenPose algorithm to judge the driver's dangerous driving behavior by detecting unsafe objects in the cab and the driver's posture, and combines the improved Retinaface face detection algorithm with the Dlib feature-point algorithm to discriminate the fatigue driving state of the driver. The experimental results show that the accuracy of the three driver danger levels (R1, R2, and R3) obtained by the proposed system reaches 95.8%, 94.5%, and 96.3%, respectively. The experimental results of this system have a specific practical significance in driver-distracted driving warnings.

Keywords: YOLOX; driver danger levels; Dlib; Image Identification

1. Introduction

With the progress of science and technology and the improvement of people's living standards, the number of automobile owners and drivers is increasing rapidly. According to the Traffic Management Bureau of the Ministry of Public Security report, there have been more than 200,000 traffic accidents in China every year in the past ten years, with more than 260,000 casualties and CNY 1.2 billion of direct economic losses [1–3]. Among them, the driver is an essential factor affecting driving safety and road smoothness, and it is necessary to analyze and pay attention to the driver's behavior [4–6]. Some research data from developed countries in Europe and the United States show that the proportion of traffic accidents caused by human factors is as high as 80% to 90% [7–9]. Therefore, it is of great significance to study a system for discriminating the degree of danger of a driver's state [10,11].

Hong S, Kwon H et al. [12] combined conventional photoelectric volumetric tracing (PPG) and electrocardiographic tracing (ECG) in the ear canal and found that fatigue detection was highly accurate. Sivaraman S, Trivedi M M et al. [13] at the University of California analyzed driving status by determining the vehicle's position concerning the rest of the vehicles through video images. The Face LAB system [14], created by the Australian National University in collaboration with Volvo, first recognizes the driver's facial features. It obtains data such as blinking frequency, pupil diameter, degree of eyelid closure, head position and rotation frequency, and mouth and eyebrow movement parameters, which are fused together to analyze and obtain the driver's state.

The physiological parameter detection method is subjective; the vehicle behavior detection method is affected by the road; and the machine vision detection method is costly. It is impossible to prejudge dangerous behavior, and the detection effect of a single judgment standard is poor. Therefore, this paper proposes a comprehensive driver hazardous state behavior discrimination system, which combines the YOLOX target detection algorithm and OpenPose driver gesture recognition algorithm to detect the driver's hazardous behavior. They are combined with driver dynamic fatigue tracking detection, driver unsafe state, and the behavior hazard classification detection system [15–19]. The system sends real-time warnings to the driver according to the degree of danger so that the driver can adjust the state and behavior in time to ensure the safety of driving, thus improving the driving safety coefficient and reducing the occurrence of traffic accidents.

2. Object Recognition in YOLOX Cab

The target detection problem is one of the more difficult problems in machine vision due to the significant disparity in the target object's shape and the influence of light intensity, occlusion, and other factors in the detection process. Deep learning-based target detection algorithms are divided into two categories: two stage and one stage. Two-stage algorithms first generate a series of candidate frames before using a convolutional neural network to classify the target, such as R-CNN, SPP-Net, Fast R-CNN, Faster R-CNN, and R-FCN, etc. Whereas, one-stage algorithms extract features directly from the network to predict object classification and location, e.g., YOLOv1, YOLOv2, RetinaNet, YOLOv3, and SSD. Although the YOLO algorithm is less effective in the detection of small target objects, since the targets in this paper, cigarettes, mugs, and phones in the cab, i.e., the images throughout the detection process, are not small objects, and also the YOLOX detection rate is speedy, the enhanced version can run at 45fps (frames per second) on the GPU, and the simplified version can even reach 155fps. In addition, YOLOX has good generalization ability and can correlate well with the background information when recognizing objects. Therefore, the algorithm satisfies the most needed real-time in this study and can significantly reduce the false detection of targets in the cab.

YOLOX can be divided into three parts: CSPDarknet, FPN, and YOLO Head, and the YOLOX network structure is shown in Figure 1.

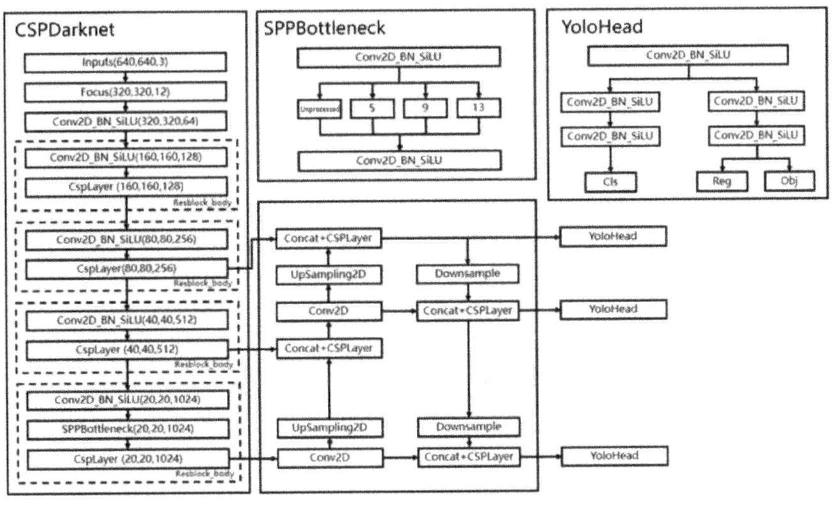

Figure 1. YOLOX network structure.

The backbone feature extraction network of YOLOX is CSPDarknet; after inputting the image, the feature extraction is first carried out in the backbone network, which is called the feature layer, storing the feature set of all the input images. The backbone of YOLOX is composed of residual convolution; the residual network has the advantages of easy optimization and improvement of accuracy. After the input image, it first passes through the Focus structure, which compresses the width and height to 1/2 of the original, extends the number of channels by four times, and then passes through the Resblock_body structure four times. The three feature layers extracted are located in the middle, lower middle, and bottom layers of the CSPDarknet, respectively. When the input is (640,640,3), the shape sizes of the three feature layers are feature1 = (80,80,256), feature2 = (40,40,512), and feature3 = (20,20,1024). The three feature layers are then fed into the FPN layer, the feature pyramid that fuses different types of shape feature layers for better feature extraction. Also, the structure of PANet is used in YOLOX to obtain the final effective feature layer by upsampling and downsampling the features.

YOLO Head is the classifier and regressor of YOLOX, and the three enhanced adequate feature layers obtained are fed into YOLO Head through the CSPDarknet and FPN network structure, each of which has width, height, and number of channels with standard convolution and activation functions. The prediction consists of three parts: Reg, Obj, and Cls. The Reg part is the regression parameter judgment of the feature points; the Obj part is the judgment of whether the feature points contain objects; and the Cls part is the kind of objects contained in the feature points.

Figure 2 shows the flow chart of the target detection system framework based on the YOLOX algorithm, which is divided into three parts: input layer, recognition layer, and entity layer.

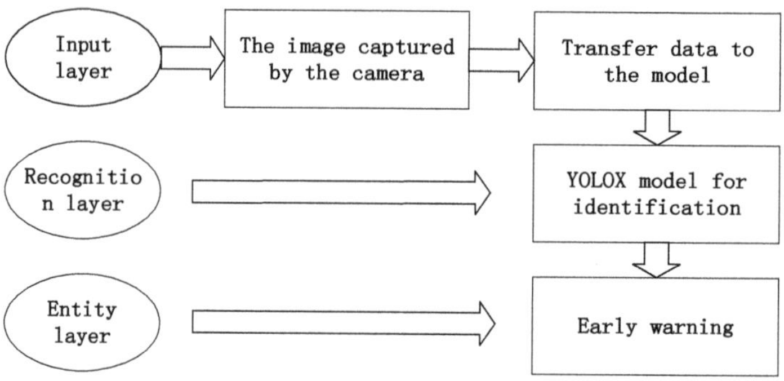

Figure 2. Target detection system framework.

Input layer: a camera that collects images is placed in the cab, the driver's state image is collected in real-time through the camera, and the image is transmitted to the YOLOX model.

Recognition layer: a series of image processing and target detection are performed through the recognition layer, and the trained YOLOX model is used to recognize the target objects (cigarettes, water cups, and telephones) in the cab.

Entity layer: by analyzing the target detection results, it can provide early warning to the driver, regulate the driver's driving behavior, and reduce accidents.

No open-source dataset is available on the market for reference for the research on driver behavior and state detection. Therefore, the dataset used in this paper includes the Kaggle driving posture area image set [20,21], the collection of different drivers in different driving states. Images and some image files are downloaded from the Internet by adding noise, flipping, and increasing contrast, 6000 images are obtained, and data annotation is performed, as shown in Figure 3. The pictures are divided into three categories: smoke, drink, and phone, and a dataset is created.

The YOLOX-S model is adopted, which pays more attention to the rate and meets the real-time requirements. The training process is divided into two stages: freezing and thawing. In the freezing stage, the epoch is set to 100, and the backbone of the model is frozen, which will not affect the feature extraction network and requires only a tiny amount of video memory. At the same time, freezing training helps to improve training speed, preventing weights from breaking in the initial stage. In the thawing stage, the epoch is also set to 100, the backbone of the model is thawed, and the feature extraction network changes. This stage takes up a lot of memory, and the parameters of the network change. The backbone network is first frozen and then unfrozen, the total number of training generations is set to 20, and multiple threads are started to read data to speed up the data reading process, the number of threads is set to 4, the graph is shown in Figure 4.

Figure 3. Data annotation.

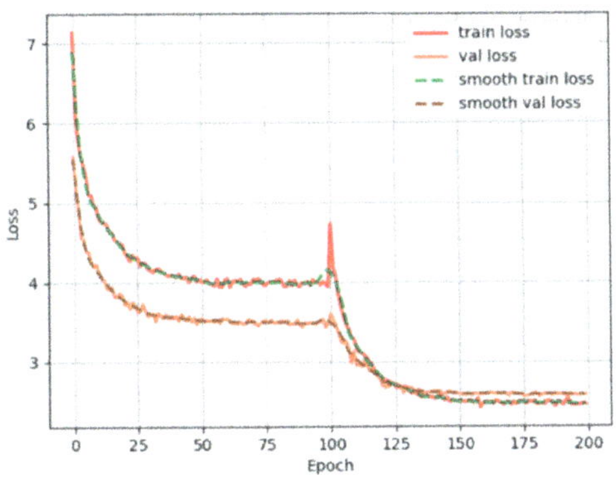

Figure 4. YOLOX training curve.

The prediction phase uses two files: yolo.py and predict.py. Modify model_path and classes_path in yolo.py. Modify model_path to the trained weight file ep200-loss2.450-val_loss2.579.pth, classes_path points to the txt file corresponding to the detection category, and predict the target image containing specific objects (water cups, cigarettes, and telephones), as shown in Figure 5.

The model precision and recall are calculated as shown in (1) and (2).

$$\text{Precision} = \frac{TP}{TP + FN} \times 100\% \quad (1)$$

$$\text{Recall} = \frac{TP}{TP + FP} \times 100\% \quad (2)$$

The test results are shown in Table 1.

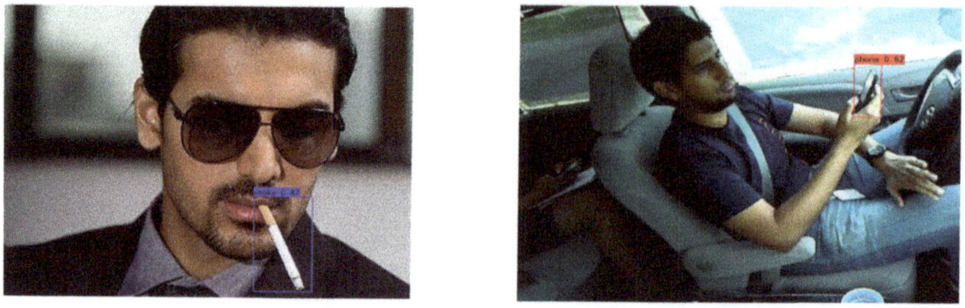

(**a**) (**b**)

Figure 5. (**a**) Cigarette test results; (**b**) phone test results.

Table 1. Test results.

	Recall Rate	Accuracy Rate
smoke	98.2%	97.8%
drink	98.5%	97.0%
phone	97.8%	96.6%
average	98.2%	97.1%

From the test results, we can conclude that the accuracy rates of the three target objects, cigarettes, water cups, and telephones, are all over 95%.

In order to verify the reasonableness and feasibility of the model, other classical target detection algorithm models are applied to this dataset, and the detection results are shown in Figure 6. Comparing the results, the YOLOX algorithm has the highest accuracy and a fast detection speed, which has obvious advantages and meets the requirements of the designed system.

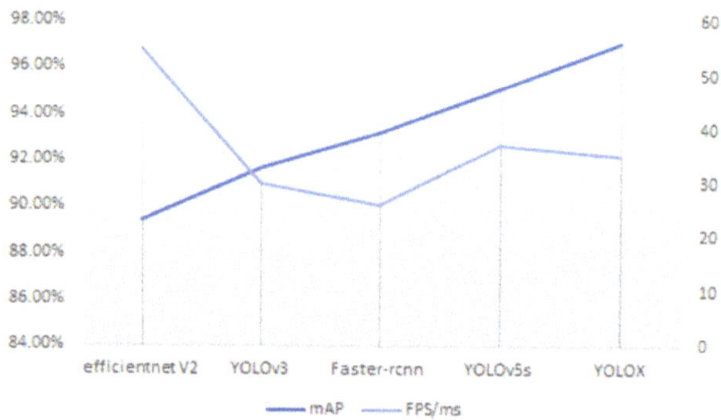

Figure 6. Detection efficiency of different models.

3. OPENPOSE Driver Posture Detection

OpenPose mainly estimates the current posture of the human body through the relative position of the critical points of the human body. By observing the positions of various essential parts of the human body, the posture of the human body can be accurately measured, and the target's posture can be predicted, such as drinking water, calling, smoking, etc. First, the posture of the driver's body part is detected. After inputting the picture, after OpenPose processing, the bone connection points are extracted. The model used in this paper for the driver's body is Body_25, which includes 25 human bone points. The joint point connection implementation diagram is shown in Figure 7.

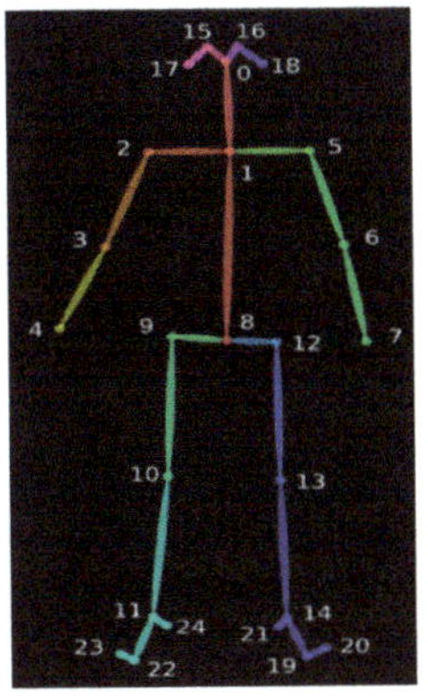

Figure 7. Joint point connection.

The OpenPose network implementation consists of the following three phases (shown in Figure 8):

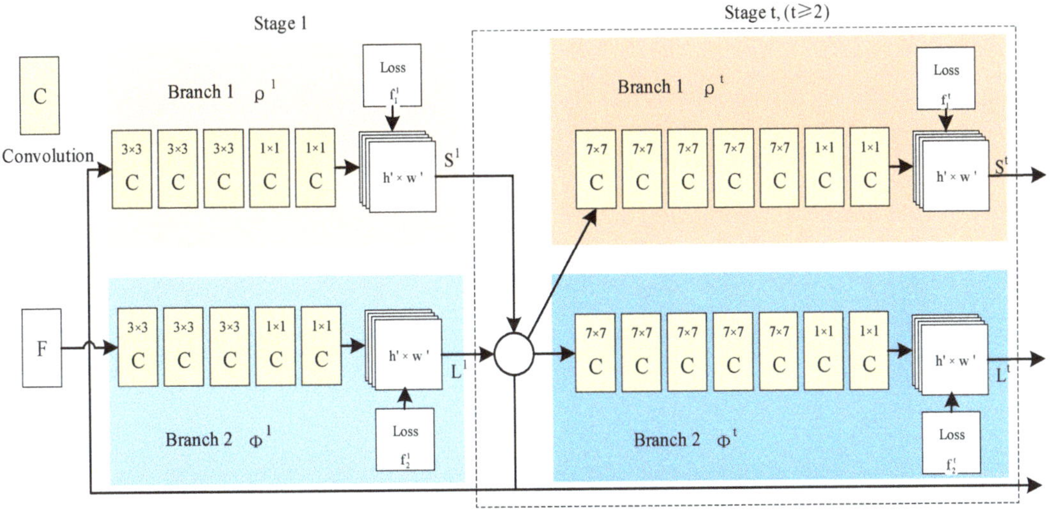

Figure 8. OpenPose schematic.

Phase 1: create a feature mapping for the input image using the first ten layers of VGGNet.

Stage 2: Create a CNN network with a two-branch multi-stage. One branch predicts a set of 2D confidence maps for body part locations (e.g., eyes, elbows, knees, etc.), and the other predicts a set of 2D vector fields for partial affinities. In the first stage (left half), the network produces an initial set of detection confidence maps S and a set of partial affinity fields L. Then, in each second stage (right half), the predictions from the two branches of the previous stage are connected to the original image features F to produce more accurate predictions. This step increases the depth of the neural network to capture more accurate predictions.

Stage 3: generate 2D key points by reasoning and parsing the confidence and affinity maps through the greedy algorithm [22].

Repeating the above steps can predict the location of the key points and their confidence maps. Finally, the skeletal connectivity map of the target human body can be obtained by connecting the critical points through the greedy algorithm.

It can be seen from the detection effect that the body posture of the driver can judge whether the driver is performing behaviors that endanger driving safety, such as making a phone call, smoking, etc. Whether the driver answers the phone or not is judged by calculating the distance from the right hand to the right ear, that is, the distance from point 4 to point 17, or the distance from the left hand to the left ear, that is, the distance from point 7 to point 18. Similarly, smoking and drinking water are judged by calculating the distance from the hands to the nose, that is, the distance from points 4 and 7 to point 0.

Through the coordinate value of the horizontal position of the hand key point and the horizontal position of the ear key point, that is, the y-axis coordinate value, it is judged whether the hand and the ear are on the same horizontal line, and thus to determine whether the driver is answering the phone (handheld phone), as shown in Figure 9. It can be seen from the common sense of making and answering calls that our hands do not coincide with the position of our ears when making calls, so this paper sets a threshold when making a judgment. If the difference between the y-axis coordinates of the hand and the ear is within the set threshold, it is initially considered to be receiving a call. After the assumption method and continuous experimentation, the threshold is set to 15, and the flowchart is shown in Figure 10.

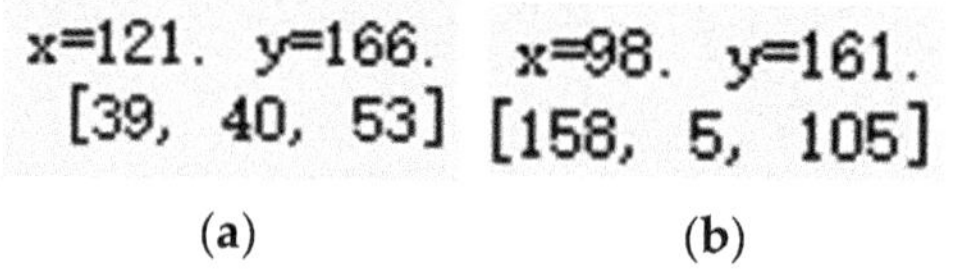

Figure 9. (**a**) Hand coordinates; (**b**) ear coordinates.

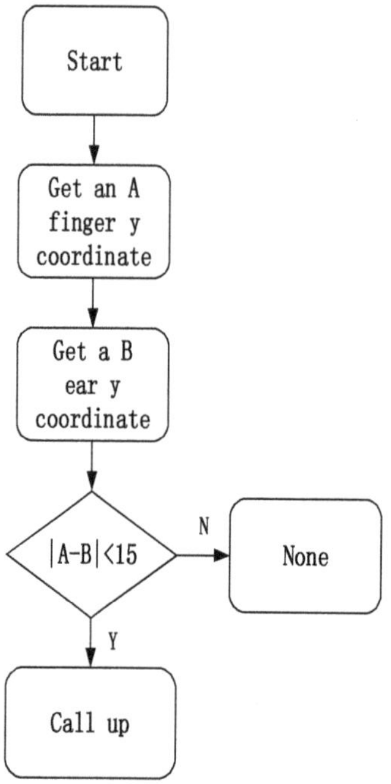

Figure 10. Corresponding flow chart of phone call status.

As shown in Figure 9a, the y-axis coordinate value of the hand is 166, and as shown in Figure 9b, the y-axis coordinate value of the ear is 161, and the difference between them is 5, which is within the set threshold range, so it was initially judged to be on the phone.

In the same way, the driver's smoking behavior is determined by the y-axis coordinate value of the hand and the y-axis coordinate value of the nose, and the threshold is set to 10 through the assumption method and continuous experimentation; the drinking behavior is similar to the smoking behavior. The y-axis coordinate value of the hand and the y-axis coordinate value of the nose can also be calculated for discrimination. However, the position of drinking water is more uncertain than smoking, so the threshold is set to 15.

Therefore, the driver must meet the following two conditions when smoking:

(1) The difference between the y-axis coordinate value of the driver's hand joint point and the y-axis coordinate value of the nose joint point does not exceed 10.
(2) The YOLOX target detection system detected the presence of cigarettes.

Drivers must meet the following two conditions when drinking water:

(1) The difference between the y-axis coordinate value of the driver's hand joint point and the y-axis coordinate value of the nose joint point does not exceed 15.
(2) There is a water cup in the detection result of the YOLOX target detection system.

The driver must meet the following two conditions when making a phone call:

(1) The difference between the y-axis coordinate value of the driver's hand joint point and the y-axis coordinate value of the ear joint point does not exceed 15.
(2) The detection result of the YOLOX target detection system has the presence of a phone.

Since the driver's state is continuous, it is impossible to complete these dangerous behaviors instantaneously, and the detection of the driver's driving state image is relatively independent; there is no contextual connection, and practical information exchange cannot be performed. Some unexpected situations, such as the vehicle turning quickly, road potholes, vehicle bumps, etc., will cause sudden jumps in the captured images. The expected effect cannot be achieved in a period, so it cannot be accurately and stably obtained—test results.

Based on this, this paper introduces a voting mechanism to reduce false detections due to image jumps, enhancing the algorithm's robustness. Based on the hardware conditions and reasonable assumptions of the camera used in this paper, a 12-frame voting queue is set up as a voter for driver behavior prediction. If, and only if, the same behavior exists in the 12-frame voting queue for more than eight frames, the output driver state will change. Otherwise, the output state will remain unchanged. Table 2 shows the comparison results of the detection accuracy with and without the voting mechanism.

Table 2. Driver risky behaviors voting detection.

Risky Behavior Classification	No Voting Queue. (Accuracy Rate)	There is a Voting Queue. (Accuracy Rate)
Drink water	190/200 = 0.95	196/200 = 0.98
Smokes	186/200 = 0.93	190/200 = 0.95
Call up	188/200 = 0.94	196/200 = 0.98

The test results show that the detection accuracy of drivers drinking water has increased from 95% to 98%; the detection accuracy of smoking has increased from 93% to 95%; and the detection accuracy of phone calls has increased from 94% to 98%.

4. Fatigue Detection

In this paper, the driver's face is detected based on the RetinaFace algorithm, proposed by Insight Face in 2019, incorporating excellent modeling ideas such as a feature pyramid network and enhanced feature extraction, and performs well on the WiderFace dataset [23].

The MobileNetV1-0.25 backbone network was chosen to significantly reduce the computation and speed up the computation rate to meet the real-time nature of face detection. The MobileNet model is a lightweight deep neural network that uses the core idea of depthwise separable convolution, whose structure is shown in the following figure. Its structure is shown in Figure 11.

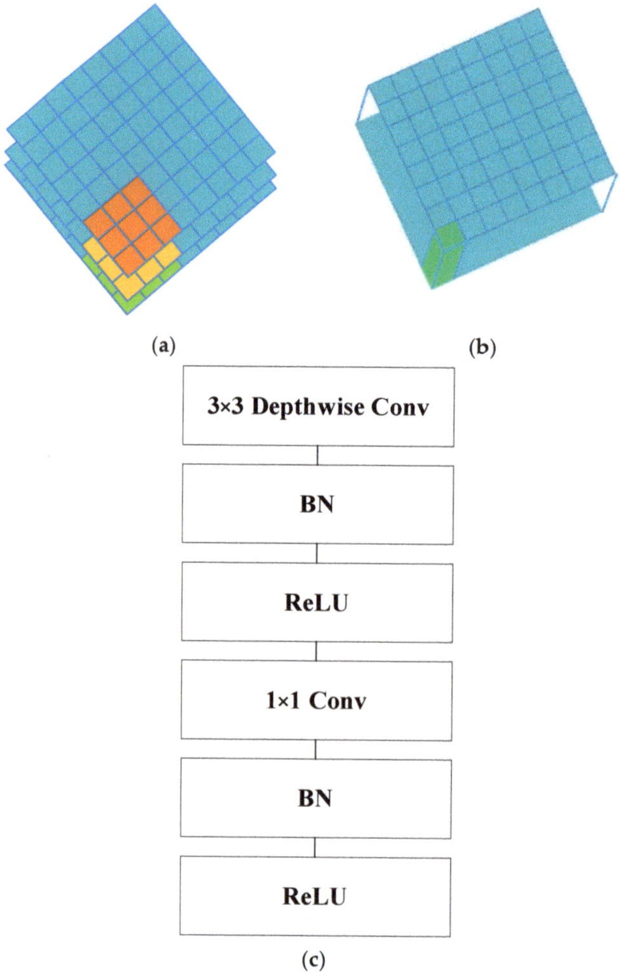

Figure 11. (**a**) Depthwise convolutional filters; (**b**) pointwise convolution filters; (**c**) depthwise separable convolution.

The prediction result of RetinaFace is divided into three, which are classification prediction result (Class Head), regression prediction result of the box (Box Head), and regression prediction result of face key points (Landmark Head), while the fatigue detection system in this paper is added with a cradle head tracking system in order to avoid affecting the accuracy of the algorithm in case of an undetectable driver's face. Meanwhile, the Dlib algorithm is used to obtain the critical points of the driver's face for fatigue detection. Therefore, the improved version of RetinaFace subtracts the regression prediction results of the face key points, significantly reducing the computation and improving the detection rate.

The three adequate feature layers obtained are equivalent to dividing the input image into grids of different sizes, and there are several a priori frames at each grid point; RetinaFace defaults to two a priori frames, and prediction results determine whether the a priori frames contain faces or not, and if they do, then the a priori frames are adjusted to obtain the prediction frames. Class Head determines whether the a priori frame at each grid point contains a face. Using 1×1 convolution, the number of channels in the feature layer is adjusted to num_anchors × 2; num_anchors is the number of a priori frames, which is defaulted to two in the RetinaFace network. Two is used to judge whether the a priori frame contains a face or not; two inside the serial number of one is more significant, as it means that the a priori frame contains a face. Two inside the serial number zero is more significant, meaning Box Head is used to adjust the center, width, and height of the a priori box to obtain the prediction box. Using 1 × 1 convolution, the number of channels of the feature layer is adjusted to num_anchors × 4, which is used to represent the adjustment parameters of each a priori box. Four represents the four adjustment parameters of the a priori box, the first two are used to adjust the center of the a priori box to obtain the center of the prediction box. The last two are used to adjust the width and height of the a priori box to obtain the width and height of the final prediction box. The process of actual frames can be divided into three steps:

(1) Calculate the degree of overlap between all actual frames and a priori frames, and use the a priori frames with IoU greater than 0.35 for prediction to obtain real frames;
(2) Coding these a priori frames that have a greater degree of overlap with the proper frame;
(3) The coding operation can be divided into classification prediction results and regression prediction results of the boxes.

The improved algorithm reduces the recognition process of face feature points in the face detection part; firstly, the face is detected and boxed out using the RetinaFace algorithm, and then combined with the Dlib algorithm to detect 68 feature points on the driver's face, which not only better accomplishes the recognition of 68 key points on the driver's face, but also improves the rate of the RetinaFace face recognition algorithm. Figure 12 below shows the partial results of face detection after the improvement; five points are recognized before the improvement of the face feature point, and after the improvement of the feature point, this prediction step is removed.

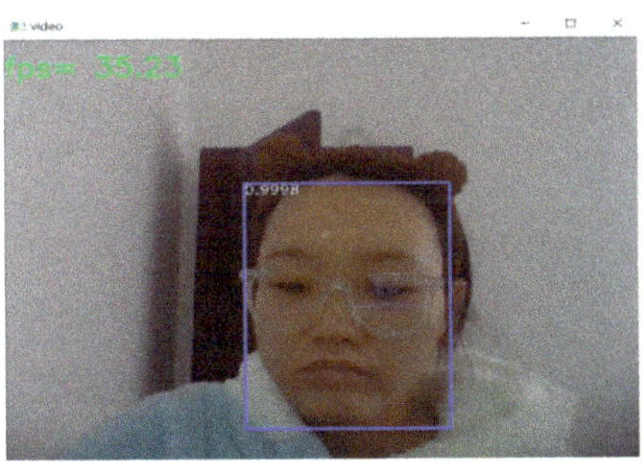

Figure 12. Face detection.

Compared to the pre-improved network, the improved face detection rate is greatly improved with guaranteed accuracy, with fps going from an average of 16 to an average of 30, as shown in Figure 13.

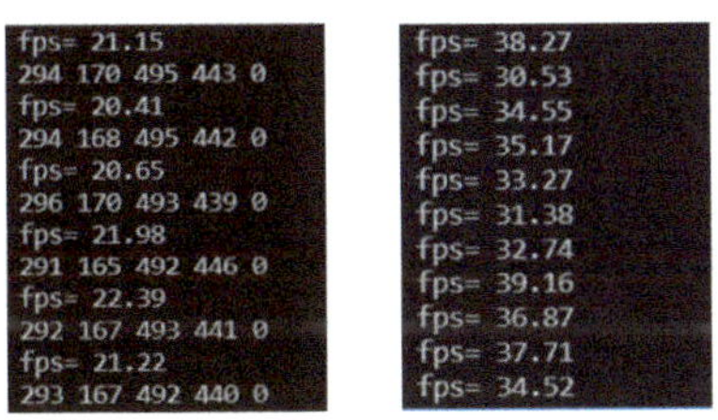

(a) (b)

Figure 13. (**a**) Pre-optimization detection rate and (**b**) optimized detection rate.

The 68 critical points of the driver's face are detected by the Dlib algorithm, as shown in Figure 14. The opening and closing of the driver's eyes are determined by the eye aspect ratio (EAR), as shown in Figure 15; P_1–P_6 are the six feature points of the eyes.

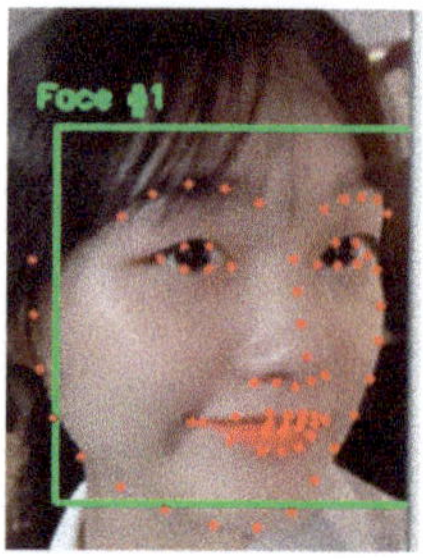

Figure 14. Keypoint detection.

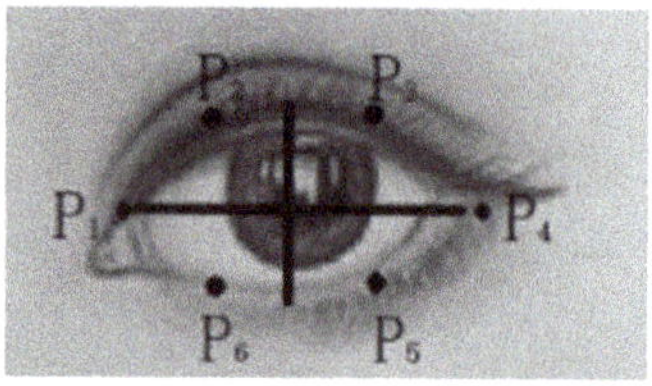

Figure 15. Eye feature points.

When the driver opens or closes his eyes, the aspect ratio will change. The formula of *EAR* is

$$EAR = \frac{||P_2 - P_6|| + ||P_3 - P_5||}{2||P_1 - P_4||} \tag{3}$$

The threshold for eye fatigue is set to 0.28; if the *EAR* is more than this, the eyes are considered open; if the *EAR* is less than it, the eyes are considered closed, as shown in Figure 16.

Figure 16. Eyes open.

Combined with the *PERCLOS* (Percentage of Eyelid Closure over the Pupil) value, the fatigue state of the driver is detected, and if Equation (3) is satisfied, it is considered that fatigue driving is performed.

$$PERCLOS = \frac{F}{T} \times 100\% \tag{4}$$

The *F* in Formula (4) represents eye closure frames, and the *T* in Formula (4) represents the total number of frames in the detection period.

Referring to the relevant data [24–26], the fatigue limit is *PERCLOS* = 20%; we use 100 frames of images as a loop. If 20 frames of closed-eye images are detected in 100 frames, it is considered fatigued and the output is tired; otherwise, the output is relaxed, as shown in Figure 17.

Figure 17. Driver status.

In order to test the effectiveness of the driver fatigue tracking detection algorithm in this paper and the influence of the dynamic tracking system after adding the gimbal to the driver fatigue detection effect. This paper conducts eIn order to ensure that the driver's face is always within the monitoring range of the camera, ensure real-time and accurate detection, and avoid false detection caused by the driver's face not being in the detection frame; the cradle head tracking system is added, as shown in Figure 18. The tracking gimbal of OpenMV first obtains the x and y coordinates of the center of the face, sends the coordinate information of the center point to OpenMV, and then controls the motion of the two servos of the gimbal by calculating the deviation between the coordinates of the center of the face and the center of the picture, to realize driving—personnel face tracking detection.

Figure 18. OpenMV Connection Diagram.

In order to test the effectiveness of the driver fatigue tracking detection algorithm, and the influence of the dynamic tracking system after adding the cradle head tracking system to the driver fatigue detection effect, this paper conducts experiments on the subjects. This is done by testing whether the system recognizes and detects when the subjects are relaxed and fatigued using the driver fatigue detection system without the cradle head tracking system and the driver fatigue detection system with the cradle head tracking system, as shown in Figure 19.

Two kinds of driver fatigue detection systems without cradle head tracking system and cradle head tracking system are tested, respectively, and the two states of driver's relaxation and fatigue are detected, respectively. Two thousand four hundred frames of images are identified and detected for each state. The detection results are shown in Tables 3 and 4. The exact number of detections is used to ensure the effectiveness of the accuracy comparison.

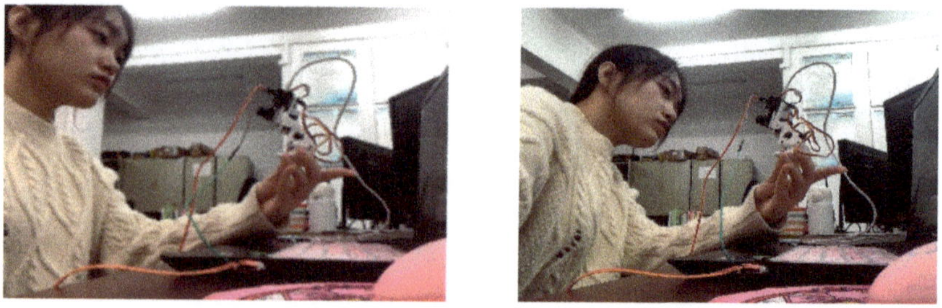

Figure 19. (**a**) Dynamic tracking of the detection process 1 (**b**) Dynamic tracking of the detection process 2.

Table 3. Fatigue test results without cradle head tracking system.

State	Test Numbers	Error Numbers	Accuracy
Relax	2400	180	92.50
Fatigue	2400	210	91.25

Table 4. Fatigue detection results when adding cradle head tracking system.

State	Test Numbers	Error Numbers	Accuracy
Relax	2400	120	95.00
Fatigue	2400	140	94.17

The experimental results show that the detection accuracy of the driver's relaxed state has increased from 92.5% when the gimbal is never added to 95% when the gimbal is added. The accuracy of the fatigue state has also increased to 94.17%, compared with that when the cradle head tracking system is not added. The accuracy of fatigue detection results is significantly improved, and dynamic tracking fatigue detection achieves better results.

5. Design of The Diver's State Risk Level Discrimination System

This paper realizes the design of the driver's state risk level discrimination system by integrating the driver's fatigue detection and the detection of several dangerous driving behaviors. Through years of driving experience and analysis of related research, we found that fatigue driving is the most influential factor in car driving [27,28]. When driving, if the driver is drowsy, it will confuse the brain, and the vision will also become blurred, not to mention the fatigue of driving for a long time. Therefore, when performing risk classification, if there is fatigue driving, it is high-risk driving. We conducted a risk-level discrimination experiment on three behaviors of smoking, drinking, and making a phone call in the laboratory (non-driving environment). The experimental results show that when a person concentrates on one thing and suddenly receives a call, his attention will be diverted immediately. He will even drop his job if he is on the phone with someone important; if on the phone and he learns something that breaks him down, he can become out of control, let alone get on with the job at hand. Simultaneously, when answering the phone, you have to look at the mobile phone screen and tap the screen with your hand for a long time, which will affect the driver's driving concentration. The reflection of the mobile phone will also interfere with the driver, making the driver unable to concentrate, resulting in traffic accidents. Therefore, we regard answering the phone as one of the most dangerous behaviors when driving, except for fatigue driving [29–31]. As far as smoking and drinking water are concerned, our lived experience tells us that drinking water has a more significant impact on driving because the cup is larger than the cigarette, it is easy to affect the driver's vision, and in some cases, the driver must raise his head when drinking water, which will significantly affect the driver's attention. Based on the above analysis, we rank the risk levels of the driver's unsafe state behaviors studied in this paper (starting from the most dangerous) as follows: fatigue driving, making phone calls, drinking water, and smoking.

We combine the above four behavior state information and the 12-frame voting queue to determine the driver's state danger level. We divide the driver's dangerous state into three dangerous states: R1, R2, and R3. R1 represents very dangerous; we are continuously alert this situation and advise to stop driving; R2 stands for medium danger; we issue a danger warning to remind the driver to modify the current state of dangerous behavior in time, and continue to drive after correcting; and R3 represents a low-risk state, no alarm is required, and only the alarm flashing light needs to appear. The status level of the driver is queried according to Table 5 below.

Table 5. Driver status level.

Action State Type	Queue Frames < 8	8 < Queue Frames< 16	16 < Queue Frames
call	R2	R2	R1
drink	R3	R2	R1
smoke	R3	R2	R2

In the experimental part, this paper collected 6000 frames of driver driving images, including 1920 frames of R1 risk level video, 2290 frames of R2 risk level video, and 1790 frames of R3 risk level video. The test results are shown in Table 6.

Table 6. Danger level detection results of driving state.

Driver's Driving State Danger Level	Number to be Checked	False Detection Rate	Detection Accuracy
R1	1920	81	95.8%
R2	2290	125	94.5%
R3	1790	66	96.3%

From the detection results of the driver's driving state risk level, it can be concluded that the driver's driving state risk level (R1, R2, and R3) has an accuracy rate of more than 90%. The accuracy rate of R3 hazard level detection even reached 96.3%, with a high accuracy rate. In experimental testing, we obtain better results, achieve the expected goals, and meet our needs for driving safety in real life.

6. Conclusions

In this paper, the YOLOX algorithm is used to detect specific target objects (cigarettes, water cups, and telephones) in the driver's cab, and combined with the OpenPose posture detection algorithm to judge the driver's dangerous behavior, and the accuracy of the detection results all reach 95%. The improved RetinaFace face detection algorithm is proposed, combined with the Dlib algorithm to determine the driver fatigue driving state according to the *PERCLOS* value, and the gimbal system is added to realize the dynamic tracking detection of the driver. A driver state danger level discrimination system is designed to classify four dangerous driving behavior states, namely, fatigue driving, drinking water, smoking, and making phone calls, and the experimental results show that the accuracy of R1, R2, and R3 danger level detection reaches 95.8%, 94.5%, and 96.3%, respectively.

With the development of deep learning and neural networks, the application in various fields is more and more comprehensive, the research scope and research data of this paper still need to be standardized and improved. In this paper, we only recognize the target objects (cigarettes, water cups, and telephones) in the driver's cab, and the target objects are still relatively single, lacking the dataset of scenes in low-light and harsh environments, and insufficient for real-vehicle driving tests. Next, we will increase the variety of scenes and target objects in complex environments to make the dataset richer to test the robustness of the algorithm, modify the original information fusion structure of the network to enhance the feature fusion ability, modify the detection layer of the original network to enhance the ability of the network to extract and localize the effective information, and conduct a large number of practical scenes to test the discriminative system.

Author Contributions: Methodology, J.G., S.Z. and W.R.; Validation, J.G., S.Z. and W.R. All authors have read and agreed to the published version of the manuscript.

Funding: This research received no external funding.

Informed Consent Statement: Informed consent was obtained from all subjects involved in the study.

Data Availability Statement: The dataset is not readily available to the public as it is privately collected and produced.

Conflicts of Interest: The authors declare no conflict of interest.

References

1. People's Daily Online. In 2020, the Number of Motor Vehicles in China Will Reach 372 Million, and the Number of Motor Vehicle Drivers will Reach 456 Million. Available online: http://en.caam.org.cn/Index/show/catid/25/id/1615.html (accessed on 1 October 2022).
2. Traffic Administration of the Ministry of Public Security. *Annual Report of Road Traffic Accident Statistics of the People's Republic of China (2020)*; Traffic Administration of the Ministry of Public Security: Beijing, China, 2021.
3. Zhang, L.; Liu, T.; Pan, F.; Guo, T.; Liu, R. Analysis of the Influence of Driver Factors on Road Traffic Accident Indicators. *China J. Saf. Sci.* **2014**, *24*, 79–84.
4. Rau, P. Drowsy driver detection and warning system for commercial vehicle drivers: Field operational test design, analysis, and progress. In Proceedings of the 19th International Conference on Enhanced Safety of Vehicles, Washington, DC, USA, 6–9 June 2005.
5. Ding, Y. *Research on the Characteristics of Drivers in Urban Traffic Accidents Based on Data Mining*; Shenyang University: Shenyang, China, 2018.
6. Connier, J.; Zhou, H.; De Vaulx, C.; Li, J.J.; Shi, H.; Vaslin, P.; Hou, K.M. Perception Assistance for the Visually Impaired Through Smart Objects: Concept, Implementation, and Experiment Scenario. *IEEE Access* **2020**, *8*, 46931–46945. [CrossRef]
7. Zhao, M. *Research on Image-Based Driver Gesture Recognition Method*; Southeast University: Nanjing, China, 2018.
8. Dinges, D.F.; Grace, R. *PERCLOS: A Valid Psychophysiological Measure of Alertness as Assessed by Psychomotor Vigilance*; US Department of Transportation: Washington, DC, USA; Federal Highway Administration: Washington, DC, USA, 1998.
9. Wang, W.; Sun, F.; Cao, Q. *Theory and Method of Driving Behavior in Road Traffic System*; Science Press: Beijing, China, 2001.
10. Patel, M.; Lal, S.K.; Kavanagh, D.; Rossiter, P. Applying neural network analysis on heart rate variability data to assess driver fatigue. *Expert Syst. Appl. Int. J.* **2011**, *38*, 7235–7242. [CrossRef]
11. Statistics Report for Traffic Accident between Ministry of Land, Infrastructure, Transport and Tourism. Available online: https://www.mlit.go.jp/en/index.html (accessed on 24 April 2023).
12. Hong, S.; Kwon, H.; Choi, S.H.; Park, K.S. Intelligent System for Drowsiness Recognition based on Ear Canal Electroencephalography with Photoplethysmography and Electrocardiography. *Inf. Sci.* **2018**, *453*, 302–322. [CrossRef]
13. Sivaraman, S.; Trivedi, M.M. Active learning based robust monocular vehicle detection for on-road safety systems. In Proceedings of the 2009 IEEE Intelligent Vehicles Symposium, Xi'an, China, 3–5 June 2009.
14. Grace, R.; Byrne, V.E.; Legrand, J.M. A Machine Vision Based Drowsy Driver Detection System for Heavy Vehicles. In Proceedings of the Ocular Measures of Driver Alertness Conference, Herndon, VA, USA, 26–27 April 1999.
15. You, F.; Tu, H.; Gong, Y. Research on the influence of mobile phone social entertainment operation on driver's visual parameters. *Transp. Syst. Eng. Inf.* **2020**, *20*, 236–243.
16. Jiang, A.; Noguchi, R.; Ahamed, T. Tree Trunk Recognition in Orchard Autonomous Operations under Different Light Conditions Using a Thermal Camera and Faster R-CNN. *Sensors* **2022**, *22*, 2065. [CrossRef] [PubMed]
17. Tong, X.; Li, Z. Improved YOLOX-S algorithm for helmet and mask detection. *J. Yibin Coll.* **2023**, *23*, 17–24.
18. Lv, Z.; Wei, X.; Ma, Z. Improved YOLOX for lightweight helmet detection. *Comput. Eng. Appl.* **2023**, *59*, 61–71.
19. Liu, Y.; Li, B.; Xi, P.; Lv, X.; Zhang, L.; Liu, Z.; Huang, Q.; Tang, R.; He, J.; Chen, L.; et al. Neuromorphic transcutaneous electrical nerve stimulation (nTENS) induces efficient tactile-related cortical networks in forearm amputees. *Sci. China Technol. Sci.* **2023**, *66*, 1451–1460.
20. Wu, Y.; Yang, T.; Tang, Y. Research on Road Object Detection Algorithm Based on Improved YOLOX. In Proceedings of the 2023 3rd International Conference on Neural Networks, Information and Communication Engineering (NNICE), Guangzhou, China, 24–26 February 2023; pp. 271–275.
21. Liu, K.; Luo, S.Y. Traffic sign recognition based on improved YOLOX-S. *Electron. Meas. Technol.* **2023**, *46*, 112–119.

22. Williams, D.J.; Shah, M. A fast algorithm for active contours. In Proceedings of the [1990] Proceedings Third International Conference on Computer Vision, Osaka, Japan, 4–7 December 1990; pp. 592–595.
23. Wang, L.; Lin, Y.; Sun, W.; Wu, Y. Improved Faster-RCNN Algorithm for Mask Wearing Detection. In Proceedings of the 2021 IEEE 4th Advanced Information Management, Communicates, Electronic and Automation Control Conference (IMCEC), Chongqing, China, 18–20 June 2021; pp. 1119–1124.
24. Zhang, Y.; Li, W.; Guo, W.; Song, Z. Lightweight pedestrian detection for multivariate scenes. *Opt. Precis. Eng.* **2022**, *30*, 1764–1774. [CrossRef]
25. Xu, F.; Xue, Y.; Zheng, C.; Hou, W.; Guo, J.; Wang, X. Piglet posture recognition based on self-attention mechanism with no anchor point. *J. Agric. Eng.* **2022**, *38*, 166–173.
26. Shao, Y.; Zhang, D.; Chu, H.; Zhang, X.; Rao, Y. A review of YOLO target detection based on deep learning. *J. Electron. Inf.* **2022**, *44*, 3697–3708.
27. Sun, Y.; Chen, Y.; Jin, X.; Yu, J.; Chen, Y. An artificial intelligence-based method for recognizing seedlings and weeds in *Brassica napus*. *Fujian J. Agric.* **2021**, *36*, 1484–1490.
28. Cao, J.; Guo, Z.; Pan, L.; Ding, X. Seatbelt wearing detection in aerial work scenarios. *J. Hunan Univ. Sci. Technol.* **2022**, *37*, 92–99.
29. Liu, X.; Zhang, Z.; Liu, J.; Zhang, M.; Flores, P.; Han, X. Detection and counting of corn kernels in the field based on multiple deep learning algorithms. *Intell. Agric.* **2022**, *4*, 49–60.
30. Sun, Q.; Chen, F.; Liu, G. Fusion of depth information for target detection in multilevel graph relational networks. *Chin. Sci. Technol. Pap.* **2022**, *17*, 1194–1200.
31. Wang, Z.; Xu, X.; Hua, Z.; Shang, Y.; Duan, Y.; Song, H. Lightweight cow estrus behavior recognition by integrating YOLO v5n and channel pruning algorithm. *J. Agric. Eng.* **2022**, *38*, 130–140.

Disclaimer/Publisher's Note: The statements, opinions and data contained in all publications are solely those of the individual author(s) and contributor(s) and not of MDPI and/or the editor(s). MDPI and/or the editor(s) disclaim responsibility for any injury to people or property resulting from any ideas, methods, instructions or products referred to in the content.

sensors

MDPI

Article

Cyclist Orientation Estimation Using LiDAR Data

Hyoungwon Chang [1], Yanlei Gu [1,*], Igor Goncharenko [1], Li-Ta Hsu [2] and Chinthaka Premachandra [3]

1 College of Information Science and Engineering, Ritsumeikan University, 1-1-1, Noji-higashi, Kusatsu 525-8577, Shiga, Japan
2 Department of Aeronautical and Aviation Engineering, The Hong Kong Polytechnic University, 11 Yuk Choi Rd, Hung Hom, Kowloon, Hong Kong
3 Department of Electronic Engineering, School of Engineering, Shibaura Institute of Technology, 3-7-5, Toyosu, Koto-ku, Tokyo 135-8548, Japan
* Correspondence: guyanlei@fc.ritsumei.ac.jp

Abstract: It is crucial for an autonomous vehicle to predict cyclist behavior before decision-making. When a cyclist is on real traffic roads, his or her body orientation indicates the current moving directions, and his or her head orientation indicates his or her intention for checking the road situation before making next movement. Therefore, estimating the orientation of cyclist's body and head is an important factor of cyclist behavior prediction for autonomous driving. This research proposes to estimate cyclist orientation including both body and head orientation using deep neural network with the data from Light Detection and Ranging (LiDAR) sensor. In this research, two different methods are proposed for cyclist orientation estimation. The first method uses 2D images to represent the reflectivity, ambient and range information collected by LiDAR sensor. At the same time, the second method uses 3D point cloud data to represent the information collected from LiDAR sensor. The two proposed methods adopt a model ResNet50, which is a 50-layer convolutional neural network, for orientation classification. Hence, the performances of two methods are compared to achieve the most effective usage of LiDAR sensor data in cyclist orientation estimation. This research developed a cyclist dataset, which includes multiple cyclists with different body and head orientations. The experimental results showed that a model that uses 3D point cloud data has better performance for cyclist orientation estimation compared to the model that uses 2D images. Moreover, in the 3D point cloud data-based method, using reflectivity information has a more accurate estimation result than using ambient information.

Keywords: cyclist; orientation estimation; LiDAR; deep neural network

1. Introduction

Understanding the behavior of bicycle riders is one of the essential factors of autonomous driving system. Non-motorized road users include pedestrians, cyclists, etc. [1]. However, while a lot of research has been conducted on pedestrian detection and behavior analysis [2,3], there is a lack of discussion on the estimation of a cyclist's behavior. As cyclists have more dynamic behavior and higher speeds than pedestrians, cyclist safety is also a critical issue that needs to be discussed. According to various traffic-related reports, a significant number of traffic incidents involve cyclists. In 2021, 43.6% of traffic accidents involved bicycle riders in Tokyo, Japan. This percentage of cyclists involved in accidents is only increasing since 2016 [4]. Furthermore, the observatory in European Union countries showed that cyclist safety is not being improved in various countries. In EU27 countries, the number of fatalities in traffic crashes involving cyclists remained constant between 2010 and 2019, while the number of fatalities in crashes involving other road users has decreased [5]. As the autonomous driving system is being developed for transportation efficiency along with safer roads, estimating cyclist orientation could be a solution for improvement on cyclist safety.

In real traffic situations, various road users including automobiles, pedestrians and bicycles move toward different directions. The easiest and most direct communication method for cyclists is to use hand signals to indicate their moving intentions. Accordingly, there are related regulations in some countries to encourage cyclist's hand signals [6,7]. However, such hand signals could be informative, but they are not efficient enough to be applied into real traffic interactions between bicycles and automobiles, as they require time for both cyclists and drivers to process the signals [8]. Thus, the autonomous driving system is expected to estimate cyclist's behavior without signals sent intentionally by the cyclist. In this research, estimation on a cyclist's body and head orientation is proposed to consider a cyclist's natural behavior, instead of hand signals. The ongoing direction of a cyclist can be represented by his or her body on a bicycle, as a cyclist rides in a fixed posture on a bicycle. Moreover, a cyclist's intention for changing direction can be represented by his or her head direction, which is the most predictable cue for a cyclist's turn [9,10]. Thus, the head and body orientations are critical factors in cyclist behavior estimation, as they indicate the future trajectory.

Most of previous research on pedestrian and cyclist orientation estimation are conducted based on the RGB-camera sensor widely used in autonomous vehicles. Schulz et al. developed a system to localize head and estimate the head orientation [11,12]. The system has eight separate classifiers corresponding to eight different head orientations. The head localization and head orientation estimation were realized by comparing the output confidence values generated by all eight classifiers. Gandhi et al. used the famous Support Vector Machines (SVM) and Histograms of Oriented Gradient (HOG) to estimate the pedestrian body orientation [13]. Gu et al. proposed using human physical model constraint and temporal constraint to accurately estimate the joined body and head orientation of pedestrians and cyclists in video [14,15]. Flohr et al. presented a probabilistic framework for the joint estimation of pedestrian head and body orientation from a mobile stereo vision platform using the pictorial structure approach [16].

Recently, deep neural networks have been widely used for vision-based orientation estimation of pedestrians and cyclists. Raza et al. presented an appearance-based pedestrian head-pose and full-body orientation prediction by using grayscale image and employing a deep learning mechanism [17]. Abadi et al. proposed to estimate the cyclist head and body orientation using joined head map information generated from Openpose [18], and then used the joined head and body orientation to predict the crossing intention of cyclist [19,20]. In order to identify the cyclist heading and predict their intentions, Garcia et al. proposed a multi-class detection with eight classes according to orientations and presented a performance comparison for cyclist detection and orientation classification between the main deep-learning-based algorithms reported in the literature, such as SSD, Faster R-CNN and R-FCN [21].

However, there is a certain weakness in using a camera for pedestrian and cyclist orientation estimation. As a camera is hugely dependent on light variation, the estimation results from images taken by a camera can be unstable. Such a problem can be solved by using LiDAR (Light Detection and Ranging) sensor. Different from the passive sensor camera, LiDAR is an active sensor, and can emit pulsed light waves into surrounding objects and receive the bounced-back wave to calculate the distance. Hence, a 3D structure of the environment can be measured and presented as 3D point cloud [22] because the ranging mechanism of LiDAR sensor is independent from light sources in surrounding environments. Therefore, LiDAR sensor is not affected by light variation such as direct sunlight or night conditions where a camera struggles. Thus, pedestrian and cyclist detection has been conducted by using LiDAR sensor [23–27]. The camera can capture higher-resolution images of objects; on the other hand, LiDAR can acquire accurate 3D information of objects and has better performance in extreme light variation. Therefore, an integration of camera image and LiDAR information was proposed for pedestrian detection [28–31]. However, there are a few discussions for cyclist orientation estimation using LiDAR sensor. The autonomous vehicle needs to work in different light conditions. Accordingly, research on the use of a LiDAR-based perception system can support the current system that uses a camera only, or be a redundancy in cases where the camera-based perception system cannot work. As there has been research on camera-based cyclist orientation estimation, this research focuses on the methods of LiDAR-based cyclist orientation.

The contribution of this research is to propose two different methods for cyclist orientation estimation. The first method is to convert LiDAR data into three gray-scaled images for cyclist orientation classification. The second method is to use 3D point cloud data to represent LiDAR data for cyclist orientation estimation. Both of the proposed methods use a model ResNet50 deep neural network for orientation classification. Finally, the performances of the above two methods are compared in order to approach the most effective usage of LiDAR sensor data for cyclist orientation estimation. In the research, a cyclist dataset including multiple cyclists with different body and head orientation is developed. The experimental results proved that the 3D point cloud data-based method has a better performance for cyclist orientation estimation compared to the 2D image-based method. Moreover, in the 3D point cloud data-based method, using reflectivity information has a more accurate estimation result than using ambient information.

The rest of the paper is organized as follows: Section 2 describes the two proposed methods. Section 3 presents the experimental result, and the paper is concluded in Section 4.

2. Cyclist Orientation Estimation Based on 2D and 3D Methods

2.1. Definition for Cyclist Body and Head Orientations

Following the conventional definition used in the research of camera-based orientation estimation [19,20], cyclist body orientation is labelled in eight classes and head orientation of body orientation is labelled in three classes in this research. The body orientation number is defined along clockwise, starting from an orientation facing the LiDAR sensor. The diagram for the eight body orientations of the cyclist in Bird's Eye View (BEV) is shown on Figure 1a. In this diagram, the cyclist is in orientation "0", and the arrow indicates the rotation direction for labeling. Figure 1b shows the example of 3D point cloud data for each body orientation taken by LiDAR sensor. Assuming that a LiDAR sensor has been equipped on a vehicle, the orientation "0" indicates that the cyclist is moving towards a vehicle and that the cyclist is moving closer to the vehicle. On the opposite direction, orientation "4" means that the cyclist is currently riding away from the vehicle. The orientations "2" and "6" indicate the situations of the cyclist riding perpendicular to the vehicle. The diagonal orientations "1" and orientation "7" indicates the situation where the cyclist is moving closer to the vehicle and there is a 45-degree angle between their moving directions. Similarly, the orientation "3" and orientation "5" refer to the behavior of the cyclist moving away from vehicle.

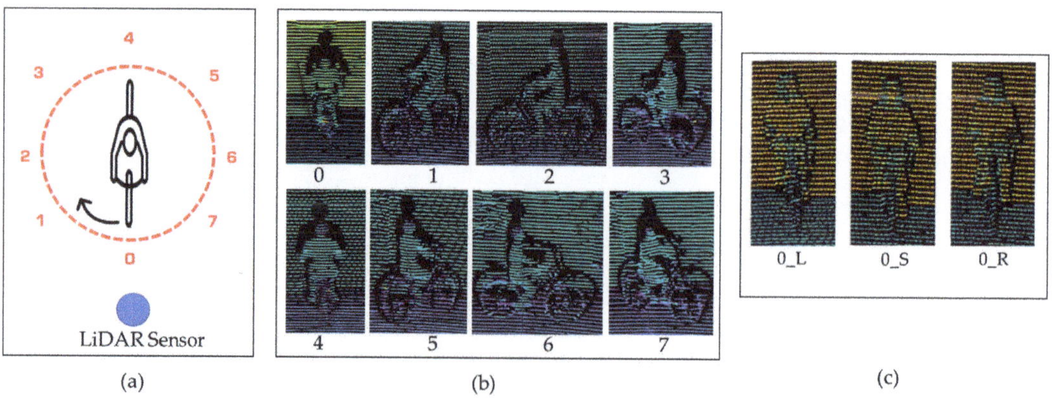

Figure 1. Demonstration of the definition of body orientation (**a**,**b**) and head orientation (**c**).

Detached from body orientation, the head orientation of a bicycle rider indicates the sight of the cyclist. Based on the head orientation, the future decisions of the cyclist are estimated. Considering the natural head rotation of a cyclist, the cyclist head orientation following the cyclist's sight is labelled in three classes, which are L (Left), S (Straight) and R (Right). This head orientation is labelled in a cyclist's perspective as well, meaning that the cyclist turns the head to the left, looks straight and turns the head to right. Thus, three head orientations are assigned to sub-classes for each of eight body orientations. This results in 24 classes that indicate the joined body and head orientation. Figure 1c demonstrates the head orientations in the event of "0" body orientation only, and the joined body and head orientations are labelled as "0_L", "0_S" and "0_R" for the three postures of bicycle riders in body orientation "0" while looking on Left (L) side, looking Straight (S) and looking on Right (R) side.

This research works under the assumption that the cyclist has been detected from the LiDAR data. Some cyclist detection algorithms exist that can possibly fulfill this requirement [21,32]. In this research, the data of the cyclist's area is manually cropped from the data collected by LiDAR sensor and used for orientation estimation.

2.2. 2D Image-Based Cyclist Orientation Estimation

The proposed 2D image-based cyclist orientation estimation aims for classification of the cyclist head and body orientation based on the images generated from LiDAR sensor data. The methodology of this system is to transform information taken by LiDAR sensor into image format and classify cyclist images into different joined body and head orientations. The flowchart of the image-based orientation estimation method is shown in Figure 2, in which the red boxes represent the main stages of the method. Moreover, the solid green boxes represent the input and output of each stage. Additionally, the images in the dotted green boxes illustrate the examples of generated images in each main stage.

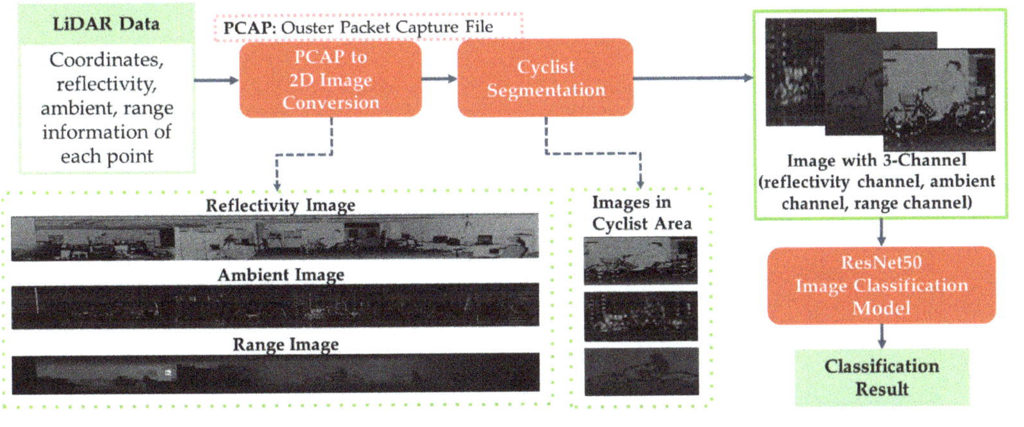

Figure 2. Flowchart of 2D image-based cyclist orientation estimation.

The first stage of the proposed method is a data preprocessing step that converts a raw datum captured from the LiDAR sensor to image. The data format of the LiDAR sensor varies depending on the LiDAR sensor manufacturer, and the LiDAR sensor used in this research records a sequence of sensor packets to a PCAP (Packet Capture) file [33]. Each sensor packet corresponds to one scan for the surrounding environment. The point cloud data in each scan contains the coordinates (x, y, z) of each point with four data layers: range, signal, reflectivity and ambient. Each data layer contains information that is not able to be captured from an RGB camera, which is a key information for data analysis in 3D space [34].

The range information of the point represents the distance from the sensor by calculating the travel time of the laser light wave. The signal information represents the strength of the light returned to the sensor for a given point. The reported reflectivity is a byproduct of range and signal that gives the user an indication of the target reflectivity. The ambient information denotes the strength of sunlight collected for a given point, also expressed in the quantity of photons detected that were not produced by the sensor's laser pulse. The signal information varies with range (objects farther away return less light), and ambient data varies with sunlight levels. On the other hand, the reflectivity data are consistent across lighting conditions and range. Therefore, reflectivity is the only piece of data that contains information about the properties of the object, which is not light [34]. To avoid information redundancy, the signal information is not employed in the system, since the signal varies with range information. Therefore, the coordinates, range, reflectivity and ambient data of each point on the point cloud are utilized for the proposed image-based cyclist orientation estimation system.

For the image conversion, the 3D data are projected into 2D. The 3D point cloud captured from LiDAR sensor is perfectly 1:1 spatial correspond [35]. Hence, the images can be generated by analyzing the coordinates of each point and assign values to each pixel. This implies that each point on the point cloud is projected as each pixel on the images. Since the LiDAR sensor used in this research has a vertical resolution of 64 and a resolution of 2048 for 360-degree, there are 131,072 points captured on each scan, which is the same as the length of returned lists of data layers. These returned lists are combined into a 3D array with a size of (2048 $\times$ 64 $\times$ 6). This array represents a single frame of the captured sequence, and each element contains values of x, y, z, range, ambient and reflectivity. Hence, this array is divided into three arrays that each represents an image of a frame. As the spatial correspondence of the points is perfect, it can be directly projected onto a 2D array with a size of (2048 $\times$ 64) based on the position on the 3D array. Each converted pixel of a 1D array contains range, ambient and reflectivity values for arrays of range image, ambient image and reflectivity image, respectively. Since the converted image arrays contain only one value per pixel, the generated images are gray-scale images with one channel. Figure 2 shows the examples of generated range, ambient and reflectivity images.

The second stage of the proposed method is to segment the cyclist area from the range, ambient and reflectivity images. This research focuses on the cyclist orientation estimation, and assumes that cyclists have been detected by other methods, for example, applying YOLO [36] on generated gray scale images [35]. Thus, the cyclist area is manually cropped in this research. Figure 3a shows the example reflectivity images of cyclists with different body orientations, and Figure 3b demonstrates the reflectivity images for three head orientations in the case of "0" body orientation.

The ultimate goal of the 2D image-based cyclist orientation estimation system is to predict and classify the body and head orientation of cyclists based on images generated from LiDAR sensor data. In this research, a Residual Neural Network (ResNet) model with 50 convolution layers is used for this goal. Compared to traditional CNNs, ResNet can overcome the "vanishing gradient" problem. Therefore, it can construct networks with thousands of convolutional layers, and outperform shallower networks. Since ResNet has deep architecture and good performance for image recognition, it is widely used in the task of image classification. The input images of the image classification model are the range image, ambient image and reflectivity image of a cyclist from LiDAR sensor data. The desired input size of the model is (224 $\times$ 224) [37]. Therefore, the images are resized to be in the same size of (224 $\times$ 224). In addition, the three images are concatenated to be in one data array, which has array size of (224 $\times$ 224 $\times$ 3), and sent to the ResNet50 model to estimate the joined body and head orientation of cyclist. The experimental result of the 2D image-based method will be represented in Section 3.

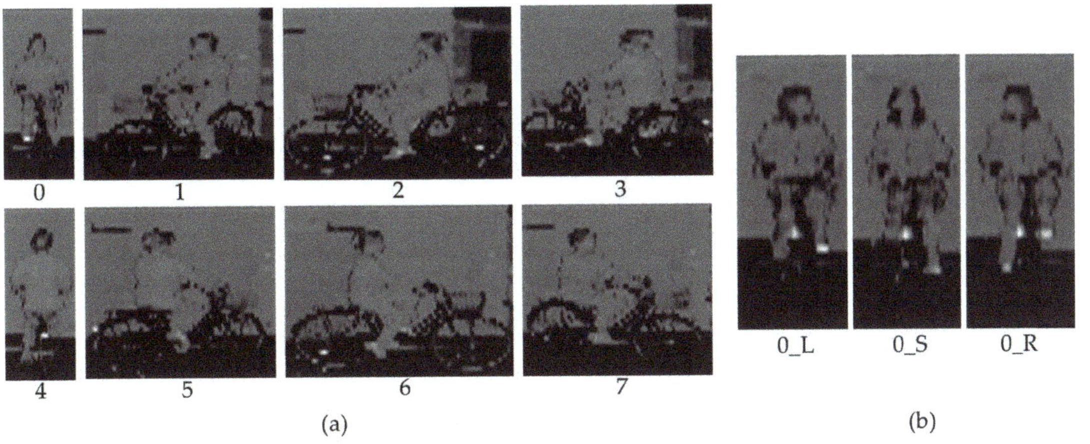

Figure 3. Reflectivity images of cyclist with different body orientations (**a**) and head orientations (**b**). (In addition to the reflectivity information, ambient and range are also used in 2D image-based cyclist orientation estimation).

2.3. *3D Point Cloud-Based Cyclist Orientation Estimation*

The basic methodology of the 3D point-cloud-based cyclist orientation estimation is similar to the 2D image-based method. However, this point-cloud-based method represents the information of a cyclist using 3D point cloud data instead of a 2D image. The flowchart of the proposed 3D point-cloud-based cyclist orientation estimation is illustrated in Figure 4. In Figure 4, the important stages are indicated in a red box. Moreover, the input and output data of each stage are indicated in green color.

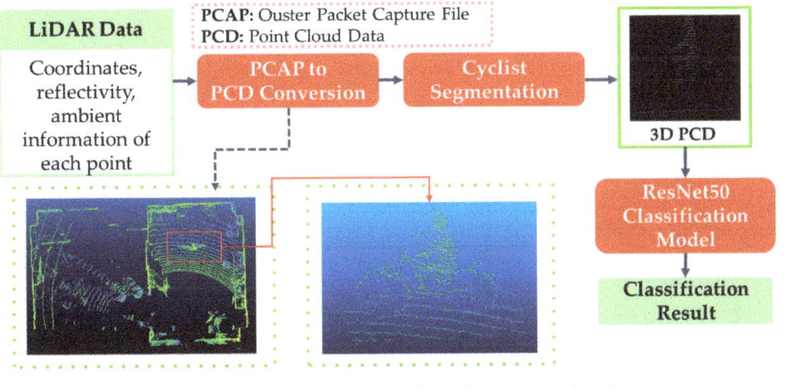

Figure 4. Flowchart of 3D point-cloud-based cyclist orientation estimation.

In order to process the point cloud data in each scan, it is necessary to convert it from a packet file of the LiDAR sensor, PCAP, to a Point Cloud Data format, PCD. The point clouds stored in PCD file contains a collection of 3D coordinates (x, y, z) with other data layers. In this proposed system, the data layers of reflectivity information and ambient information are chosen to be utilized, while excluding range information. This is because the range information represents the 3D shape of an object using the distance of the points from the sensor to the object, with the 3D shape of the object also able be represented by coordinates (x, y, z). Therefore, using a range data layer is a redundancy in the information. Thus, reflectivity and ambient information are used in the proposed system. Hence, there are two types of input to the system, which are two arrays containing 3D coordinates with reflectivity and ambient each, respectively. After each array is used as an input to the orientation classification model, the results of each case can be compared to understand the best data layer for this 3D point-cloud-based method. Finally, the arrays are saved as a PCD file by the point cloud library [38].

After the PCD conversion stage, cyclist segmentation is performed to acquire point cloud data in the cyclist area. This research also assumes that cyclists have been detected by other methods. For example, it is possible to convert a 3D point cloud into RGB-map in Bird's Eye View (BEV) and implement image object detection algorithm, YOLO, on the BEV image [21]. Following this idea, the point cloud data of the cyclist is segmented from BEV in the software CloudCompare. Figure 4 shows the BEV of point cloud data and the front view of the segmented cyclist area.

After segmentation process, the point cloud data are no longer in a vertical size of 64 and a horizontal size of 2048. Instead, the size of the point cloud data is dependent on the posture of the cyclist, especially the body orientation. On the other words, the number of points on the point cloud is also inconsistent across the body orientation. As the point cloud data is an input to the classification model, the size of the data should be equal. Therefore, the point cloud data are normalized in a vertical size of 224, a horizontal size of 224 and depth of 50, followed by the input configuration of classification model.

The segmented and normalized point cloud data of a cyclist is returned as an output from the previous data preprocessing stages. Hence, they are input to the point cloud data classification model for cyclist body and head orientation estimation. The purpose of this research is to compare the cyclist orientation estimation results between converting LiDAR data to a 2D image and using the 3D point cloud data from the sensor. Hence, the same classification model for the image-based method is used in the point-cloud-based method, which is ResNet50. The point cloud data in 3D array is input to ResNet50 in a size of (224 × 224 × 50). Since elements of points on the point cloud data indicate their reflectivity or ambient value, the values of points are processed as a value of a pixel in the case of images. For a clear comparison between the two different usages of LiDAR sensor data, the same architecture is used with a slight modification on input configuration for the 3D point-cloud-based method. The experimental result of the 3D point-cloud-based method will be represented in Section 3.

3. Experiments

3.1. Data Collection

In this research, the LiDAR sensor data of a cyclist with different body and head orientations is required. The LiDAR sensor that is used for the experiment is Ouster OS1-64. The LiDAR sensor used in the experiment has 64 channels of resolution, which indicates the number of beams that are sent at once. Moreover, it has a horizontal resolution of 2048, which means that the 64 channel beams are sent 2048 times per frame. Thus, it records 131,072 points per frame, and it can capture 10 frames per second. The data are collected indoor with a bicycle in a fixed position. The LiDAR sensor is installed about 7 m away from the bicycle's position. Hence, the bicycle is rotated clockwise within the target body orientation. In the data collection, a total of 12 students participated as cyclists. Participants are asked to ride a bicycle by pedaling on a bicycle in a fixed position.

Twelve participants are asked to do the pedaling, and they are asked to turn their head three times per body orientation. To capture a clear posture, the cyclists are supposed to turn their head completely to each direction and maintain a solid head orientation while the data are taken. The cyclists' data are taken for 5 s per head and body orientation, which indicates 50 frames per joined body and head orientation. Finally, there are about 14,000 frames taken for joined body and head orientation in total. Table 1 is shown below to represent the number of frames taken for each joined body and head orientation with the total number of each body orientation.

Table 1. Number of frames in dataset for joined body and head orientation.

Head Orientation	Body Orientation								Total
	0	1	2	3	4	5	6	7	
Left	590	591	591	592	589	591	589	592	
Straight	589	589	589	589	592	590	594	591	
Right	589	592	590	589	591	592	591	592	
Sub-total	1768	1772	1770	1770	1772	1773	1774	1775	14,174

The datasets of cyclists are divided into training datasets and validation datasets for cross validation. The 4-fold cross validation is used in evaluation. The training process has an epoch of 40, and the validation process compares the predicted labels of a cyclist's orientation with the ground truth labels for evaluation. For the experiments, a computer with an Intel Core i9-9900K CPU running at 3.60 GHz, an NVIDIA GeForce RTX 2070-Super GPU, and a RAM memory of 16 GB is adopted.

3.2. Experimental Results

This research presents three classification results which are the results of the 2D image-based method, the 3D point cloud with reflectivity-value-based method and 3D point cloud with ambient-value-based method. The main evaluation method of the classification model is prediction accuracy, as it is a direct and quantitative measurement of the classification. However, such an evaluation method by comparing accuracy can lead to unclear classification errors between the classes. Therefore, a confusion matrix is additionally used to present the performance of the classification. The confusion matrix is a table that records the correctness of predicted labels with true labels in percentage. Since there are multiple numbers of orientation classes, the multi-class confusion matrix is used as a performance indicator.

Table 2 shows the accuracy of classification for the cyclist joined body and head orientation estimation of the two proposed methods. The confusion matrix in Figure 5 illustrates the classification results of the 2D-based method for joined body and head orientation estimation. In Figures 6 and 7, the classification results of the 3D-based method with ambient and reflectivity as input are shown, respectively. Obviously, the classification accuracy of joined body and head orientation is higher in the 3D point-cloud-based method in general. Moreover, using reflectivity resulted in better accuracy than using ambient.

Table 2. Cyclist Joined Body and Head Orientation Estimation Accuracy of Different Usage of LiDAR data.

	2D Image Based Method	3D Point Cloud Based Methods	
		Ambient	Reflectivity
Accuracy	47.69%	50.96%	60.52%

Figure 5. Confusion matrix of cyclist joined body and head orientation estimation using 2D images.

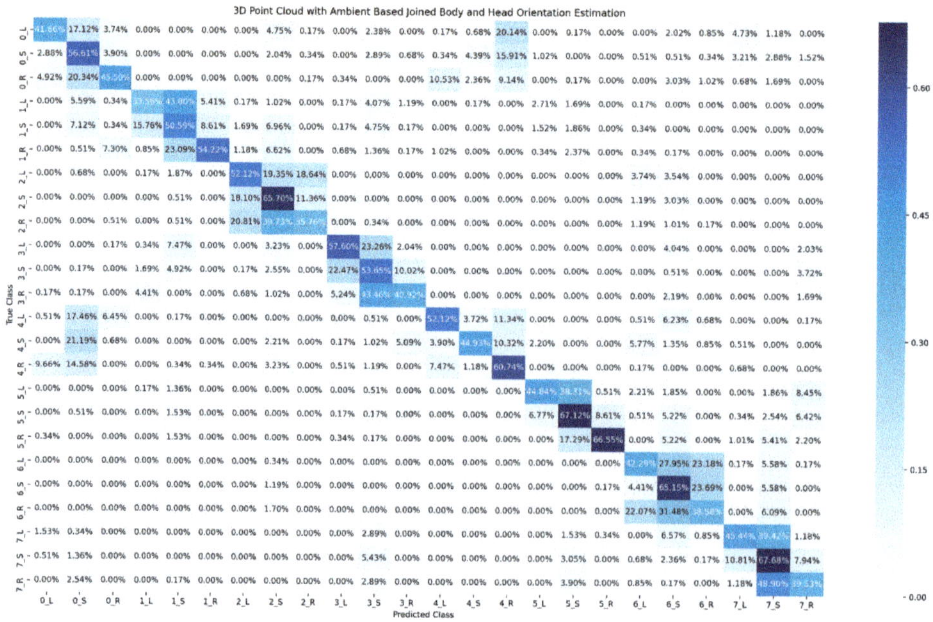

Figure 6. Confusion matrix of cyclist joined body and head orientation estimation using 3D point cloud data with ambient.

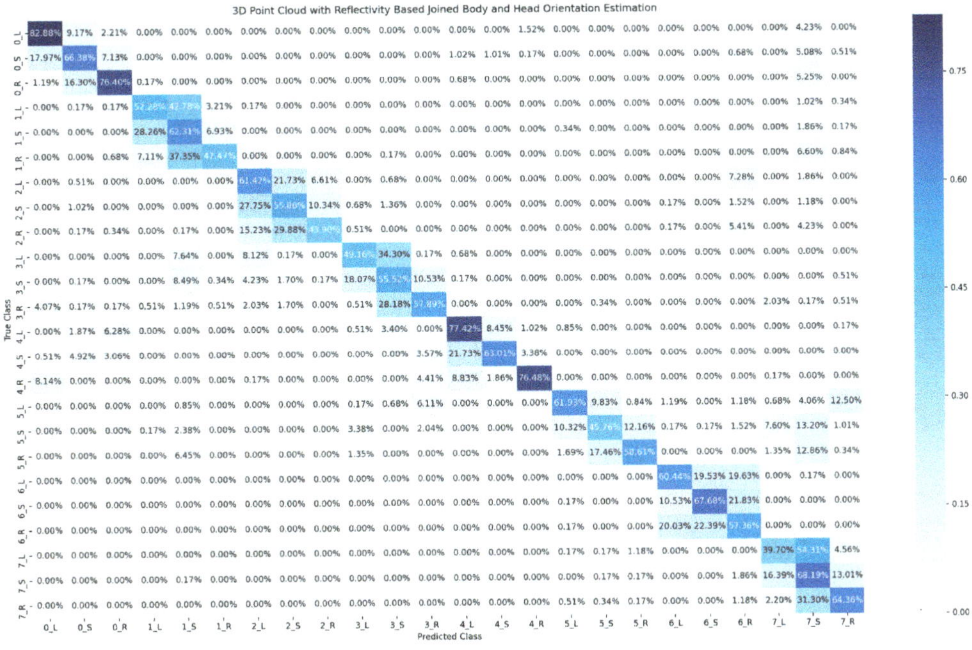

Figure 7. Confusion matrix of cyclist joined body and head orientation estimation using 3D point cloud data with reflectivity.

Furthermore, the results of the 3D-based method are accumulated from 24 joined body and head orientation classes into 8 body orientation classes in Figures 8 and 9. Once the estimation result of joined body and head orientation estimation is summarized into a body orientation estimation result, it is obvious that there are much higher correction rates and stability than in joined body and head orientation estimation. Since the joined body and head orientation estimation aims for more detailed behavior of cyclists, the head orientation is added as a sub-class to the body orientation. Hence, there are more classes for classification, numbering 24 in total. This increase in number of classes for more detailed posture analysis resulted in a decrease in estimation accuracy. Moreover, the head orientation is not well estimated since the accuracy dropped when head orientation is added as a sub-class to the body orientation. More specifically, for the estimation accuracy calculated after the accumulation for eight body orientation, the 3D point cloud data with ambient have 81.62% accuracy and 3D point cloud data with reflectivity have 90.34% accuracy. In this research, all experiments were performed in an indoor environment with artificial illumination, which is better than the illumination condition during night in an outdoor environment. However, using ambient information cannot have a better performance than using reflectivity information for body orientation estimation.

In order to have analysis on head orientation specifically, the results of the 3D-based method are accumulated from 24 joined body and head orientation classes into 3 head orientation classes in Figure 10. The confusion matrix of three head classes shows more unstable estimation results than when they are accumulated into eight body orientation classes. The head orientation estimation correction rate of 3D point cloud data with ambient shows 56.79% accuracy, while 3D point cloud data with reflectivity shows a correction rate of 63.42%. The estimation resulted in better accuracy when the head orientation is eliminated from the joined body and head orientation estimation. Therefore, there is a challenge for head orientation estimation with both the proposed 2D image-based method and the 3D point-cloud-based method.

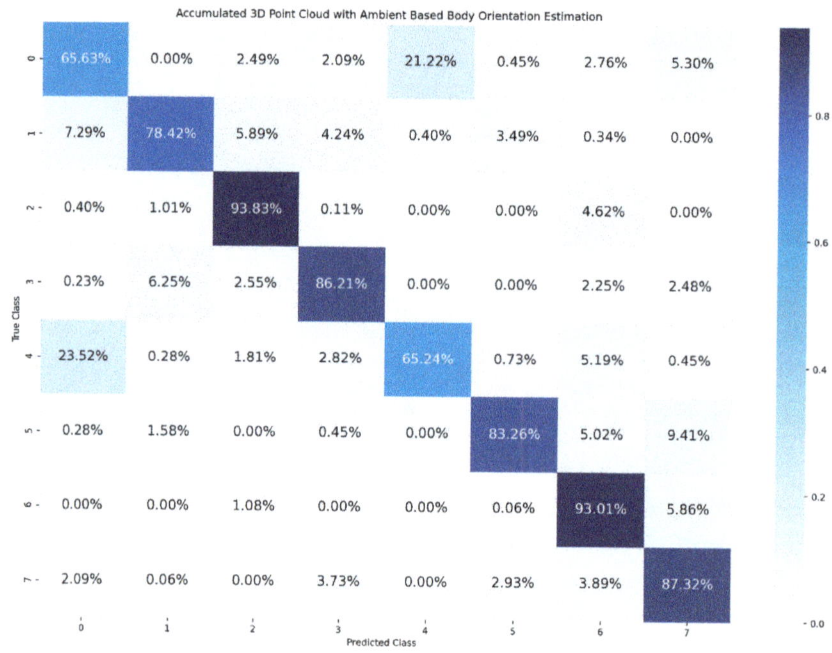

Figure 8. Confusion matrix of body orientation estimation accumulated from joined body and head orientation estimation using 3D point cloud data with ambient.

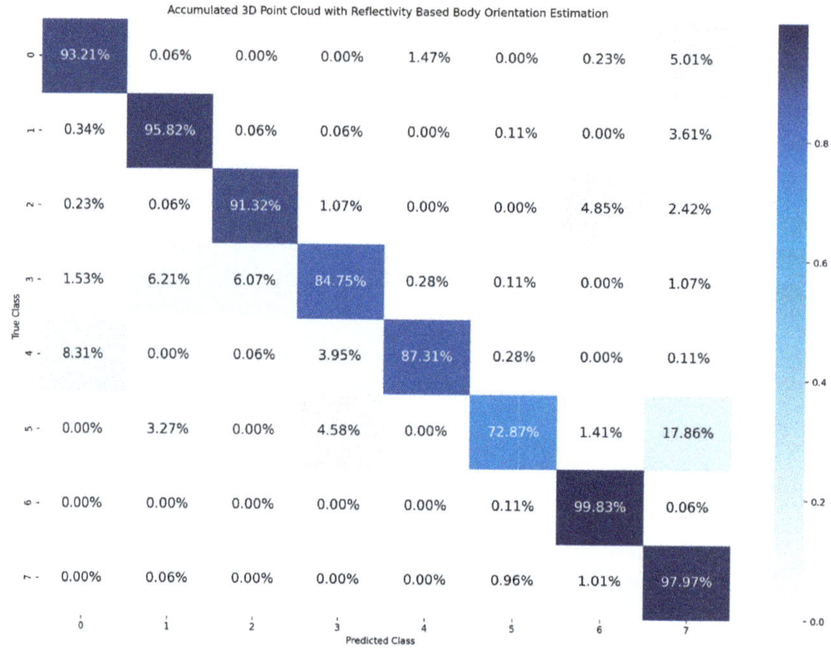

Figure 9. Confusion matrix of body orientation estimation accumulated from joined body and head orientation estimation using 3D point cloud data with reflectivity.

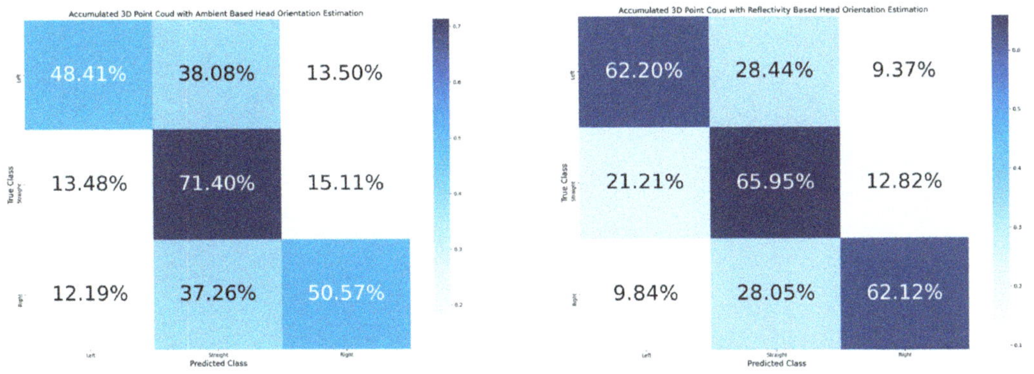

Figure 10. Left: Confusion matrix of head orientation estimation accumulated from joined body and head orientation estimation using 3D point cloud data with ambient. **Right**: Confusion matrix of head orientation estimation accumulated from joined body and head orientation estimation using 3D point cloud data with reflectivity.

3.3. Discussions

In this research, the cyclist head orientation is labeled into three classes: L (Left), S (Straight) and R (Right). The three classes correspond to the three cases: cyclist turns head to the left, looks straight and turns head to the right. In fact, head movement is the most reliable indicator for detecting cyclist intention when they are about to make a turn. Figure 11 demonstrates one of the most dangerous scenarios in the real traffic situation. The cyclist is planning to turn right and cross the road, as indicated by the green arrow trajectory line. Usually, cyclists maximally turn their heads (about 90 degrees, as indicated by the red arrow) to check the situation behind them before they turn their bicycle. To simulate this situation, head orientations L and R are defined as a 90-degree difference from the body orientation. The head orientation S means that the body and head orientations are the same (0-degree difference). In fact, there are more complicated cases in real traffic situations. This research uses three cases to test the feasibility of head orientation estimation.

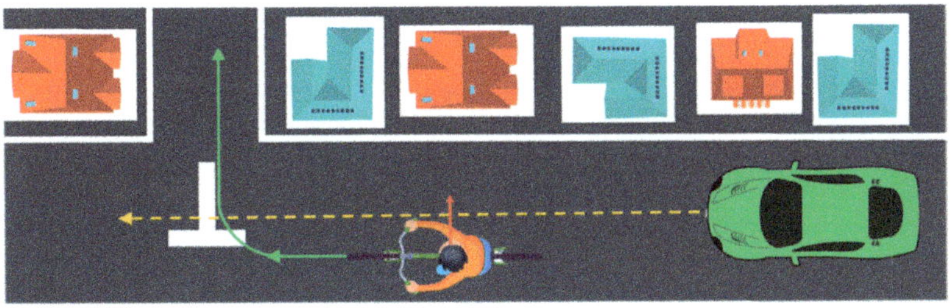

Figure 11. Demonstration of a scenario where a cyclist is turning his or her head before crossing the road.

The experimental result indicates that the accuracy of the head orientation is relatively low and needs to be improved. Recently, the deep-learning-based super-resolution technique is not only used for image processing, but also extended for point cloud processing. It is possible to employ the super-resolution [39,40] techniques to increase the resolution of a cyclist to improve the accuracy of the head orientation estimation.

As we can see from Figures 8 and 9, misclassification often happens between body orientation 0 and body orientation 4. This is because the LiDAR data of these two classes are similar. However, when the body orientation estimation is performed based on an RGB camera in the ideal light condition, the misclassifications between body orientation 0 and orientation 4 are few [20] because the cyclist's face can be clearly represented in an RGB image and used as the main feature to distinguish the two classes. However, the resolution of LiDAR sensor is low and cannot represent the face clearly. In fact, distinguishing these two classes is significant for cyclist safety; one possible solution is to use super-resolution techniques [39,40] to increase the resolution of LiDAR data. Another way is to track the trajectory of the cyclist in order to understand whether the cyclist is moving towards the vehicle or riding away from the vehicle.

In the experiment, the distance between cyclists and LiDAR sensor is around 7 m. Theoretically, when cyclists go far from the sensor, the classification capability of the developed system should decrease. This problem can be solved by increasing the resolution of LiDAR sensor data, e.g., using super-resolution [39,40] techniques.

The paper proposes two different methods for cyclist orientation estimation, aiming to illustrate the comparison between the two proposed methods. The dataset used in this research only contains cyclists without carrying bags. However, cyclists sometimes may carry bags or other objects in real traffic situations. When the proposed method is used in real applications, the dataset for training should be enriched to include different types of cyclists, e.g., those carrying backpacks or sling bags. By adding the different types of sample data, the retrained model is expected to recognize the orientation of cyclists with carrying bags.

4. Conclusions

In this research, a system for cyclist body and head orientation estimation using LiDAR sensor data is proposed. The ultimate goal of the proposed system is to approach the most effective usage of LiDAR sensor data for cyclist body orientation estimation. The first method suggested an approach to convert LiDAR sensor data into three different gray-scaled images by utilizing the data layers of range, ambient and reflectivity. Hence, the method used the images together as a three-channel image as an input to orientation classification model. On the other hand, the second method proposed using the LiDAR sensor as a 3D point cloud data with each point containing an ambient layer or a reflectivity layer for the orientation classification. The evaluation of the proposed system is based on the classification accuracy and confusion matrix. The results of the experimentation proved that the proposed 3D point-cloud-based cyclist orientation system leads to better prediction results for the cyclist joined body and head orientation estimation than the 2D image-based method. The 2D image-based method resulted in 47% accuracy, the 3D point cloud data with ambient in 51% accuracy and the 3D point cloud data with reflectivity in 60% accuracy. Therefore, the best usage of LiDAR data for cyclist orientation estimation is to utilize the data into 3D point cloud data with a reflectivity layer of each point. Moreover, there is a challenge regarding joined body and head orientation estimation which resulted in much less prediction accuracy than body orientation alone.

In the future, the super-resolution technique will be adopted to improve the accuracy of orientation estimation, especially for the head orientation estimation.

Author Contributions: Conceptualization, Y.G.; methodology, Y.G. and H.C.; software, H.C. and Y.G.; validation, H.C.; writing—original draft preparation, Y.G. and H.C.; investigation, Y.G. and H.C., I.G., L.-T.H. and C.P.; supervision, Y.G.; project administration, Y.G.; funding acquisition, Y.G. All authors have read and agreed to the published version of the manuscript.

Funding: This research was funded by JSPS (Japan Society for the Promotion of Science) KAKENHI Grant Number JP21K12019. Y.G. is the research representative of the research grant.

Institutional Review Board Statement: Not applicable.

Informed Consent Statement: Not applicable.

Data Availability Statement: The data are not publicly available due to restrictions on privacy.

Conflicts of Interest: The authors declare no conflict of interest.

References

1. European Commission. ITS & Vulnerable Road Users. Available online: https://transport.ec.europa.eu/transport-themes/intelligent-transport-systems/road/action-plan-and-directive/its-vulnerable-road-users_en (accessed on 24 January 2023).
2. Dollar, P.; Wojek, C.; Schiele, B.; Perona, P. Pedestrian detection: An evaluation of the state of the art. *IEEE Trans. Pattern Anal. Mach. Intell.* **2011**, *34*, 743–761. [CrossRef] [PubMed]
3. Hamed, M.M. Analysis of pedestrians' behavior at pedestrian crossings. *Saf. Sci.* **2001**, *38*, 63–82. [CrossRef]
4. Tokyo Metropolitan Police Department. Status of Bicycle Traffic Accidents in Tokyo. Available online: https://www.keishicho.metro.tokyo.lg.jp/about_mpd/jokyo_tokei/tokei_jokyo/bicycle.html (accessed on 24 January 2023).
5. European Commission. *Facts and Figures Cyclists. European Road Safety Observatory*; European Commission, Directorate General for Transport: Brussels, Belgium, 2020.
6. RCUK. Essential Guide to Road Cycling Hand Signals and Calls. Available online: https://roadcyclinguk.com/how-to/technique/essential-guide-road-cycling-hand-signals-calls.html (accessed on 24 January 2023).
7. Bicycle Law USA. Bike Hand Signals. Available online: https://www.bikelaw.com/2021/03/bike-hand-signals/ (accessed on 24 January 2023).
8. Walker, I. Signals are informative but slow down responses when drivers meet bicyclists at road junctions. *Accid. Anal. Prev.* **2005**, *37*, 1074–1085. [CrossRef] [PubMed]
9. Westerhuis, F.; De Waard, D. Reading cyclist intentions: Can a lead cyclist's behaviour be predicted? *Accid. Anal. Prev.* **2017**, *105*, 146–155. [CrossRef] [PubMed]
10. Hemeren, P.E.; Johannesson, M.; Lebram, M.; Eriksson, F.; Ekman, K.; Veto, P. The use of visual cues to determine the intent of cyclists in traffic. In Proceedings of the 2014 IEEE International Inter-Disciplinary Conference on Cognitive Methods in Situation Awareness and Decision Support, San Antonio, TX, USA, 3–6 March 2014; pp. 47–51.
11. Schulz, A.; Damer, N.; Fischer, M.; Stiefelhagen, R. Combined head localization and head pose estimation for video–based advanced driver assistance systems. In Proceedings of the Pattern Recognition: 33rd DAGM Symposium, Frankfurt, Germany, 31 August–2 September 2011; pp. 51–60.
12. Schulz, A.; Stiefelhagen, R. Video-based pedestrian head pose estimation for risk assessment. In Proceedings of the 2012 15th International IEEE Conference on Intelligent Transportation Systems, Anchorage, AK, USA, 16–19 September 2012; pp. 1771–1776.
13. Gandhi, T.; Trivedi, M.M. Image based estimation of pedestrian orientation for improving path prediction. In Proceedings of the 2008 IEEE Intelligent Vehicles Symposium, Eindhoven, The Netherlands, 4–6 June 2008; pp. 506–511.
14. Gu, Y.; Kamijo, S. Bicyclist recognition and orientation estimation from on-board vision system. *Int. J. Automot. Eng.* **2015**, *6*, 67–73.
15. Gu, Y.; Hsu, L.T.; Xie, L.; Kamijo, S. Accurate estimation of pedestrian orientation from on-board camera and inertial sensors. *IEICE Trans. Fundam. Electron. Commun. Comput. Sci.* **2016**, *99*, 271–281. [CrossRef]
16. Flohr, F.; Dumitru-Guzu, M.; Kooij, J.F.; Gavrila, D.M. A probabilistic framework for joint pedestrian head and body orientation estimation. *IEEE Trans. Intell. Transp. Syst.* **2015**, *16*, 1872–1882. [CrossRef]
17. Raza, M.; Chen, Z.; Rehman, S.U.; Wang, P.; Bao, P. Appearance based pedestrians' head pose and body orientation estimation using deep learning. *Neurocomputing* **2018**, *272*, 647–659. [CrossRef]
18. Cao, Z.; Simon, T.; Wei, S.E.; Sheikh, Y. Realtime multi-person 2d pose estimation using part affinity fields. In Proceedings of the 2017 IEEE Conference on Computer Vision and Pattern Recognition, Honolulu, HI, USA, 21–26 July 2017; pp. 7291–7299.
19. Abadi, A.D.; Gu, Y.; Goncharenko, I.; Kamijo, S. Detection of Cyclists' Crossing Intentions for Autonomous Vehicles. In Proceedings of the 2022 IEEE International Conference on Consumer Electronics, Las Vegas, NV, USA, 7–9 January 2022; pp. 1–6.
20. Abadi, A.D.; Gu, Y.; Goncharenko, I.; Kamijo, S. Detection of Cyclist's Crossing Intention based on Posture Estimation for Autonomous Driving. *IEEE Sens. J.* **2023**. [CrossRef]
21. Garcia-Venegas, M.; Mercado-Ravell, D.A.; Pinedo-Sanchez, L.A.; Carballo-Monsivais, C.A. On the safety of vulnerable road users by cyclist detection and tracking. *Mach. Vis. Appl.* **2021**, *32*, 109. [CrossRef]
22. Velodyne, Inc. What Is Lidar? Available online: https://velodynelidar.com/what-is-lidar/ (accessed on 29 January 2023).
23. Szarvas, M.; Sakai, U.; Ogata, J. Real-time pedestrian detection using LIDAR and convolutional neural networks. In Proceedings of the 2006 IEEE Intelligent Vehicles Symposium, Tokyo, Japan, 13–15 June 2006; pp. 213–218.
24. Premebida, C.; Ludwig, O.; Nunes, U. Exploiting LIDAR-based features on pedestrian detection in urban scenarios. In Proceedings of the 2009 12th International IEEE Conference on Intelligent Transportation Systems, St. Louis, MO, USA, 4–7 October 2009; pp. 1–6.
25. Ogawa, T.; Sakai, H.; Suzuki, Y.; Takagi, K.; Morikawa, K. Pedestrian detection and tracking using in-vehicle lidar for automotive application. In Proceedings of the 2011 IEEE Intelligent Vehicles Symposium, Baden-Baden, Germany, 5–9 June 2011; pp. 734–739.
26. Saleh, K.; Hossny, M.; Hossny, A.; Nahavandi, S. Cyclist detection in lidar scans using faster r-cnn and synthetic depth images. In Proceedings of the 2017 IEEE 20th International Conference on Intelligent Transportation Systems, Yokohama, Japan, 16–19 October 2017; pp. 1–6.
27. Wang, W.; Chang, X.; Yang, J.; Xu, G. LiDAR-based dense pedestrian detection and tracking. *Appl. Sci.* **2022**, *12*, 1799. [CrossRef]
28. Premebida, C.; Ludwig, O.; Nunes, U. LIDAR and vision-based pedestrian detection system. *J. Field Robot.* **2009**, *26*, 696–711. [CrossRef]

29. Dimitrievski, M.; Veelaert, P.; Philips, W. Behavioral pedestrian tracking using a camera and lidar sensors on a moving vehicle. *Sensors* **2019**, *19*, 391. [CrossRef] [PubMed]
30. Pang, L.; Cao, Z.; Yu, J.; Liang, S.; Chen, X.; Zhang, W. An efficient 3D pedestrian detector with calibrated RGB camera and 3D LiDAR. In Proceedings of the 2019 IEEE International Conference on Robotics and Biomimetics, Dali, China, 6–8 December 2019; pp. 2902–2907.
31. Alfred Daniel, J.; Chandru Vignesh, C.; Muthu, B.A.; Senthil Kumar, R.; Sivaparthipan, C.B.; Marin, C.E.M. Fully convolutional neural networks for LIDAR–camera fusion for pedestrian detection in autonomous vehicle. *Multimed. Tools Appl.* **2023**, 1–24. [CrossRef]
32. Simony, M.; Milzy, S.; Amendey, K.; Gross, H.M. Complex-yolo: An euler-region-proposal for real-time 3d object detection on point clouds. In Proceedings of the European Conference on Computer Vision Workshops, Munich, Germany, 8–14 September 2018; pp. 1–14.
33. Ouster Inc. Ouster Sensor Documentations. Available online: https://static.ouster.dev/sensor-docs/ (accessed on 24 January 2023).
34. Ouster Inc. Firmware 2.1.1: Better Perception Performance with Improved Reflectivity and Signal Multiplier Mode. Available online: https://ouster.com/blog/firmware-2-1-1-better-perception-performance-with-improved-reflectivity-and-signal-multiplier-mode/ (accessed on 24 January 2023).
35. Ouster Inc. Object Detection and Tracking using Deep Learning and Ouster Python SDK. Available online: https://ouster.com/blog/object-detection-and-tracking-using-deep-learning-and-ouster-python-sdk/ (accessed on 24 January 2023).
36. Ultralytics. Ultralytics YOLOv8 Docs. Available online: https://docs.ultralytics.com/ (accessed on 29 January 2023).
37. Simonyan, K.; Zisserman, A. Very deep convolutional networks for larg, e-scale image recognition. *arXiv* **2014**, arXiv:1409.1556.
38. Rusu, R.B.; Cousins, S. 3D is here: Point cloud library (PCL). In Proceedings of the 2011 IEEE International Conference on Robotics and Automation, Shanghai, China, 9–13 May 2011; pp. 1–4.
39. Wang, X.; Yu, K.; Wu, S.; Gu, J.; Liu, Y.; Dong, C.; Qiao, Y.; Change Loy, C. Esrgan: Enhanced super-resolution generative adversarial networks. In Proceedings of the European Conference on Computer Vision, Workshops, Munich, Germany, 8–14 September 2018; pp. 1–16.
40. Li, R.; Li, X.; Fu, C.W.; Cohen-Or, D.; Heng, P.A. Pu-gan: A point cloud upsampling adversarial network. In Proceedings of the IEEE/CVF International Conference on Computer Vision, Seoul, Republic of Korea, 27 October–2 November 2019; pp. 7203–7212.

Disclaimer/Publisher's Note: The statements, opinions and data contained in all publications are solely those of the individual author(s) and contributor(s) and not of MDPI and/or the editor(s). MDPI and/or the editor(s) disclaim responsibility for any injury to people or property resulting from any ideas, methods, instructions or products referred to in the content.

sensors

MDPI

Article

Design and Calibration of Plane Mirror Setups for Mobile Robots with a 2D-Lidar

James E. Kibii [1], Andreas Dreher [2], Paul L. Wormser [2] and Hartmut Gimpel [2,3,*]

1 Department of Electrical Engineering and Information Technology, HTWG Konstanz, University of Applied Sciences, 78462 Konstanz, Germany
2 Department of Mechanical Engineering, HTWG Konstanz, University of Applied Sciences, 78462 Konstanz, Germany
3 Institute for Optical Systems, HTWG Konstanz, University of Applied Sciences, Alfred-Wachtel-Straße 8, 78462 Konstanz, Germany
* Correspondence: hartmut.gimpel@htwg-konstanz.de

Abstract: Lidar sensors are widely used for environmental perception on autonomous robot vehicles (ARV). The field of view (FOV) of Lidar sensors can be reshaped by positioning plane mirrors in their vicinity. Mirror setups can especially improve the FOV for ground detection of ARVs with 2D-Lidar sensors. This paper presents an overview of several geometric designs and their strengths for certain vehicle types. Additionally, a new and easy-to-implement calibration procedure for setups of 2D-Lidar sensors with mirrors is presented to determine precise mirror orientations and positions, using a single flat calibration object with a pre-aligned simple fiducial marker. Measurement data from a prototype vehicle with a 2D-Lidar with a 2 m range using this new calibration procedure are presented. We show that the calibrated mirror orientations are accurate to less than 0.6° in this short range, which is a significant improvement over the orientation angles taken directly from the CAD. The accuracy of the point cloud data improved, and no significant decrease in distance noise was introduced. We deduced general guidelines for successful calibration setups using our method. In conclusion, a 2D-Lidar sensor and two plane mirrors calibrated with this method are a cost-effective and accurate way for robot engineers to improve the environmental perception of ARVs.

Keywords: Lidar; environmental perception; mirror; calibration procedure; mobile robot; ground detection

1. Introduction

1.1. Lidar Sensors on Vehicles

Light detection and ranging (Lidar) sensors generally measure the distance to an object by emitting laser light towards the object and then observing the time of flight until some of the backscattered light returns to the sensor. This Lidar measurement principle can be turned into different kinds of optical sensors: either a laser range sensor statically directs its light into a single direction and, therefore, measures a single distance value for any object in this direction or a 2D laser scanner, also called single-layer laser scanner or 2D-Lidar, additionally uses one mechanical axis of rotation to spin the emitted laser light, up to a full 360° rotation. This kind of 2D-Lidar has been commercially available for several decades. During the last few years, 3D-Lidar sensors have also become available. These can roughly be separated into two groups: The group of spinning 3D-Lidar sensors, also called multiple-layer laser scanners, still uses a macroscopic mechanical rotation axis to spin a multitude of emitted laser light beams, generally using a motor. They can have a field of view (FOV) of up to 360° along their scanning direction and typically around 40° in the perpendicular direction. The other group consists of 3D flash Lidar sensors, some of which are also called 3D solid-state Lidar sensors. These sensors avoid the macroscopic mechanical spinning movement by a multitude of different alternatives. A common property of these 3D flash

Lidar sensors is their camera-like FOV, typically 40° in both axes. A recent overview of 3D-Lidar sensors for automotive applications can be found in [1].

Lidar sensors are commonly used on different kinds of vehicles for environmental perception [2] because of their advantages over Radar and camera sensors: Lidar sensors usually have a higher spatial resolution than Radar sensors, because of the shorter wavelength of light compared to radio-frequency Radar waves. Lidar sensors also incorporate a well-adapted light source and, thus, can still operate reasonably well under challenging conditions for optical sensors, such as darkness, bright sunshine, or moderate fog and rain. Compared to triangulating optical sensors, for example stereo cameras, Lidar sensors have the advantage that the distance resolution stays nearly constant even at the maximum object distance.

Even though 3D-Lidar sensors have attracted more attention, especially for their use in autonomous passenger vehicles, there are currently many application areas where 2D spinning Lidars are still predominantly used. One example of this are autonomous guided vehicles (AGVs), also called autonomous mobile robots (AMRs) [3]. The use of 2D-Lidars on AGVs is still reasonable because they already supply extremely useful data and their price is currently still much lower than a 3D-Lidar. Additionally, the smaller amount of point cloud data generated by these sensors makes data processing less challenging; it can be achieved with less expensive hardware and, therefore, further reduces costs.

The data from 2D-Lidars on AGVs are used for several purposes: firstly, for localization of the vehicle inside a known room; secondly, for navigation purposes, so the vehicle can plan and find a path to where it is ordered to move; thirdly, Lidar data are often used for mapping. This means that during the operation of the AGV, data are stored and used to continuously generate or update a map of the room in which the AGV moves. This is performed with various simultaneous localization and mapping (SLAM) algorithms. A fourth purpose is collision avoidance: the Lidar data help the AGV not to collide with fixed or moving objects.

A problem that arises when using 2D-Lidar sensors on AGVs is the lack of ground detection. To achieve long-distance sight, as well as to acquire and perceive map data on a fixed height above the ground, the scanning plane of a 2D-Lidar sensor is generally oriented parallel to the ground on which the AGV drives. Therefore, the ground itself cannot be detected by the 2D-Lidar sensor. This becomes especially problematic when the ground is missing, for example when the AGV approaches a staircase to a lower floor. In this case, the AGV could fall down those stairs and damage itself, as well as its environment. Another problem is obstacles on the floor that are not high enough to be detected by the scanning laser beam, but are still big enough to seriously stop or damage the AGV when it simply tries to drive over the obstacle.

This problem of missing ground detection can be solved by adding a separate sensor to detect the ground surface. This can be an ultrasonic distance sensor or an additional optical distance sensor such as a short-range laser range sensor. However, these additional sensors add further cost to the AGV hardware. In addition, the data quality and usefulness of the above-mentioned additional ground detection sensors is much weaker than the 2D-Lidar data. However, the ultimate solution of adding a second 2D-Lidar for ground detection or using a 3D Lidar both add significant cost.

Therefore, reshaping the FOV of a 2D-Lidar sensor to simultaneously achieve a wide range and ground detection is an important issue for improving 2D-Lidars on vehicles.

1.2. Previous Work on Lidar Sensors with Reshaped FOV

Several research groups have already added plane mirrors next to a Lidar sensor to reshape the FOV of the sensor. By this approach, a single Lidar sensor can additionally be used for ground detection, avoiding the need for a second sensor. This seems especially reasonable when a 2D-Lidar sensor with a 360° FOV used on an AGV is mounted at a position, e.g., at the very front of the AGV, where some of its FOV would otherwise be blocked by the presence of the vehicle. This use of mirrors to reshape the FOV of an optical sensor is similar to what has been done before with cameras looking simultaneously into several directions [4,5] or generating an omnidirectional FOV with catadioptric optics in front of a camera objective lens [6,7].

A conference paper from the year 2001 [8] is probably the first publication on the topic of Lidar sensors with additional plane mirrors. Two plane mirrors are attached to a 2D-Lidar sensor mounted at the front of an automobile. The resulting Lidar data were presented, as well as possible future applications sketched. A method for the calibration of the mirror positions was not mentioned in this publication.

This idea was not described in more detail until in 2017, when a research paper by Abiko et al. [9] used plane mirrors attached to a 2D spinning Lidar mounted on an unmanned aerial vehicle to help a quadcopter fly at a constant distance below a road bridge. This setup is very similar to the one presented here. However, Abiko et al. did not deal with the task of geometrical calibration of the mirrors either.

A later research paper [10] by Matsubara et al. used fixed plane mirrors next to a spinning 3D Lidar sensor on an autonomous vehicle to extend the sensor FOV to the ground. They also attached plane mirrors onto a manipulator arm on a vehicle [11] to improve its stair climbing ability. The work of this group differs from ours because they used a 3D spinning Lidar as opposed to a 2D spinning Lidar in our paper. In addition, they also did not deal with the calibration of the mirrors.

In a paper by Aalerud et al. [12], the FOV of a 3D spinning Lidar was reshaped by means of eight plane mirrors. The goal was to generate a similar FOV to that of a 3D flash Lidar, but using a 3D spinning Lidar, and to improve the spatial resolution. This group also presented mechanical design considerations for the placement of the mirrors and performance figures of the resulting prototype, and they stressed the need for a calibration method [12] (p. 22) to increase the precision of the measurement.

We know of only two publications that addressed the calibration process of the involved mirror positions and orientations: In a patent [13] by Waymo LLC, a manufacturing company of 3D Lidars, the need for such a calibration was mentioned briefly. In their context, the mirrors are far (meters) away from the Lidar sensor and the calibration is meant to track the position variations of the mirrors during vibrations that occur on a driving vehicle. Presumably, this ongoing calibration uses fiducial markers attached to the mirrors.

Recently, a paper [14] by Pełka et al. described a calibration procedure for a 3D flash Lidar where the FOV is reshaped by six plane mirrors into a shape that is similar to that of a 3D spinning Lidar. Their presented calibration method does not use any fiducials on the mirrors and relies on "pairs of the nearest neighborhood points that were reflected by a different mirror" [14] (p. 9). This approach cannot succeed with data from 2D-Lidar sensors, as in our proposed configuration, because there can be no more than one of these point pairs at the position where two scan lines of the 2D-Lidar cross each other. From our experience, one single point pair is insufficient for calibration without additional information such as fiducials or similar aids.

Therefore, to our knowledge, so far, no calibration procedure has been presented to acquire the position and orientation of plane mirrors attached to a 2D-Lidar sensor.

1.3. Aim

The aim of this work was to develop an easy-to-perform calibration procedure for a setup of a 2D-Lidar sensor with two attached plane mirrors. Accurate calibration is necessary because, usually, the position and orientation values that can be taken from the computer-aided design (CAD) software used to design the mechanical parts that hold the mirrors are not precise enough to measure accurate point cloud data. Production tolerances in the mechanical parts themselves, as well as individual variations during the fixation process of the optical mirror part, e.g., by gluing, are causes for deviations in orientation and position from the ideal CAD values.

We aimed to develop a calibration method that uses a simple calibration object and that can be used cost effectively in industrial production processes, as well as by researchers or amateurs when building a mobile robot with a 2D-Lidar and a two-mirror setup. This goal is complicated by the fact that the mechanical system to calibrate has many degrees of freedom, but point cloud data acquired with a 2D-Lidar on calibration objects of a simple geometry do not provide many features to be used for the calibration algorithm.

Additionally, we explored different design possibilities for the geometric arrangement of two plane mirrors around a 2D-Lidar sensor in order to add ground-detecting capabilities to the vehicle. We give several examples for geometric setups that are suitable for different purposes and boundary conditions that may be encountered on an AGV.

2. Materials and Methods

2.1. Prototype Vehicle

The presented measurements were conducted with a prototype vehicle that consisted of a "PiCar-S" [15] by SunFounder, not using any of the included sensors, and a "Raspberry Pi 4B" computer [16] running Linux (Raspberry Pi OS) and Python 3 [17]. A 2D scanning Lidar "TIM310-S01" [18] by Sick AG was connected via USB to the vehicle's computer. This Lidar has an angular FOV of 270° with a resolution of 1°, a maximum measurement distance of 2 m on targets with 10% remission, a scanning frequency of 15 Hz, and an IR (infrared) laser wavelength of 850 nm. The computer and the Lidar sensor were attached on top of the vehicle, while two rechargeable lithium polymer batteries and a step-up voltage converter were attached at the bottom of the vehicle for power supply.

Two plane rectangular first surface mirrors [19] by Edmund Optics measuring 100 × 75 mm with a 3 mm tick float glass substrate and a "protected silver" reflection coating were attached at either side of the Lidar sensor. The nominal reflectivity of these mirrors at an 850 nm wavelength was specified as >98%. In the prototype vehicle, only a fraction of the existing mirror surface is actually used for reflecting the Lidar's laser beam; the mirror dimensions are only this large because the mirrors were bought as off-the-shelf parts. A metal reflection coating was chosen so that mirror reflectivity does not strongly depend on the angle of incidence of the light for a large range of incidence angles. For holding these mirrors at the intended positions, mechanical parts were designed in the CAD software "Creo Parametric 8" [20] and produced as fused deposition modeling parts on a 3D printer out of polylactic acid (PLA) polymer material. The mirrors were attached to the mechanical parts by "Fixogum" rubber cement.

A picture of the described prototype vehicle is shown in Figure 1. This vehicle also has a simple driving algorithm implemented in Python to perform manual and automatic driving on the base of the Lidar data, which is not the topic of this article. A video of the driving vehicle can be seen in the Supplementary Material to this article.

2.2. Geometric Position and Orientation of Mirrors

Extensive CAD simulations were performed regarding the different possibilities to position and orient the two plane mirrors next to the Lidar sensor. These simulations were performed with the parametric CAD software "Creo Parametric" [20]. In Figure 2a, a top view projection of the geometry used is sketched. The green circles drawn around the mirror edges have a diameter of 27 mm; this corresponds to the light receiving aperture of the Lidar sensor used here. Whenever the red line of an observation direction touches a green circle, this corresponds to a power loss of the received Lidar light of 50%, because only half of the receiving aperture of the Lidar sensor can then receive the light reflected at this mirror edge. For this reason, the CAD model was parametrized such that these observation direction lines always keep a distance of 13.5 mm from the mirror edges, as well as from the housing of the Lidar sensor.

Figure 1. Prototype vehicle. As the construction base, the "PiCar" can be seen, consisting of four rubber tires, a servomotor for steering the two front wheels, and a "Raspberry Pi" computer for data processing and navigation. A mechanical platform holds the 2D-Lidar sensor (blue) on top of the computer board. Attached to the Lidar sensor is a larger mechanical part (black), which holds the two plane glass mirrors at either side of the Lidar sensor. Lithium polymer batteries and the power supply board are not visible as they are attached underneath the vehicle.

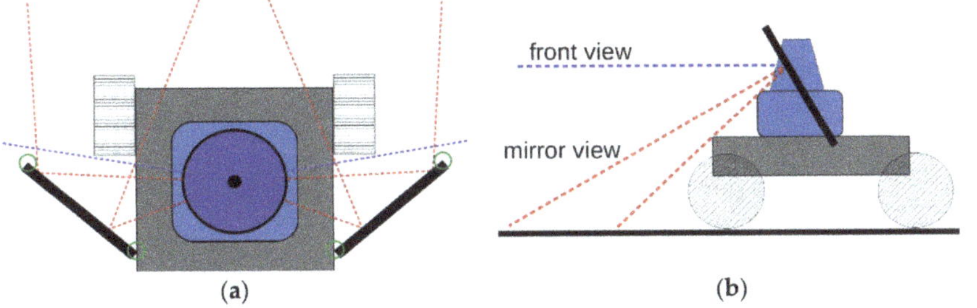

Figure 2. (a) Top view and (b) side view of the geometry for the geometric design simulations. In the center of the image, the 2D-Lidar sensors can be seen (blue). The black lines represent the two plane mirrors. The green circles at their ends have a diameter corresponding to the light receiving aperture of the Lidar sensor. The red lines represent the light beams, or observation directions, of the Lidar sensor. They all originate from the center point inside the Lidar.

The resulting reshaped Lidar FOV can be seen in Figure 2b. Observation angles that are not reshaped by any mirror make up the front view of the vehicle, whereas each of the two mirrors reflects a piece of the original 2D FOV towards the ground. As a result, two line-shaped pieces of the FOV observe the ground. These two line-shaped mirror view FOVs should cross each other somewhere on the ground to ensure a gapless observation of the ground in front of the vehicle.

The Cartesian coordinate system used has its origin in the central rotation axis of the Lidar sensor at the height of the center of the receiving aperture of the Lidar. The y-axis points into the direction of movement of the vehicle, the x-axis towards the right mirror (seen from the top), and the z-axis towards the top of the vehicle.

The definitions used to describe the geometric setup of the two mirrors, as well as the resulting key performance figures of the geometries are visualized in Figure 3. All CADs were mirror-symmetrical around the yz-plane. The angle α describes the FOV of the scanning Lidar that is not disturbed by the mirrors; this part of the FOV generally points to the front of the vehicle. The angle β is the tilt angle of both mirrors around the global z-axis, and γ is their inclination angle around the global x-axis. The length L_1 is the total mechanical width of the mirror setup; L_2 is the minimum distance between the front of the vehicle and the detectable part of the ground; L_3 is the maximum distance between vehicle and the detectable ground. Finally, L_4 is the distance from the center of the Lidar to the front of the vehicle measured along the y-axis.

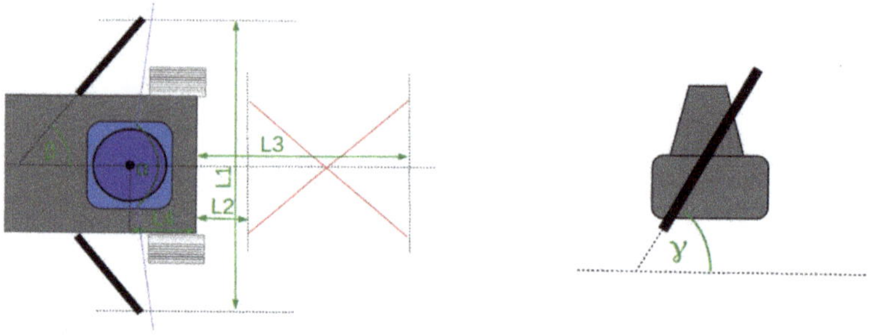

(a) **(b)**

Figure 3. Sketch of the geometry parameters of the mirror configuration (green). The mirror FOV on the ground is sketched in red color. Top view (**a**) and side view (**b**). Note that the γ angle in the side view is defined with respect to the local "front" direction of the mirror and not in the global coordinate system.

The following additional geometric quantities were not changed during the CAD simulations, because they were determined by the prototype vehicle and the properties of the Lidar sensor: the width of the vehicle was 135 mm; the height of the Lidar (Cartesian origin) above the ground was 160 mm; the diameter of the nearly cylindrical Lidar housing in the xy-plane was 38 mm.

For the experiments to calibrate the geometric position of the mirrors in Section 3.3, one fixed geometry configuration was used. The corresponding CAD values of all geometric parameters during those experiments are listed in Table 1. This geometric setup was chosen as a good compromise between most of the other explored geometric setups.

Table 1. Geometry parameters used in experiments with the prototype vehicle. Definitions of the geometry parameters are shown in Figure 3. Alternative geometry setups, as well as a visualization of the geometry setup are presented in Section 3.2.

Geometry Parameter	α	β	γ	L_1	L_2	L_3	L_4
CAD value used in prototype vehicle	110.2°	25.1°	73.5°	214.4 mm	81.1 mm	419.8 mm	100 mm

For the data in Section 3.1, the CAD model of the prototype vehicle was used to verify that the angle of incidence of the laser beam on the two mirrors stayed below 60° for all Lidar rotation positions that contributed to the Lidar measurement data. Therefore, no significant power loss of the laser beam during reflection on the metallic mirror coatings was expected.

During the CAD, no special care was taken to avoid a direct back reflection of the Lidar sensor's emitted laser light into the receiving aperture of the Lidar. This problem was reported in [10] (p. 4), but was not observed with the present setup.

2.3. Point Cloud Transformation Algorithm

The 2D-Lidar sensor returns a distance value for each rotation angle inside its 270° FOV. Depending on the angular position of the Lidar rotor, this distance value needs to be transformed differently into a triplet of x-, y-, and z-coordinates, the so-called point cloud data.

For Lidar viewing angles around the forward direction of the vehicle, the laser beam was not deflected by any mirror, so the returned distance value can be directly taken as the object distance for this viewing angle. These data values only undergo a simple coordinate transformation from the raw angle–distance value pair in polar coordinates into a 3D Cartesian coordinate vector. In this case, the resulting point cloud data remain on the x-y plane per definition.

For Lidar viewing angles where the laser beam and the returned light were reflected by one of the mirrors, a slightly more complex transformation needs to be calculated. The resulting point cloud data after the transformation depend on the distance and angular orientation of the mirror that reflects the light. The algorithm of this point cloud transformation as presented in Formulas (1) to (4) was taken from [12] (p. 5), but the formula (2) for $\boldsymbol{r}$ was corrected. For a visualization of the underlying 3D geometry, the figure in the given reference can be used. First, the Lidar viewing angle and measured distance value are transformed into a Cartesian coordinate vector $\boldsymbol{d}$ of the apparent object point. The geometric plane of the mirror surface is described with a support vector $\boldsymbol{w}$ somewhere on this plane and a normal vector $\boldsymbol{n}$. The vector to the reflection point $\boldsymbol{I}$ on the mirror is calculated as

$$\boldsymbol{I} = \frac{\boldsymbol{n} \cdot \boldsymbol{w}}{\boldsymbol{n} \cdot \boldsymbol{d}} \boldsymbol{d} \tag{1}$$

The unit vector $\boldsymbol{r}$ of the reflected light beam is

$$\boldsymbol{r} = \frac{\boldsymbol{d}}{|\boldsymbol{d}|} - 2\left(\frac{\boldsymbol{d}}{|\boldsymbol{d}|} \cdot \frac{\boldsymbol{n}}{|\boldsymbol{n}|}\right) \frac{\boldsymbol{n}}{|\boldsymbol{n}|} \tag{2}$$

and the vector $\boldsymbol{v}$ from this reflection point to the transformed data coordinate point is

$$\boldsymbol{v} = (|\boldsymbol{d}| - |\boldsymbol{I}|)\, \boldsymbol{r} \tag{3}$$

This finally makes the vector $\boldsymbol{d}'$ the transformed data point

$$\boldsymbol{d}' = \boldsymbol{I} + \boldsymbol{v} \tag{4}$$

There are some Lidar viewing angles where the laser beam is emitted in a direction that gets close to an edge of either of the mirrors. The corresponding viewing angles of the Lidar were precalculated, and Lidar distance data from those directions were not used for further data analysis. The interval of these omitted viewing angles was calculated such that at least 50% of the receiving optical aperture of the Lidar always had a clear beam path either on one of the mirrors or missing both of the mirrors.

2.4. Calibration Setup and Procedure

Our calibration procedure consisted of the following easy steps, which are summarized in Figure 4 and detailed below:

1. Determine the distance between each mirror and the Lidar along a fixed direction;
2. Position the flat calibration object in front of the Lidar;
3. Acquire and save Lidar distance data on this flat calibration object;
4. Calculate the position and orientation of both mirrors that fit the data best.

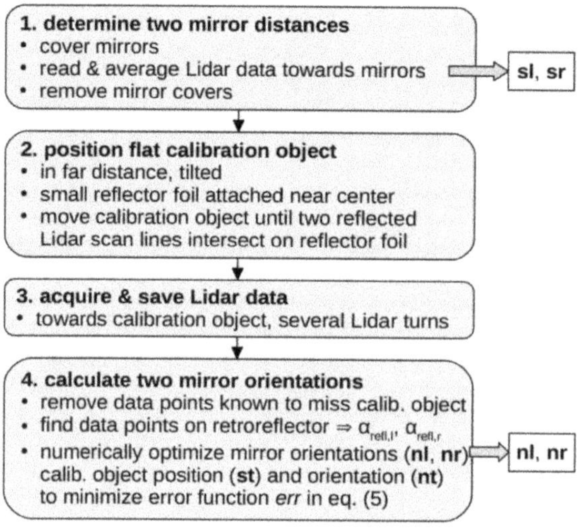

Figure 4. Flow chart of the calibration procedure. These four steps were conducted to calibrate the positions (**sl**, **sr**) and orientations (**nl**, **nr**) of the two plane mirrors next to the 2D-Lidar sensor. Details on each step are presented in Section 2.4.

In the first step, the mirrors were covered with white flat paper, e.g., using adhesive film. Lidar distance data were acquired for several hundred rotation periods of the spinning Lidar. For each mirror, the arithmetic mean value of the measured distances along one fixed direction towards this mirror was taken. In our geometry, both distances dr and dl to the two mirrors were determined in the x-direction. The paper was then removed from the mirrors after this measurement.

For the second step, a flat calibration object was used, which consisted of a flat sheet of material that had a homogenous, light, and diffuse surface. Our experiments were performed with a 4 mm-thick sheet of polymer material with size 700 × 700 mm. which was entirely coated with white paint on its front side; see Figure 5. Additionally, a small square piece of retro-reflection foil was attached roughly to the center of this calibration object. The size of this reflector foil was chosen such that it corresponded to the angular resolution of the Lidar sensor in the intended calibration distance. For our experiment, a 10 mm square piece of white-color "Diamond Grade" reflective sheeting [21] was used, but any other retroreflecting foil should work equally well. The position of the retroreflector can later be recognized easily in the acquired Lidar data because the sensor used in this experiment additionally returns the brightness value for every returned distance value. This brightness value is much higher on the reflector material than on the surrounding white surface. In the case of using a different Lidar sensor, which does not return brightness values, this reflector foil can be omitted, and a hole of a similar size can be made in the flat calibration object instead.

Figure 5. Photograph of the calibration setup. A white flat surface is positioned in front of the prototype vehicle. The position of the vehicle is elevated in order to have the Lidar sensor detect a larger part of the surface. The calibration surface is tilted around the x-axis for better calibration results. Note the 10 mm square piece of reflector foil that is attached near the center of the white surface and is necessary for the calibration procedure to succeed. (Background of room blurred for privacy.)

The calibration target needs to be positioned with the retroreflection foil at the crossing point where the laser beam passes twice, once reflected by one of the mirrors and also reflected by the second mirror. To achieve this, the Lidar was turned on and the light lines of the Lidar sensor's IR laser on the calibration target were observed with an IR observation tool while positioning the calibration target. In our experiments, a hand-held IR viewer was used for this purpose, but a smartphone camera with an imperfect IR blocking filter can alternatively be used to visualize this in a dark room.

As the third step, Lidar data with this well-positioned calibration object were acquired and saved to a data file. Usually, several hundred rotation periods of the spinning Lidar are used; this typically corresponds to a few minutes of data acquisition time.

For step four, we transferred this acquired Lidar data from the vehicle's computer to a different computer for convenience, but the following data processing steps could in general also be performed on the Raspberry Pi computer on the vehicle itself. The acquired data were used in the algorithm detailed below to accurately determine the exact orientation angles of both mirrors in the setup. This calibration algorithm was implemented in Python [17] and mainly relies on an implementation of the downhill simplex algorithm, which is available in Python's SciPy library [22]. This algorithm minimizes the value of a scalar function of multiple arguments. A problem-specific scalar error function was defined as detailed below. The free variables in this error function were the two independent orientation angles of both mirrors, as well as the distance and orientation of the flat calibration target. These free variables were parametrized with support (**s**) and normal (**n**) vectors as follows:

- Normal vector of the right mirror surface: $\mathbf{nr} = (nr_x, nr_y, -1)$;
- Normal vector of the left mirror surface: $\mathbf{nl} = (nl_x, nl_y, -1)$;
- Normal vector of the flat calibration target: $\mathbf{nt} = (nt_x, -1, nt_z)$;
- Support vector of the flat calibration target: $\mathbf{st} = (st_x, st_y, st_z)$; this vector was also taken to be the position vector of the retroreflector foil on the target.

Setting one coordinate value of the above normal vectors to -1 does not restrict the generality of the chosen parametrization because none of these planes are oriented perfectly perpendicular to any of the coordinate axes. The minus sign was chosen to describe the expected orientation in the given coordinate system.

Additionally, the following fixed parameters were used as determined in Step 1:

- Support vector of right mirror surface: **sr** = (dr, 0, 0);
- Support vector of left mirror surface: **sl** = (−dl, 0, 0).

The problem-specific error function *err* to be minimized during the optimization of the above mirror orientation parameters is defined as the sum of three positive contributions:

$$err = err_{RMS} + err_r + err_l \tag{5}$$

The first contribution err_{RMS} to this error function in Formula (5) is defined as the root-mean-squared (RMS) distance of the transformed point cloud (see Section 2.3) to the calibration target plane. For this RMS value, the data from several turns of the spinning Lidar sensor were used. Only data points that were known either to hit the calibration target directly or to be reflected properly by one of the two mirrors were considered. The transformed point cloud data still depends on the two mirror orientations. The distance of a point $\boldsymbol{p}$ to a plane $E_{\boldsymbol{s},\boldsymbol{n}}$ defined by a support vector $\boldsymbol{s}$ and a normal vector $\boldsymbol{n}$ is calculated according to Euclidean geometry [23] as

$$dist(\boldsymbol{p}, E_{\boldsymbol{s},\boldsymbol{n}}) = \left|(\boldsymbol{p} - \boldsymbol{s})\cdot\frac{\boldsymbol{n}}{|\boldsymbol{n}|}\right| \tag{6}$$

Before the other two contributions err_r and err_l to the error function in Formula (5) can be evaluated, two angle parameters need to be calculated first. The parameter $\alpha_{\text{refl,r}}$ is the angular position of the Lidar's rotor when viewing the retroreflector foil on the target via a reflection on the right mirror. The corresponding parameter $\alpha_{\text{refl,l}}$ is the angular position of the Lidar when viewing the reflector foil via a reflection on the left mirror. Both angular positions can be calculated from the saved calibration dataset by searching for data points that have an extraordinarily high brightness value returned by the Lidar sensor. If the previous positioning (Step 2 of the calibration procedure) succeeded, then the dataset included several of those points for reflections over both mirrors. Only data points observed under $\alpha_{\text{refl,r}}$ were considered for the calculation of the error function contribution err_r and, accordingly, for err_l. Both of these error function contributions were then calculated as the RMS distance value of all the considered data points with a high brightness value to the position of the support vector **st** of the calibration target.

Since both types of error function contributions were measured in units of length, the calculations were performed in millimeters during our experiments, and since they also were of a similar size, no scaling prefactors were used in the definition of the error function in Formula (5). For other Lidar sensors or different angular resolutions, it could be useful to add a scalar prefactor in front of the err_{RMS} term in Formula (5) in order to scale both types of contributions to a similar magnitude.

Eventually, with this problem-specific error function defined, the minimization algorithm was used to determine the optimal normal vectors of both mirror surfaces that gave the smallest possible value for this error function. Initial values for all optimization parameters were taken to be CAD values or even rough measurement values acquired on the prototype vehicle with a measuring tape. We allowed the optimization algorithm a maximum number of 100,000 iterations, and it normally took a few minutes on a common laptop computer for it to finish.

The described marking of the intersection point of the two reflected Lidar FOV sections with a piece of bright retroreflector foil proved to be a very useful additional constraint during the parameter optimization. Measuring the mirror distance to the Lidar beforehand with an independent measurement (Step 1) also proved to be a useful and necessary addition to the procedure, as it reduced the number of free parameters to be optimized. Leaving out either of these two described extra steps resulted in a highly unstable calibration and often also in the wrong calibration results.

3. Results

3.1. Prototype Vehicle Performance

Three relevant performance properties of our prototype vehicle were evaluated experimentally as follows: First, we checked how the reflection of the Lidar's laser light at the mirrors affected the acquired distance data. For this purpose, two fixed observation angles of the spinning Lidar were chosen. In one of these directions, the Lidar's light was reflected off the plane mirror and directed towards a flat object of homogenous brightness. In the second chosen direction, the Lidar sensor measured data from a direction where the same flat object was observed at mostly the same distance. For both directions, 1000 distance values were acquired. The measurement results were:

- Direct view (no mirror): mean distance 403.2 mm, standard deviation 1.9 mm;
- View reflected by mirror: mean distance 400.4 mm, standard deviation 2.1 mm.

The resulting standard deviation of the measured distance without any mirror is typical for the 2D-Lidar sensor used. When reflecting the light with a mirror, the standard deviation increased by 10%. This small increase did not significantly deteriorate the Lidar's distance measurement performance. It was probably due to the loss of light intensity during the reflection on the mirror. The mirror datasheet states a reflectivity of $R > 98\%$ for an angle of incidence of 45°; it is plausible that for the used angle of incidence (<60°) in the prototype setup, this power loss is higher. Additionally, the optical power loss is experienced twice during Lidar operation, once by the emitted laser light beam and once again by the returning light.

To compare this small observed increase in distance noise by a mirror reflection to other environmental effects on the distance noise, a third measurement was performed on a different target object with much lower surface roughness:

- Direct view (no mirror), less object roughness: mean dist. 404.5 mm, st. dev. 4.5 mm.

This confirmed the conclusion that the distance noise increase that was due to the mirror reflection was much smaller than distance noise increases by other environmental influences, such as this smoother surface of the observed object. Therefore, we concluded that the mirrors used in the prototype setup were suitable for the chosen purpose, because they only increased the distance noise of the measured distance values by 10%.

A second check performed was to verify the actual size and position of the reshaped FOV on the ground in front of the prototype vehicle. The geometric parameters used in this vehicle, as well as the resulting theoretical FOV size on the ground from the CAD calculations are listed in Table 1. A hand-held IR viewer and a measuring tape were used to determine experimentally the positions where the Lidar light can be seen on the ground in front of the vehicle. The results are presented in Table 2.

Table 2. Experimental results for the FOV position on the ground in front of the prototype vehicle. Definitions of the position values are in Figure 3; theoretical CAD values are identical to Table 1.

FOV Position Value on Ground in Front of Vehicle	Theoretical CAD Value	Measurement Result
min. distance L_2, left side of vehicle	81.1 mm	53 mm
min. distance L_2, right side of vehicle	81.1 mm	62 mm
max. distance L_3, left side of vehicle	419.8 mm	368 mm
max. distance L_3, right side of vehicle	419.8 mm	425 mm

The results in Table 2 underline the need for a proper calibration of the mirror positions and orientations on the vehicle. The measurement results were similar to what the CAD calculation predicted, but they still differed by significant amounts up to 50%. Additionally, it can be seen that the left and the right mirror mounted onto the vehicle differed in their orientation, because the left and right measured FOV positions were not identical.

In a third experimental check, obstacles were placed on a flat and horizontal ground in front of the prototype vehicle, and the recorded Lidar data were visualized. Here, the intention was to find out for what size of obstacles the prototype setup can be useful. Little polymer cubes were used as obstacles, with sizes between 5 mm and 30 mm. A picture of the measurement setup can be seen in Figure 6a and the resulting Lidar distance data from a single Lidar rotation period in Figure 6b.

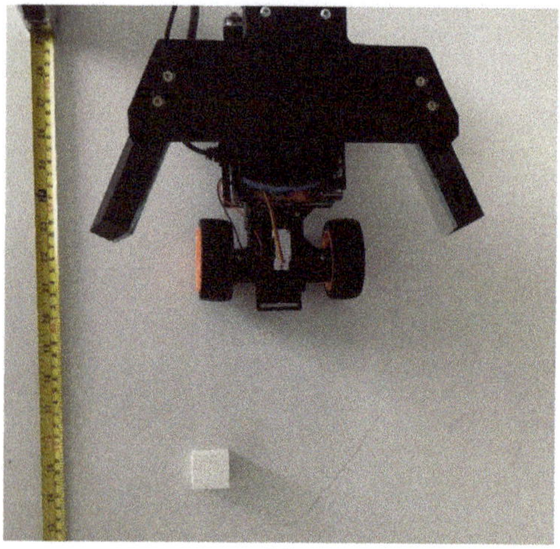

(**a**)

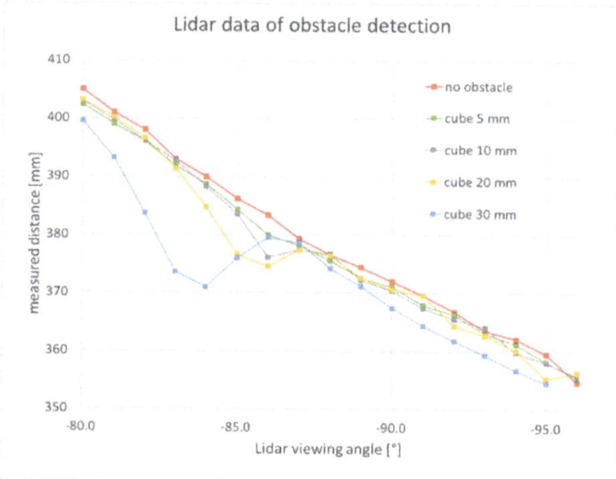

(**b**)

Figure 6. (**a**) Measurement setup with a cubical obstacle placed in front of the prototype vehicle. The Lidar setup with two mirrors acquires distance data in this setup to check what sizes of obstacles can be reliably detected. The presented setup is also a typical example for obstacles that severely threaten the prototype vehicle and keep it from driving on, if not detected and avoided. (**b**) Measurement data: Lidar distance raw values over Lidar rotation position for several sizes of cube obstacles in front of the prototype vehicle.

In all the resulting Lidar raw distance data of Figure 6b, it can be seen that the distance value decreased for increasing Lidar rotation angles. This is only an artefact of the geometric configuration where the same flat ground in front of the vehicle was further away when observed by mirror reflections under different Lidar rotation angles. The slight deviation of the distance values from a straight line was due to the inherent distance noise of the Lidar data. The dip in these decreasing curves was caused by the obstacle cube. It can be seen that cubes of sizes between 30 mm and 10 mm caused distance variations, which can readily be distinguished from the distance noise. Thus, obstacles of these sizes could be detected by the presented setup in the Lidar data taken during just one single Lidar rotation period. The dip in the data for a cube size of 5 mm was very slight and could therefore not be reliably distinguished from random distance noise. Therefore, cubes of size 5 mm could not be reliably detected with the present setup during a single rotation period of the Lidar. (This experimental conclusion assumed that a straight line is fit to the data points in Figure 6b; some acceptance corridor around this line is defined such that the small distance deviations from this line due to noise were not considered, but the visible larger distance deviations from this line due to the obstacles can be recognized.) Nevertheless, the distance data could be averaged over several turns of the Lidar rotation, thus decreasing the distance noise of the averaged distance values. This could probably even make the detection of 5 mm-sized objects possible with the present setup, but would result in a longer reaction time, because the necessary data acquisition for this averaging obstacle detection algorithm would take the time of several Lidar turns. For the presented prototype vehicle, this averaging strategy was not pursued, because the vehicle was capable of driving over obstacles of this size without noticeable damage. In the Supplementary Material to this article, some videos show how the vehicle drives and avoids obstacles.

3.2. Geometric Design Results

To explore the wide range of geometric possibilities when using a 2D-Lidar and two mirrors for ground detection, several different geometry configurations were simulated in the CAD software. All simulated configurations used the same vehicle and the same 2D-Lidar sensor. However, the mounting position and orientation of both mirrors was changed, while still keeping the Lidar sensor mounted at a position on the central yz-plane. For most geometry configurations, also the size of the two plane mirrors was kept unchanged, with the exception of the geometry aiming for a minimum vehicle width. The mounting position of the Lidar sensor on top of the vehicle was also changed. A prerequisite was that each FOV covered the ground at least over the width of the vehicle without blind spots.

The five different design goals were:

1. Each mirror covers the near ground on the opposite side;
2. Each mirror covers the near ground on the same side;
3. Angular FOV in the forward direction (parameter α) as large as possible;
4. Total width of the vehicle (parameter L_1) as small as possible;
5. Width of the mirrors as small as possible.

For Goals 1, 2, and 4, the position of the Lidar was fixed centrally on the vehicle, while for Goals 3 and 5, it could be moved forward and backward.

In each of the above design goals, there were still many other geometric degrees of freedom available that led to a multitude of possible geometric setups. Therefore, for each of these design goals, one typical geometric solution was chosen that seemed to achieve the aim without going to unrealistic extremes in any other geometric parameter. The characteristic geometry parameters of each of our chosen CAD solutions are listed in Table 3. A top view of each CAD solution is shown in Figure 7.

Table 3. Resulting geometry parameters for typical CADs aiming for each of the five different design goals. (See Figure 3 for definitions of the used geometry parameters.).

Design Goal	α	β	γ	L_1	L_2	L_3	L_4
1	110.2°	25.1°	73.5°	214.4 mm	81.1 mm	419.8 mm	100 mm
2	145.4°	45.0°	73.0°	220.8 mm	96.7 mm	337.1 mm	100 mm
3	180.0°	46.2°	67.0°	223.3 mm	59.6 mm	237.0 mm	34.8 mm
4	135.6°	29.8°	73.0°	159.3 mm	85.0 mm	239.0 mm	100 mm
5	144.6°	38.0°	70.0°	167.9 mm	182.8 mm	290.4 mm	34.8 mm

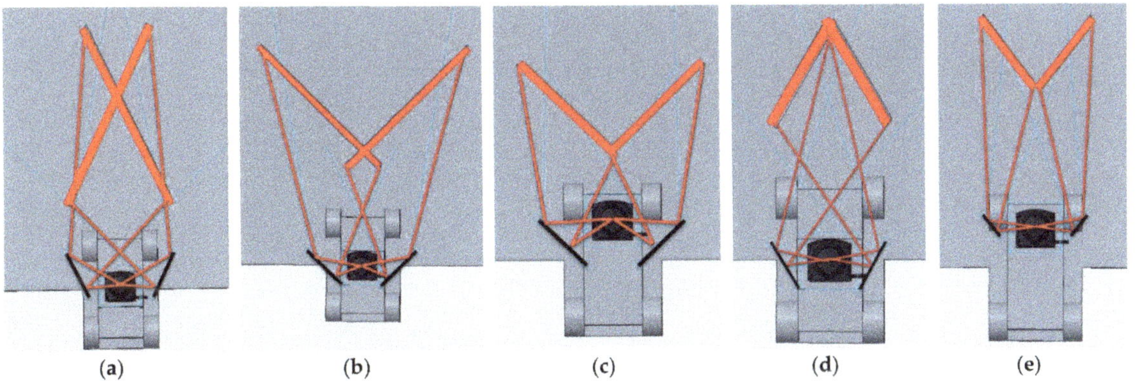

Figure 7. Top views of typical resulting CADs for the five design goals 1 to 5. The displayed geometries in (**a**–**e**) correspond to the listed geometry parameters 1 to 5 in Table 3. The thick red lines show where the Lidar laser scans the ground in front of the vehicle, and the thin red lines show the laser beam path from the Lidar center towards both ends of the thick red lines on the ground.

The CAD variant for Design Goal 1 (Figure 7a) was identical to the prototype vehicle used here. This geometry was used throughout the measurements for this article because it is a good compromise between all the other presented geometries.

In general, there are two different strategies for reshaping the FOV of a 2D-Lidar sensor in order to achieve ground detection. This can be seen when comparing Design 1 to Design 2. The first possibility, as in Design 1 (Figure 7a), is to make the laser beam that hits the mirrors at the outward end reflect mostly into the driving direction of the vehicle, i.e., along the y-axis. This results in the right mirror illuminating the near ground at the left side in front of the vehicle and vice versa. The second possibility, as in Design 2 (Figure 7b), is to make the laser beam that hits the mirrors at the inwards end reflect mostly into the driving direction of the vehicle. This results in the right mirror illuminating the near ground at the right side in front of the vehicle and vice versa. When comparing the presented two geometry results, it can be seen that Geometry 1 generated a FOV on the ground that was much nearer to the front of the vehicle (L_2 is smaller) and roughly had the width of the vehicle along its full extent in front of the vehicle. In contrast, Geometry 2 created an FOV on the ground with the nearest position further away from the front of the vehicle, but with a width that increased with the distance from the front of the vehicle.

The resulting geometry (Figure 7c) for Design Goal 3 proved that it was also possible to keep a large unaffected viewing angle α towards the driving direction of the vehicle, in the displayed example of 180°. For mounting positions of the 2D-Lidar at the very front of a vehicle, as is common for AGVs, this is the maximum useful viewing angle, because towards the back, the vehicle itself normally blocks the Lidar's view. Therefore, the additional FOV on the ground can in this case be achieved without losing any relevant FOV towards the front.

Design Goal 4 was to minimize the total width L_1 of the mirror setup. This can be necessary when space for adding plane mirrors next to a Lidar is limited. The corresponding result (Figure 7d) had a total width L_1 of around 160 mm. Compared to the width of the current and typical 2D-Lidar sensor housing of around 70 mm, we considered this size increase by a factor of 2 acceptable.

In all previous designs presented, the size of the mirrors was kept constant. However, as can be seen in Figure 7, the 100 mm length of the reflecting surface of the mirrors was not fully used in most cases. This was the motivation for decreasing the size of the mirror parts in Design Goal 5. The result is presented in Figure 7e, where the length of the used mirror surfaces was reduced to 43 mm. This significantly reduced the weight of the resulting glass mirror setup. At the same time, it was possible to achieve a very narrow setup with L_1 of only 168 mm in the presented example.

3.3. Calibration Results and Verification

The calibration procedure presented in Section 2.4 was experimentally tested for several different distances and orientations of the simple flat calibration object. The aim of these experiments was to check if the numeric optimization converged to a realistic result of the mirror orientations. For realistic calibration outputs, we additionally verified that the resulting geometry parameters also yielded realistic results for independent point cloud data, not already used during the calibration.

The flat calibration target with retro-reflector foil attached was positioned in the following different distances and orientations in front of the prototype vehicle. The given values were only rough estimates; more precise values were acquired as the output of the calibration procedure, but were not relevant to the following conclusions:

- Setup A: target tilted ~45° around x-axis, ~0° around z-axis, distance ~300 mm;
- Setup B: target tilted ~30° around x-axis, ~0° around z-axis, distance ~300 mm;
- Setup C: target tilted ~45° around x-axis, ~20° around z-axis, distance ~300 mm;
- Setup D/E/F: same orientation as Setup A/B/C, larger distance ~400 mm;
- Setup G/H/I: same orientation as Setup A/B/C, larger distance ~600 mm.

For each of the above calibrations setups, the following quality criteria, chosen in analogy to the criteria used in [12] (p. 11), were evaluated:

- The RMS absolute distance between the acquired points and the location of the calibration target plane; see Equation (6). All Lidar rotation angles that produced data on the calibration target, as well as the data from several turns of the spinning Lidar were considered for the calculation of this criterion.
- The arithmetic mean intensity value of the acquired points on the calibration target, in received signal strength indicator (RSSI) values as given by the Lidar sensor.
- The number of successfully acquired data points, given as a percentage of the number of measured data points. (The Lidar sensor can mark return data as invalid.)
- The standard deviation of the measured object distances into one fixed direction directly in front of the vehicle.

The percentage of acquired data points resulted in 100% for all our measurements on the white calibration target; therefore, this value is not stated individually in the following results.

The measured mean (± standard deviation) values for the mirror distances were:

- Distance right mirror: dr = (83.3 ± 1.9) mm;
- Distance left mirror: dl = (82.5 ± 1.9) mm.

For each of these calibrations setups the results are given in Table 4. Additionally, for each setup, the calibrated mirror positions from the first line were used to calculate the quality criteria for the dataset of the second line. This was meant to be a first verification check. However, the resulting quality criteria values always were the same as in the first line. Therefore, this did not yield meaningful results, and these are not shown in Table 4.

For the ease of comprehension, the resulting values for the left and right mirror normal vectors (**nr**, **nl**) were transformed into an angle β, describing the tilt of the mirror around the global z-axis, and an angle δ, describing the tilt of the mirrors around the global x-axis. This β angle is defined identically to the β angle in the above CAD definitions and, thus, can be compared directly. The resulting (global) δ angle is defined differently as the (local) γ angle used above in the CAD definitions. Both angles, β and δ, are presented in Table 4. For the ease of comparison, the theoretical CAD values of the mirror angles were transformed to the global (δ) angle and displayed there as well.

The calibration setups A, B, D, E, and G produced δ angles that were more than a few degrees away from the respective CAD values. Those calibration setups were therefore considered unreliable, because the real prototype setup is known to have mirror angles that differ no more than a few degrees from their CAD values. The calibration setup F resulted in double the RMS distance value of the total point cloud compared to the distance standard deviation of only the point in the front direction. This was also taken as an indication that the calibration process did not succeed. In the results of Setup B, it can additionally be seen that the RMS distance of the point cloud to the calibration plane was unrealistically small, being only around 0.5 mm. This value is even smaller than the standard deviation of the distances in one fixed direction (more than 0.8 mm in this case); this emphasizes that these results were not plausible.

We often observed during the optimization process that led to implausible results that the optimization algorithm tilts the two mirrored point clouds in such a way that the fitted calibration plane is parallel to the viewing direction of the Lidar. As a result, the present distance noise in the data did not result in an RMS distance of the point cloud to the calibration plane any more, since this distance noise now occurred along the extension of the plane. An unrealistically low value of the RMS distance to the plane is a good indicator of this happening. The same problem also occurred when the calibration procedure was executed without the mirror distances fixed or without the additional constraint of the two Lidar directions of the retroreflection foil to coincide. In general, these far-off calibration results were due to the calibration problem being numerically under-constrained.

Table 4. Calibration results of the different calibration setups. The resulting quality criteria, as well as the resulting orientation angles of the two mirrors are given. Two lines per calibration setup represent two independent measurements to illustrate the stability of the results. The first line shows results for a calibration performed on a dataset acquired for the stated geometric setup and the results calculated from this dataset. The second line shows these results with the calibration and evaluation procedure repeated on a second independent dataset for the very same geometric setup.

Calibration Setup	RMS Distance	Mean Intensity(RSSI)	Distance st. dev. Front	Right Mirror		Left Mirror	
				β_r	δ_r	β_l	δ_l
CAD (theory)	-	-	-	−25.07°	34.96°	25.07°	34.96°
A	1.05 mm 1.22 mm	142.46 142.29	1.25 mm 1.19 mm	−26.73°. −25.34°	15.25° 25.29°	24.99° 23.76°	15.82° 26.18°
B	0.55 mm 0.40 mm	143.54 140.50	0.97 mm 0.79 mm	−26.83° −26.95°	9.82° 6.07°	24.97° 25.12°	10.25° 6.32°
C	1.18 mm 1.30 mm	156.25 155.25	1.17 mm 1.19 mm	−24.57° −24.67°	33.37° 32.39°	23.36° 23.47°	34.48° 33.49°
D	1.17 mm 1.18 mm	159.73 158.44	0.98 mm 0.89 mm	−25.37° −25.07°	29.51° 30.58°	23.83° 23.66°	30.55° 31.61°
E	1.07 mm 1.03 mm	158.68 158.25	0.70 mm 0.76 mm	−25.52° −25.14°	28.86° 31.19°	23.73° 23.48°	30.00° 32.33°
F	1.53 mm 1.35 mm	175.41 174.06	0.74 mm 0.83 mm	−23.45° −23.61°	37.37° 37.03°	23.63° 23.56°	37.25° 36.97°
G	1.65 mm 1.05 mm	178.87 180.81	1.18 mm 1.11 mm	−24.44° −24.63°	28.91° 22.29°	23.10° 24.32°	30.08° 22.84°
H	1.11 mm 1.03 mm	168.35 168.86	1.06 mm 1.06 mm	−24.60° −24.79°	33.30° 32.44°	24.01° 24.10°	34.21° 33.35°
I	1.23 mm 1.30 mm	170.79 170.69	0.90 mm 0.69 mm	−25.00° −24.93°	32.18° 32.14°	23.72° 23.75°	33.22° 33.16°

The calibration setups C, H, and I all produced calibration results that we considered reasonable. In these cases, the numerical optimization converged to a realistic geometric scenario. Looking at the details of those successful calibration setups, those with maximal tilt of the calibration target (Setups C, F, and I) were the most successful (except for Setup F). At the largest distance, the setup with a target tilt around only one axis (Setup H) also succeeded. All successful calibration setups had in common that the Lidar points covered a large width on the calibration plane. This large width can be achieved by a large distance between the target and the Lidar or, alternatively, by tilting the calibration plane such that the Lidar points have a larger distance on one side of the calibration plane. Furthermore, a rotated calibration plane was observed to have an additional advantage (e.g., Setup G versus H) over calibration plane orientations that are symmetric to the yz-plane.

To evaluate the measurement error of the determined mirror angles, the mean value of those mirror angles was calculated, as well as the standard deviation, but only for the three successful calibration setups (C, H, I). This gave (mean ± standard deviation):

- Left mirror: $\beta_l = 23.74° \pm 0.29°$, $\delta_l = 33.62° \pm 0.57°$;
- Right mirror: $\beta_r = -24.76° \pm 0.18°$, $\delta_r = 32.64° \pm 0.55°$.

This shows that the presented calibration method was capable of yielding reliable angles of both mirrors when taking care that the calibration target was positioned at a distance of ~600 mm from the Lidar sensor and was additionally tilted. In this case, the uncertainty of the mirror angles was around 0.5°. This angle uncertainty translated into a systematic transversal distance uncertainty of around 5 mm at an object distance of 600 mm (where the calibration target was used) and of 18 mm in the maximum measurement distance of 2 m of the current Lidar sensor. This is already smaller than the specified longitudinal systematic distance uncertainty given by the Lidar manufacturer of 40 mm and, therefore, considered good enough for the purpose of precise environmental perception in front of a robot vehicle. Additionally, this resulting angle uncertainty of only 0.5° was much smaller than the error of the angles taken directly from the CAD.

An example of the 3D point cloud data generated from such a calibrated mirror setup on a Lidar sensor is shown in Figure 8.

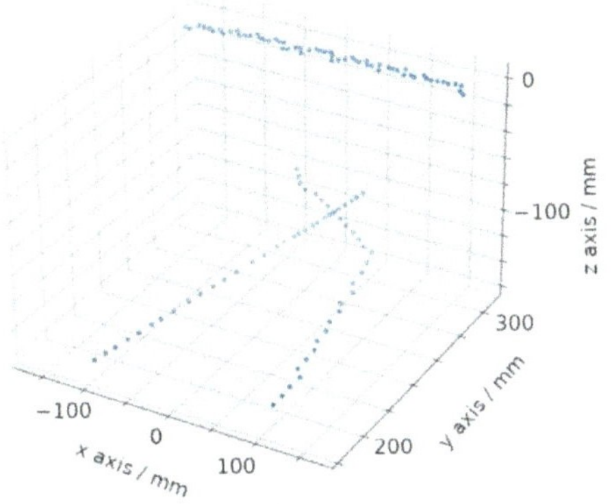

Figure 8. Example of resulting 3D point cloud data acquired by the Lidar sensor with two mirrors. The prototype vehicle had a flat wall in front for the data acquisition. Data points on this wall, as well as on the floor surface can be seen.

In order to make sure that the above calibration results not only produced precise measurement data when used on the same geometric object that was used during calibration, but also resulted in precise data when used on different objects, the following verification experiments were conducted: A first point cloud dataset was acquired and saved. This dataset was used to calibrate the mirror positions and orientations as described above. The resulting mirror angles and positions from this calibration were then taken as fixed during the following steps. Then, a second dataset was taken with the calibration plane positioned at a different position or in a different orientation. The calibration algorithm was used for a second time, but only the position and orientation of the calibration target were set as free for the optimizer during this second run. As a result, we could check how accurate the mirror calibrations values from the first (calibration) run could transform point cloud data from a different obstacle setup in the second (verification) run. To quantify the results of this verification experiments, the following Table 5 shows the resulting RMS deviations of the calibration plane distance after the verification run, as well as the best-fit orientation angles of the calibration plane after the verification run.

The presented verification data in Table 5 show that the RMS distances were only in the range of 1 mm. This showed that, during the verification run, the parameters still properly converged and no significant additional distance error was present. The presented plane angles in Table 5 fit well to the known orientations of the plane object in the verification dataset. The observed small deviations from those angle values were due to the theoretical values only being known to an uncertainty of around ±7°, as well as a different sign of the β_{target} angle, which was compensated for in the table. Therefore, the selected calibrations C, H, and I proved to be correct also when used on point cloud data not seen during the calibration process. As a result, we could show that the presented calibration method is suitable for the calibration of the mirror positions and orientations for use with a 2D-Lidar sensor.

Table 5. Verification results for the three successful calibration setups (C, H, I). The column with the verification setup also shows the approximate real angles of the used plane object. The RMS distance variation of the point cloud data on the plane object is given after the verification run. Additionally, the resulting orientation angles of the plane object of the verification run are given. (The plane object used for the verification runs was the same as the calibration plane object used for the calibration runs.)

CalibrationSetup	Verification Setup	RMS Distance	Distancest. dev. Front	Plane Object Angle 90° + β_{target}	Plane Object Angle δ_{target}
C	D (~45°,~0°)	1.27 mm	0.98 mm	39.60°	−2.15°
H	I (~45°~,20°)	1.57 mm	0.69 mm	36.56°	21.67°
H	A (~45°,~0°)	1.55 mm	1.25 mm	39.42°	1.38°
I	H(~30°, ~0°)	1.10 mm	1.06 mm	25.97°	−0.60°

4. Discussion and Conclusions

This research paper first described a prototype vehicle that used a 2D-Lidar setup with two mirrors to detect the position of the ground it drives on. It can use this information to avoid driving over obstacles or over edges. The design of this vehicle is not new in itself; similar designs and performance have already been presented earlier [8,9]; however, this prototype vehicle was the basis for the following new experiments.

This article explored a variety of different geometric design possibilities for reshaping the FOV of a 2D-Lidar with two plane mirrors. A few of these design variants were also mentioned earlier [8], but to our knowledge, this is the first systematic overview of different geometric setups in order to detect the ground under an AGV with a 2D-Lidar and plane mirrors. For several typical geometrical configurations, the resulting FOV shape on the ground was demonstrated. The geometric setup can be optimized for minimum mirror weight, minimum vehicle width, a short distance of the ground FOV in front of the vehicle, or a large width of the FOV in front of the vehicle. We suggest that the driving properties of the vehicle, such as driving speed and turning radius, are most important for determining which of those FOV shapes is most appropriate for the intended use. A closer FOV on the ground is advantageous for slower vehicles and a wider FOV on the ground for vehicles with a smaller turning radius.

As our main result, we presented a new method to calibrate the positions and angular orientations of the plane mirrors attached to a 2D-Lidar (Section 2.4). To the best knowledge of the authors, this is the first calibration method presented for this kind of mirror setup around 2D-Lidars. Earlier publications either did not deal with this problem (and, presumably, just used coarse CAD values for the mirror angles) or focused exclusively on the calibration of plane mirrors next to a 3D-Lidar sensor. Comparing our calibrated mirror angles values in Section 3.3 with the theoretical CAD values from Table 4, one can see a difference of around 2° between them. This underlines the need for such a calibration procedure in order to obtain precise point cloud data when using mirror setups on a Lidar sensor. A lack of calibration leads to systematic measurement errors of the point cloud data otherwise.

In this publication, we presented our systematic experiments to find the best parameters to make the presented calibration method succeed (Section 3.3). We found that for optimal calibration results, the calibration plane should have a large distance and ideally also be tilted around both axes. For the 2D-Lidar sensor used in this experiment, which has a 2 m measurement range, a distance of 600 mm from the calibration target was shown to be sufficient. We could show that with the presented prototype vehicle, the resulting accuracy of the point cloud data was improved after performing the calibration, compared to using the angular mirror positions from the CAD program directly without any calibration. It was additionally shown that the resulting point cloud data after the geometric transformation with these calibrated mirror orientations had a sufficient accuracy compared to the typical inherent distance noise of Lidar sensors. The distance error of the point cloud data did not significantly change after the calibration process. Finally, the presented calibration procedure yielded mirror pose results that also still worked fine on point cloud data not seen during the calibration process. We therefore concluded that the presented calibration procedure is a useful and robust process.

The presented calibration procedure is easy to perform in several aspects: The necessary steps during the calibration procedure are easy to conduct. Only materials that are easily available to both industrial manufacturing companies of AGVs, as well as to robot researchers are needed for the calibration procedure. Importantly, the shape of the main calibration object used is a simple plane. This allows for easy upsizing of this calibration procedure to 2D-Lidars with a longer measuring range by simply using larger calibration planes. Manufacturing any other shape of calibration object would probably pose much bigger problems in the necessary dimensions of several meters.

Additionally, the presented calibration procedure is general in the sense that it could easily be adopted for a different mobile robot using a different 2D-Lidar sensor. In case a different Lidar sensor should not have the possibility to read out intensity values in addition to the distance values, we outlined a variation of the calibration procedure in Section 2.4 that substitutes the reflector foil in our setup with a hole in the calibration plane object. We predict this procedure variant to succeed for Lidar sensors without intensity data output. The plastic material of our calibration object is not relevant: the calibration could be performed also with any plane object of a different material. Otherwise, our developed calibration methodology does not make use of any geometric or other peculiarity of our prototype vehicle or 2D-Lidar sensor. Therefore, we predict that this calibration method will succeed as well on mobile robots of a smaller or bigger size, as well as using different 2D-Lidar sensors or different calibration plane materials.

As a limitation of our experiments, the rather short measuring range of 2 m of the Lidar sensor should be mentioned. For using our calibration method on Lidar sensors with a larger measurement range, it would be necessary to increase the distance and the transverse dimensions of the plane measurement target. However, such distances and large transverse extensions could probably easily be realized with the proposed calibration setup, with a wall of a room or even an outside wall of a building taken as flat calibration target. We hypothesize the presented calibration method to be easily upscalable.

Further research using a 2D-Lidar sensors with a much larger measurement range, e.g., 15 m, is desirable, because it could enable the presented technique to be used in mobile robots that have a higher velocity than indoor AGVs, e.g., for outdoor settings where objects at further distances are relevant for robot navigation and autonomous vehicles. As a second area of future research, performance figures on the obstacle detection rate and overall functionality of a mobile robot in its real operation area with an obstacle avoidance algorithm using our ground detection and calibration method would be of interest. A third area of further research is the development of an open-source software library to assist robot developers by automating the calibration procedure described in Section 2.4 and by performing the real-time transformation of the point cloud data using the calibrated mirror parameters. Additionally, a very interesting topic for further research is to generate a fully analytical parametrization of the mirror geometry in this setup. Analyzing these analytical expressions could lead to a deeper insight into why the two additional constraints included in our calibration procedure are necessary: the mirror distance measured directly when covered with paper and the reflector foil attached at the center of the calibration target.

To draw a conclusion, we showed that a 2D-Lidar sensor with additional plane mirrors significantly improved the FOV of an ARV to achieve both environmental distance perception and ground detection without loss of data accuracy and without great expense. The manufacturing costs and engineering effort of such a setup are much smaller than for a 3D-Lidar sensor or other competing sensors, and still, this setup delivers the necessary point cloud data to enable robots to perceive the ground they are driving on or an obstacle they are flying close to precisely. The presented calibration procedure ensures that the geometric information perceived by the autonomous robot has a high precision. In the presented prototype vehicle, a measurement uncertainty of around 15 mm was achieved, which is comparable to the performance of current 3D-Lidar sensors. This improves ARVs' capabilities to navigate and act in spaces that are not exclusively prepared for robot activity and where human activity can lead to unexpected obstacles such as objects on the floor, as well as unavoidable obstacles such as downward stairs. It hereby enables the development and production of more cooperative autonomous robots, for example for transport in hospitals, assistance in nursing homes, and a multitude of novel fields of application.

Supplementary Materials: The following Supporting Information can be downloaded at: https://www.mdpi.com/article/10.3390/s22207830/s1, Videos S1–S3: Driving vehicle without and with different obstacles.

Author Contributions: Conceptualization, H.G.; methodology, H.G.; software, J.E.K.; validation, H.G.; formal analysis, J.E.K. and H.G.; investigation, J.E.K., A.D., and P.L.W.; resources, H.G.; data curation, J.E.K., A.D., and P.L.W.; writing—original draft preparation, H.G.; writing—review and editing, J.E.K., A.D., and P.L.W.; visualization, J.E.K., A.D., P.L.W., and H.G.; supervision, H.G.; project administration, H.G.; funding acquisition, H.G. All authors have read and agreed to the published version of the manuscript.

Funding: The article processing charge was funded by the Open Access Publication Fund of the HTWG Hochschule Konstanz University of Applied Sciences.

Institutional Review Board Statement: Not applicable.

Data Availability Statement: Not applicable.

Acknowledgments: The authors want to thank Manuel Greiter at the HTWG for his support with setting up the mobile vehicle used in this work, as well as the Sick AG in Waldkirch, Germany, for their kind donation of the Lidar sensor.

Conflicts of Interest: Hartmut Gimpel has given professional training courses for Sick AG.

References

1. Roriz, R.; Cabral, J.; Gomes, T. Automotive LiDAR Technology: A Survey. *IEEE Trans. Intell. Transp. Syst.* **2021**, *23*, 6282–6297. [CrossRef]
2. Li, Y.; Ibanez-Guzman, J. Lidar for Autonomous Driving: The Principles, Challenges, and Trends for Automotive Lidar and Perception Systems. *IEEE Signal Process. Mag.* **2020**, *37*, 50–61. [CrossRef]
3. Ito, S.; Hiratsuka, S.; Ohta, M.; Matsubara, H.; Ogawa, M. Small Imaging Depth LIDAR and DCNN-Based Localization for Automated Guided Vehicle. *Sensors* **2018**, *18*, 177. [CrossRef] [PubMed]
4. Endres, F.; Sprunk, C.; Kümmerle, R.; Burgard, W. A Catadioptric Extension for RGB-D Cameras. In Proceedings of the 2014 IEEE/RSJ International Conference on Intelligent Robots and Systems, Chicago, IL, USA, 14–18 September 2014; pp. 466–471.
5. Anselment, C.; Weber, H.; Aschenbrenner, J.; Roser, C.; Schnurr, N.; Strohmeier, D. Device for Deviating and for Enlarging the Field of View. European Patent EP 2863253 B1, 19 August 2015.
6. Barone, S.; Carulli, M.; Neri, P.; Paoli, A.; Razionale, A.V. An Omnidirectional Vision Sensor Based on a Spherical Mirror Catadioptric System. *Sensors* **2018**, *18*, 408. [CrossRef] [PubMed]
7. Baker, S.; Nayar, S.K. A Theory of Single-Viewpoint Catadioptric Image Formation. *Int. J. Comput. Vis.* **1999**, *35*, 175–196. [CrossRef]
8. Willhoeft, V.; Fuerstenberg, K. Multilevel-Extension for Laserscanners. In Proceedings of the ITSC 2001. 2001 IEEE Intelligent Transportation Systems. Proceedings (Cat. No.01TH8585), Oakland, CA, USA, 25–29 August 2001; pp. 446–450.
9. Abiko, S.; Sakamoto, Y.; Hasegawa, T.; Yuta, S.; Shimaji, N. Development of Constant Altitude Flight System Using Two Dimensional Laser Range Finder with Mirrors. In Proceedings of the 2017 IEEE International Conference on Advanced Intelligent Mechatronics (AIM), Munich, Germany, 3–7 July 2017; pp. 833–838.
10. Matsubara, K.; Nagatani, K. Improvement in Measurement Area of Three-Dimensional LiDAR Using Mirrors Mounted on Mobile Robots. In *Field and Service Robotics*; Springer: Berlin/Heidelberg, Germany, 2019.
11. Matsubara, K.; Nagatani, K.; Hirata, Y. Improvement in Measurement Area of 3D LiDAR for a Mobile Robot Using a Mirror Mounted on a Manipulator. *IEEE Robot. Autom. Lett.* **2020**, *5*, 6350–6356. [CrossRef]
12. Aalerud, A.; Dybedal, J.; Subedi, D. Reshaping Field of View and Resolution with Segmented Reflectors: Bridging the Gap between Rotating and Solid-State LiDARs. *Sensors* **2020**, *20*, 3388. [CrossRef] [PubMed]
13. Chen, M.; Pitzer, B.; Droz, P.-Y.; Grossman, W. Mirrors to Extend Sensor Field of View in Self-Driving Vehicles. U.S. Patent Application US 2020/0341118 A1, 29 October 2020.

14. Pełka, M.; Będkowski, J. Calibration of Planar Reflectors Reshaping LiDAR's Field of View. *Sensors* **2021**, *21*, 6501. [CrossRef] [PubMed]
15. SunFounder PiCar-S Kit V2.0 for Raspberry Pi. Available online: https://www.sunfounder.com/products/raspberrypi-sensor-car (accessed on 22 April 2022).
16. Raspberry Pi 4 Model B. Available online: https://www.raspberrypi.com/products/raspberry-pi-4-model-b/ (accessed on 22 April 2022).
17. Python.Org. Available online: https://www.python.org/ (accessed on 28 April 2022).
18. 2D LiDAR Sensors | TiM3xx | SICK (TIM310-S01 Is Not Sold Anymore). Available online: https://www.sick.com/de/en/detection-and-ranging-solutions/2d-lidar-sensors/tim3xx/c/g205751 (accessed on 22 April 2022).
19. Metallic Mirror Coatings | Edmund Optics. Available online: https://www.edmundoptics.eu/knowledge-center/application-notes/optics/metallic-mirror-coatings/ (accessed on 22 April 2022).
20. Creo Parametric 3D Modeling Software | PTC. Available online: https://www.ptc.com/en/products/creo/parametric (accessed on 22 April 2022).
21. Diamond Grade Reflective Sheeting | 3M United States. Available online: https://www.3m.com/3M/en_US/p/c/films-sheeting/reflective-sheeting/b/diamond-grade/ (accessed on 28 April 2022).
22. Scipy.Optimize.Fmin—SciPy v1.8.0 Manual. Available online: https://docs.scipy.org/doc/scipy/reference/generated/scipy.optimize.fmin.html (accessed on 28 April 2022).
23. Plane (Geometry)—Distance from a Point to a Plane. *Wikipedia*. Available online: https://en.wikipedia.org/wiki/Plane_(geometry) (accessed on 13 April 2022).

sensors

MDPI

Article

Quantifying the Foregone Benefits of Intelligent Speed Assist Due to the Limited Availability of Speed Signs across Three Australian States

Sujanie Peiris *, Stuart Newstead, Janneke Berecki-Gisolf and Brian Fildes

Accident Research Centre, Monash University, 21 Alliance Ln, Clayton, VIC 3800, Australia
* Correspondence: sujanie.peiris@monash.edu; Tel.: +61-3-9905-9979

Abstract: By being able to communicate the speed limit to drivers using speed sign recognition cameras, Intelligent Speed Assist (ISA) is expected to bring significant road safety gains through increased speed compliance. In the absence of complete digital speed maps and due to limited cellular connectivity throughout Australia, this study estimated the forgone savings of ISA in the event that speed signs are solely relied upon for optimal advisory ISA function. First, speed-related fatalities and serious injuries (FSI) in the Australian states of Victoria, South Australia, and Queensland (2013–2018) were identified, and published effectiveness estimates of ISA were applied to determine the potential benefits of ISA. Subsequently, taking into account speed sign presence across the three states, the forgone savings of ISA were estimated as FSI that would not be prevented due to absent speed signage. Annually, 27–35% of speed-related FSI in each state are unlikely to be prevented by ISA because speed sign infrastructure is absent, equating to economic losses of between AUD 62 and 153 million. Despite a number of assumptions being made regarding ISA fitment and driver acceptance of the technology, conservative estimates suggest that the benefits of speed signs placed consistently across road classes and remoteness levels would far outweigh the costs expected from the absence of speed signs. The development and utilisation of a methodology for estimating the foregone benefits of ISA due to suboptimal road infrastructure constitutes a novel contribution to research. This work provides a means of identifying where infrastructure investments should be targeted to capitalise on benefits offered by advanced driver assist technologies.

Keywords: intelligent speed assist; advisory; roads; infrastructure; speed signs; speed sign sensing

1. Introduction

Speeding is the largest behavioural contributor to road traffic deaths and injuries in Australia and other developed countries [1], with speeding (i.e., travelling above the sign-posted speed limit) being cited as the most common road safety infringement [2]. Speed regulation remains a highly controversial area of road safety, with significant media resistance to positive speed management strategies, policies, and practice [3–5], even though poor speed compliance contributes to both crash risk and severity [6]. Due to the limited success of many speed reduction strategies and targeted interventions, advanced driver assist technologies, such as Intelligent Speed Assist (ISA), have the potential to influence driver speeding and are being considered as viable speed management interventions.

Given the prevalence of speeding in Australia and overseas, the ability to mitigate speeding is likely to have significant road safety benefits. In certain jurisdictions, such as Queensland, up to 94% of drivers are estimated to engage in low-level speeding (i.e., travelling up to 10 km/h above the limit) [7], while in other states this estimate is 78% or 88% [8,9]. It is therefore not surprising that a large proportion of fatal and serious injury crashes involve individuals who are exceeding the speed limit. Estimates suggest that, in NSW, up to 38% of all casualty crashes and 76% of fatal crashes could be attributed to low-level speeding [9]. These figures are reinforced by those in other areas of the world, with low- and medium-level speeding found to correspond to a major fraction of fatal crashes in Europe [10].

The exact contribution of speeding to crashes, however, is a contentious topic as it is difficult to quantify. Crash databases in most jurisdictions, both in Australia and internationally, do not record detailed information regarding speed involvement as a contributing or causative factor to the crash. Of those that do, reference to speed involvement is only general, and caution should be exercised in using these data to ascertain the proportion of crashes that involve speeding as a causal factor [11]. This is because crash data are typically compiled by attending police officers who do not reconstruct crashes to determine pre-crash speed estimates. Such estimates should be based on Newtonian mechanics, considering vehicle damage profiles and resting positions of the vehicles, and are thus resource intensive.

Estimates of speed involvement using the GIDAS study [12], which examined 10,742 fully documented and reconstructed crashes (Germany, 2011–2016), identified that 24.2% of fatal crashes, 18.5% of fatal and serious injury crashes, and 15.2% of all injury crashes involved speed as a contributing factor. Overall estimates suggest that 10 to 15% of all crashes and 30% of all fatal crashes in Europe are a direct result of speeding or inappropriate speed [13]. In the USA during 2017, speeding was implicated in 26% of all road trauma fatalities [14], but this figure was as high as 35% in some states [15]. In New Zealand, up to 30% of fatal crashes were attributed to speeding, although this figure was estimated to be as high as 60% when corrective factors from another study were applied [11]. Estimates from event data recorders analysed from South Australian crashes suggest that speeding contributes to 18% of all serious and fatal road trauma [16], but the contribution of speed to fatalities and serious injuries from crashes in Victoria is thought to be as high as 30% and 25%, respectively [17]. While scientific evidence that links fatal and serious injury crashes with inadvertent or unintentional speeding is sparse, it is highly likely that at least a proportion of those crashes attributed to speeding involve those inadvertently speeding [1,18,19]. Given the relationship between travel speed and crash risk, and travel speed and crash severity, the ability to mitigate speed-related crashes, whether it be low- or high-level speeding, inadvertent, or intentional, is likely to have significant road safety gains [20,21].

Given the limited success of physical speed countermeasures such as chicanes and speed bumps [22–29], advanced in-vehicle driver assist technologies that can influence travel speed, such as ISA, are expected to bring significant reductions in road trauma [30–33] by informing drivers of the posted speed limit for the road they are travelling on. ISA technology commonly uses vehicle-mounted sign recognition cameras and a GPS-linked speed limit database to display the legal speed limit on the dashboard. There are, at present, three forms of ISA that are designed to aid drivers in observing the speed limit: (1) advisory ISA, which notifies the driver when the travel speed of the vehicle exceeds the legal speed limit through an audio or visual warning; (2) voluntary or supportive ISA, which enables the driver to choose if the vehicle should adjust to the speed limit if/when the speed limit is exceeded; and (3) mandatory or limiting ISA, which physically regulates the vehicle's travel speed based on the speed limit detected.

It is expected that if drivers are constantly aware of the speed limit, and notified if and when it is being exceeded, that they are more likely to adhere to the speed limit. Based on this premise, it is estimated that the most intervening variant of ISA (limiting ISA, if fitted to every vehicle) could reduce crash likelihood by up to 30% and road deaths by 20% in Europe [34], while these figures could be as high as 30% and 28%, respectively, in Australia [35]. Despite the large safety gains expected from supportive or limiting ISA, advisory ISA, because of its limited intrusiveness and greater potential for public acceptance, is the most widely promoted form of ISA at present, with all newly manufactured vehicles (from May 2022) in Europe required to be fitted with the technology [36]. The mandate is unlikely to be unique to Europe, with ANCAP, Australia's independent vehicle safety testing authority, promoting ISA fitment and assessing vehicles based on the presence and performance of the technology [37].

The ability of ISA to function, however, is dependent on the technology being able to identify local speed limits using cellular connectivity (which allows the vehicle to access digital speed maps) or, in its absence, being able to access physical speed signs which vehicle-mounted cameras can detect using embedded sensors [38]. The European Commission has specified that ISA systems that employ a combination of camera systems, possess a Global Navigation Satellite System (GNSS), and use up-to-date digital maps are *state of the art*, having *the greatest real-world performance and reliability* [39]. In recognition of the limitations associated with accessing digital speed maps, and in support of the recent technology mandate in the EU, all member states of the EU were encouraged to facilitate the correct placement of numerical speed limit signs on streets and roads with clear indications of when a speed zone starts and ends [39]. Cellular networks enable continuous data connectivity, supporting a number of autonomous driver functions by delivering information about traffic and road conditions over the wireless network. In the case of ISA systems, access to mobile data would enable vehicles to acquire base map information and receive live updates regarding posted speed limits. Such connectively, however, is absent or limited in many regions of Australia, and, as an alternative, speed sign infrastructure is required for ISA to function [40]. In particular locations, such as in remote and regional Australia, cellular connectively is absent [41] and vehicles fitted with ISA are solely reliant on physical speed signs. Reflecting this, existing ANCAP test protocols for ISA do not specify the need for vehicles to be able to access digital speed maps, provided that the speed recognition cameras can detect speed signs and relay the appropriate information to the driver [42].

1.1. Speed Signage and ISA

The lack of regulation regarding speed sign availability (frequency and/or placement) and consistency (position, size, angle, and design) is problematic throughout most Australian jurisdictions [43], making camera detection of the signs difficult and, at times, impossible [44,45]. A recent review of speed management information in Australia identified that speed limit information is *widely utilised, diversely managed, dependent on legislative requirements and organisational structure, and spread throughout jurisdictions*, with limited guidance available at a national level for jurisdictions on speed management [46]. Regional and remote areas in Australia are particularly problematic, since the presence of speed signs is sporadic at best and the lack of digital or cellular connectivity in these areas means that even if vehicles are fitted with ISA, the technology will be of no use in assisting drivers to keep to speed limits. Ironically, regional and remote areas are also often responsible for a disproportionate number of deaths and serious injuries [47,48], and are thus likely to benefit most from having infrastructure that is supportive of driver assist technologies [49], including ISA. While digital HD speed data maps and GPS-linked speed limit databases are available, especially in urban city regions, limited cellular coverage across areas of Australia [40] means that camera detection of physical speed signs will currently be relied upon for ISA functioning in rural and remote areas. Speed signs are, therefore, a key input for sensing technology which allows ISA to function. In the absence of speed signs, on-board sensors, radars, and speed sign recognition cameras will not detect the data that are needed for ISA to be able to advise drivers of speeds or for ISA to regulate speed. Since speed sign availability is limited and, by law, the driver is often responsible for knowing the default speed in unmarked areas [50], the lifesaving potential of ISA in countries such as Australia may not always be realised and the benefits of ISA may potentially be overestimated.

1.2. Study Aims

Speed compliance encouraged through the use of ISA is expected to bring significant road safety gains in terms of speed-related crash mitigation. The overall objective of this study was to quantify the proportion of fatalities and serious injuries from ISA-sensitive crashes that will not be prevented due to the absence of speed signage. This required first quantifying the proportion of crashes that will likely be mitigated through speed compliance aided by ISA (independent of ISA-supportive infrastructure being present). Subsequently, the proportion of fatalities and serious injuries from speed-involved crashes that will not be prevented by ISA (i.e., the forgone savings) needed to be quantified, assuming that speed signage is the principal support for all vehicles fitted with ISA.

This objective was achieved by: (1) quantifying the proportions of crashes in Victoria, South Australia, and Queensland that were attributed to exceeding the speed limit (i.e., were ISA-sensitive) and the proportion of these that would be reduced by ISA based on technology effectiveness estimates published in the literature; (2) determining how many of these crashes occurred in areas that are not fitted with adequate speed signage; and (3) calculating the foregone saving of ISA in terms of crash-related injuries and fatalities and their correlated cost estimates. It is anticipated that the results of this study can inform appropriate investments into speed signage placement, or equivalent infrastructure, to ensure that ISA technologies perform as intended and provide life-saving benefits to road users.

2. Methods

Data on crashes reported by police from the Australian states of Victoria, South Australia, and Queensland were used in this study. Based on the assumption that crash trends and other parameters related to travel and driver behaviour remain similar over time and to ensure an adequate sample for analysis, data from 2013–2018 were used. The lost savings of ISA, assuming it were to be fitted to each vehicle, were estimated as follows:

1. Speed-related crashes (considered here to be *ISA-sensitive*) were identified from crash data (2013–2018 inclusive) using an established methodology [51] and mapped onto the road network with respect to remoteness level and road class.
2. ISA effectiveness estimates (published by Doecke, et al. [52]) were applied to ISA-sensitive crashes to quantify the number of fatalities and serious injuries likely prevented if ISA had been fitted to the crashed vehicles.
3. Using published literature that allowed speed sign availability to be approximated based on road class and remoteness level [53], ISA-sensitive crashes that would vs. would not be benefited by ISA were identified.
4. The lost benefits of ISA were quantified by estimating the number of fatalities and serious injuries that are unlikely to have been prevented due to the lack of speed signs at the crash locations.

A schematic summarising the overall methodology is presented in Figure 1, with each of the above steps outlined in greater detail below.

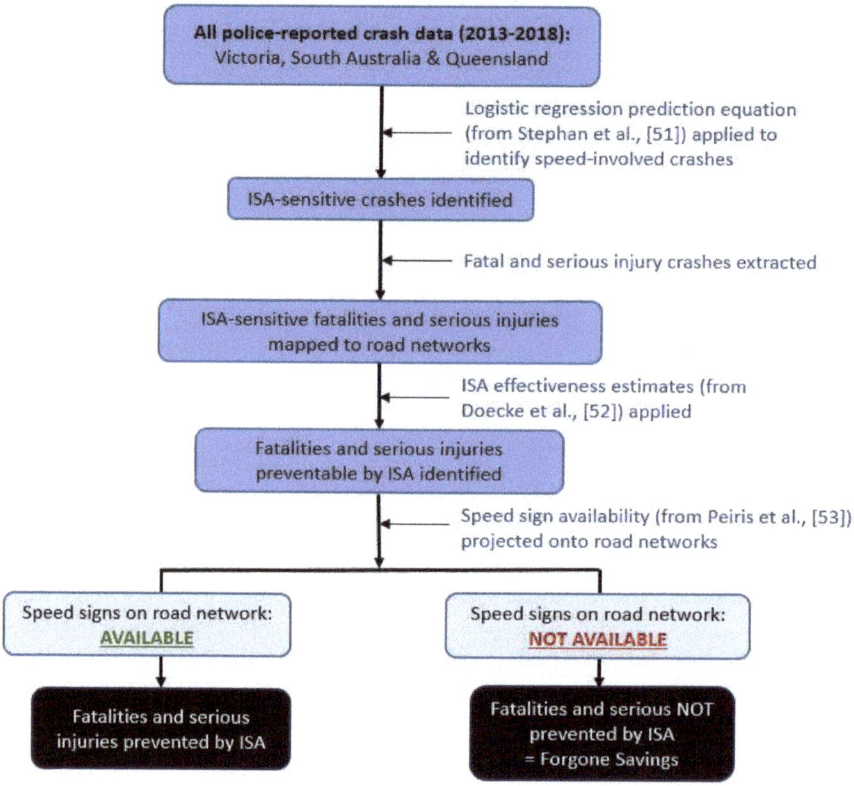

Figure 1. Schematic depicting the methodology which was undertaken to identify speed-related impacts that would be prevented by ISA and those which would not be prevented (forgone savings) due to the absence of speed signs.

2.1. *Identifying ISA-Sensitive Crashes from Crash Data*

An analysis originally conducted in 2003 by Diamantopoulou, et al. [54] to determine the characteristics of speed-involved crashes, and updated in 2022 by Stephan, Budd, Newstead and Young [51], was used here to predict the ISA-sensitive crashes based on a number of available fields in the crash data. The re-analysis by Stephan and colleagues involved conducting a logistic regression using in-depth South Australian crash data [55–57] to identify factors associated with speed involvement in South Australian mass crash data (2008 to 2012). Stepwise logistic regression using likelihood ratio tests was used to identify main and interactive effects of model variables that contributed significantly to the model. The equation was generalised across locations, motor vehicle types, and speed zones to identify speed-involved crashes. The reader is referred to the study by Stephan, Budd, Newstead and Young [51] for an in-depth explanation of how the model was developed. For the current study, the prediction equation generated by Stephan, Budd, Newstead and Young [51] was applied to Victoria, South Australia, and Queensland mass crash data (2013–2018) to identify crashes likely to involve speeding.

There is some conjecture in the evidence regarding the contribution of speed in impacts [11,17,58–61]. The logistic regression prediction equation developed by Stephan, Budd, Newstead and Young [51] generates a probability of speed involvement for each crash. In doing so, every fatal and serious injury crash is ranked based on the model's prediction of how likely each crash was to have been speed-related. In order to dichotomise the results and determine if a crash was speed-related or not, three specific thresholds were chosen: 10%, 15%, and 30%. While using a 10% threshold for speed involvement was likely to produce the best specificity for the speed prediction model, thresholds of 15% and 30% were also selected for greater sensitivity and to provide a range of positive predictive values. Due to the low specificity of the prediction equation (which generally produced very high probabilities of speed involvement across the crash population in each state), choosing three empirical cut-offs to dichotomise crashes as being speed-related or not provided a realistic range of outcome estimates. The most conservative estimate was used for the cost-benefit analysis.

In order to identify the speed-related crashes that were likely to be influenced by speed sign availability, the crashes in each state that were identified as being speed-related were plotted onto the road network using ArcGis Pro software. The crashes were then disaggregated by road class specific to each state and the remoteness levels across Australia, these being Major Cities, Inner Regional Australia, Outer Regional Australia, Remote Australia, and Very Remote Australia according to the Aria Index [62].

2.2. Applying ISA Effectiveness Values to the Crash Data

A literature review was conducted to identify the most suitable estimates of ISA crash reduction effectiveness. Since the benefits and limitations of ISA are likely to be country-specific and Australian crash data were used here to identify ISA-sensitive crashes, only ISA effectiveness values derived from Australian-run ISA trials using light vehicles were reviewed.

The effectiveness values for advisory ISA used in this study were derived from the 2009–2010 ISA trial funded by NSW RTA [52], in which 104 light vehicles were fitted with advisory ISA and speed data recorders. As part of the 3-month trial, vehicles communicated remotely with a centralised database so that up-to-date speed, location, and speed zone information could be constantly relayed. The road network within the trial area extended approximately 2500 km and incorporated more than 4000 speed signs across 932 speed zones, ensuring speed sign availability for ISA function. Baseline speed compliance of drivers prior to ISA installation was measured using a GPS-based speed and location data recorder fitted to all study vehicles at least one month prior to ISA device installation, logging the speed and location of each vehicle every ten seconds and producing a total of 7.5 million speed records (with the speed behaviour of drivers being measured before, during, and after the ISA device was operating in the vehicle). Modelling based on the results of this trial (using baseline speed compliance measurements) suggested that fatal and serious injuries in NSW would be reduced by 8.4% and 5.9%, respectively, if all vehicles in NSW were fitted with the technology. The reader is referred to Doecke, Kloeden and Woolley [52] for a detailed overview of the study, results, and analysis.

Since ISA changes the shape of the distribution of speed, using a speed-specific risk curve applied to the distribution of speeds in different speed zones allows the change in risk associated with speed distribution changes resulting from ISA fitment to be more accurately quantified. The reduction in crash risk using this approach was estimated by Doecke, Kloeden and Woolley [52] by applying crash risk curves for travel speeds originally generated by Kloeden, McLean, Moore and Ponte [56] in 1997, and developed further by the researchers in 2001 [57] and 2002 [63], to the speed distributions generated from the trial. Using the curves, the casualty crash risk reductions estimated for each speed zone were multiplied by the average number of fatal and serious injury crashes (2004–2008) within each speed zone across the Australian states [64] in order to determine annual crash reductions that could be achieved if all vehicles in Australia were fitted with advisory ISA.

The ISA effectiveness estimates generated by Doecke, Kloeden and Woolley [52] were applied in the current study to the fatalities and serious injuries resulting from the subset of crashes identified as being speed-involved (see Table 1 below).

Table 1. Estimated annual percentage (%) of reductions in excessive-speed crashes produced by advisory ISA based on crash data (2013–2018) using effectiveness estimates derived by Doecke, Kloeden and Woolley [52].

Crash Type	Victoria	South Australia	Queensland
Fatal	18.7	19.5	18.8
Serious injury	19.1	19.3	19.6

2.3. *Quantifying the Foregone Savings of ISA*

As a premise for the calculations made in this study, it was assumed that, in Australia, the benefits of ISA will not be realised in regions where speed signs are unavailable. Current ISA technology fitted to new vehicles is highly reliant on speed sign recognition cameras communicating detected speed limits to drivers.

In order to approximate where ISA was most likely to function or fail based on speed sign availability, a methodology study was conducted in 2021 [65] to estimate speed sign availability in Victoria. In brief, this involved obtaining crash location data from respective state authorities to identify where (using GPS coordinates) each crash occurred. Fatal and serious injury (FSI) crashes were then mapped in Victoria (2013–2018) according to road class and remoteness level (as defined by the ABS 2016 [56]). Based on this distribution, locations were randomly selected to check for speed sign availability. Ten random locations (if <5% of FSI crashes occurred on the particular road class and remoteness level) were selected or twenty random locations (if >5% of FSI crashes occurred on the particular road class and remoteness level) were selected for the assessment per road class and remoteness level.

Speed sign data generated and made available by Data VIC [66] were overlaid on the Victorian road network map in ArcGis Pro software, and the distance from each random point to available speed signs was measured for both directions of travel in ArcGIS and verified using Google Maps. The distances were then averaged for each road type and remoteness level to identify the availability and frequency of speed sign locations based on road class and remoteness level. It was assumed that, in the absence of digital speed maps, drivers would have to travel these distances on average before vehicle-mounted speed sign detection cameras could identify a speed sign and inform drivers of the appropriate location-specific speed limit. The results of this analysis can be found in Peiris, Berecki-Gisolf, Newstead, Chen and Fildes [53].

Based on the findings from the methodology study [53], it was assumed that if 50% or more of the random locations identified did not have a speed sign available on the roads in that road class and remoteness level, speed sign availability for that road class and remoteness would be considered largely unavailable. Therefore, a vehicle travelling above the speed limit on that class of road would not benefit from ISA in terms of reduced crash risk or severity. Based on this assumption, six classes of road (disaggregated by remoteness level) were assumed to have no speed sign availability in Victoria. Since the road class hierarchy is similar in South Australia, the same assumption was made regarding speed sign availability across the road classes and remoteness levels (Table 2). In Queensland, however, since there are only five major road classes, a conservative approach was used to estimate speed sign availability. In Queensland, four classes of roads, that are each roughly equivalent to those in Victoria and South Australia, were assumed to not have speed sign availability (Table 3).

Table 2. Assumptions made regarding speed sign availability across remoteness levels (Major City, Regional, and Remote) and road classes in Victoria and South Australia.

Road Class	Major City	Regional	Remote
Freeway	✓	✓	✓
Highway	✓	✓	✓
Arterial	✓	✓	✓
Sub-Arterial	✓	✗	✗
Collector	✓	✓	✗
Local	✗	✗	✗

(✓) speed signs were considered to be available; (✗) speed signs were considered to be unavailable.

Table 3. Assumptions made regarding speed sign availability across remoteness levels (Major City, Regional, and Remote) and road classes in Queensland.

Road Class	Major City	Regional	Remote
Freeway	✓	✓	✓
Highway	✓	✓	✓
Secondary	✓	✓	✓
Local Connector	✓	✓	✗
Streets	✗	✗	✗

(✓) speed signs were considered to be available; (✗) speed signs were considered to be unavailable.

ISA-sensitive crashes that occurred on road classes where speed signs were not available, as identified in Tables 2 and 3, were assumed to not benefit from ISA technology; this information was used to calculate the estimated potential forgone savings of ISA.

When estimating the economic benefits and forgone savings due to ISA, one of two approaches can be used to value the economic cost of road crashes (these being the willingness-to-pay and human capital approach). Here, a hybrid of the approaches (hybrid human capital), derived by BITRE [67], was used to estimate cost of death and injury from forgone ISA-sensitive crashes.

3. Results

The total number of fatalities (F) and serious injuries (SI) resulting from crashes reported to police over the six-year period analysed, and the subset of those where speed was identified as a likely causal factor using the logistic regression prediction equation developed by Stephan, Budd, Newstead and Young [51], is presented in Table 4.

Table 4. Number of fatalities (F) and serious injuries (SI) from crashes identified as being speed-related. Results are shown using various thresholds (top 10%, 15%, and 30% of crashes) and ranked according to estimated likelihood of speed involvement.

	Total		F and SI in Speeding-Related Crashes at Various Thresholds					
	F	SI	F (10%)	SI (10%)	F (15%)	SI (15%)	F (30%)	SI (30%)
Victoria	1505	23,301	696	4024	757	5289	958	9048
South Australia	454	2701	135	364	152	565	246	1094
Queensland	1480	31,249	582	3726	678	5251	846	9784

Applying the estimated percentage (%) of casualty crash risk reductions per year produced by advisory ISA, as derived by Doecke, Kloeden and Woolley [52] and summarised by Wall, Creef, Boland, Vecovski, Prendergast, Stow, Fernandes, Beck, Doecke and Woolley [64], to the crashes identified as resulting from excessive speed, the safety benefits likely to be observed are summarised in Table 5. These estimates are based on the assumptions that all vehicles are fitted with advisory ISA, that the effectiveness rates specified are accurate, that ISA is functional in all instances, and that ISA is equally effective in reducing speeds amongst all driver demographics (i.e., comparable levels of speed compliance to those observed from the NSW trials), independent of crash location or circumstance.

Table 5. Estimated annual reductions in fatalities and serious injuries from excessive-speed crashes, as identified from crash data in each state, based on crash reduction estimates specified by Wall, Creef, Boland, Vecovski, Prendergast, Stow, Fernandes, Beck, Doecke and Woolley [64].

State	Threshold *	Total F and SI Prevented	Fatalities Prevented	Serious Injuries Prevented
Victoria	10	149.6	21.7	127.9
	15	191.7	23.6	168.1
	30	317.5	29.9	287.7
South Australia	10	16.1	4.4	11.7
	15	23.1	4.9	18.2
	30	43.2	8	35.2
Queensland	10	140.0	18.2	121.7
	15	192.8	21.2	171.5
	30	346.1	26.5	319.6

* Thresholds refer to the top 10%, 15%, and 30% of crashes most likely to have involved speeding based on the calculated probabilities.

Based on the generalised assumptions regarding speed sign availability across road classes and remoteness levels (Tables 3 and 4), the lost benefits of ISA were estimated for each of the probability thresholds in each state (Table 6). That is, speed-related crashes that occurred on local roads in major cities, sub-arterial and local roads in regional areas, and sub-arterial, collector, and local roads in remote areas within Victoria and South Australia; and in Queensland, speed-related crashes that occurred on streets in major cities and regional areas, and streets and local connector roads in remote areas, were assumed not to be influenced by ISA due to the absence of speed signs.

Table 6. Annual fatalities (F) and serious injuries (SI) likely to be prevented and unlikely to be prevented in speed-related crashes (due to absent speed signs).

State and Threshold *		Total F and SI	F		SI		Total	
			Prevented	Not Prevented	Prevented	Not Prevented	Prevented (%)	Not Prevented (%)
Victoria	10	149.6	15.7	6.0	92.3	35.7	108.0 (72.2)	41.6 (27.8)
	15	191.7	17.0	6.5	121.0	47.0	138.1 (72.0)	53.6 (28.0)
	30	317.5	21.9	7.9	208.7	78.9	230.6 (72.6)	86.9 (27.4)
South Australia	10	16.1	3.2	1.2	7.4	4.3	10.6 (65.8)	5.5 (34.3)
	15	23.1	3.5	1.5	12.2	6.0	15.6 (67.7)	7.5 (32.4)
	30	43.2	5.9	2.1	24.9	10.3	30.8 (71.4)	12.3 (28.6)
Queensland	10	140.0	14.4	3.9	88.2	33.5	102.6 (73.3)	37.4 (26.7)
	15	192.8	17.0	4.2	123.5	48.1	140.5 (72.9)	52.3 (27.1)
	30	346.1	21.2	5.3	229.9	89.7	251.2 (72.6)	95.0 (27.4)

* Thresholds refer to the top 10%, 15%, and 30% most likely to involve speeding based on the calculated probabilities.

Table 7 presents the estimated contribution of ISA to overall road safety in the three states in terms of the percentage of fatalities likely or unlikely to be prevented (due to absent speed signage) based on average annual road tolls (2013–2018). The 15% threshold estimates of ISA effectiveness were used to quantify the likely benefits of ISA on the overall road toll (measured in terms of fatalities). It is estimated that between 1.7 and 2.6% of the total road toll will continue to occur in each state because ISA will not be able to prevent these speed-related crashes due to absent speed signs.

Table 7. Percentage of ISA-sensitive fatalities prevented and not prevented, with respect to the annual average road toll, due to the presence/absence of speed signs.

State	Road Toll (Fatalities)	Fatalities in ISA-Sensitive Crashes		% Road Toll	
		Prevented	Not Prevented	Prevented	Not Prevented
Victoria	251	17	6.5	6.8	2.6
South Australia	76	3.5	1.5	4.6	2.0
Queensland	247	17	4.2	6.9	1.7

Estimating Economic Benefits and Forgone Savings Due to ISA

In the absence of a nationally agreed method for valuing road trauma, the Australian Government Office of Best Practice Regulation (OBRP) more recently published an updated means of estimating the base costs of road trauma. This method, *willingness-to-pay*, which is based on values of statistical life and statistical life year, was used by Economic Connections and published by the Australian Automobile Association (2017) to value the cost of a fatality at AUD 4.339 million and a serious injury at AUD 239,000, indexed to the year 2015 [68].

The monetary values of deaths and serious injuries estimated to be saved (due to functioning ISA) and foregone (due to ISA not being able to function in the absence of speed sign infrastructure) were expressed in monetary terms by multiplying the expected crash savings by the dollar values estimated for a fatality and serious injury by Economic Connections [68] (Table 8).

Table 8. The estimated costs and foregone savings * (AUD million) due to fatalities and serious injuries resulting from ISA-sensitive crashes where ISA could have intervened (based on speed sign availability).

State and Threshold		Prevented		Not Prevented	
		F	SI	F	SI
Victoria	10	55.25	21.11	28.96	11.20
	15	59.82	22.87	37.97	14.75
	30	77.07	27.80	65.49	24.76
South Australia	10	11.26	4.22	2.32	1.35
	15	12.32	5.28	3.83	1.88
	30	20.76	7.39	7.81	3.23
Queensland	10	50.67	13.72	27.68	10.51
	15	59.82	14.78	38.75	15.09
	30	74.60	18.65	72.14	28.15

* All costs associated with death and serious injuries calculated here are expressed in 2022 AUD.

Estimates suggest that, in Victoria alone, deaths and serious injuries costing AUD 40 to 69 million annually could potentially be prevented by placement of speed sign infrastructure. These estimates are proportionately less in South Australia. In Queensland, lost lives and serious injuries resulting from high-speed crashes which could have been prevented by ISA would cost close to an estimated AUD 30 to 76 million.

4. Discussion

In the absence of cellular connectivity, speed sign sensing technology is solely reliant on speed signs being readily detectable. This study set out to estimate the lost benefits of ISA technology in three Australian states due to the absence of adequate speed sign infrastructure. By estimating the proportion of roads that are unlikely to support ISA (due to inadequate speed sign availability), and hence unlikely to prevent ISA-sensitive fatalities and serious injuries, the forgone benefits of ISA were estimated. It was assumed that roads that did not have speed signage or up-to-date digital speed maps that vehicles could access were correctly identified and would produce no road trauma savings for ISA-equipped vehicles. To do this, the presence and frequency of speed signs across road classes and remoteness levels in Victoria (based on a publicly available digital speed sign map) were quantified in a previously published desktop study [53], and these results were projected to South Australia and Queensland based on broad assumptions regarding speed sign availability across road class and remoteness levels. The number of fatalities and serious injuries that ISA may be able to mitigate was estimated by applying the most relevant advisory ISA effectiveness estimates available to speed-related (i.e., ISA-sensitive) fatal and serious injury crashes identified using historical crash data (disaggregated by road class and remoteness levels). Despite the fact that a large proportion of individuals suffer minor injuries in speed-involved crashes [69], and that ISA is likely to reduce these injuries, estimates of minor injuries prevented by ISA were not derived here due to minor injury data not being reliably collected or routinely validated within crash databases [11,58,60]. Therefore, the lost savings of ISA were calculated as lives lost and individuals seriously injured in speed-related crashes where ISA may have been of assistance in mitigating the crash but, ultimately, would not have been due to absent speed signage.

The estimates made in this paper regarding the benefits of ISA (and therefore the forgone savings) were based on a number of assumptions. These include that all light vehicles will be fitted with ISA in future and that crash trends remain similar over time across jurisdictions. Further, it was assumed that ISA is equally effective at reducing speeding amongst all driver demographics for the following reasons: (1) ISA fitment will induce comparable levels of driver compliance across the jurisdiction to those measured in the NSW ISA trial; (2) ISA functionality will be dependent on physical speed sign infrastructure; and (3) speed sign infrastructure remained unchanged over the study period. To our knowledge, no researchers to date have published a means of quantifying the ADAS technology benefits that are potentially foregone due to poor or absent road infrastructure. This work provides a novel contribution to efforts to better prepare roads for vehicles fitted with lifesaving ADAS technologies, such as ISA.

4.1. Effectiveness of ISA

Based on the stated assumptions, advisory ISA was estimated to potentially have the capacity to annually prevent 21–30 fatalities and 128–288 serious injuries in Victoria, 7–8 fatalities and 12–35 serious injuries in South Australia, and 18–27 fatalities and 121–320 serious injuries in Queensland. However, due to speed sign absence, approximately 28% of the fatal and serious injuries that ISA would be able to prevent in Victoria would not be prevented. Projecting the speed sign availability assumptions to South Australia and Queensland (based on road class and remoteness level), it was approximated that 29–34% of fatalities and serious injuries in South Australia and 27% of fatalities and serious injuries that could be prevented by ISA in Queensland would not be prevented. Annually, across the three states alone, it was estimated that between 11 and 15 lives would be lost and between 74 and 179 serious injuries would continue to occur in speed-related crashes that would otherwise be preventable with advisory ISA. These figures were not projected to a national level due to the inability to project assumptions regarding speed sign availability across road classes in other states for lack of comparative data. However, it is likely that the forgone benefits of ISA will be significantly higher in states that have more regional and remote areas of road travel exposure than Victoria does. The monetary values of the lost benefits of ISA, per annum, vary from AUD 30 to 69 million in Victoria and AUD 29 to 76 million in Queensland, though in South Australia they are proportionately less. Determining the percentage of total road trauma forgone nationally due to the absence of speed signage would be informative and help better direct infrastructure resources to regions where speed-related crashes are frequent but where ISA would be incapable of assisting drivers to be speed compliant.

While these figures should be interpreted with caution due to the number of assumptions made during the analysis, it is likely that the estimated foregone trauma savings due to absent speed signs are an underestimate of actual losses due to a number of factors including:

- The conservative estimates of crashes that were considered likely to involve speeding;
- The effectiveness rates of ISA employed here;
- Potential overestimates of speed sign availability in South Australia and Queensland since Victoria was used in the original research to generate assumptions about speed sign availability. Victoria may have more speed signs per road class than the other states.

4.2. Estimates of Speed Involvement in Crashes

The probability that a vehicle was exceeding the speed limit prior to impact was estimated using a logistic regression prediction equation [70]. While there are a number of limitations associated with the model (discussed below), the ranking and selection process is likely to have resulted in conservative estimates of speed involvement.

The logistic regression equation used in this study was based on in-depth South Australian crash data, which had been re-analysed and applied to Victorian mass crash data to identify factors associated with speed involvement. As such, there are a number of limitations associated with this model. The model was developed using SA in-depth crash data, which was restricted to casualty crashes that occurred in 60 km/h zones in urban areas involving passenger vehicles travelling at free speed for 1995 to 1997 [55,56]. The model developed based on this data was applied to all Victorian crashes to identify driver and vehicle characteristics associated with speed involvement (based on crash data from the years 2008 to 2012). Implicit is the assumption that factors that were associated with speed-involved passenger vehicle crashes in 60 km/h urban areas in South Australia were equally generalisable to Victorian crashes across all areas, speed zones, and vehicle types. As a consequence, it is likely that speed-related impacts, that are typically over represented in remote and regional areas of Australia [47], are under-represented in the number of fatal and serious injury crashes that were identified as being speed-involved in this analysis.

The application of this model to crash data from a different time period (2013–2018) and different state (such as Queensland) has not been validated. While results from the model should be interpreted with caution, three thresholds for identifying the probability of speed involvement were used. Based on the thresholds used, 19–40%, 15–42%, and 13–32% of fatal and serious injury crashes in Victoria, South Australia, and Queensland, respectively, were estimated to involve excessive speed as a crash-contributing factor. These figures are comparable to those published in the literature suggesting that, in New Zealand for example, close to 30% of fatalities and 20% of serious injuries are likely to involve excessive speeds (based on police reported crash data) [11]. Doecke and Woolley [71] identified that, in Victoria, over 30% of fatal crashes were likely to be speed-related (based on crash data), while in South Australia close to 40% of fatal crashes were speed-related. In Queensland, it was estimated that 22% of fatal crashes and close to 7% of serious injury crashes involved excessive speed, although the authors noted that limitations in data recording were likely to make these estimates conservative. When Event Data Recorder (EDR) data from South Australia were used to estimate the contribution of speed to fatal and serious injury crashes more recently, it was estimated that 18% of all fatal and serious injury crashes were a result of speeding [16]. Therefore, the estimates of speed-related fatal and serious injury crashes generated using the logistic regression equation here appear reasonable.

4.3. The Effectiveness Rates of ISA

It is important to note that outcome measures or effectiveness estimates derived in ISA trials vary widely in terms of study settings, methodology, and results. For example, the speed reduction effects of ISA can be used to calculate the reduction in crash likelihood and severity. Using Nilsson's [20] or Elvik's power model [6], the changes in mean speed induced by ISA can be related to changes in crash frequency. Alternatively, Kloeden's relative risk curves [55–57] can be used to predict the difference in casualty crashes for a change in the speed distribution (across speed zones) induced by the ISA technology. ISA effectiveness can also be reported in the form of other measures, such as time spent exceeding the speed limit, maximum or mean speeds pre-ISA vs. during-ISA, or numbers of speeding infringements issued pre-ISA compared to during or post-ISA. Ultimately, however, ISA effectiveness (independent outcome measures reported) is dependent on the pre-ISA travel speeds of the subjects taking part in the trial (which, in turn, depend on where the trial was conducted, i.e., on-road vs. test track, rural vs. urban, etc.), the demographics of the participants and selection biases, and the types of vehicles that the ISA devices were fitted to during the trial (i.e., private vs. commercial). ISA fitted to a less speed-compliant group of drivers who engage in inadvertent speeding, for example, is likely to have a more pronounced effect on mean travel and speed distributions and, therefore, provide a higher estimated benefit [72]. Factors that are likely to influence the estimated effectiveness of ISA are presented schematically in Figure 2.

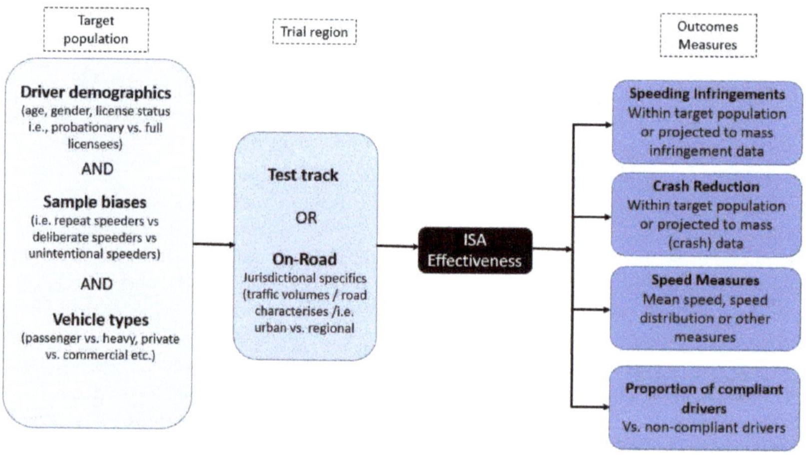

Figure 2. Schematic of how outcome measures of ISA effectiveness are influenced by the target population and trial region in ISA trials.

Estimates of ISA effectiveness in terms of crash risk reduction generated from mean speed measures are readily available in the literature [73,74]. However, these estimates imply that ISA has a uniform effect on the reduction of speeds across all speed zones. The ISA effectiveness estimates used in this study were generated from the reduction in speed distribution attributed to ISA during the NSW ISA trial [64] (Table 1), similar to effectiveness estimates cited in other studies [75,76]. The magnitudes in crash risk reduction estimates generated by Wall, Creef, Boland, Vecovski, Prendergast, Stow, Fernandes, Beck, Doecke and Woolley [64], and used here, are comparable to crash reduction estimates cited in a number of other similar ISA trials run globally. For example, the long-term use of supportive ISA systems has been estimated to reduce injury crashes by up to 25% and fatal crashes by up to 32% [77], limiting ISA has been estimated to be up to 26.5% and 28.3% effective in reducing serious injury and fatal crash risk, and advisory ISA has been estimated to be up to 8.3% and 11% effective for reducing serious injury and fatal crash risk, respectively [76].

An independent investigation by TRL [12,78] suggested that, on average, 19% of fatalities would likely be prevented with the introduction of mandatory ISA (which could be over-ridden by applying the accelerator pedal). Carsten and Tate [32] estimated that fitting a *simple* mandatory ISA system to every vehicle would reduce injury crashes by 20% and fatal crashes by 37%, but a more advanced system capable of communicating dynamic speed changes via cellular connectivity would reduce injury crashes by 36% and fatal crashes 59%. Not surprisingly, the estimated crash reduction benefits of advisory ISA are typically less due to the less intrusive nature of the technology.

It is also noteworthy that the crash risk reduction estimates generated by Wall, Creef, Boland, Vecovski, Prendergast, Stow, Fernandes, Beck, Doecke and Woolley [64], and used here, were generated based on the change in the distribution of speeds (i.e., before vs. during ISA installation). Kloeden's risk curves [56] relative to travel speed were used by the authors in the ISA study to estimate the reductions in crash risk. The equations for risk relative to travel speed (generated for 50 km/h and 60 km/h speeds) and mean speed (for 80 km/h or above) were applied to the speed distributions observed. While these risk curves are widely accepted, have been previously used for identifying the benefits of ISA [52,75], and are based on Australian conditions, there are limitations that may have affected the ISA effectiveness estimates derived from the NSW ISA trial. For example, Kloeden's risk curves [56] may not apply to the full range of speed zones, environments, or driving/road conditions that subjects in the ISA trial were exposed to. Further, Kloeden's et al.'s research used free travel speed to estimate crash risks, but this may not have been the case when applying the curves to the speed data generated from the ISA trial. More broadly, the observed speeding behaviour from the NSW ISA trial may not directly translate to Victoria, Queensland, or South Australia (to which the effectiveness values were applied), in which case ISA may not have an equivalent effect to that observed and quantified from the ISA trial. Further to this, the distribution of crashes per speed zone in Victoria, South Australia, and Queensland in the crash data which are used here may not be the same distribution of crashes per speed zone in the crash data that were used by the NSW ISA trial researchers to derive the effectiveness estimates.

Given that low-level speeding is socially acceptable in Australia and globally [1], it is difficult to accurately quantify the difference ISA will make to road safety. Driver acceptance (including usability) of ISA was not considered in this study. The actual effectiveness of ISA is likely to be highly dependent on how receptive or reactive drivers in future will be to warnings issued when ISA is functional. With advisory ISA, drivers are not obliged to reduce their speed when speed warnings are issued, and hence the level of speed compliance induced by advisory ISA is uncertain. When the beliefs underlying intensions to override ISA (to exceed the speed limit) were investigated using the Theory of Planned Behaviour, Rowe, et al. [79] found that driver attitudes strongly predicted these intentions. While the behavioural response of drivers to ISA is not within the scope of this study, it is important to state that since only advisory ISA was considered here, there is potential for drivers to ignore ISA-issued warnings and for the crash benefits (and therefore foregone benefits) of ISA to be severely overstated. Effectiveness estimates of limiting or mandatory ISA, therefore, are likely to be more reflective of actual ISA effectiveness for these ISA types, since this removes the behavioural or decision-making component associated with the technology. Data from international trials of supportive ISA systems suggest that speeding can be reduced by up to 53%, with potential reductions in serious injury crashes of up to 25% [80], or 50% if limiting ISA is considered [32]. However, given the European mandate for advisory ISA, only the forgone benefits of advisory ISA were considered here; therefore, effectiveness rates of advisory ISA were employed for this study. It is also noteworthy that, even if ISA is effective in encouraging speed compliance, limitations in driver reactions may, in some settings, only reduce the severity of a crash rather than prevent a crash. Further to this, ISA may be more effective in certain speed zones than others, as demonstrated across all of the Australian-run ISA trials [51,64,72,76]. Despite this, and the likely institutional and regulatory barriers associated with mandating the technology, ISA has the potential to encourage speed compliance and improve speed-related behaviour. The long-term benefits of deploying ISA, however, can only be realised if the infrastructure that supports its function is also made available.

4.4. Availability of Road Sign Infrastructure

In Australia, other than in the state of Victoria, road asset data pertaining to the availability of speed signs are not publicly available. It is therefore difficult to quantify the density of available speed signs across the nation. The availability of speed signs in Victoria was quantified in a previously published study based on a sampling methodology reflective of the distribution of FSI crashes [53]. Here, it was assumed that the same speed sign availability trends would be present across road classes and remoteness levels in South Australia and Queensland. This was considered to be a conservative means of gauging speed sign availability, since the position or visibility of the speed signs were not considered. That is, provided a speed sign was present, it was assumed that it would be detected by speed sign recognition cameras. While this method is a not a scientifically rigorous means of quantifying speed sign availability in the other states, in the absence of speed sign data, the projection allowed for speed sign availability to be estimated in order to quantify the readiness of roads for ISA.

To better understand the required investment in speed signage, in context of the potential cost saving from ISA, estimates are made of costs associated with installing speed signs across the road classes and remoteness levels deemed to have inadequate speed signs in this analysis (see Tables 2 and 3). Assuming that these road classes have no speed signs at present, then the cost of installing new speed signs at 5 km intervals within major Cities, 15 km intervals in regional areas, and at 20 km intervals within remote areas would amount to approximately AUD 3.3 million in Victoria, AUD 4.0 million in South Australia, and AUD 3.7 million in Queensland. These costs are based on current estimates of installing a sign, pole, and foundation (including installation/labour) in Victoria. Assuming no maintenance or on-going costs are associated with the upkeep of static speed signs, this results in a one-off AUD 11 million investment across the three states to prevent annual losses (due to death and serious injury associated with road trauma) of approximately AUD 62 million. A 2011 analysis of costs associated with the upgrade and maintenance of the digital road network (again using Victoria's cost estimates as the basis and projected nationally) suggested that it would cost approximately AUD 15.6 million to digitally map speeds in all Australian states and require a further AUD 2.4 million per year to update and maintain the maps (accounting for inflation, this is AUD 19.7 million and AUD 3 million, respectively, at current currency values). Given the somewhat restricted cellular connectivity throughout the nation, relying on digital speed maps alone for assisting drivers in being speed compliant is unlikely to be best practice, and investments into the strategic placement of static speed signs are recommended.

The assumption in this study that digital speed maps will have no effect on improving ISA function is unlikely to adversely impact the findings made here. While digital speed maps are available in most major city areas, and connectivity in these regions is high, changes in speed limit due to weather, road works, or traffic incidents are unlikely to be updated frequently enough for ISA systems to observe updated speed limits. Therefore, at present, speed signs will be heavily relied upon for adequate ISA function.

4.5. Future Research

This study was a macro-level analysis of the availability of physical speed signs (which constitute the required input data for onboard sensors and speed recognition cameras which, in turn, determines ISA function). The findings from this study suggest that the availability of speed signs in Victoria, and assuming similar speed sign availability in South Australia and Queensland, is inadequate to fully support advisory ISA in the absence of cellular connectively. Future research is recommended to identify speed sign availability in other states in order to determine if equal investments in infrastructure are necessary for the full potential of ISA to be realised across Australia. The effect that ISA is likely to have on mitigating minor injuries (not taken into account here) should also be considered so as to more accurately justify the investments that should be made across road networks to support ISA. Further, extending the scope of this work, the availability of cellular connectively can be taken into account so that the preparedness of the road network (for not only ISA but also for other autonomous vehicle technologies) can be evaluated, justifying more readily the investments in physical and digital infrastructure that should be made to gain maximum benefits from advanced driver assist technologies.

5. Conclusions

The annual forgone benefits of ISA, due to the absence of fixed speed signs estimated here, warrant investments into speed sign infrastructure, particularly since the annual economic costs of the lost benefits are likely to far outweigh that of one-off infrastructure investments in the installation of speed signs. In the interim, data collection and data sharing of current speed sign availability across Australian jurisdictions is recommended, as this will provide more realistic estimates of road-readiness for technologies such as ISA and allow for more targeted and strategic investments. Currently, the true extent of infrastructure deficiencies remains unknown. The estimates of the forgone benefits of ISA generated here use the best available data, a sound methodology, and a transparent approach. In doing so, the need for investments into speed sign infrastructure likely to bring significant additional road safely gains in terms of lives lost and injuries prevented is identified.

Author Contributions: Conceptualization, S.N. and S.P.; methodology, S.P. and S.N.; formal analysis, S.P.; investigation, S.P. and J.B.-G.; resources, J.B.-G. and B.F.; data curation, S.P.; writing—original draft preparation, S.P. and J.B.-G.; writing—review and editing, S.P., J.B.-G., S.N., and B.F.; supervision, B.F., S.N., and J.B.-G.; project administration, B.F. and J.B.-G. All authors have read and agreed to the published version of the manuscript.

Funding: This research received no external funding.

Institutional Review Board Statement: Not applicable.

Informed Consent Statement: Not applicable.

Data Availability Statement: Not applicable.

Acknowledgments: The authors would like to thank Karen Stephan for her assistance with the application of the logistic regression model used to identify speed-involved crashes.

Conflicts of Interest: The authors declare no conflict of interest.

References

1. Job, S.; Sakashita, C.; Mooren, L.; Grzebieta, R. Community perceptions and beliefs regarding low-level speeding and suggested solutions. In Proceedings of the Annual Meeting of the Transportation Research Board, Washington, DC, USA, 8–12 January 2012.
2. Grabar, H. The American Addiction to Speeding: How We Became Obsessed with Driving Fast, No Matter the Cost. Available online: https://slate.com/business/2021/12/speed-limit-americas-most-broken-law-history.html (accessed on 25 February 2022).
3. Mooren, L.; Grzebieta, R.; Job, S. Speed-the biggest and most contested road killer. *J. Australas. Coll. Road Saf.* **2014**, *25*, 13–18.
4. Litras, P.; Spits, S. Mark Webber Attacks 'Nanny State'. 2010. Available online: https://www.smh.com.au/sport/motorsport/mark-webberattacks-nanny-state-20100328-r585.html (accessed on 25 February 2022).
5. Matthew-Wilson, C. Speed limits do not alter behaviour. *Otago Daily Times*. 22 July 2019.
6. Elvik, R.; Christensen, P.; Amundsen, A. *Speed and Road Accidents: An Evaluation of the Power Model*; The Institute for Transport Economics (Transportøkonomisk Institutt): Oslo, Norway, 2004; p. 2004.
7. Alavi, H.; Keleher, S.; Nieuwesteeg, M. Quantifying the contribution of low-level speeding to trauma in Victoria. In Proceedings of the Australasian Road Safety Research Policing Education Conference, Melbourne, VIC, Australia, 12–14 November 2014.
8. Gavin, A.; Walker, E.; Fernandes, R.; Graham, A.; Job, R.; Sergeant, J. Creation and validation of a tool to measure the real population risk of speeding. In Proceedings of the 2011 Australasian Road Safety Research, Policing and Education Conference, Perth, WA, Australia, 6–9 November 2011.
9. Gavin, A.; Walker, E.; Murdoch, C.; Graham, A.; Fernandes, R.; Job, R.F.S. Is a focus on low level speeding justified? Objective determination of the relative contributions of low and high level speeding to the road toll. In Proceedings of the 2010 Australasian Road Safety Research, Policing and Education Conference, Canberra, ACT, Australia, 31 August–3 September 2010.
10. Viallon, V.; Laumon, B. Fractions of fatal crashes attributable to speeding: Evolution for the period 2001–2010 in France. *Accid. Anal. Prev.* **2013**, *52*, 250–256. [CrossRef] [PubMed]

11. Job, S.; Brodie, C. Understanding the role of speeding and speed in serious crash trauma: A case study of New Zealand. *J. Road Saf.* **2022**, *33*, 5–25. [CrossRef]
12. Unger, T.; Schubert, A. *Qualitative Assessment of Intelligent Speed Adaptation (ISA) Systems*; Verkehrsunfallforschung an der TU Dresden GmbH—VUFO: Dresden, Germany, 22 October 2018; p. 47.
13. European Commission. *Road Safety Thematic Report: Speeding*; European Commission: Brussels, Belgium, 2020.
14. National Centre for Statistics and Analysis. Speeding: 2017 Data; Vol. DOT HS 812 687. 2019. Available online: https://crashstats.nhtsa.dot.gov/Api/Public/ViewPublication/812687 (accessed on 10 January 2022).
15. Turner, B.; Machata, K.; Hels, T.; Lassarre, S.; Salathé, M.; Yannis, G.; Hollo, P.; Toth, V.; Shingo, D.; van Schagen, I.; et al. *Speed and Crash Risk*; International Traffic Safety Data and Analysis Group: Paris, France, 2018; p. 82.
16. Doecke, S.; Elsegood, M.; Ponte, G. *The Contribution of Various Levels of Speeding to Fatal and Serious Road Trauma*; CASR Report CASR189; The Centre for Automotive Safety Research: Adelaide, SA, Australia, 2021; p. 21.
17. *Department of Transport Victorian Road Safety Strategy 2021–2030*; Department of Transport: Melbourne, VIC, Australia, 2020; p. 44.
18. Blincoe, K.M.; Jones, A.P.; Sauerzapf, V.; Haynes, R. Speeding drivers' attitudes and perceptions of speed cameras in rural England. *Accid. Anal. Prev.* **2006**, *38*, 371–378. [CrossRef] [PubMed]
19. Richard, C.M.; Divekar, G.; Brown, J.L. *Motivations for Speeding-Additional Data Analysis*; National Highway Traffic Safety Administration: Washington, DC, USA, 2016; p. 117.
20. Nilsson, G. *Traffic Safety Dimensions and the Power Model to Describe the Effect of Speed on Safety*; Lund Institute of Technology: Lund, Sweden, 2004.
21. Cameron, M.H.; Elvik, R. Nilsson's Power Model connecting speed and road trauma: Applicability by road type and alternative models for urban roads. *Accid. Anal. Prev.* **2010**, *42*, 1908–1915. [CrossRef] [PubMed]
22. Aydin, M.M.; Gunay, B.; Akgol, K. Performance Comparison of Various Chicane Types: A Driving Simulator Study. *Int. J. Civ. Eng.* **2019**, *17*, 1753–1765. [CrossRef]
23. Sołowczuk, A.B.; Kacprzak, D. Identification of the Determinants of the Effectiveness of On-Road Chicanes in the Village Transition Zones Subject to a 50 km/h Speed Limit. *Energies* **2021**, *14*, 4002. [CrossRef]
24. Kosakowska, K. Evaluation of the impact of speed bumps on the safety of residents-selected aspects. *Transp. Res. Procedia* **2022**, *60*, 418–423. [CrossRef]
25. Kyriakou, C.; Christodoulou, S.E.; Dimitriou, L. Do vehicles sense, detect and locate speed bumps? *Transp. Res. Procedia* **2021**, *52*, 203–210. [CrossRef]
26. Yeo, J.; Lee, J.; Cho, J.; Kim, D.-K.; Jang, K. Effects of speed humps on vehicle speed and pedestrian crashes in South Korea. *J. Saf. Res.* **2020**, *75*, 78–86. [CrossRef] [PubMed]
27. Blewden, M.; Mackie, H.; Thorne, R. *Effectiveness and Implementation of Raised Safety Platforms*; Austroads Publication No. AP-R642-20; Austroads Ltd.: Sydney, NSW, Australia, 2020.
28. Cairney, P.; Fildes, B.; Newstead, S.; Oxley, J.; Thompson, L.; Davy, S. *Short-Term Evaluation of Raised Safety Platforms: Evaluation Methodology and Survey Design*; Report prepared for VicRoads; ARRB Group CN9887: Melbourne, VIC, Australia, 2018.
29. Lawrence, B.; Fildes, B.; Cairney, P.; Davy, S.; Sobhani, A. Evaluation of raised safety platforms (RSP) on-road safety performance. *Sustainability* **2022**, *14*, 138. [CrossRef]
30. Carsten, O.; Lai, F.; Chorlton, K.; Goodman, P.; Carslaw, D.; Hess, S. *Speed Limit Adherence and Its Effect on Road Safety and Climate Change*; University of Leeds: Leeds, UK, 2008; p. 127.
31. Lai, F.; Carsten, O.; Tate, F. How much benefit does Intelligent Speed Adaptation deliver: An analysis of its potential contribution to safety and environment. *Accid. Anal. Prev.* **2012**, *48*, 63–72. [CrossRef]
32. Carsten, O.M.; Tate, F. Intelligent speed adaptation: Accident savings and cost–benefit analysis. *Accid. Anal. Prev.* **2005**, *37*, 407–416. [CrossRef] [PubMed]
33. Guo, W.; Blythe, P.T.; Edwards, S.; Pavkova, K.; Brennan, D. Effect of intelligent speed adaptation technology on older drivers' driving performance. *IET Intell. Transp. Syst.* **2015**, *9*, 343–350. [CrossRef]
34. *European Transport Safety Council Briefing: Intelligent Speed Assistance (ISA)*; European Transport Safety Council: Brussels, Belgium, 2018.
35. Doecke, S.; Raftery, S.; Elsegood, M.; Mackenzie, J. *Intelligent Speed Adaptation (ISA): Benefit Analysis Using EDR Data from Real World Crashes*; CASR Report CASR176; Centre for Automotive Safety Research, The University of Adelaide: Adelaide, SA, Australia, 2021.
36. SAE. European Union Approves Mandatory Vehicle Speed-Limiting Feature. Available online: https://www.sae.org/news/2019/04/eu-to-mandate-intelligent-speed-assistance (accessed on 23 March 2022).
37. ANCAP. ANCAP. ANCAP Assessment Protocol: Safety Assist v8.0.2. In *Assessment of Speed Assist Systems*; Australasian New Car Assessment Program: Manuka, ACT, Australia, 2018; pp. 1–24.
38. Seidl, M.; Hynd, D.; McCarthy, M.; Martin, P.; Hunt, R.; Mohan, S.; Krishnamurthy, V.; O'Connell, S. *Depth Cost-Effectiveness Analysis of the Identified Measures and Features Regarding the Way Forward for EU Vehicle Safety*; TRL Limited: Brussels, Belgium, 2017; p. 234.
39. *European Commission Commission Delegated Regulation (EU) 2019/2144 Supplementary Material*; C(2021) 4455 Final 9; European Commission: Brussels, Belgium, 2021.

40. Germanchev, A.; Eastwood, B.; Hore-Lacy, W. *Infrastructure Changes to Support Automated Vehicles on Rural and Metropolitan Highways and Freeways Road Audit (Module 2)*; Austroads Ltd.: Sydney, NSW, Australia, October 2019; p. 90.
41. NPerf 3G/4G/5G Coverage Map, Australia. Available online: https://www.nperf.com/en/map/AU/-/2445.Telstra/signal/?ll=-35.32633026307483&lg=139.75708007812503&zoom=4 (accessed on 1 August 2021).
42. Terrell, M.; Australasian New Car Assessment Program (ANCAP), Manuka, ACT, Australia. Personal Communication, 2 March 2022.
43. Raj, A.C.; Sozio, A.; Velden, M.v.d. *Guidance and Readability Criteria for Traffic Sign Recognition Systems Reading Electronic Signs*; Austroads Ltd.: Sydney, NSW, Australia, 2020; p. 57.
44. Transurban. *NSW Partially Automated Vehicle Trials. Stage One—Sydney Orbital Network*; Transurban: Sydney, NSW, Australia, 2018; p. 26.
45. Transurban. *Victorian Connected and Automated Vehicle Trials Program: Phase One*; Transurban: Melbourne, VIC, Australia, 2018; p. 26.
46. Bilton, P.; Williams, A.; Plath, E.; Shah, S. *Best Practice Speed Information Management*; AP-R671-22; Austroads Ltd.: Sydney, NSW, Astrailia, 2022; p. 101.
47. Austroads. *Guide to Road Safety Part 5: Road Safety for Regional and Remote Areas*; Austroads: Sydney, NSW, Australia, 2019; p. 66.
48. Austroads. *National View on Regional and Remote Road Safety*; Austroads: Sydney, NSW, Australia, 2019; p. 163.
49. Peiris, S.; Newstead, S.; Berecki-Gisolf, J.; Chen, B.; Fildes, B. Quantifying the lost safety benefits of ADAS technologies due to inadequate supporting road infrastructure. *Sustainability* **2022**, *14*, 2234. [CrossRef]
50. VicRoads. *Traffic Engineering Manual: Volume 3—Additional Network Standards & Guidelines. Speed Zoning Guidelines, ANS&G Part 2.11: Speed Zoning Guidelines (Edition 1)*; VicRoads: Carlton, VIC, Australia, 2017; pp. 1–98.
51. Stephan, K.; Budd, L.; Newstead, S.; Young, K. *Road Safety Benefits of Intelligent Speed Adaptation. Stage 1: Characteristics of Drivers and Vehicles Involved in Speed-Related Chrashes in Victoria*; Monash University Accident Research Centre: Melbourne, VIC, Australia, 2022; p. 40.
52. Doecke, S.; Kloeden, C.; Woolley, J. *NSW Intelligent Speed Adaptation (ISA) Trial: Modelling the Effects of Advisory ISA on the Australian Driving Population*; Centre for Automotive Safety Research: Adelaide, SA, Australia, 2011.
53. Peiris, S.; Berecki-Gisolf, J.; Newstead, S.; Chen, B.; Fildes, B. Development of a Methodology for Estimating the Availability of ADAS-Dependent Road Infrastructure. *Sustainability* **2021**, *13*, 9512. [CrossRef]
54. Diamantopoulou, K.; Hoareau, E.; Oxley, P.; Cameron, M. *The Feasibility of Identifying Speeding-Related and Fatigue-Related Crashes in Police-Reported Mass Crash Data*; Monash University Accident Research Centre: Melbourne, VIC, Australia, 2003; p. 77.
55. Kloeden, C.; McLean, A.; Moore, V.; Ponte, G. *Travelling Speed and the Risk of Crash Involvement Volume 2-Case and Reconstruction Details*; NHMRC Road Accident Research Unit, The University of Adelaide: Canberra, ACT, Australia, 1997; pp. 1–318.
56. Kloeden, C.; McLean, A.; Moore, V.; Ponte, G. *Travelling Speed and the Risk of Crash Involvement Volume 1-Findings*; NHMRC Road Accident Research Unit, The University of Adelaide: Canberra, ACT, Australia, 1997; pp. 1–69.
57. Kloeden, C.N.; Ponte, G.; McLean, A.J. *Travelling Speed and the Risk of Crash Involvement on Rural Roads*; CR 204; Road Accident Research Unit; The University of Adelaide: Adelaide, SA, Australia, 2001; pp. 1–69.
58. Doecke, S.D.; Dutschke, J.K.; Baldock, M.R.; Kloeden, C.N. Travel speed and the risk of serious injury in vehicle crashes. *Accid. Anal. Prev.* **2021**, *161*, 106359. [CrossRef]
59. Klinjun, N.; Kelly, M.; Praditsathaporn, C.; Petsirasan, R. Identification of Factors Affecting Road Traffic Injuries Incidence and Severity in Southern Thailand Based on Accident Investigation Reports. *Sustainability* **2021**, *13*, 12467. [CrossRef]
60. Doecke, S.; Kloeden, C.N. The accuracy of determining speeding directly from mass crash data and using the NSW Centre for Road Safety method. *J. Australas. Coll. Road Saf.* **2014**, *25*, 35.
61. NSW Centre for Road Safety Driving too fast. Available online: https://roadsafety.transport.nsw.gov.au/speeding/index.html (accessed on 15 January 2022).
62. ABS. The Australian Statistical Geography Standard (ASGS) Remoteness Structure: Map of the 2016 Remoteness Areas for Australia. Available online: https://www.abs.gov.au/websitedbs/D3310114.nsf/home/remoteness+structure (accessed on 13 September 2019).
63. Kloeden, C.N.; McLean, J.; Glonek, G.F.V. *Reanalysis of Travelling Speed and the Risk of Crash Involvement in Adelaide South Australia*; Australian Transport Safety Bureau: Canberra, ACT, Australia, 2002.
64. Wall, J.; Creef, K.; Boland, P.; Vecovski, V.; Prendergast, M.; Stow, J.; Fernandes, R.; Beck, J.; Doecke, S.; Woolley, J. Road Safety Benefits of Intelligent Speed Adaptation for Australia. In Proceedings of the 2011 Australasian Road Safety Research, Policing and Education conference, Perth, WA, Australia, 6–9 November 2011.
65. Peiris, S.; Berecki-Gisolf, J.; Chen, B.; Fildes, B. Road Trauma in Regional and Remote Australia and New Zealand in Preparedness for ADAS Technologies and Autonomous Vehicles. *Sustainability* **2020**, *12*, 4347. [CrossRef]
66. Department of Transport, DATAVIC. Speed Sign Data, Creative Commons Attribution 4.0 International. Victoria, Australia. Funded by the State Government of Victoria. 2021. Available online: http://data.vicroads.vic.gov.au/Metadata/Speed_Zones%20-%20Open%20Data.html (accessed on 11 July 2019).
67. Risbey, T.; Cregan, M.; De Silva, H. Social Cost of Road Crashes. In the Proceedings of the Proc of 33rd Australasian Transport Research Forum, Canberra, ACT, Australia, 29 September–1 October 2010; pp. 1–16.

68. Economic Connections Pty Ltd. *Cost of Road Trauma in Australia*; Australian Automobile Association: Canberra, Australia, 2017; p. 20.
69. *Transport for NSW Road Traffic Casualty Crashes in NSW: Statistical Statement for the Year Ended 31 December 2019*; Centre for Road Safety: Macquarie Park, NSW, Austraila, 2020; p. 125.
70. Stephan, K.L.; Young, K.L.; Newstead, S.V.; Lenne, M.G.; Cavallo, A.; Duck, N.J.; Imberger, K.; Healy, D. The effectiveness of an advisory intelligent speed adaptation (ISA) system for Victorian repeat speeders. In Proceedings of the 2014 Australasian Road Safety Research, Policing and Education Conference, Melbourne, VIC, Australia, 12–14 November 2014; pp. 1–11.
71. Doecke, S.; Woolley, J. *Cost Benefit Analysis of Intelligent Speed Assist. Adelaide: Centre for Automotive Safety Research*; CASR Report CASR093; The University of Adelaide: Adelaide, SA, Australia, 2010.
72. Regan, M.A.; Triggs, T.J.; Young, K.L.; Tomasevic, N.; Mitsopoulos, E.; Stephan, K.; Tingvall, C. *On-Road Evaluation of Intelligent Speed Adaptation, Following Distance Warning and Seatbelt Reminder Systems: Final Results of the TAG SafeCar Project*; Monash University Accident Research Centre: Melbourne, VIC, Australia, 2006; p. 312.
73. Driscoll, R.; Page, Y.; Lassarre, S.; Ehrlich, J. LAVIA–an Evaluation of the Potential Safety Benefits of the French Intelligent Speed Adaptation Project. In the Proceedings of the Annual Proceedings/Association for the Advancement of Automotive Medicine, Melbourne, VIC, Australia, 15–17 October 2007; p. 485.
74. Young, K.; Stephan, K.; Newstead, S. *Benefits and Costs of Intelligent Speed Adaptation for the Australian Fleet: Literature Review*; Monash University Accident Research Centre: Melbourne, VIC, Australia, 2015; p. 20.
75. Tate, F.; Carsten, O. *ISA-UK Intelligent Speed Adaptation*; University of Leeds and MIRA: Leeds, UK, 2008; p. 98.
76. Doecke, S.; Woolley, J.; Anderson, R. Cost Benefit Analysis of Intelligent Speed Adaptation. In Proceedings of the 2010 Australasian Road Safety Research, Policing and Education Conference, Canberra, ACT, Australia, 31 August–3 September 2010; p. 10.
77. Hjalmdahl, M.; Almqvist, S.; Varhelyi, A. Speed regulation by in-car active accelerator pedal: Effects on speed and speed distribution. *IATSS Res.* **2002**, *26*, 60–66. [CrossRef]
78. Barrow, A.; Edwards, A.; Smith, L.; Khatry, R.; Kalaiyarasan, A.; Hynd, D. *Effectiveness Estimates for Proposed Amendments to the EU's General and Pedestrian Safety Regulations*; Transport Research Laboratory: Berkshire, UK, 2017; p. 120.
79. Rowe, R.; Maurice-Smith, M.; Mahmood, M.; Shuja, A.; Gibson, D. Understanding intentions to override intelligent speed assistance prior to widespread availability: An application of the theory of planned behaviour. *Accid. Anal. Prev.* **2021**, *151*, 105975. [CrossRef] [PubMed]
80. Hjälmdahl, M.; Várhelyi, A.; Hydén, C.; Risser, R.; Draskoczy, M. *The Effects of Active Accelerator in Cars-Results from a Large Scale Trial with Intelligent Speed Adaptation*; European Transport Conference: Cambridge, UK, 9–11 September 2002.

MDPI AG
Grosspeteranlage 5
4052 Basel
Switzerland
Tel.: +41 61 683 77 34

Sensors Editorial Office
E-mail: sensors@mdpi.com
www.mdpi.com/journal/sensors

Disclaimer/Publisher's Note: The title and front matter of this reprint are at the discretion of the Guest Editors. The publisher is not responsible for their content or any associated concerns. The statements, opinions and data contained in all individual articles are solely those of the individual Editors and contributors and not of MDPI. MDPI disclaims responsibility for any injury to people or property resulting from any ideas, methods, instructions or products referred to in the content.

www.ingramcontent.com/pod-product-compliance
Lightning Source LLC
Chambersburg PA
CBHW061610220926
51850CB00005B/473
* 9 7 8 3 7 2 5 8 6 4 4 0 9 *